FEDERAL EXECUTIVE TEAM

Director, Climate Change Science Program .... William J. Brennan
Director, Climate Change Science Program Office .... Peter A. Schultz
Lead Agency Principal Representative to Climate Change Science Program;
Senior Advisor for Global Change Programs, U.S. Geological Survey .... Thomas R. Armstrong
Product Lead, U.S. Geological Survey .... Joan J. Fitzpatrick
Synthesis and Assessment Product Advisory Group Chair; Associate Director, Environmental
Protection Agency National Center for Environmental Assessment .... Michael W. Slimak
Synthesis and Assessment Product Coordinator, Climate Change Science Program Office .... Fabien J.G. Laurier
Special Advisor, National Oceanic and Atmospheric Administration .... Chad A. McNutt

OTHER AGENCY REPRESENTATIVES

National Aeronautics and Space Administration .... Waleed Abdalati
National Oceanic and Atmospheric Administration .... John A. Calder
National Oceanic and Atmospheric Administration .... Kathleen Crane
National Science Foundation .... David J. Verardo
U.S. Department of Energy .... Wanda R. Ferrell

EDITORIAL AND PRODUCTION TEAM

Graphic Design (USGS)
Plates 1 and 2 .... Amber Swallow
Illustrations .... Loretta Ulibarri, Mary Berger, Sharon Powers
Layout .... Margo VanAlstine
Copy editor (USGS, contractor through ATA Services) .... Mary-Margaret Coates

USGS-CCSP ARCTIC PALEOCLIMATE FEDERAL ADVISORY COMMITTEE

A four-member Federal Advisory Committee oversaw the scientific development of this Synthesis and Assessment Product (SAP) at the request of the U.S. Geological Survey. These Federal Advisory Committee members also served as part of the group of chapter lead authors and contributing authors who wrote this report.

Richard B. Alley, Pennsylvania State University
Julie Brigham-Grette, University of Massachusetts
Gifford H. Miller, University of Colorado
Leonid Polyak, Ohio State University

This Synthesis and Assessment Product, described in the U.S. Climate Change Science Program (CCSP) Strategic Plan, was prepared in accordance with Section 515 of the Treasury and General Government Appropriations Act for Fiscal Year 2001 (Public Law 106-554) and the information quality act guidelines issued by the Department of the Interior and the U.S. Geological Survey pursuant to Section 15). The CCSP Interagency Committee relies on the Department of the Interior and the U.S. Geological Survey certifications regarding compliance with Section 515 and Department guidelines as the basis for determining that this product conforms with Section 515. For purposes of compliance with Section 515, this CCSP Synthesis and Assessment Product is an "interpreted product" as that term is used in U.S. Geological Survey guidelines and is classified as "highly influential." This document does not express any regulatory policies of the United States or any of its agencies, or provide recommendations for regulatory action.

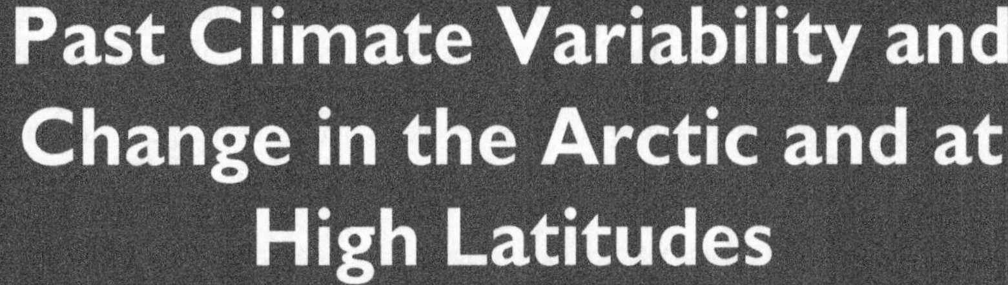

Past Climate Variability and Change in the Arctic and at High Latitudes

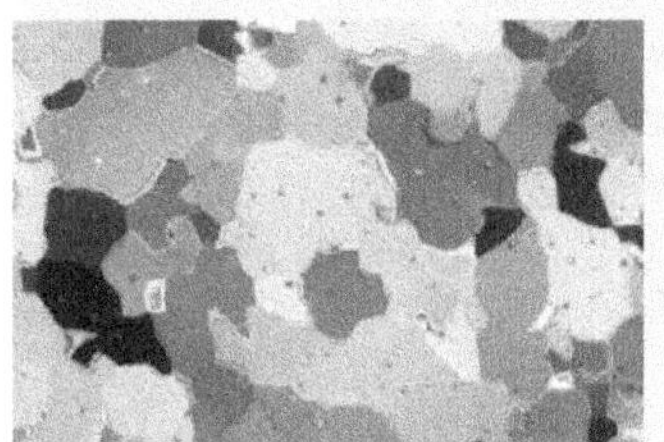

Final Report
Synthesis and Assessment Product 1.2

U.S. Climate Change Science Program
and the Subcommittee on Global Change Research

LEAD AGENCY:
U.S. Geological Survey

CONTRIBUTING AGENCIES:
National Oceanic and Atmospheric Administration
National Science Foundation

January 2009

Members of Congress:

On behalf of the National Science and Technology Council, the U.S. Climate Change Science Program (CCSP) is pleased to transmit to the President and the Congress this Synthesis and Assessment Product (SAP) *Past Climate Variability and Change in the Arctic and at High Latitudes*. This is part of a series of 21 SAPs produced by the CCSP aimed at providing current assessments of climate change science to inform public debate, policy, and operational decisions. These reports are also intended to help the CCSP develop future program research priorities.

The CCSP's guiding vision is to provide the Nation and the global community with the science-based knowledge needed to manage the risks and capture the opportunities associated with climate and related environmental changes. The SAPs are important steps toward achieving that vision and help to translate the CCSP's extensive observational and research database into informational tools that directly address key questions being asked of the research community.

This SAP summarizes the current knowledge of the past climate of the Arctic and discusses its relevance to key questions about present and future changes of relevance to policy makers and stakeholders. It was developed in accordance with the Guidelines for Producing CCSP SAPs, the Information Quality Act (Section 515 of the Treasury and General Government Appropriations Act for Fiscal Year 2001 (Public Law 106-554)), and the guidelines issued by the Department of Interior and the U.S. Geological Survey pursuant to Section 515.

We commend the report's authors for both the thorough nature of their work and their adherence to an inclusive review process.

Sincerely,

Carlos M. Gutierrez Secretary of Commerce Chair, Committee on Climate Change Science and Technology Integration	Samuel W. Bodman Secretary of Energy Vice Chair, Committee on Climate Change Science and Technology Integration	John H. Marburger III Director, Office of Science and Technology Policy Executive Director, Committee on Climate Change Science and Technology Integration

AUTHOR TEAMS FOR THIS REPORT

Executive Summary	**Lead Authors**	Richard B. Alley, Pennsylvania State University, University Park, PA Julie Brigham-Grette, University of Massachusetts, Amherst, MA Gifford H. Miller, University of Colorado, Boulder, CO Leonid Polyak, Ohio State University, Columbus, OH James W.C. White, University of Colorado, Boulder, CO
Chapter 1	**Lead Author**	Joan J. Fitzpatrick, U.S. Geological Survey, Denver, CO
	Contributing Authors	Richard B. Alley, Pennsylvania State University, University Park, PA Julie Brigham-Grette, University of Massachusetts, Amherst, MA Gifford H. Miller, University of Colorado, Boulder, CO Leonid Polyak, Ohio State University, Columbus, OH Mark Serreze, University of Colorado, Boulder, CO
Chapter 2	**Lead Authors**	Richard B. Alley, Pennsylvania State University, University Park, PA Joan Fitzpatrick, U.S. Geological Survey, Denver, CO
	Contributing Authors	Julie Brigham-Grette, University of Massachusetts, Amherst, MA Gifford H. Miller, University of Colorado, Boulder, CO Daniel Muhs, U.S. Geological Survey, Denver, CO Leonid Polyak, Ohio State University, Columbus, OH
Chapter 3	**Lead Authors**	Gifford H. Miller, University of Colorado, Boulder, CO Julie Brigham-Grette, University of Massachusetts, Amherst, MA
	Contributing Authors	Lesleigh Anderson, U.S. Geological Survey, Denver, CO Henning Bauch, GEOMAR, University of Kiel, Germany Mary Anne Douglas, University of Alberta, Edmonton, Alberta, Canada Mary E. Edwards, University of Southampton, United Kingdom Scott Elias, Royal Holloway, University of London, United Kingdom Bruce Finney, Idaho State University, Pocatello, ID Svend Funder, University of Copenhagen, Denmark Timothy Herbert, Brown University, Providence, RI Larry Hinzman, University of Alaska, Fairbanks, AK Darrell Kaufman, Northern Arizona University, Flagstaff, AZ Glen MacDonald, University of California, Los Angeles, CA Alan Robock, Rutgers University, Rutgers, NJ Mark Serreze, University of Colorado, Boulder, CO John Smol, Queen's University, Kingston, Ontario, Canada Robert Spielhagen, GEOMAR, University of Kiel, Germany Alexander P. Wolfe, University of Alberta, Edmonton, Alberta, Canada Eric Wolff, British Antarctic Survey, Cambridge, United Kingdom

AUTHOR TEAMS FOR THIS REPORT—Continued

Chapter 4 **Lead Authors** James W.C. White, University of Colorado, Boulder, CO
Richard B. Alley, Pennsylvania State University, University Park, PA

Contributing Authors Anne Jennings, University of Colorado, Boulder, CO
Sigfus Johnsen, University of Copenhagen, Denmark
Gifford H. Miller, University of Colorado, Boulder, CO
Steven Nerem, University of Colorado, Boulder, CO

Chapter 5 **Lead Author** Richard B. Alley, Pennsylvania State University, University Park, PA

Contributing Authors John T. Andrews, University of Colorado, Boulder, CO
Garry K.C. Clarke, University of British Columbia, Vancouver, British Columbia, Canada
Kurt M. Cuffey, University of California, Berkeley, CA
Svend Funder, University of Copenhagen, Denmark
Shawn J. Marshall, University of Calgary, Alberta, Canada
Jerry X. Mitrovica, University of Toronto, Ontario, Canada
Daniel R. Muhs, U.S. Geological Survey, Denver, CO
Bette Otto-Bliesner, National Center for Atmospheric Research, Boulder, CO

Chapter 6 **Lead Author** Leonid Polyak, Ohio State University, Columbus, OH

Contributing Authors John Andrews, University of Colorado, Boulder, CO
Julie Brigham-Grette, University of Massachusetts, Amherst, MA
Dennis Darby, Old Dominion University, Norfolk, VA
Arthur Dyke, Geological Survey of Canada, Ottowa, Ontario, Canada
Svend Funder, University of Copenhagen, Copenhagen, Denmark
Marika Holland, National Center for Atmospheric Research, Boulder, CO
Anne Jennings, University of Colorado, Boulder, CO
James Savelle, McGill University, Montreal, Quebec, Canada
Mark Serreze, University of Colorado, Boulder, CO
Eric Wolff, British Antarctic Survey, Cambridge, United Kingdom

Chapter 7 **Lead Authors** Richard B. Alley, Pennsylvania State University, University Park, PA
Julie Brigham-Grette, University of Massachusetts, Amherst, MA
Gifford H. Miller, University of Colorado, Boulder, CO
Leonid Polyak, Ohio State University, Columbus, OH
James W.C. White, University of Colorado, Boulder, CO

ACKNOWLEDGMENTS

Many individuals made significant contributions and improvements to this document at various stages in its journey toward completion and publication.

The product lead offers special thanks to the lead authors and to the CCSP Product Coordinator of this report, whose patience and professionalism during all stages of writing, review, and production made its completion possible, and to Patricia Jellison of the U.S. Geological Survey Office of Global Change for her much-valued assistance navigating the final requirements for its approval and publication.

We are grateful to the following colleagues who provided extensive feedback on early drafts through the peer review process: Becky Alexander, University of Washington; Roger Barry, University of Colorado; Cecilia Bitz, University of Washington; Rosanne D'Arrigo, Lamont-Doherty Earth Observatory, Columbia University; Daniel Kirk-Davidoff, University of Maryland; David Reusch, The Pennsylvania State University; Jose Rial, University of North Carolina; Vladimir Romanovsky, University of Alaska, Fairbanks; William Ruddiman, University of Virginia; Gavin Schmidt, Goddard Institute for Space Studies, NASA; and Henriette Skourup, National Space Institute, Denmark. Details of the peer review process can be found at *http://www.usgs.gov/peer_review/docs/sap1-2_pr_results.pdf.*

We are also grateful to all individuals who provided valuable comments through the public comment period. Documents related to the public comment process can be found at *http://www.climatescience.gov/Library/sap/sap1-2/default.php.*

We also extend our thanks to the following: Arlene Compher and Scott Horvath of the USGS Office of Communications, for their invaluable assistance with web-enabled communications; Patricia Campbell (USGS), for ensuring Section 508 compliance with our web release; Angela Peavy, USGS Director's Office Liaison to the Department of the Interior, for her assistance in expediting matters relating to our Federal Advisory Committee Act (FACA) papers; Anne Waple (CCSP), Deborah Riddle, National Oceanographic and Atmospheric Administration (NOAA), and Sara Veasey (NOAA), for their excellent publishing guidance; Frances Pierce (USGS), for her timely help with our FACA requirements; and to the USGS SAP leads for SAP 3.4 and 4.2, Jack McGeehin and Colleen Charles, for their advice and support throughout the production of this document.

We extend our appreciation to all the participants of our initial scoping meeting and to everyone who has provided suggestions for improvement of text and figures along the way.

RECOMMENDED CITATIONS

For the Report as a whole:

CCSP, 2009: *Past Climate Variability and Change in the Arctic and at High Latitudes.* A report by the U.S. Climate Change Science Program and Subcommittee on Global Change Research [Alley, R.B., J. Brigham-Grette, G.H. Miller, L. Polyak, and J.W.C. White (coordinating lead authors)]. U.S. Geological Survey, Reston, VA, 257 pp.

For the Executive Summary:

Alley, R.B., J. Brigham-Grette, G.H. Miller, L. Polyak, and J.W.C. White, 2009: Executive Summary. In: *Past Climate Variability and Change in the Arctic and at High Latitudes.* A report by the U.S. Climate Change Science Program and Subcommittee on Global Change Research. U.S. Geological Survey, Reston, VA, pp. 1-4.

For Chapter 1:

Fitzpatrick, J.J., R.B. Alley, J. Brigham-Grette, G.H. Miller, L. Polyak, and M. Serreze, 2009: Preface: Why and how to use this synthesis and assessment report. In: *Past Climate Variability and Change in the Arctic and at High Latitudes.* A report by the U.S. Climate Change Science Program and Subcommittee on Global Change Research. U.S. Geological Survey, Reston, VA, pp. 5-10.

For Chapter 2:

Alley, R.B., J.J. Fitzpatrick, J. Brigham-Grette, G.H. Miller, D. Muhs, and L. Polyak, 2009: Paleoclimate concepts. In: *Past Climate Variability and Change in the Arctic and at High Latitudes.* A report by the U.S. Climate Change Science Program and Subcommittee on Global Change Research. U.S. Geological Survey, Reston, VA, pp. 11-30.

For Chapter 3:

Miller, G.H., J. Brigham-Grette, L. Anderson, H. Bauch, M.A. Douglas, M.E. Edwards, S. Elias, B. Finney, S. Funder, T. Herbert, L. Hinzman, D.K. Kaufman, G. MacDonald, A. Robock, M. Serreze, J. Smol, R. Spielhagen, A.P. Wolfe, and E. Wolff, 2009: Temperature and precipitation history of the Arctic. In: *Past Climate Variability and Change in the Arctic and at High Latitudes.* A report by the U.S. Climate Change Science Program and Subcommittee on Global Change Research. U.S. Geological Survey, Reston, VA, pp. 31-90.

For Chapter 4:

White, J.W.C., R.B. Alley, A. Jennings, S. Johnsen, G.H. Miller, and S. Nerem, 2009: Past rates of climate change in the Arctic. In: *Past Climate Variability and Change in the Arctic and at High Latitudes.* A report by the U.S. Climate Change Science Program and Subcommittee on Global Change Research. U.S. Geological Survey, Reston, VA, pp. 91-110.

For Chapter 5:

Alley, R.B., J.T. Andrews, G.K.C. Clarke, K.M. Cuffey, S. Funder, S.J. Marshall, J.X. Mitrovica, D.R. Muhs, and B. Otto-Bleisner, 2009: History of the Greenland ice sheet. In: *Past Climate Variability and Change in the Arctic and at High Latitudes.* A report by the U.S. Climate Change Science Program and Subcommittee on Global Change Research. U.S. Geological Survey, Reston, VA, pp. 111-158.

For Chapter 6:

Polyak, L., J.T. Andrews, J. Brigham-Grette, D. Darby, A. Dyke, S. Funder, M. Holland, A. Jennings, J. Savelle, M. Serreze, and E. Wolff, 2009: History of Arctic sea ice. In: *Past Climate Variability and Change in the Arctic and at High Latitudes.* A report by the U.S. Climate Change Science Program and Subcommittee on Global Change Research. U.S. Geological Survey, Reston, VA, pp. 159-184.

For Chapter 7:

Alley, R.B., J. Brigham-Grette, G.H. Miller, L. Polyak, and J.W.C. White, 2009: Key findings and recommendations. In: *Past Climate Variability and Change in the Arctic and at High Latitudes.* A report by the U.S. Climate Change Science Program and Subcommittee on Global Change Research. U.S. Geological Survey, Reston, VA, pp. 185-190.

TABLE OF CONTENTS

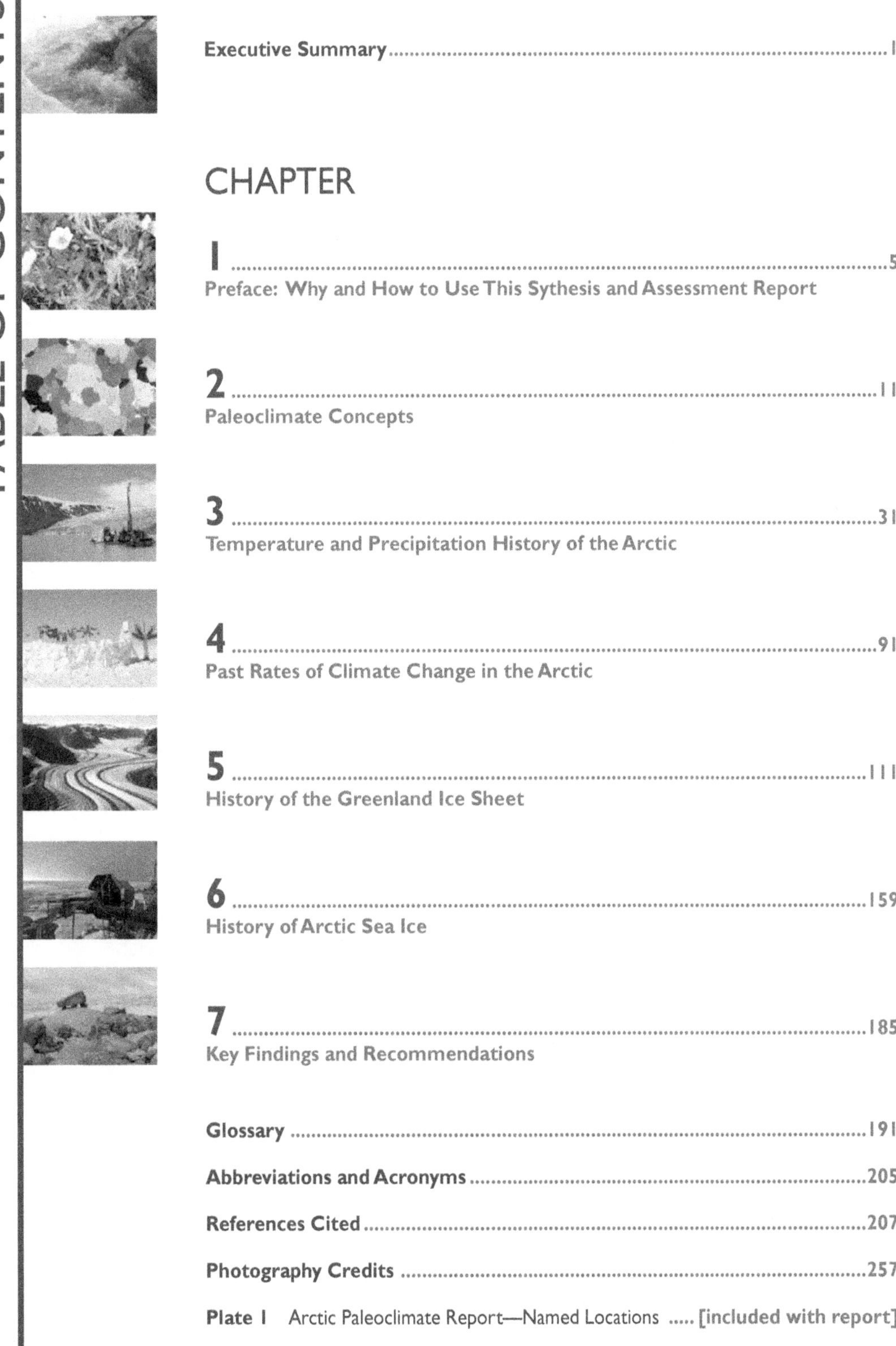

EXECUTIVE SUMMARY

Lead Authors: Richard B. Alley*, Pennsylvania State University, University Park, PA; Julie Brigham-Grette*, University of Massachusetts, Amherst , MA; Gifford H. Miller*, University of Colorado, Boulder, CO; Leonid Polyak*, Ohio State University, Columbus, OH; James W.C. White, University of Colorado, Boulder, CO

* SAP 1.2 Federal Advisory Committee Member

Introduction

Paleoclimate records play a key role in our understanding of Earth's past and present climate system and in our confidence in predicting future climate changes. Paleoclimate data help to elucidate past and present active mechanisms of climate change by placing the short instrumental record into a longer term context and by permitting models to be tested beyond the limited time that instrumental measurements have been available.

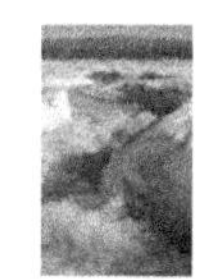

Recent observations in the Arctic have identified large ongoing changes and important climate feedback mechanisms that multiply the effects of global-scale climate changes. Ice is especially important in these "Arctic amplification" processes, which also involve the ocean, the atmosphere, and the land surface (vegetation, soils, and water). As discussed in this report, paleoclimate data show that land and sea ice have grown with cooling temperatures and have shrunk with warming ones, amplifying temperature changes while causing and responding to ecosystem shifts and sea-level changes.

MAJOR QUESTIONS AND RELATED FINDINGS

1) How have temperature and precipitation changed in the Arctic in the past? What does this tell us about Arctic climate that can inform projections of future changes?

The Arctic has undergone dramatic changes in temperature and precipitation during the past 65 million years (m.y.) (the Cenozoic Era) of Earth history. Arctic temperature changes during this time exceeded global average temperature changes during both warm times and cold times, supporting the concept of Arctic amplification.

At the beginning of the Cenozoic Era, 65 million years ago (Ma), there was no sea ice on the Arctic Ocean, and neither Greenland nor Antarctica supported an ice sheet. General cooling since that time is attributed mainly to a slow decrease in greenhouse gases, especially carbon dioxide, in the atmosphere. Ice developed during this slow, "bumpy" cooling, first as mountain glaciers and as seasonal sea ice with the first continental ice sheet forming over Antarctica as early as 33 Ma ago. Following a global warm period about 3.5 Ma in the middle Pliocene, when extensive deciduous forests grew in Arctic regions now occupied by tundra, further cooling crossed a threshold about 2.6 Ma, allowing extensive ice to develop on Arctic land areas and thus initiating the Quaternary ice ages. This ice has responded to persistent features of Earth's orbit over tens of thousands of years, growing when sunshine shifted away from the Northern Hemisphere and melting when northern sunshine returned. These changes were amplified by feedbacks such as greenhouse-gas concentrations that rose and fell as the ice shrank and grew, and by the greater reflection of sunshine caused by more-extensive ice. Human civilization has developed during the most recent of the relatively warm interglacials, the Holocene (about 11.5 thousand years ago (ka) to the present). The penultimate warm interval, about 130–120 ka, received somewhat more Northern-Hemisphere summer sunshine than the Holocene owing to differences in Earth's orbital configuration. Because this more abundant summer sunshine warmed the Arctic summer about 5°C above recent temperatures, the Greenland Ice Sheet was substantially smaller than its current size and almost all glaciers melted completely at that time.

The Last Glacial Maximum peaked approximately 21 ka when the Arctic was about 20°C colder than at present. Ice recession was well underway by 16 ka, and most of the Northern Hemisphere ice sheets melted by 7 ka. Summer sunshine rose steadily from 20 ka to a maximum (10% higher than at present due to the Earth's orbit) about 11 ka ago, and has been decreasing since then. The extra energy received in summer in the early Holocene resulted in warmer summers throughout the Arctic. Summer temperatures were 1°–3°C above 20th century averages, enough to completely melt many small glaciers in the Arctic and to slightly shrink the ice sheet on Greenland. Summer sea-ice limits were significantly less than their 20th century average. As summer sunshine decreased in the second half of the Holocene, glaciers re-established or advanced, and sea ice became more extensive. Late Holocene cooling reached its nadir during the Little Ice Age (about 1250–1850 AD), when most Arctic glaciers reached their maximum Holocene extent. The Little Ice Age temperature minimum may also have been augmented by multiple large volcanic eruptions that lofted a reflective aerosol layer into the stratosphere at that time. Subsequent warming during the 19th and 20th centuries has resulted in Arctic-wide glacier recession, the northward advance of terrestrial ecosystems, and the reduction of perennial (year-round) sea ice in the Arctic Ocean. These trends will continue if greenhouse gas concentrations continue to increase into the future.

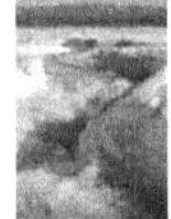

2) How rapidly have temperature and precipitation changed in the Arctic in the past? What do these past rates of change tell us about Arctic climate that can inform projections of future changes?

As discussed with the previous question, climate changes on numerous time scales for various reasons, and it has always done so. In general, longer-lived changes are somewhat larger but much slower than shorter-lived changes.

Processes linked to continental drift (plate tectonics) have affected atmospheric and oceanic currents and the composition of the atmosphere for tens of millions of years; in the Arctic, a global cooling trend has switched conditions from being ice-free year-round near sea level to icy conditions more recently. Within the icy times, variations in Arctic sunshine in response to features of Earth's orbit have caused regular cycles of warming and cooling of tens of thousands of years that were roughly half the size of the continental-drift-linked changes. This "glacial-interglacial" cycling was amplified by colder times, bringing reduced greenhouse gases and greater reflection of sunlight, especially from expanded ice-covered regions. This glacial-interglacial cycling has been punctuated by sharp-onset, sharp-end (in as little as 1–10 years) millennial oscillations, which near the North Atlantic were roughly half as large as the glacial-interglacial cycling but which were much smaller Arctic-wide and beyond. The current warm period of the glacial-interglacial cycling has been influenced by cooling events from single volcanic eruptions, slower but longer lasting changes from random fluctuations in frequency of volcanic eruptions and from weak solar variability, and perhaps by other classes of events. Very recently, human effects have become evident, not yet showing both size and duration that exceed peak values of natural fluctuations further in the past, but with projections indicating that human influences could become anomalous in size and duration and, hence, in speed.

3) What does the paleoclimate record tell us about the past size of the Greenland Ice Sheet and its implications for sea level changes?

The paleo-record shows that the Greenland Ice Sheet has consistently lost mass and contributed to sea-level rise when the climate warmed, and has grown and contributed to sea-level fall when the climate cooled. These changes occurred even at times when offsetting effects from elsewhere in the climate system caused the net sea-level change around Greenland to be negligible, and so these changes in the ice sheet cannot have been caused primarily by sea-level change. In contrast, no changes in the ice sheet have been documented independent of temperature changes. Moreover, snowfall has increased with major climate warmings, but the ice sheet lost mass nonetheless; increased accumulation in the ice sheet center was not sufficient to counteract increased melt and flow near the edges. Most of the documented changes (of both ice sheet and forcings) spanned multi-millennial periods, but limited data show rapid responses to rapid forcings have also occurred. In particular, regions near the ice margin have been observed to respond within a few decades or less. However, major changes of the ice sheet are thought to take centuries to millennia, and this is supported by the limited data.

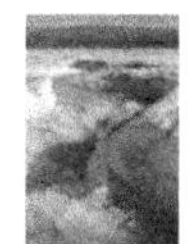

The paleo-record does not yet give any strong constraints on how rapidly a near-complete loss of the ice sheet could occur, although the paleo-data indicate that onset of shrinkage will be essentially immediate after forcings begin. The available evidence suggests such a loss requires a sustained warming of at least 2–7°C above mean 20th century values, but this threshold is poorly defined. The paleo-archives are sufficiently sketchy that temporary ice sheet growth in response to warming, or changes induced by factors other than temperature, could have occurred without being recorded.

4) What does the paleoclimate record tell us about past changes in Arctic sea ice cover, and what implications does this have for consideration of recent and potential future changes?

Although incomplete, existing data outline the development of Arctic sea-ice cover from the ice-free conditions of the early Cenozoic. Some data indicate that sea ice has covered at least part of the Arctic Ocean for the last 13–14 million years, and it has been most extensive during the last several million years in relationship with Earth's overall cooler climate. Other data argue against the development of perennial (year-round) sea ice until the most recent 2–3 million years. Nevertheless, episodes of considerably reduced ice cover, or even a seasonally ice-free Arctic Ocean, probably punctuated even this latter period. Warmer climates associated with the orbitally-paced interglacials promoted these episodes of diminished ice. Ice cover in the Arctic began to diminish in the late 19th century and this shrinkage has accelerated during the last several decades. Shrinkages that were both similarly large and rapid have not been documented over at least the last few thousand years, although the paleoclimatic record is sufficiently sparse that similar events might have been missed. Orbital changes have made ice melting less likely than during the previous millennia since the end of the last ice age, making the recent changes especially anomalous. Improved reconstructions of sea-ice history would help clarify just how anomalous these recent changes are.

RECOMMENDATIONS

- Paleoclimatic data on the Arctic are generated by numerous international investigators who study a great range of archives throughout the vast reaches of the Arctic. The value of this diversity is evident in this report. Many of the key results of this report rest especially on the outcomes of community-based syntheses, including the CAPE Project, and multiply replicated, heavily sampled archives such as the central Greenland deep ice cores. Results from the ACEX deep coring in Arctic Ocean sediments were appearing as this report was being written. These results are quite valuable and will become more so with synthesis and replication, including comparison with land-based and marine records. The number of questions answered, and raised, by this one new data set shows how sparse the data are on many aspects of Arctic paleoclimatic change. Future research should maintain and expand the diversity of investigators, techniques, archives, and geographic locations, while promoting development of community-based syntheses and multiply replicated, heavily sampled archives. Only through breadth and depth can the remaining uncertainties be reduced while confidence in the results is improved.

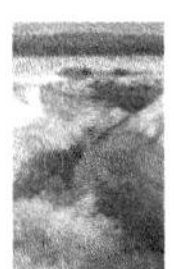

- The questions asked of this study by the CCSP are relevant to public policy and require answers. The answers provided here are, we hope, useful and informative. However, we recognize that despite the contributions of many community members to this report, in many cases a basis was not available in the refereed scientific literature to provide answers with the accuracy and precision desired by policy-makers. Future research activities in Arctic paleoclimate should address in greater detail the policy-relevant questions motivating this report.

- Paleoclimatic data provide very clear evidence of past changes in important aspects of the Arctic climate system. The ice of the Greenland Ice Sheet, smaller glaciers and ice caps, the Arctic Ocean, and in soils is shown to be vulnerable to warming, and Arctic ecosystems are strongly affected by changing ice and climate. National and international studies generally project rapid warming in the future. If this warming occurs, the paleoclimatic data indicate that ice will melt and associated impacts will follow, with implications for ecosystems and economies. The results presented here should be utilized by science managers in the design of monitoring, process, and model-projection studies of Arctic change and linked global responses.

Preface—Why and How to Use This Synthesis and Assessment Report

Lead Author: Joan Fitzpatrick, U.S. Geological Survey, Denver, CO

Contributing Authors: Richard B. Alley*, Pennsylvania State University, University Park, PA; Julie Brigham-Grette*, University of Massachusetts, Amherst, MA; Gifford H. Miller*, University of Colorado, Boulder, CO; Leonid Polyak*, Ohio State University, Columbus, OH; Mark Serreze, University of Colorado, Boulder, CO

* SAP 1.2 Federal Advisory Committee Member

INTRODUCTION

The U.S. Climate Change Science Program (CCSP), a consortium of Federal agencies that investigates climate, has established a Synthesis and Assessment Program as part of its Strategic Plan. A primary objective of the CCSP is to provide the best science-based knowledge possible to support public discussion and government- and private-sector decisions about the risks and opportunities associated with changes in climate and in related environmental systems (U.S. Climate Change Science Program, 2007). The CCSP has identified an initial set of 21 Synthesis and Assessment Products (SAPs) that address the highest-priority research, observation, and information needed to support decisions about issues related to climate change. This assessment, SAP 1.2, focuses on the evidence for and record of past climate change in the Arctic. This SAP is one of three reports that addresses the climate-variability-and-change research element and Goal 1 of the CCSP Strategic Plan to improve knowledge of Earth's past and present climate and environment, including its natural variability, and improve understanding of the causes of observed variability and change.

The development of an improved understanding of natural, long-term cycles in climate is one of the primary goals of the climate research element and Goal 1 of the CCSP (U.S. Climate Change Science Program, 2007). The Arctic region of Earth, by virtue of its sensitivity to the effects of climate change through strong climate feedback mechanisms, has a particularly informative paleoclimate record. Because mechanisms operating in the Arctic and at high northern latitudes are also linked to global climate mechanisms, an examination of how Arctic climate has changed in the past is also globally informative.

MOTIVATION FOR THIS REPORT

Why Does the Past Matter?

The pre-instrumental context of Earth's climate system provided by paleodata strengthens the interlocking web of evidence that supports scientific results regarding climate change.

Paleoclimate records* play a key role in our understanding of Earth's past and present **climate** system and in predicting future **climate changes**. Paleoclimate data help to elucidate past and present active mechanisms by permitting computer-based **models** to be tested beyond the short period (less than 250 years) of instrumental records. **Paleorecords** also provide estimates of past polar **amplification** of (more intense response to) climate change.

This important role of paleoclimate records is recognized, for example, by inclusion of paleoclimate as Chapter 6 of the 11-chapter Fourth Assessment Report of Working Group I (AR4-I) of the **Intergovernmental Panel on Climate Change (IPCC)** (IPCC, 2007), and by the extensive references to paleoclimatic data in climate change reports of the U.S. National Research Council, such as *Climate Change Science: An Analysis of Some Key Questions* (Cicerone et al., 2001).

The pre-instrumental context of Earth's climate system provided by paleodata strengthens the interlocking web of evidence that supports scientific results regarding climate change. For example, in considering whether fossil-fuel burning is an important contributor to the recent rise in atmospheric carbon-dioxide concentrations, researchers must determine and quantify global sources and sinks of carbon in Earth's overall carbon budget. But one can also ask whether the change of atmospheric carbon-dioxide concentrations observed in the instrumental record for the past 100 years falls inside or outside the range of natural variability as revealed in the paleo-record and, if inside, whether the timing of changes in **carbon dioxide** levels matches any known natural cycles that can explain them. Answers to such questions must come from paleoclimate data, because the instrumental record is much too short to characterize the full range of natural fluctuations.

Testing and validation of climate models requires the use of several techniques, as described in Chapter 8 of IPCC (2007) The specific role of paleoclimate information is described there: "Simulations of climate states from the more distant past allow models to be evaluated in regimes that are significantly different from the present. Such tests complement the 'present climate' and 'instrumental period climate' evaluations, because 20th century climate variations have been small compared with the anticipated future changes under **forcing** scenarios derived from the IPCC *Special Report on Emission Scenarios* (SRES) (IPCC, 2000)."

*For bold terms, refer to Glossary; for italic terms, refer to Plate 1; for geologic ages, refer to Plate 2.

Why the Arctic?

During the past century the planet has warmed, overall, by 0.74°C (0.56°–0.92°C) (IPCC, 2007). Above land areas in the *Arctic*, air temperatures have warmed as much as 3°C (exceeding 4°C in winter; Serreze and Francis, 2006) during the same period of time. Instrumental records indicate that in the past 30 years, average temperatures in the *Arctic* have increased at almost twice the rate of the planet as a whole. Attendant changes include reduced **sea ice**, reduced **glacier** extent, increased coastal erosion, changes in vegetation and wildlife habitats, and **permafrost** degradation. Global climate models incorporating the current trend of increasing **greenhouse gases** project continued warming in the near future and a continued amplification of global signals in the Arctic. The sensitivity of the Arctic to changed forcing is due to powerful **positive feedbacks** in the Arctic climate system. These **feedbacks** produce large effects on Arctic climate while also having significant impacts on the global climate system.

This high degree of sensitivity makes the paleoclimate history of the Arctic especially informative when one considers the issue of modern climate change. Summaries of recent changes in the Arctic environment (e.g., ACIA, 2005; Richter-Menge et al., 2006) are based primarily on observations and instrumental records. This report uses paleoclimate records to provide a longer-term context for recent Arctic warming; that context allows us to better understand the potential for future climate changes. Paleoclimate records provide a way to:

- define the range of past natural variability in the Arctic and estimate the sensitivity of the Arctic's response to global climate change,
- evaluate the past rates of Arctic climate change (and thereby provide a long-term context for current rates of change),
- identify past Arctic warm states that are potential **analogues** of future conditions,
- quantify the effects of abrupt **perturbations** (such as large injections of volcanic ash into the atmosphere) and threshold behaviors, and
- gain insights into how the Arctic has behaved during past warm times by identifying critical feedbacks and their mechanisms.

Above land areas in the Arctic, air temperatures have warmed as much as 3°C (exceeding 4°C in winter; Serreze and Francis, 2006) during the past century.

FOCUS AND SCOPE OF THIS SYNTHESIS REPORT (GEOGRAPHIC AND TEMPORAL)

The content of this report follows from the prospectus developed early in its planning (this prospectus is available at the **CCSP** website, *http://www.climatescience.gov*, and it is focused on four topical areas in which the paleo-record can most strongly inform discussions of climate change. These topics, each addressed in a separate chapter of this synthesis report, are:

- the history of past changes in Arctic temperature and precipitation,
- past rates of change in the Arctic,
- the paleo-history of the *Greenland Ice Sheet*, and
- the paleo-history of sea ice in the Arctic.

In general, the temporal scope of this report covers the past 65 million years (**m.y.**) from the early **Cenozoic** (65 **Ma**, million years ago) to the recent **Holocene** (today). Each chapter presents information in chronological sequence from oldest to youngest. The degree of detail in the report generally increases as one moves forward in time because the amount and detail of the available information increases as one approaches the present. The geographic scope of this report, although focused on the Arctic, includes some sub-Arctic areas especially in and near the *North Atlantic Ocean* in order to make use of many relevant paleo-records from these regions.

The specific questions posed in the report are as listed below:

This report summarizes paleoclimate data on past rates of change in the Arctic and subarctic on all relevant time scales, and it characterizes in particular detail the records of past abrupt changes that have had widespread effects.

1) How have temperature and precipitation changed in the Arctic in the past? What does this tell us about Arctic climate that can inform projections of future changes?

This report documents what is known of high-latitude temperature and precipitation during the past 65 million years at a variety of time scales, using sedimentary, biological, and geochemical **proxies**—indirect recorders—obtained largely from ice cores, lake sediment, marine sediment, and tree rings, but also from sediment found in river and coastal bluffs and elsewhere. Sedimentary deposits do not record climate data in the same way that a modern scientific observer does, but climatic conditions control characteristics of many sediments, so these sedimentary characteristics can serve as proxies for the climate that produced them (e.g., Bradley, 1999). (See Chapter 2 for a discussion of proxies.) Some of the many proxies routinely used are:

- the character of organic matter,
- the isotopic geochemistry of minerals or ice,
- the abundance and types of macrofossils and microfossils, and
- the occurrence and character of specific chemicals (**biomarkers**) that record the presence or absence of certain species and of the conditions under which those species grew.

Historical records taken from diaries, notebooks, and logbooks are also commonly used to link modern data with **paleoclimate reconstructions**.

The **proxy** records document large changes in the Arctic. As described in Chapter 3, comparison of Arctic paleoclimatic data with records from lower latitude sites for the same time period shows that temperature changes in the Arctic were greater than temperature changes elsewhere (changes were "amplified"). This **Arctic amplification** occurred for climate changes with different causes. Physical understanding shows that this amplification is a natural consequence of features of the Arctic climate system.

2) How rapidly have temperature and precipitation changed in the Arctic in the past? What do these past rates of change tell us about Arctic climate that can inform projections of future changes?

The climate record of Earth shows changes that operate on many time scales—tens of millions of years for continents to rearrange themselves, to weeks during which particles from a major volcanic eruption spread in the stratosphere and block the sun. This report summarizes paleoclimate data on past rates of change in the Arctic and subarctic on all relevant time scales, and it characterizes in particular detail the records of past abrupt changes that have had widespread effects. This section of the report has been coordinated with CCSP Synthesis and Assessment Product 3.4, the complete focus of which is on global aspects of abrupt climate change.

The data used to assess rates of change in Chapter 4 are primarily the same as those used to assess the magnitudes of change in Chapter 3. However, as discussed in Chapter 3, the existence of high-time-resolution records that cannot always be synchronized exactly to other records, and additional features of the paleoclimatic record, motivate separate treatment of these closely related features of Arctic climate history.

Faster or less expected changes have larger effects on natural and human systems than do slower, better anticipated changes (e.g., National Research Council, 2002). Comparison of projected rates of change for the future (IPCC, 2007) with those experienced in the past can thus provide insights to the level of impacts that may occur. Chapter 4 summarizes rates of Arctic change in the past, compares these with recent Arctic changes and to non-Arctic changes, and assesses processes that contribute to the rapidity of some Arctic changes.

3) What does the paleoclimate record tell us about the past size of the Greenland Ice Sheet and its implications for sea level changes?

Paleoclimate data allow us to reconstruct the size of the *Greenland Ice Sheet* at various times in the past, and they provide insight into the climatic conditions that produced those changes. This report summarizes those paleoclimate data and what they suggest about the mechanisms that caused past changes and might contribute to future changes.

An **ice sheet** leaves tracks—evidence of its passage—on land and in the ocean; those tracks show how far it extended and when it reached that extent, (e.g., Denton et al., 2005). On land, **moraines** (primarily rock material), which were deposited in contact with the edges of the ice, document past ice extents especially well. Beaches now raised out of the ocean following retreat of ice that previously depressed the land surface, and other geomorphic indicators, also preserve important information. Moraines and other ice-contact deposits in the ocean record evidence of extended ice; isotopic ratios of shells that grew in the ocean may reveal input of meltwater, and iceberg-rafted debris identified in sediment cores can be traced to source regions supplying the icebergs (e.g., Hemming, 2004). The history of ice thickness can be traced by use of moraines or other features on rock that projected above the level of the ice sheet, by the history of land rebound following removal of ice weight, and by indications (especially total gas content) in ice cores (Raynaud et al., 1997). Models can also be used to assimilate data from coastal sites and help constrain inland conditions. This report integrates these and other sources of information that describe past changes in the *Greenland Ice Sheet*.

Paleoclimate data allow us to reconstruct the size of the Greenland Ice Sheet at various times in the past, and they provide insight into the climatic conditions that produced those changes.

Changes in glaciers and ice sheets, especially the *Greenland Ice Sheet*, have global repercussions. Complete melting of the *Greenland Ice Sheet* would raise global sea level by 7 meters (m); even partial melting would flood the world's coasts (Lemke et al., 2007). Freshwater from melting ice-sheets delivered to the oceans in sensitive regions—the *North Atlantic Ocean*, for example—could contribute to changes in extent of **sea ice**, ocean circulation, and climate and could produce strong regional and possibly global effects (Meehl et al., 2007).

4) What does the paleoclimate record tell us about past changes in Arctic sea-ice cover, and what implications does this have for consideration of recent and potential future changes?

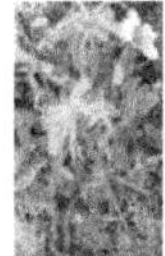

This report documents past periods when the extent of Arctic sea ice was reduced, and it evaluates the scope, causes, and effects of these reductions (e.g., **CAPE**, 2006). The extent of past sea ice and patterns of sea-ice drift are recorded in sediments preserved on the sea floor. Sea-ice extent can also be reconstructed from fossil assemblages preserved in ancient beach deposits along many *Arctic* coasts (Brigham-Grette and Hopkins, 1995; Dyke et al., 1996).

Sea ice fundamentally affects the climate and oceanography of the Arctic (e.g., Seager et al., 2002), its ecosystems, and human use.

Recent advances in tapping the Arctic **paleoceanographic archives**, notably the first deep-sea drilling in the central *Arctic Ocean* (Shipboard Scientific Party, 2005) and the 2005 Trans-Arctic Expedition (Darby et al., 2005), have provided new, high-quality material with which to identify and characterize warm, reduced-ice events of the past, which may serve as analogues for possible future conditions (e.g., Holland et al., 2006). Sea ice fundamentally affects the climate and oceanography of the Arctic (e.g., Seager et al., 2002), its ecosystems, and human use. The implications of reduced sea ice extend throughout the Arctic and beyond, and they bear on such issues as national security and search-and-rescue (National Research Council, 2007).

Italicized locations in the text can be found on a location map (Plate 1) that is provided to help keep the reader spatially oriented.

Report and Chapter Structure

This report is organized into five primary technical chapters. The first of these (Chapter 2) provides a conceptual framework for the information presented in the succeeding chapters, each of which focuses on one of the topics described above. Chapter 2 also contains information on the standardized use of time scales and geological terminology in this report.

A common timeline (Plate 2) is also provided to assist the reader to stay temporally oriented amid the multiple types of event nomenclatures utilized by the paleoclimatic and geologic communities.

Each of the topical chapters (Chapters 3 through 6) answers, in this order, the questions "Why, how, what, and so what?" The "Why" or opening introductory segment for each chapter outlines the relevance of the topic to the issue of modern climate change. The "How"' segment discusses the sources and types of data compiled to build the paleoclimate record and the strengths and weaknesses of the information. The "What" segment is the paleo-record information itself, presented in chronological order, oldest to most recent. The final "So what" segment discusses the significance of the material contained in the chapter and its relevance to current climate change. Each technical chapter is preceded by an abstract that outlines the principal conclusions contained in the body of the chapter itself. Bolded words in the text indicate entries in the technical glossary at the end of this report. Italicized locations in the text can be found on a location map (Plate 1) that is provided to help keep the reader spatially oriented. This figure includes locations referred to in the text that are located above 64° north latitude. A common timeline (Plate 2) is also provided to assist the reader to stay temporally oriented amid the multiple types of event nomenclatures utilized by the paleoclimatic and geologic communities.

The Synthesis and Assessment Product Team

Four of the Lead Authors of this report were constituted as a Federal Advisory Committee (FAC) that was charged with advising the U.S. Geological Survey and the CCSP on the scientific and technical content related to the topic of the paleoclimate history of the *Arctic* as described in the **SAP** 1.2 prospectus. (See Public Law 92-463 for more information on the Federal Advisory Committee Act; see the GSA website *http://fido.gov/facadatabase/* for specific information related to the SAP 1.2 Federal Advisory Committee.) The FAC for **SAP** 1.2 acquired input from more than 30 contributing authors in five countries. These authors provided substantial content to the report, but they did not participate in the Federal Advisory Committee deliberations upon which this SAP was developed.

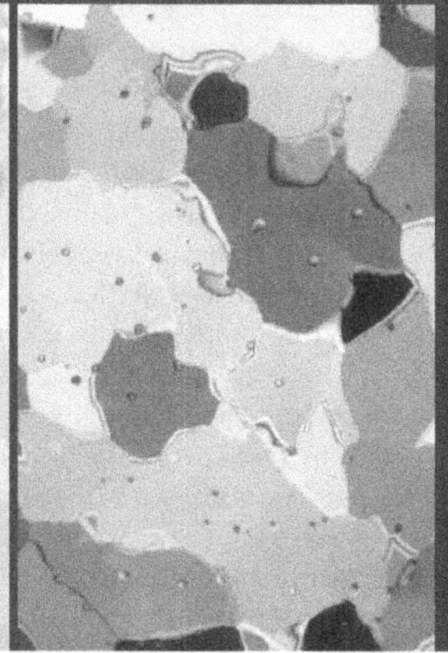

CHAPTER 2

Paleoclimate Concepts

Lead Authors: Richard B. Alley*, Pennsylvania State University, University Park, PA; Joan Fitzpatrick, U.S. Geological Survey, Denver, CO

Contributing Authors: Julie Brigham-Grette*, University of Massachusetts, Amherst, MA; Gifford H. Miller*, University of Colorado, Boulder, CO; Daniel Muhs, U.S. Geological Survey, Denver, CO; Leonid Polyak*, Ohio State University, Columbus, OH

*SAP 1.2 Federal Advisory Committee Member

ABSTRACT

Interpretation of paleoclimate records requires an understanding of Earth's climate system, the causes (forcings) of climate changes, and the processes that amplify (positive feedback) or damp (negative feedback) these changes. Paleoclimatologists reconstruct the history of climate from proxies, which are those characteristics of sedimentary deposits that preserve paleoclimate information. A great range of physical, chemical, isotopic, and biological characteristics of lake and ocean sediments, ice cores, cave formations, tree rings, the land surface itself, and more are used to reconstruct past climate. Ages of climate events are obtained by counting annual layers, measuring effects of the decay of radioactive atoms, assessing other changes that accumulate through time at rates that can be assessed accurately, and using time-markers to correlate sediments with others that have had their ages measured more accurately. Not all questions about the history of Earth's climate can be answered through paleoclimatology: in some cases the necessary sediments are not preserved, or the climatic variable of interest is not recorded in the sediments. Nonetheless, many questions can be answered from the available information.

An overview of the history of Arctic climate during the past 65 million years (m.y.) shows a long-term irregular cooling for tens of millions of years. As ice became established in the Arctic, it grew and shrank for tens of thousands of years in regular cycles. During at least the most recent of these cycles, shorter lived, large, and rapid fluctuations occurred, especially around the North Atlantic Ocean. The last 11,000 years or so have remained generally warm and relatively stable, but with small climate changes of varying spacing and size. Assessment of the causes of climate changes, and the records of those causes, shows that reduction in atmospheric carbon-dioxide concentration and changes in continental positions were important in the cooling trend throughout tens of millions of years. The cycling in ice extent was paced by features of Earth's orbit and amplified by the effects of the ice itself, changes in carbon dioxide and other greenhouse gases, and additional feedbacks. Abrupt climate changes were linked to changes in the circulation of the ocean and the extent of sea ice. Changes in the Sun's output and in Earth's orbit, volcanic eruptions, and other factors have contributed to the natural climate changes since the end of the last ice age.

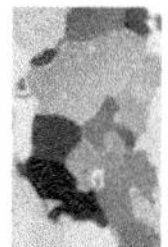

Indications of past climate, called climate proxies, are preserved in geological records; they tell us that Earth's climate has rarely been static.

2.1 INTRODUCTION

Most people notice the weather. Day to day, week to week, and even year to year, changes in such parameters as minimum and maximum daily temperatures, precipitation amounts, wind speeds, and flood levels are all details about the weather that nearly everyone shares in daily conversations. When all else fails, most people can talk about the weather.

Evaluating longer-term trends in the weather (tens to hundreds of years or even longer) is the realm of climate science. "**Climate**"* is the average weather, usually defined as the average of the past 30 years. "**Climate change**" is the long-term change of the average weather, and climate change is the focus of this assessment report. While most people accept that the weather is always changing on the time scale of recent memory, geologists reconstruct climate on longer time scales and use these reconstructions to help understand why climate changes. This improved understanding of Earth's climate system informs our ability to predict future climate change. Reconstructions of past climate also allow us to define the range of natural climate variability throughout Earth's history. This information helps scientists assess whether climate changes observable now may be part of a natural cycle or whether human activity may play a role. The relevance of climate science lies in the recognition that even small shifts in climate can and have had sweeping economic and societal effects (Ladurie, 1971; Lamb, 1997).

Climate change includes long-term trends lasting tens of millions of years and abrupt shifts occurring in as little as a decade or less, both of which have resulted in large-scale reorganizations of oceanic and atmospheric circulation patterns.

Indications of past climate, called climate proxies, are preserved in geological records; they tell us that Earth's climate has rarely been static. For example, during the past 70 million years (**m.y.**), of Earth history, large changes have occurred in average global temperature and in temperature differences between tropical and polar regions, as well as in ice-age cycles during which more than 100 m of sea level was stored on land in the form of giant continental **ice sheets** and then released back to the ocean by melting of that ice. Climate change includes long-term trends lasting tens of millions of years, and abrupt shifts occurring in as little as a decade or less, both of which have resulted in large-scale reorganizations of oceanic and atmospheric circulation patterns. As we discuss in the following

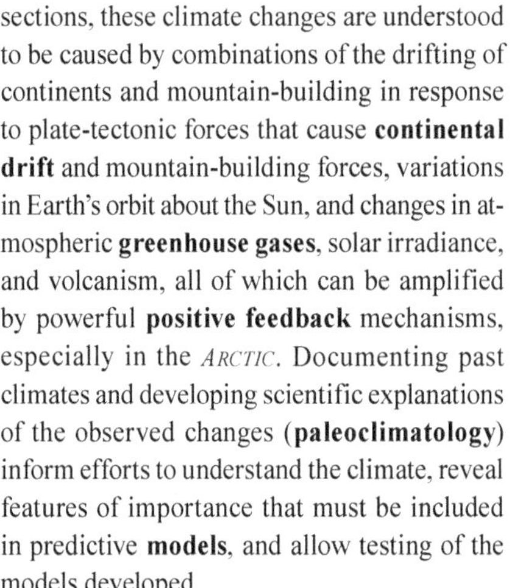

sections, these climate changes are understood to be caused by combinations of the drifting of continents and mountain-building in response to plate-tectonic forces that cause **continental drift** and mountain-building forces, variations in Earth's orbit about the Sun, and changes in atmospheric **greenhouse gases**, solar irradiance, and volcanism, all of which can be amplified by powerful **positive feedback** mechanisms, especially in the *Arctic*. Documenting past climates and developing scientific explanations of the observed changes (**paleoclimatology**) inform efforts to understand the climate, reveal features of importance that must be included in predictive **models**, and allow testing of the models developed.

An overview of key climate processes is provided here, followed by a summary of techniques for reconstructing past climatic conditions. Additional details pertaining to specific aspects of the *Arctic* climate system and its history are presented in the subsequent chapters.

2.2 FORCINGS, FEEDBACK, AND VARIABILITY

An observed change in climate may depend on more than one process. Tight linkages and interactions exist between these processes, as described below, but it is commonly useful to divide these processes into three categories: internal variability, **forcings**, and **feedbacks**. (For additional information, see Hansen et al., 1984, Peixoto and Oort, 1992; or **IPCC**, 2007 among other excellent sources.)

Internal variability is familiar to weather watchers: if you don't like the weather now, wait for tomorrow and something different may arrive. Even though the Sun's energy, Earth's orbit, the composition of the atmosphere, and many other important controls are the same as yesterday, different weather arrives because complex systems exhibit fluctuations within themselves. This variability tends to average out over longer time periods, so climate is less variable than weather; however, even the 30-year averages typically used in defining the climate vary internally. For example, without any external cause, a given 30-year period may have one more El Niño event in the Pacific Ocean, and thus slightly warmer average temperatures, than the previous 30-year period.

*For bold terms, refer to Glossary; for italic terms, refer to Plate 1; for geologic ages, refer to Plate 2.

Forced changes are caused by an event outside the climate system. If the Sun puts out more energy, Earth will warm in response. If fewer volcanoes than average erupt during a given century, then less sunlight than normal will be blocked by particles from those volcanoes, and Earth's surface will warm in response. If burning fossil fuel raises the carbon-dioxide concentration of the atmosphere, then more of the planet's outgoing radiation will be absorbed by that **carbon dioxide**, and Earth's surface will warm in response. Depending on often-random processes, different forcings may combine to cause large climate swings or offset to cause climate changes to be small.

When one aspect of climate changes, whether in response to some forcing or to internal variability, other parts of the climate system respond, and these responses may affect the climate further; if so, then these responses are called feedbacks. How much the temperature changes in response to a forcing of a given magnitude (or in response to the net magnitude of a set of forcings) depends on the sum of all of the feedbacks. Feedbacks can be characterized as positive, serving to amplify the initial change, or negative, acting to partially offset the initial change.

As an example, some of the sunshine reaching Earth is reflected back to space by snow without warming the planet. If warming (whether caused by an El Niño, increased output from the Sun, increased carbon dioxide concentration in the atmosphere, or anything else) melts snow and ice that otherwise would have reflected sunshine, then more of the Sun's energy will be absorbed, causing additional warming and the melting of more snow and ice. This additional warming is a feedback (usually called the ice-albedo feedback). This ice-albedo feedback is termed a positive feedback, because it amplifies the initial change.

2.2.1 The Earth's Heat Budget—A Balancing Act

On time scales of hundreds to thousands of years, the energy received by the Earth from the Sun and the energy returned to space balance almost exactly; imbalance between incoming and outgoing energy is typically less than 1% over periods as short as years to decades (Figure 2.1). This state of near-balance is maintained by the very strong **negative feedback** linked to thermal radiation. All bodies "glow" (send out radiation), and warmer bodies glow more brightly and send out more radiation than cooler ones. (Watching the glow of a burner on an electric stove become visible as it warms shows this effect very clearly.) Some of the Sun's energy reaching Earth is reflected without causing warming, and the rest is absorbed to warm the planet. The warmer the planet, the more energy it radiates back to space. A too-cold planet (that is, a planet colder than the temperature at which it would be in equilibrium) will receive more energy than is radiated, causing the planet to warm, thus increasing radiation from the planet until the incoming and outgoing energy balance.

When one aspect of climate changes, whether in response to some forcing or to internal variability, other parts of the climate system respond, and these responses may affect the climate further; if so, then these responses are called feedbacks.

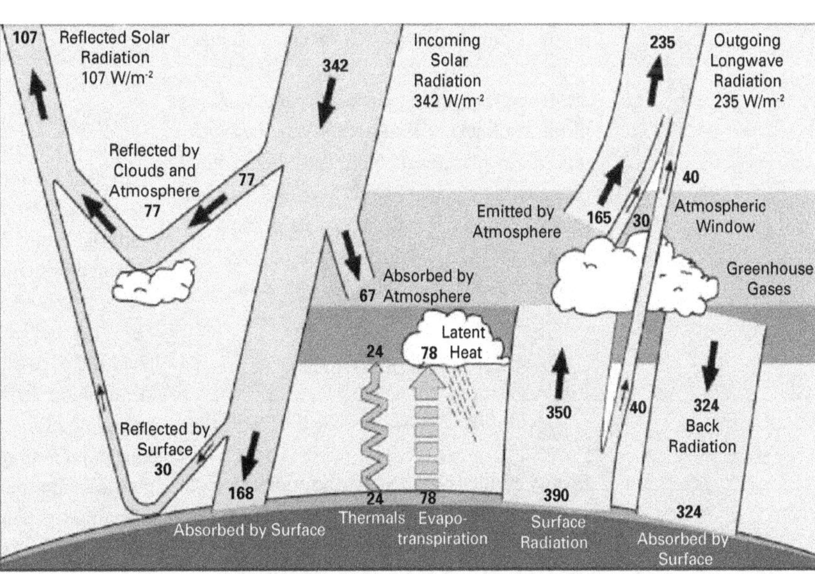

Figure 2.1. Earth's energy budget is a balance between incoming and outgoing radiation. (Numbers are in watts per square meter of the Earth's surface, and some estimates may be uncertain by as much as 20%.) Incoming shortwave radiation from the Sun entering Earth's atmosphere (342 W/m^2) may be reflected by clouds, or absorbed or reflected as longwave radiation by the Earth. The greenhouse effect involves the absorption and reradiation of energy by atmospheric greenhouse gases and particles, resulting in a downward flux of infrared (longwave) radiation from the atmosphere to the surface (back radiation) causing higher surface temperatures. In this figure, Earth is in energy balance with the total rate of energy lost from Earth (107 W/m^2) of reflected sunlight plus 235 W/m^2 of infrared (longwave) radiation equal to the 342 W/m^2 of incident sunlight (Kiehl and Trenberth, 1997). [Copyright 1997 American Meteorological Society, reproduced by permission of American Meteorological Society.]

Similarly, a too-warm planet will radiate more energy than is received from the Sun, producing cooling to achieve balance. Greenhouse gases in the atmosphere block some of the outgoing radiation, transferring some of the energy from the blocked radiation to other air molecules to warm them, or radiating the energy up or down. The net effect is to cause the lower part of the atmosphere (the **troposphere**) and the surface of the planet to be warmer than they would have been in the absence of those greenhouse gases. The global average temperature can be altered by changes in the energy from the Sun reaching the top of our atmosphere, in the reflectivity of the planet (the planet's albedo), or in strength of the greenhouse effect.

Equatorial regions receive more energy from space than they emit to space, polar regions emit more energy to space than they receive, and the atmosphere and ocean transfer sufficient energy from the equatorial to the polar regions to maintain balance (for additional information see Nakamura and Oort, 1988, Peixoto and Oort, 1992, and Serreze et al., 2007).

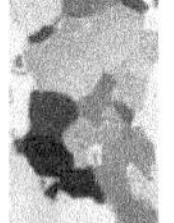

Important **forcings** described later in this section include changes in the Sun; cyclical features of Earth's orbit (**Milankovitch forcing**); changes in greenhouse gas concentrations in Earth's atmosphere; the shifting shape, size, and positions of the continents (**plate tectonics**); biological processes; volcanic eruptions; and other features of the climate system. Other possible forcings, such as changes in cosmic rays or in blocking of sunlight by space dust, cannot be ruled out entirely but do not appear to be important.

2.2.2 Solar Irradiance Forcing

2.2.2a Effects of the Aging of the Sun

Energy emitted by the Sun is the primary driver of Earth's climate system. The Sun's energy, or **irradiance**, is not constant, and changes in solar irradiance force changes in Earth's climate. Our understanding of the physics of the Sun indicates that during Earth's 4.6-billion-year history, the Sun's energy output should have increased smoothly from about 70% of modern output (see, for example, Walter and Barry, 1991). (Direct paleoclimatic evidence of this increase in solar output is not available.) During the last 100 m.y., changes in solar irradiance are calculated to have been less than 1%, or less than 0.000001% per century. Therefore, the effects of the Sun's aging have no bearing on climate change over time periods of millennia or less. For reference, the 0.000001% per century change in output from aging of the Sun can be compared with other changes, for example:

- maximum changes of slightly under 0.1% over 5 to 6 years as part of the sunspot cycle (Foukal et al., 2006);
- the estimated increase from the year 1750 to 2005 in solar output averaged across sunspot cycles, which also is slightly under 0.1% (Forster et al., 2007; see below); and
- the warming effect of carbon dioxide added to the atmosphere from 1750 to 2005. This addition is estimated to have had the same warming effect globally as an increase in solar output of ~0.7% (Forster et al., 2007), and thus it is much larger than changes in solar irradiance during this same time interval.

2.2.2b Effects of Short-Term Solar Variability

Earth-based observations and, in recent years, more-accurate space-based observations document an 11-year solar cycle that results from changes within the Sun. Changes in solar output associated with this cycle cause peak solar output to exceed the minimum value by slightly less than 0.1% (Beer et al., 2006; Foukal et al., 2006; Camp and Tung, 2007). A satellite thus measures a change from maximum to minimum of about 0.9 W/m^2, out of an average of about 1,365 watts per square meter (W/m^2). This value is usually recalculated as a "radiative forcing" for the lower atmosphere. It is divided by 4 to account for spreading of the radiation around the spherical Earth and multiplied by about 0.7 to allow for the radiation that is directly reflected without warming the planet (Forster et al., 2007). The climate response to this sunspot cycling has been estimated as less than 0.1°C (Stevens and North, 1996) to almost 0.2°C (Camp and Tung, 2007). As discussed by Hegerl et al. (2007), the lack of any trend in solar output over longer times than this sunspot cycling, as measured by satellites, excludes the Sun as an important contributor to the strong warming during the interval of satellite observations, but the solar variability may have contributed weakly to temperature trends in the early part of the 20th century.

Over longer time frames, indirect proxies of solar activity (historical sunspot records, tree-rings and ice-cores) also exhibit 11-year solar cycles as well as longer-term variability. Common longer cycles are about 22, 88, and 205 years (e.g., Frohlich and Lean, 2004). The historical climate record suggests that periods of low solar activity may be linked to climate anomalies. For example, the solar minima known as the "Dalton Minimum" and the "Maunder Minimum" (1790–1820 AD, and 1645–1715 AD, respectively) correspond with the relatively cool conditions of the **Little Ice Age**, suggesting a role for changes in solar activity in the climate anomalies (along with other influences; see Chapter 3). However, the magnitude of radiative forcing that can be attributed to variations in solar irradiance remains debated (e.g., Baliunas and Jastrow, 1990; Bard et al., 2000; Fleitmann, et al., 2003; Frolich and Lean, 2004; Amman et al., 2007; Muscheler et al., 2007). An extensive summary of estimates of solar increase since the Maunder Minimum is given by Forster et al. (2007), which lists a preferred value of a radiative forcing of ~0.2 W/m^2, although the report also lists older estimates of just less than 0.8 W/m^2, still well below the estimated radiative forcing of the human-caused increase in atmospheric carbon dioxide (~1.7 W/m^2) (IPCC, 2007).

2.2.3 Orbital Forcing and Milankovitch Cycles

Irregularities in Earth's orbital parameters, often referred to as "Milankovitch variations" or "**Milankovitch cycles**," after the Serbian mathematician who suggested that these irregularities might control ice-age cycles, result in systematic changes in the seasonal and geographic distribution of incoming solar radiation **(insolation)** for the planet (Milankovitch, 1920, 1941). The Milankovitch cycles have almost no effect on total sunshine reaching the planet over time spans of years or decades; they have only a small effect on total sunshine reaching the planet over tens of thousands of years and longer; but they have large effects on north-south and summer-winter distribution of sunshine.

These Milankovitch variations (Figure 2.2) are due to three types of changes:

1. the eccentricity (out-of-roundness) of Earth's orbit around the Sun varies from nearly circular to more elliptical and back over about 100 thousand years (**k.y.**) (E in Figure 2.2);
2. the **obliquity** (how far the North Pole is tilted away from "straight up" out of the plane containing Earth's orbit about the Sun) tilts more and then less over about 41 k.y. (T in Figure 2.2); and
3. the **precession** (the wobble of Earth's rotational axis) moves Earth from its position closest to the Sun in the Northern-Hemisphere summer (the southern winter) to its position farthest from the Sun in the northern summer (the southern winter) and back again in cycles of about 19–23 k.y. (P in Figure 2.2) (e.g., Loutre et al., 2004).

These orbital features are linked to the influence of the gravity of Jupiter and the moon, among others, acting on Earth itself and on the bulge at the equator caused by Earth's rotation. These features are relatively stable and can be calculated for periods of millions of years with high accuracy. Paleoclimatic records show the influence of these changes very clearly (e.g., Imbrie et al., 1993).

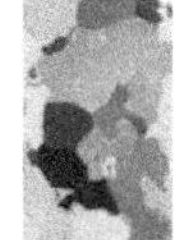

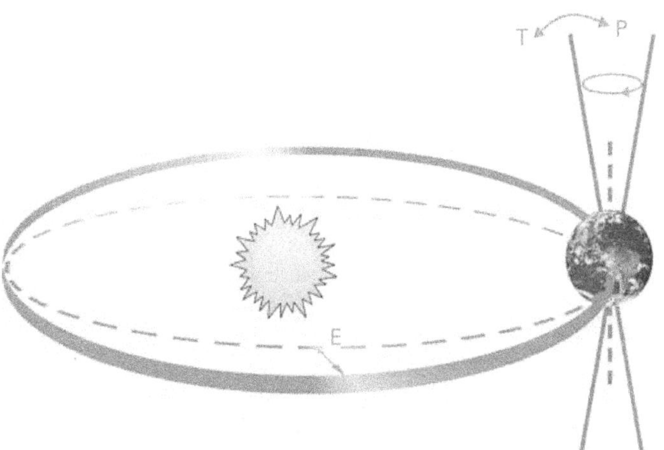

Figure 2.2. Earth's orbital variations (Milankovitch cycles) control the amount of sunlight received (insolation) at a given place on Earth's surface (Rahmstorf and Schellnhuber, 2006; Jansen et al., 2007, in IPCC, 2007, Figure 3.2). E = variation in the eccentricity of the orbit (owing to variations in the minor axis of the ellipse) with an approximate 100 k.y. periodicity; P = precession, changes in the direction of the axis tilt at a given point of the orbit, which has an approximate 19 to 23 k.y. periodicity; T = changes in the tilt (obliquity) of Earth's axis, which has an approximate 41 k.y. periodicity.

In the Arctic, the most important orbital controls are the tilt of Earth's axis and the precession or wobble of Earth's rotational axis

The variations in eccentricity (orbital "out of roundness" or departure from circularity) affect the total sunshine received by the planet in a year, but by less than 0.5% between extremes (hence giving very small changes of less than 0.001% per century). The other orbital variations have essentially no effect on the total solar energy received by the planet as a whole. However, large variations do occur in energy received at a particular latitude and season (with offsetting changes at other latitudes and in other seasons); changes have exceeded 20% in 10,000 years (which is still only 0.2% per century, again with offsetting changes in other latitudes and seasons so that the total energy received is virtually constant).

In the Arctic, the most important orbital controls are the tilt of Earth's axis (T in Figure 2.2), where high tilt angles result in much more high-latitude insolation than do low tilt angles, and the precession or wobble of Earth's rotational axis (P in Figure 2.2). When Earth is closest to the Sun at the summer solstice, insolation is significantly greater than when Earth is at its greatest distance from the Sun at the summer solstice. For example, 11 thousand years ago (**ka**), Earth was closest to the Sun at the Northern Hemisphere summer solstice, but the summer solstice has been steadily moving toward the greatest distance from the Sun since then, such that at present Northern Hemisphere summer occurs when Earth is almost the greatest distance from the Sun, resulting in 9% less insolation in Arctic midsummers today than at 11 ka (Figure 2.3). On the basis of this orbital consideration alone, Arctic summers should have been cooling during this interval in response to Earth's precession.

2.2.4 Greenhouse Gases in the Atmosphere

Roughly 70% of the incoming solar radiation is absorbed by the planet, warming the land, water, and air (Forster et al., 2007). Earth, in turn, radiates energy to balance what it receives, but at a longer wavelength than that of the incoming solar radiation. Greenhouse gases are those gases present in the atmosphere that allow incoming shortwave radiation to pass largely unaffected, but that absorb some of Earth's outgoing longwave radiation band (Figure 2.1). Greenhouse gases play a key role in keeping the planetary temperature within the range conducive to life. In the absence of greenhouse gases in Earth's atmosphere, the planetary temperature would be about –19°C (–2°F); with them, the average temperature is about 33°C (about 57°F) higher (with constant albedo; Hansen et al., 1984; Le Treut et al., 2007). The primary pre-industrial greenhouse gases include, in order of importance, water vapor, carbon dioxide, **methane**, **nitrous oxide**, and tropospheric ozone. Concentrations of these gases are directly affected by **anthropogenic** (human) activities, with the exception of water vapor as discussed below. Purely anthropogenic recent additions to greenhouse gases include a suite of halocarbons and fluorinated sulfur compounds (Ehhalt et al., 2001).

Typically, carbon dioxide is a less important greenhouse gas than water vapor near Earth's surface. Changing the carbon-dioxide concentration of the atmosphere is relatively easy, but changing the atmospheric concentration of water vapor to any appreciable degree is difficult except by changing the temperature.

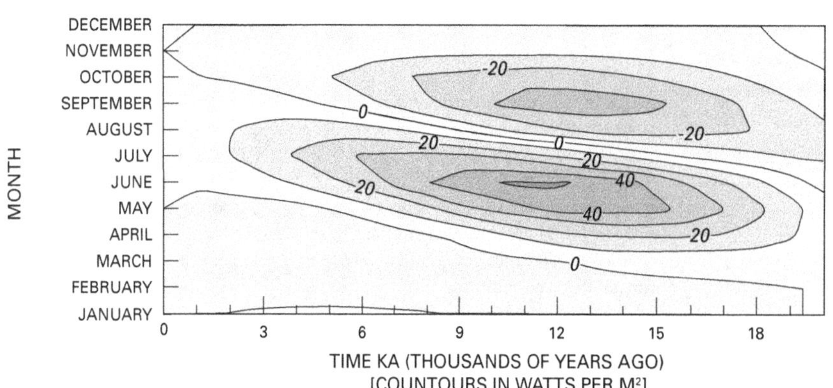

Figure 2.3. Milankovitch-driven monthly insolation anomalies (deviations from present), 20–0 ka at 60°N. Y axis, calendar months. Contours and numbers depict a history of insolation values. Contours in watts per square meter (W/m^2) (data from Berger and Loutre, 1992). Midsummer insolation values at 11 ka exceeded 40 W/m^2, whereas current values are less than 10 W/m^2.

Natural fluxes of water vapor into and out of the atmosphere are very large, equivalent to a layer of water across the entire surface of Earth of about 2 centimeters per week (cm/week) (e.g., Peixoto and Oort, 1992); human perturbations to these fluxes are relatively very small (Forster et al., 2007). However, the large ocean surface and moisture from plants provide important water sources that can yield more water vapor to warmer air; relative humidity tends to remain nearly constant as climate changes, so warming for any reason introduces more water vapor to the air and increases the greenhouse effect in a positive feedback (Hansen et al., 1984; Pierrehumbert et al., 2007). Hence, discussions of forcing of changes in climate focus especially on carbon dioxide, and to a lesser degree on methane and other greenhouse gases, rather than on water vapor (Forster et al., 2007).

Carbon dioxide concentrations in the atmosphere are tied into an extensive natural system of terrestrial, atmospheric, and oceanic sources and sinks called the global carbon cycle (see Prentice et al. (2001) in the **IPCC** 3rd Assessment Report for a comprehensive discussion). The possible effect of increasing **CO_2** levels in the atmosphere was first recognized by Arrhenius (1896). By the 1930s, mathematical models linking greenhouse gases and climate change (Callendar, 1938) projected that a doubling of atmospheric CO_2 concentration would increase the mean global temperature by 2°C and would warm the poles considerably more. (Le Treut et al. (2007) provides a detailed historical perspective on the recognition of Earth's greenhouse effect.) By the 1970s, **CH_4**, **N_2O** and **CFCs** were widely recognized as important additional anthropogenic greenhouse gases (Ramanathan, 1975).

The direct relationship between climate change and greenhouse gases such as CO_2 and methane is clearly described by the recent **Intergovernmental Panel on Climate Change** report (IPCC, 2007). Information summarized there highlights the likelihood that changes in concentrations of greenhouse gases will especially affect the Arctic (Figure 2.4) and focuses attention on greenhouse gases as well as other influences on the Arctic, as discussed in this report especially in Chapter 3 (Temperature and Precipitation History of the Arctic).

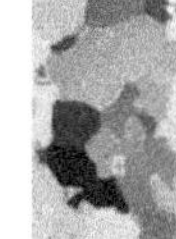

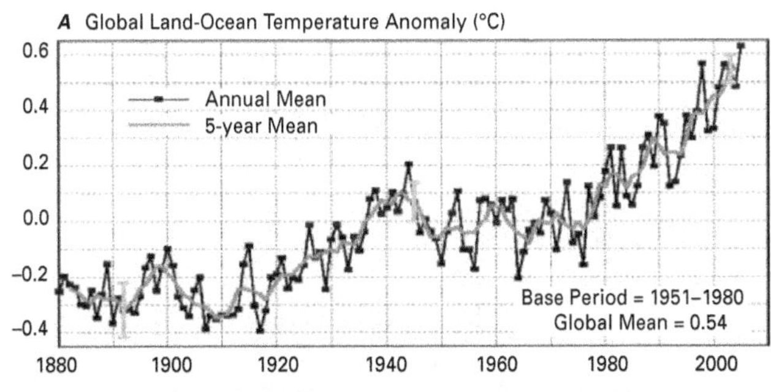

B 2001–2005 Mean Surface Temperature Anomaly (°C)

−2.0 −1.6 −1.2 −0.8 −0.4 −0.2 0.2 0.4 0.8 1.2 1.6 2.1

Figure 2.4. Mean surface temperature anomalies for Earth relative to 1951–1980. *A*) the global average. *B*) temperature anomalies 2000–2005. High northern latitudes show the largest anomalies for this time period (Hansen et al., 2006). [Copyright 2006 National Academy of Sciences, United States. Reproduced by permission of National Academy of Sciences.]

2.2.5 Plate Tectonics

The drifting of continents (explained by the theory of plate tectonics) moves land masses from equator to pole or the reverse, opens and closes oceanic "gateways" between land masses thus redirecting ocean currents, raises mountain ranges that redirect winds, and causes other changes that may affect climate. These changes can have very large local to regional effects (moving a continent from the pole to the equator obviously will greatly change the climate of that continent). Moving continents around may have some effect on the average global temperature, in part through changes in the planet's albedo (Donnadieu et al., 2006).

During millions of years, the atmospheric concentration of carbon dioxide is controlled primarily by the balance between carbon-dioxide removal through chemical reactions with rocks near Earth's surface, and carbon-dioxide release from volcanoes or other pathways involving melting or heating of rocks that sequester carbon dioxide.

Processes linked to continental rearrangement can strongly affect global climate by altering the composition of the atmosphere and thus the strength of the greenhouse effect, especially through control of the carbon-dioxide concentration of the atmosphere (e.g., Berner, 1991; Royer et al., 2007). Duirng millions of years, the atmospheric concentration of carbon dioxide is controlled primarily by the balance between carbon-dioxide removal through chemical reactions with rocks near Earth's surface, and carbon-dioxide release from volcanoes or other pathways involving melting or heating of rocks that sequester carbon dioxide. Because higher temperatures cause carbon dioxide to react more rapidly with Earth-surface rocks, atmospheric warming tends to speed removal of carbon dioxide from the air and thus to limit further warming, in a negative feedback (Walker et al., 1981). Because the tectonic processes causing continental drift control the rate of volcanism, and can change over millions of years, changes in atmospheric carbon-dioxide concentration can be forced by the planet beneath.

2.2.6 Biological Processes

Biological processes can both absorb and release carbon dioxide, such that evolutionary changes have contributed to atmospheric changes. For example, some carbon dioxide taken from the air by plants is released by their roots into the soil, by respiration while living and by decay after death. Thus, plants speed the reaction of atmospheric carbon dioxide with rocks (Berner, 1991; Beerling and Berner, 2005). This process could not have occurred on the early Earth before the evolution of plants with roots.

Plants are composed in part of carbon dioxide removed from the atmosphere, and burning (oxidation) of plants releases most of this carbon dioxide back to the atmosphere (minus the small fraction that reacts with rocks in the soil). When plants are buried without burning and altered to form fossil fuels, the atmospheric carbon-dioxide level is reduced; later, natural processes may bring the fossil fuels back to the surface to decompose and release the stored carbon dioxide. (Humans are greatly accelerating these natural processes; fossil fuels that required hundreds of millions of years to accumulate are being burned in hundreds of years.) Rapid burial favors preservation of organic matter, whereas dead things left on the surface will decompose. Thus, changes in rates of sediment deposition linked to continental rearrangement are among the processes that may affect the formation and breakdown of fossil fuels and thus the strength of the atmospheric greenhouse effect.

Continents move more or less as rapidly as fingernails grow, so that a major reshuffling of the continents requires about 100 million years, and the opening or closing of an oceanic gateway may require millions of years (e.g., Livermore et al., 2007). Major evolutionary changes have required millions of years or longer (e.g., d'Hondt, 2005). Thus, those changes in the greenhouse effect that modified Earth's climate or were linked to continental drift or biological evolution have been highly influential over time spans of tens of millions of years, but they have had essentially no effect over shorter intervals of centuries or millennia. (Note that if one considers hundreds of thousands of years or longer, an increase in volcanic activity may notably increase carbon dioxide in the atmosphere, causing warming. However, volcanic release of carbon dioxide is small enough that in a few millennia or less the changes in volcanic release have not notably affected the carbon-dioxide concentration of the atmosphere. The main short-term effect of an increase in volcanic eruptions is to cool the planet by blocking the Sun, as discussed next.)

2.2.7 Volcanic Eruptions

Volcanic eruptions are an important natural cause of climate change on seasonal to multidecadal time scales. Large explosive volcanic eruptions inject both particles and gases into the atmosphere. Particles are removed by gravity in days to weeks. Sulfur gases, in contrast, are converted rapidly to sulfate aerosols (tiny droplets of sulfuric acid) that have a residence time in the stratosphere of about 3 years and are transported around the world and poleward by circulation within the stratosphere. Tropical eruptions typically influence both hemispheres, whereas eruptions at middle to high latitudes usually affect only the hemisphere of eruption (Shindell et al., 2004; Fischer et al., 2007). Consequently, the *ARCTIC* is affected primarily by tropical and Northern Hemisphere eruptions.

The radiative and chemical effects of the global volcanic aerosol cloud produce strong responses in the climate system on short time scales (see Figure 4.6) (Briffa et al., 1998; deSilva and Zielinski, 1998; Oppenheimer, 2003). By scattering and reflecting some solar radiation back to space, the aerosols cool the planetary surface, but by absorbing both solar and terrestrial radiation, the aerosol layer also heats the stratosphere. A tropical eruption produces more heating in the tropics than in the high latitudes and thus a steeper temperature gradient between the pole and the equator, especially in winter. In the Northern Hemisphere winter, this steeper gradient produces a stronger jet stream and a characteristic stationary tropospheric wave pattern that brings warm tropical air to Northern Hemisphere continents and warms winter temperatures. Because little solar energy reaches the Arctic during winter months, the transfer of warm air from tropical sources to high latitudes has more effect on winter temperatures than does the radiative cooling effect from the aerosols. However, during the summer months, radiative cooling dominates, resulting in anomalously cold summers across most of the Arctic. The 1991 Mt. Pinatubo eruption in the Philippines resulted in volcanic aerosols covering the entire planet, producing global-average cooling, but winter warming over the Northern Hemisphere continents in the subsequent two winters (Stenchikov et al., 2004, 2006).

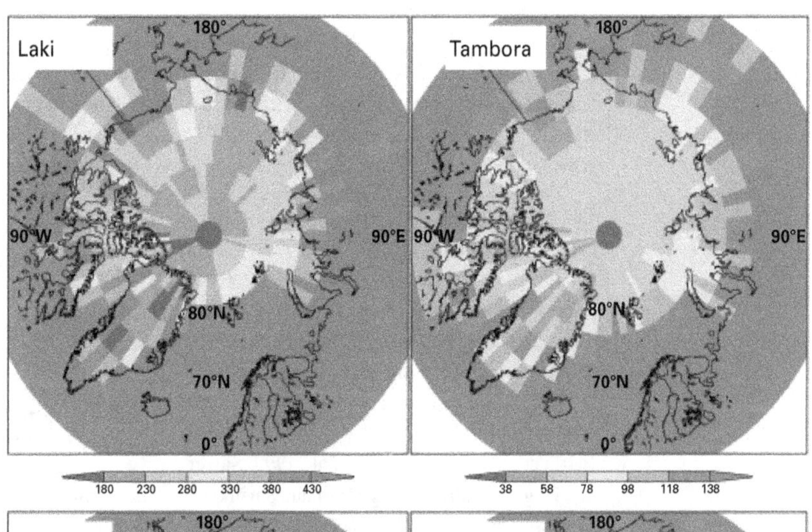

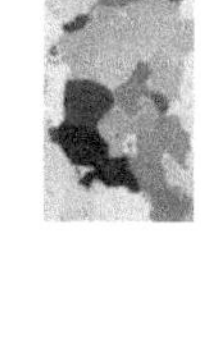

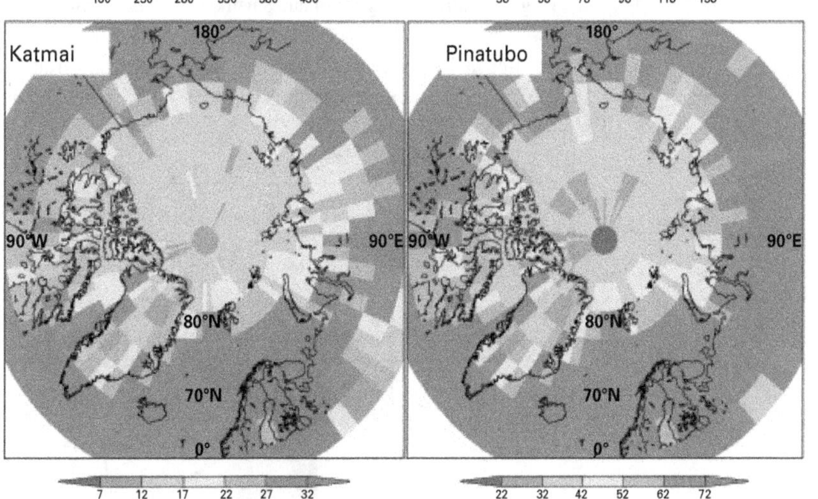

Figure 2.5. Simulated spatial distribution of volcanic sulfate aerosols (kg/km²) produced by the Laki (1783), Katmai (1912), Tambora (1815), and Pinatubo (1991) eruptions in the Arctic (region shown, 66°–82°N. and 50°–35°W.). Blue = smaller than average deposits; yellow, orange, and red = increasingly larger than average deposits (from Gao et al., 2007). Volcanic evidence derived from 44 ice cores; analysis used the NASA Goddard Institute for Space Studies (GISS) ModelE climate model. [Copyright 2007 American Geophysical Union. Reproduced by permission of American Geophysical Union.]

Three large historical Northern Hemisphere eruptions have been studied in detail: the 939 AD *Eldgjá (Iceland)*, 1783–1784 AD *Laki (Iceland)*, and 1912 AD Novarupta (Katmai, *Alaska*) eruptions. All caused cooling of the Arctic during summer but no winter warming (Thordarson et al., 2001; Oman et al., 2005, 2006).

When widespread stratospheric volcanic aerosols settle out, some of the sulfate falls onto the Antarctic and *Greenland Ice Sheets* (Figure 2.5). Measurements of those sulfates present in ice cores can be used to estimate the Sun-blocking effect of the eruption. Large volcanic eruptions, especially those within a few decades of each other, are thought to have promoted cooling during the Little Ice Age (about 1280–1850 AD) (Anderson et al., 2008). A comprehensive review of the effects of volcanic eruptions on climate and of records of past volcanism is provided by Robock (2000, 2007).

The effects of volcanic eruptions are clearly evident in ice-core records (e.g., Zielinski et al., 1994); major eruptions cooled *Greenland* about 1°C for about 1 or 2 years as recorded in *Greenland* ice cores (e.g., Stuiver et al., 1995) (Figure 2.6). Tree-ring records also support the connection between climate and volcanic eruptions (LaMarche and Hirschbeck, 1984; Briffa et al., 1998; D'Arrigo et al., 1999; Salzer and Hughes, 2007). The growth and shrinkage of the great ice-age ice sheets, and the associated loading and unloading of Earth, may have affected the frequency of volcanic eruptions somewhat (e.g., Maclennan et al., 2002), but in general the recent timing of explosive volcanic eruptions appears to be random There is no mechanism for a volcano in, say, *Alaska* to synchronize its eruptions with a volcano in Indonesia; hence, volcanic eruptions in recent millennia appear to have introduced unavoidable climatic "noise" as opposed to controlling the climate in an organized way.

2.2.8 Other Influences

Paleoclimatic records discount some speculative mechanisms of climate change. For example, about 40,000 years ago natural fluctuations reduced the strength of Earth's magnetic field essentially to zero for about one millennium. The cosmic-ray flux into the Earth system increased greatly, as recorded by a large peak in beryllium-10 in sedimentary records. However, the climate record does not change in parallel with changes in beryllium-10, indicating that the cosmic-ray increase had little or no effect on climate (Muscheler et al., 2005). Large changes in concentration of extraterrestrial dust between Earth and Sun might lead to changes in solar energy reaching Earth and thus to changes in climate; however, the available sedimentary records show no significant changes in the rate of infall of such extraterrestrial dust (Winckler and Fischer, 2006).

The climate is a complex, integrated system, and it operates through strong linked feedbacks, internal variability, and numerous forcings. On time scales of centuries or less, however, many of the drivers of past climate change—such as drifting continents, biological evolution, aging of the Sun, and features of Earth's orbit—have no discernible influence on the climate. Small variations in climate appear to have been caused by small variations in the Sun's output, occasional short-lived cooling caused by explosive volcanic eruptions, and greenhouse-gas changes have affected the planet's temperature.

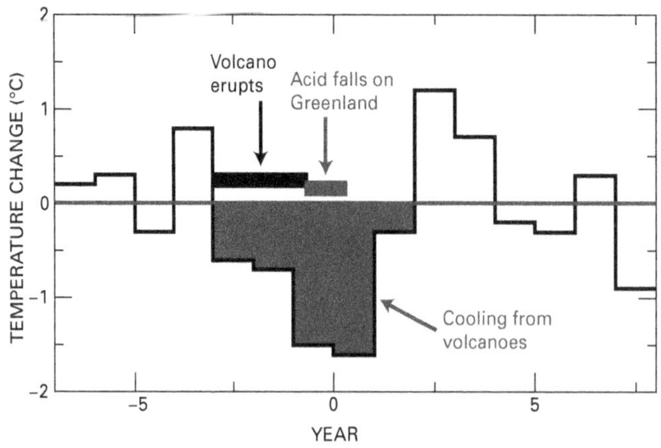

Figure 2.6. Temperature response (derived from stable isotopes) in Greenland snow to large volcanic eruptions reconstructed from the GISP2 ice core (modified from Stuiver et al., 1995). [Copyright 1995, reproduced by permission of Elsevier.]

2.3 READING THE HISTORY OF CLIMATE THROUGH PROXIES

A modern historian trying to understand our human story cannot go back in time and replay an important event. Instead, the historian must rely on indirect evidence: eyewitness accounts (which may not be highly accurate), artifacts, and more. It is as if the historical figures, who cannot tell their tale directly, have given their proxies to other people and other things to deliver the story to the modern historian.

Historians of climate—paleoclimatologists—are just like other historians: they read the indirect evidence that the past sends by **proxy**. All historians are aware of the strengths and weaknesses of proxy evidence, of the value of weaving multiple strands of evidence together to form the complete fabric of the story, of the necessity of knowing when things happened as well as what happened, and of the ultimate value of using history to inform understanding and guide choices.

Some of the proxy evidence used by paleoclimatologists would be familiar to more-traditional historians. Written accounts of many different activities often include notes on the weather, on the presence or absence of ice on local water bodies, and on times of planting or harvest and the crops that grew or failed. If care is taken to account for the tendency of people to report the rare rather than the commonplace, and to include the effects of changes in husbandry and other issues, written records can contribute to knowledge of climate back through written history. However, human accounts are lacking for almost all of Earth's history. The paleoclimatologist is forced to rely on evidence that is less familiar to most people than are written records. Remarkably, these natural proxies may reveal even more than the written records.

2.3.1 Climate's Proxies

Much of the history of a civilization can be reconstructed from the detritus its people left behind. Similarly, paleoclimate records are typically developed through analysis of sediment, broadly defined. "Sediment" may include the ice formed as years of snowfall pile up into an ice sheet, the mud accumulating at the bottom of the sea or a lake, the annual layers of a tree, the thin sheets of mineral laid one on top of another to form a stalagmite in a cave, the piles of rock bulldozed by a **glacier**, the piles of desert sand shaped into dunes by the wind, the odd things collected and stored by packrats, and more (e.g., Crowley and North, 1991; Bradley, 1999; Cronin, 1999). For a sediment to be useful, it must do the following:

1. preserve a record of the conditions when it formed (i.e., subsequent events cannot have erased the original story and replaced it with something else);
2. be interpretable in terms of climate (the characteristics of the deposit must uniquely relate to the climate at the time of formation); and
3. be "datable" (i.e., there must be some way to determine the time when the sediment was deposited).

Here, we first present one well-known paleoclimatic indicator as an example, then discuss general issues raised by that example, and follow with a discussion of many types of paleoclimatic indicators.

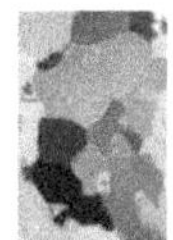

Long records of Earth's climate are commonly reconstructed from climate proxies preserved in deep-ocean sediments. One of the best-known proxy records of climate change is that recorded by benthic (bottom-dwelling) **foraminifers**, microscopic organisms that live on the sea floor and secrete calcium-carbonate shells in equilibrium with the sea water. The isotopes of oxygen in the carbonate are a function of both the water temperature (which often does not change very rapidly with time or very steeply with space in the deep ocean) and changes in global ice volume. Global ice volume determines the relative abundances of the isotopes oxygen-16 and oxygen-18 in seawater. Snow has relatively less of the heavy oxygen-18 than its seawater source. Consequently, as ice sheets grow on land, the ocean becomes enriched in the heavy oxygen-18, and this enrichment is recorded by the oxygen isotope composition of foraminifer shells. The proportion of the heavy and light isotopes of oxygen is usually expressed as **$\delta^{18}O$**; positive $\delta^{18}O$ values represent extra amounts of the heavy isotope of oxygen, and negative values represent samples with less of the heavy isotope than average seawater. Positive $\delta^{18}O$ reflects **glacial** times (colder, more ice), whereas more negative $\delta^{18}O$ reflects **interglacial** (warmer, less ice) times in Earth's history. Although the $\delta^{18}O$ of foraminifer shells does not reveal where the glacial ice was located, the record does provide a globally integrated value of the amount of glacial ice on land, especially if appropriate corrections are made for temperature changes by use of other indicators. In the absence of changes in global ice volume, changes in **benthic foraminifer** $\delta^{18}O$ reflect changes in ocean temperatures: more positive $\delta^{18}O$ values indicate colder water, and more negative $\delta^{18}O$ values indicate warmer water.

In the absence of changes in global ice volume, changes in benthic foraminifer $\delta^{18}O$ reflect changes in ocean temperatures: more positive $\delta^{18}O$ values indicate colder water, and more negative $\delta^{18}O$ values indicate warmer water.

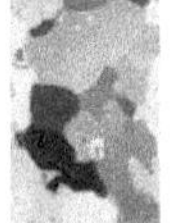

Written documents have sometimes been erased and rewritten, in a deliberate attempt to distort history or because the paper was more valuable than the original words. Paleoclimatologists are continually watching for any signs that a climate record has been "erased" and "rewritten" by events since deposition of the sediment. Occasionally, this vigilance proves to be important. For example, water may remove isotopes carrying paleoclimatic information from shells and replace them with other isotopes telling a different story (e.g., Pearson et al., 2001). However, except for the very oldest deposits from early in Earth's history, it is usually possible to tell whether a record has been altered, and this problem should not affect any of the conclusions presented in this report.

Finding the link between climate and some characteristic of the sediment is then required. The climate is recorded in myriad ways by physical, biological, chemical, and isotopic characteristics of sediments.

Physical indicators of past climate are often easy to read and understand. For example, a sand dune can form only if dry sand is available to be blown around by the wind, without being held down by plant roots. Except near beaches (where fluctuations in water level reveal bare sand), a dry climate is needed to keep grass off the sand so the sand can blow around. Today in northwestern Nebraska, the huge dune field of the Sand Hills is covered in grass (Figure 2.7). The dunes formed during drier conditions in the past, but wetter conditions now allow grass to grow on top (e.g., Muhs et al., 1997). Similarly, the sediments left by glaciers are readily identified, and those sediments in areas that are ice free today attest to changing climate. A very different physical indicator of past climate is the temperatures measured in boreholes. Just as a Thanksgiving turkey placed in an oven takes a while to warm in the middle, the two-mile-thick ice sheet of *GREENLAND* has not finished warming from the ice age, and the cold temperatures at depth reveal how cold the ice age was (Cuffey and Clow, 1997).

Many paleoclimate records are based directly on living things. **Tundra** plants are quite different from those living in temperate forests. If pollen, seeds, and twigs found in deep layers of a sediment core came from tundra plants, and those found in shallow layers came from temperate-forest plants, a formerly cold time that has warmed is indicated. Trees grow more rapidly and add thicker rings when climatic conditions are more favorable. In very dry regions, this feature allows trees to be used in reconstruction of rainfall; in cold regions, growth may be more closely linked to temperature (Fritts, 1976; Cook and Kairiukstis, 1990).

Chemical analysis of sediments may reveal additional information about past climates. As one example, some single-celled organisms in the ocean change the chemistry of their cell walls in response to changing temperature: they use more-flexible molecules to offset the increase in brittleness caused by colder temperatures. These molecules are sturdy and persist in sediments after the organism dies, so the history of the ratio of stiffer to less-stiff molecules in a sediment core provides a history of the temperature at which the organisms grew. (In this case, the organisms are **prymnesiophyte algae**, the chemicals are **alkenones**, and the frequency of carbon double bonds controls the stiffness (Muller et al., 1998); other such indicators exist).

Isotopic ratios are among the most commonly used proxy indicators of past climates. Consider just one example, providing one of the ways to determine the past concentration of carbon dioxide. All carbon atoms have 6 protons in their nuclei, most have 6 neutrons (making carbon-12), but some have 7 neutrons (carbon-13) and a few have 8 neutrons (radioactive carbon-14). The only real difference between carbon-12 and carbon-13 is that carbon-13 is a bit heavier. The lighter carbon-12 is "easier" for plants to use, so growing plants preferentially incorporate carbon from carbon dioxide containing only carbon-12 rather than carbon-13. However, if

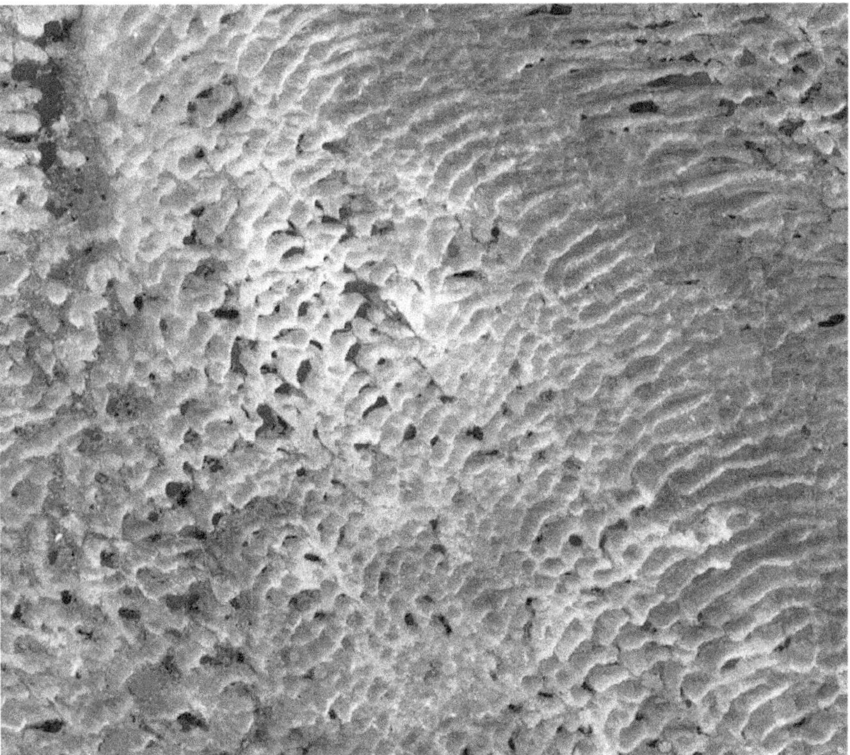

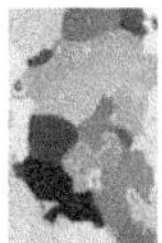

Figure 2.7. The Sand Hills of western Nebraska. The Sand Hills cover 51,400 km^2 (about a quarter of the state) and are the largest sand-dune deposit in the United States. They derive from Pleistocene glacial outwash eroded from the Rocky Mountains and now stabilized by vegetation. The hills are characterized by crowded crescent-shaped (barchan) dunes, general absence of drainage, and numerous tiny lakes filling the closed depressions between dunes. [Photo credit: NASA/GSFC/METI/ERSDAC/JAROS, and U.S./Japan ASTER Science Team. This ASTER simulated natural color image was acquired September 10, 2001, covers an area of about 57.9 × 61.6 km, and is centered near 42.1°N. and 102.2°W.]

carbon dioxide is scarce in the environment, the plants cannot be picky and must use what is available. Hence, the carbon-12:carbon-13 ratio in plants provides an indicator of the availability of carbon dioxide in the environment. The sturdy cell-wall chemicals described in the previous paragraph can be recovered and their carbon isotopes analyzed, providing an estimate of the carbon-dioxide concentration at the time the algae grew (e.g., Pagani et al., 1999).

Much of the science of paleoclimatology is devoted to calibration and interpretation of the relation between sediment characteristics and climate (see National Research Council, 2006). The relationship of some indicators to climate is relatively straightforward, but other relationships may be complex. The width of a tree ring, for example, is especially sensitive to water availability in dry regions, but it may also be influenced by changes in shade from neighboring trees, an attack of beetles or other pests that weaken a tree, the temperature of the growing season, and more. Extensive efforts go into calibration of paleoclimatic indicators against the climatic variables. Because paleoclimatic data cannot be collected everywhere, additional work is devoted to determining which areas of the globe have climates that can be reconstructed from the available paleoclimatic data. Wherever possible, multiple indicators are used to reconstruct past climates and to assess agreement or disagreement (National Research Council, 2006). Conclusions about climate typically rest on many lines of evidence.

2.3.2 The Age of the Sediments

History requires "when" as well as "what." Many techniques reveal the "when" of sediments, sometimes to the nearest year. In general, more-recent events can be dated more precisely.

Climate records that have been developed from most trees, and from some ice cores and sediment cores, can be dated to the nearest year by counting annual layers. The yearly nature of tree rings from seasonal climates is well known. A lot of checking goes into demonstrating that layers observed in ice cores and special sediment cores are annual, but in some cases the layering clearly is annual (Alley et al., 1997), allowing quite accurate counts. The longest-lived trees may be 5,000 years old; use of overlapping living and dead wood has allowed extension of records to more than 10,000 years (Friedrich et al., 2004); and the longest annually layered ice cores recovered to date extend beyond 100,000 years (Meese et al., 1997). However, relatively few records can be absolutely dated in this way.

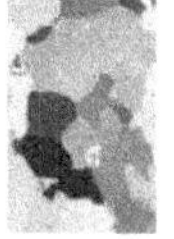

Other techniques that have been used for dating include measuring the damage that accumulates from cosmic rays striking things near Earth's surface (those rays produce beryllium-10 and other isotopes), observing the size of lichen colonies growing on rocks deposited by glaciers, and identifying the fallout of particular volcanic eruptions that can be dated by historical accounts or annual-layer counting.

Most paleoclimatic dating uses the decay of radioactive elements. Radiocarbon is commonly used for samples containing carbon from the most recent 40,000 years or so (very little of the original radiocarbon survives in older samples, causing measurements difficulties and allowing even trace contamination by younger materials to cause large errors in estimated age, so other techniques are preferred). Many other isotopes are used for various materials and time intervals, extending back to the formation of Earth. Intercomparison with annual-layer counts, with historical records, and between different techniques shows that quite high accuracy can be obtained, so that it is often possible to have errors in age estimates of less than 1%. (That is, if an event is said to be 100,000 years old, the event can be said with high confidence to have occurred sometime between 99,000 years and 101,000 years ago.)

2.4 CENOZOIC GLOBAL HISTORY OF CLIMATE

As emphasized in the Summary for Policymakers of **IPCC** (2007) and in the body of that report, a paleoclimatic perspective is important for understanding Earth's climate system and its forcings and feedbacks. Arctic records, and especially Arctic ice-core records, have provided key insights. The discussion that follows briefly discusses selected features in the history of Earth's climate and the forcings and feedbacks of those climate events. This discussion does not treat all of the extensive literature on these topics, but it is provided here as a primer to help place the main results of this report in context. (Kump et al. (2003) contains a more-complete yet accessible introduction to this topic.)

This report focuses on the **Cenozoic** Era, which began about 65 **Ma** with the demise of the dinosaurs and continues today (see section 2.5 for a discussion of the chronology used in this report). During most of this 65 m.y. interval, deep-sea records of foraminifer $\delta^{18}O$ (a powerful paleoclimatic indicator, described above in section 2.3.1), which integrate the sedimentary record in several ocean basins, show that Earth was warmer than at present and supported a smaller volume of ice (Figure 2.8). Yet, following the peak warming of the early **Eocene**, about 50–55 Ma, global temperatures generally declined (Miller et al., 2005). Although this record is not specific about Arctic climate change, the record indicates that the global gradient (or difference) in temperature between polar regions and the tropics was smaller when global climate was warmer, and that this gradient increased as the high latitudes progressively cooled (Barron and Washington, 1982). Changes in the gradient cause changes in atmospheric and oceanic circulation. The overall cooling trend of the past 55 m.y. was punctuated by intervals during which the cooling was reversed and the oceans warmed, only to cool rapidly again at a later time. Examples of such accelerated cooling include rapid increases in foraminifer $\delta^{18}O$ about 34 Ma and again about 23 Ma, which are thought to reflect the rapid buildup of ice in Antarctica in only a few hundred thousand years (Zachos et al., 2001). The **Paleocene-Eocene thermal maximum** (about 55 Ma) represents a major interval of global warming when CO_2 levels are estimated to have risen abruptly (Shellito et al.,

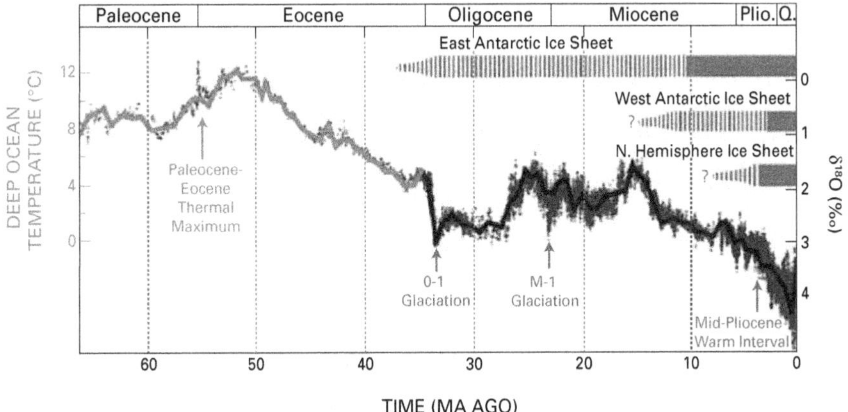

Figure 2.8. Global compilation of more than 40 deep sea benthic $\delta^{18}O$ isotopic records taken from Zachos et al. (2001), updated with high-resolution Eocene through Miocene records from Billups et al. (2002), Bohaty and Zachos (2003), and Lear et al. (2004). Lefthand y-axis refers only to solid red portion of the curve, when the oceans were ice free. Dashed blue bars = times when glaciers came and went or were smaller than now; solid blue bars = ice sheets of modern size or larger. [Figure and text modified from IPCC Chapter 6, Paleoclimate, Jansen et al., 2007, in IPCC, 2007, Figure 6.1.]

2003, Higgins and Schrag, 2006), perhaps owing to the rapid release of methane from sea-floor sediments (Bralower et al., 1995).

The style and tempo of global climate change during the past 5.3 m.y. is depicted well by the foraminifer $\delta^{18}O$ record of Lisiecki and Raymo (2005) (Figure 2.9; see section 2.3.1 for a discussion of this proxy). This composite record provides a well-dated stratigraphic tool against which other records from around world can be compared. The foraminifer $\delta^{18}O$ record reflects changes in both global ice volume and ocean bottom-water temperature change, and with the same sense—an increase in global ice or a decrease in ocean temperatures pushes the indicator in the same direction. The foraminifer $\delta^{18}O$ record indicates low-magnitude climate changes from 5.3 until about 2.7 Ma, when the amplitude of the foraminifer $\delta^{18}O$ signal increased markedly. This shift in foraminifer $\delta^{18}O$ amplitude coincides with widespread indications of onset of northern continental **glaciation** (see Chapter 3, Temperature and Precipitation History of the Arctic). The oxygen isotope fluctuations since 2.7 Ma are commonly used as a global index of the frequency and magnitude of glacial-interglacial cycles. In addition to the fluctuations, the data show that within the past 3 m.y., average ocean temperatures have been dropping. Global circulation models constrained by extensive paleoclimatic data targeting the late **Pliocene** interval from 3.3 to 3.0 Ma suggest that global temperatures were warmer by as much as 2°C or 3°C at that time (see Jiang et al., 2005; IPCC, 2007).

The large fluctuations in foraminifer $\delta^{18}O$ beginning about 2.7 Ma exhibited clear periodicities matching those of the Milankovitch forcing (those periodicities are also present in smaller, older fluctuations). A 41 k.y. periodicity was especially apparent, as well as the 19–23 k.y. periodicity. More recently, within the last 0.9 m.y. or so, the variations in $\delta^{18}O$ became even bigger, and while the 41 k.y. and 19–23 k.y. periodicities continued, a 100 k.y. periodicity became dominant. The reasons for this shift remain unclear and are the focus of much research (Clark et al., 2006; Ruddiman, 2006; Huybers, 2007; Lisiecki and Raymo, 2007).

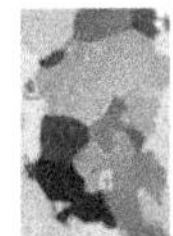

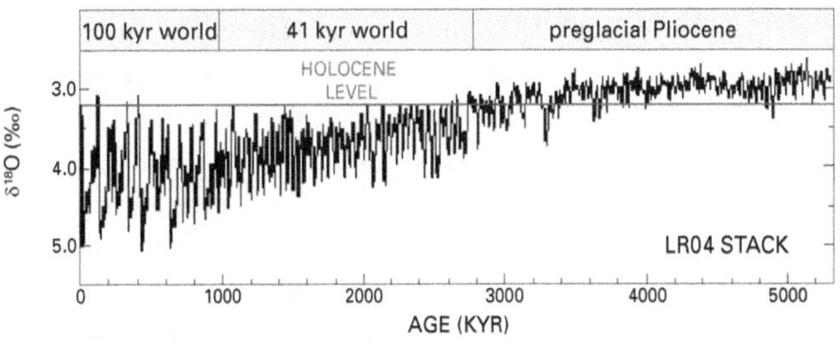

Figure 2.9. Composite stack of 57 benthic oxygen isotope records (a proxy for temperature) from a globally distributed network of marine sediment cores. This foraminifer $\delta^{18}O$ record indicates low-magnitude climate changes from about 5.3–2.7 Ma, when the amplitude of the foraminifer $\delta^{18}O$ signal increased markedly. [Data from Lisiecki and Raymo (2005) and associated website: *http://lorraine-lisiecki.com/stack.html*.]

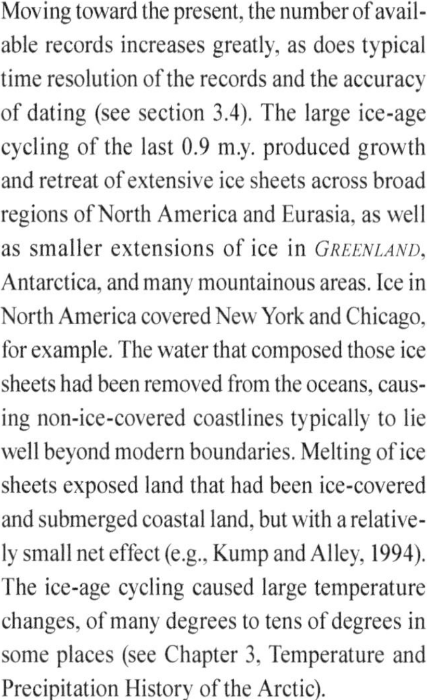

Moving toward the present, the number of available records increases greatly, as does typical time resolution of the records and the accuracy of dating (see section 3.4). The large ice-age cycling of the last 0.9 m.y. produced growth and retreat of extensive ice sheets across broad regions of North America and Eurasia, as well as smaller extensions of ice in *GREENLAND*, Antarctica, and many mountainous areas. Ice in North America covered New York and Chicago, for example. The water that composed those ice sheets had been removed from the oceans, causing non-ice-covered coastlines typically to lie well beyond modern boundaries. Melting of ice sheets exposed land that had been ice-covered and submerged coastal land, but with a relatively small net effect (e.g., Kump and Alley, 1994). The ice-age cycling caused large temperature changes, of many degrees to tens of degrees in some places (see Chapter 3, Temperature and Precipitation History of the Arctic).

Climate changed in large abrupt jumps (see section 4.4.3) during the most recent of the glacial intervals and probably during earlier ones.

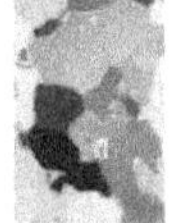

Climate changed in large abrupt jumps (see section 4.4.3) during the most recent of the glacial intervals and probably during earlier ones. In records from near the *NORTH ATLANTIC* such as *GREENLAND* ice cores, roughly half of the total difference between glacial and interglacial conditions was achieved (as recorded by many climate-change indicators) in time spans of decades to years. Changes away from the *NORTH ATLANTIC* were notably smaller, and in the far south the changes appear to see-saw (southern warming with northern cooling). The "shape" of the climate records is interesting: northern records typically show abrupt warming, gradual cooling, abrupt cooling, near-stability or slight gradual warming, and then they repeat (see Figure 5.9).

The most recent interglacial interval has lasted slightly more than 10,000 years. Generally warm conditions have prevailed compared with the average of the last 0.9 m.y. However, important changes have been observed. These changes include broad warming and then cooling in only millennia, abrupt events probably linked to the older abrupt changes, and additional events with various spacings and sizes that have a range of causes, which will be described more in Chapters 3 (Temperature and Precipitation History of the Arctic) and 4 (Past Rates of Climate Change in the Arctic).

2.5 CHRONOLOGY

In any discussion of past climate periods, we must use a time scale understandable to all readers. Beyond the historical period, then, we must use time periods that are within the realm of geology. In this report, we use two sets of terminology for prehistoric time periods, one for the longer history of Earth and one for much more recent Earth history, approximately the past 2.6 m.y. (the Quaternary Period). For the longer period of Earth history, we use the terminology and time scale adopted by the International Commission on Stratigraphy (Ogg, 2004). This time scale is well established and has been widely accepted throughout the geologic community. The **Quaternary** Period is the youngest geologic period in this time scale, and it constitutes the past approximately 2.6 m.y. (*http://www.stratigraphy.org/gssp.htm*; Jansen et al., 2007) (Figure 2.10). The Quaternary Period is of particular interest in this report, because this time interval is characterized by dramatic changes—between glacial and interglacial—in climate.

Some problems are associated with the use of time scales within the Quaternary Period. These problems are common to all geologic dating, but they assume additional importance in the Quaternary because the focus during this geologically short, recent period is on relatively short-lived events. Very few geologic records for the Quaternary Period are continuous, well dated, and applicable to all other records of climate change. Furthermore, many geologic deposits preserve records of events that are **time-transgressive** or **diachronous**. That is, a particular geologic event is recorded earlier at one geographic location and later at another.

A good example of time-transgression is the most recent deglaciation of mid-continent North America, the retreat of the ***LAURENTIDE ICE SHEET***. Although this retreat marked a major shift in a climate state, from a glacial period to an interglacial period, by its very nature it occurred at different times in different places. In midcontinental North America, the *LAURENTIDE ICE SHEET* had begun to retreat from its southernmost position in central Illinois after about 22.6 ka, but it was still present in what is now northern Illinois until after about 15.1 ka, and it was still present in Wisconsin and Michigan until after about 12.9 ka (Johnson et al., 1997) (radiocarbon ages were converted using the algorithm of Fairbanks et al., 2005), and in north-central *Labrador* until about 6 ka (Dyke and Prest, 1987). Thus, the geologic record of when the present "interglacial" period began is older in central Illinois than it is in northern Michigan, which in turn is older than it is in southern *Canada*. Time transgression as a concept also applies to phenomena other than geologic processes. Migration of plant communities (**biomes**) as a result of climate change is not an instantaneous process throughout a wide geographic region. Thus, many records of climate change that reflect changes in plant communities will take place at different times in a region as taxa within that community migrate.

ERATHEM/ERA	SYSTEM, SUB-SYSTEM, PERIOD, SUBPERIOD		SERIES/ EPOCH	AGE ESTIMATE OF BOUNDARY
Cenozoic	Quaternary		Holocene	
				11,477 yr
			Pleistocene	
				2.588 Ma
	Tertiary	Neogene	Pliocene	
				5.332 Ma
			Miocene	
				23.03 Ma
		Paleogene	Oligocene	
				33.9 Ma
			Eocene	
				55.8 Ma
			Paleocene	
				65.5 Ma

Figure 2.10. Cenozoic time periods as used in this report (modified from Ogg, 2004). [Copyright 2005, International Commission on Stratigraphy.]

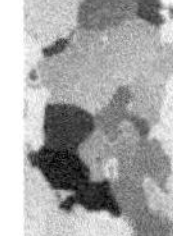

Another difficulty is not with the geologic records themselves but with the terms used in different regions to describe them. For example, "Sangamon" is the name of the last interglacial period in the mid-continent of North America (Johnson et al., 1997) and the term "Eemian" is used for the last interglacial period in Europe. However, North American workers apply the term Sangamon primarily to rock-stratigraphic records (tills deposited by glaciers and old soils called paleosols). The Sangamon interglacial is considered to have lasted several tens of thousands of years, because no glacial ice was present in the mid-continent between the last major glacial event ("Illinoian") and the most recent one ("Wisconsinan"). In contrast, the term Eemian, used by European workers, is often applied to pollen records and is reserved for a period of time, perhaps less than 10,000 years, when climate conditions were as warm as or warmer than present.

Nevertheless, it is crucial that at least some terminology is used as a common basis for discussion of geologic records of climate change during the Quaternary. In this report, we have chosen to use the stages of the oxygen isotope record from foraminifers in deep-sea cores as our terminology for discussing different intervals of time within the Quaternary Period. The identification of glacial-interglacial changes in deep-sea cores, and the naming of stages for them, began with a landmark report by Emiliani (1955). The oxygen isotope composition of carbonate in foraminifer skeletons in the ocean shifts as climate shifts from glacial to interglacial states (see section 2.4.1, above). These shifts are due both to changes in ocean temperature and changes in the isotopic composition of seawater. The latter changes result from the shifts in oxygen isotopic composition of seawater, in turn a function of ice volume on land. Because the temperature and ice-volume influences on foraminiferal oxygen-isotope compositions are in the same direction, the record of glacial-interglacial changes in deep-sea cores is particularly robust.

The oxygen isotope record of glacial-interglacial cycles has been studied and well documented in hundreds of deep-sea cores. The same glacial-interglacial cycles are easily identified in cores from all the world's oceans (Bassinot, 2007). It is, therefore, truly a continuous and global record of climate change within the Quaternary Period. Furthermore, a variety of geologic records of climate change show the same glacial-interglacial cycles that can be compared and correlated with the deep-sea record. These geologic records include glacial records (e.g., Booth et al., 2004; Andrews and Dyke, 2007), ice cores (e.g., **NGRIP**, 2004; Jouzel et al., 2007), cave carbonates (e.g., Winograd et al., 1992, 1997), and eolian sediments (e.g., Sun et al., 1999). Furthermore, deep-sea cores themselves sometimes contain, in addition to foraminifers, other records of climate change such as pollen from past vegetation (e.g., Heusser et al., 2000) or eolian (wind-deposited) sediments that record glacial and interglacial climates on land (e.g., Hovan et al., 1991).

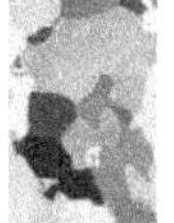

The time scales that have been developed for the oxygen isotope record are important to understand. The mostly widely used time scales are those that have been developed by use of "stacked" deep-sea core records (i.e., multiple core records, from more than one ocean) that are in turn, "tuned" or "dated" by a combination of identification of dated paleomagnetic events and an assumed forcing of climate change by changes in the parameters related to Earth-Sun orbital geometry, precession, and obliquity.

Initially, dated paleomagnetic events were used with an assumed constant sedimentation rate to provide a first estimate of the timing of the main variations in the climate. The timing closely matched the known periodicities in Earth-Sun orbital geometry, to a degree that provided very high confidence that those known periodicities were affecting the climate. Then, this result was used to fine-tune the dating by adjusting the sedimentation rates to allow closer match between the data and the orbital periodicities. The practice is often referred to as "astronomical" or "orbital" tuning. The strategy behind "stacking" multiple records is to eliminate possible local effects on a core and present a smoothed, global record. Several highly similar time scales have been developed using this approach. The most commonly cited are the SPECMAP studies of Imbrie et al. (1984) and Martinson et al. (1987) (Figure 2.11), and the more recent work of Lisiecki and Raymo (2005).

However, there are disadvantages to using the astronomically tuned oxygen isotope records. Very few deep-sea cores are dated directly, except in the upper parts that are within the range of radiocarbon dating, or at widely spaced depths where paleomagnetic events are recorded. In addition, after the initial tests, the astronomical tuning approach assumes that the orbital parameters, particularly precession and obliquity, are the primary forcing mechanisms behind climate change on glacial-interglacial time scales in the Quaternary Period. Challenges to this assumption are based on directly dated cave calcite records

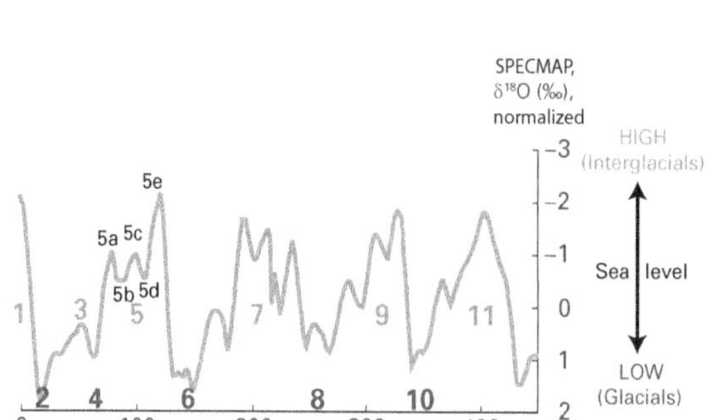

Figure 2.11. Marine Isotope Stage nomenclature and chronology used in this report (after Imbrie et al., 1984; Martinson et al., 1987). Red numbers = interglacial intervals; blue numbers = glacial intervals.

(Winograd et al., 1992, 1997) and emergent coral reef terraces (Szabo et al., 1994; Gallup et al., 2002; Muhs et al., 2002), although in general the assumption appears to be more-or-less accurate. Additional assumptions, including that response is proportional to forcing, are inherent in tuning.

Recognizing the assumptions inherent in the SPECMAP time scale, we use this time scale and the marine oxygen isotope stage terminology in this report for four reasons:

1. the wide acceptance and use in the scientific community,
2. the continuous nature of the record,
3. the global aspect of the record, and
4. the ability to subdivide the periods of time under consideration.

Regarding the latter, for example, the marine record can accommodate the problem in the use of "Sangamon," as used in North America compared with "Eemian," in Europe. The Sangamon interglacial, as used by North Americans, includes all of **Marine Isotope Stage** 5 (**MIS** 5), as well as perhaps parts of MIS 4. However, the Eemian, as used by most European workers, would include only MIS 5e or 5.5, an interval within the greater MIS 5.

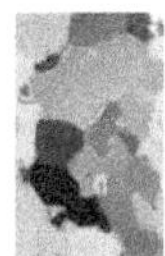

2.6 SYNOPSIS

Earth's climate is a complex, interrelated system of air, water, ice, land surface, and living things responding to the Sun's energy. Scientific understanding of this system has been increasing rapidly, and the broad outline is now quite well known, although many details remain obscure and further discoveries are guaranteed.

The climate system can be forced to change, but it also varies internally without external forcing. Both forced and unforced variations interact with various feedback processes that may either amplify or reduce the resulting climate change.

The climate system can be forced to change, but it also varies internally without external forcing. Both forced and unforced variations interact with various feedback processes that may either amplify or reduce the resulting climate change, often with interesting patterns in space and time.

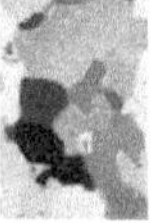

Changes in the energy emitted by the Sun, the amount of that energy reaching Earth, the amount of that energy reflected by Earth, and the greenhouse effect of the atmosphere are important in controlling global climate. Changes in continental positions, ocean currents, wind patterns, clouds, vegetation, ice, and more affect regional climates as well as contribute to the global picture. The Sun has brightened slowly for billions of years, and its brightness shows very small fluctuations measured in years to centuries. Features of Earth's orbit change the latitudinal and seasonal distribution of sunshine, and they have a small effect on total sunshine reaching the planet over tens of thousands of years. Great tectonic forces in the Earth rearrange continents and promote or reduce volcanic activity and growth of mountain ranges. All three affect greenhouse-gas concentrations and other features of the climate over millions of years or longer, and they interact with changes in the biosphere in response to biological evolution. And, these general statements omit many interesting and increasingly well-understood features of the system.

Many deposits of the Earth system—muds and cave formations and tree rings and ice layers and many more—have characteristics that reflect the climate at the time of formation, that are preserved after formation, and that reveal their age of formation. Careful consideration of these deposits underlies paleoclimatology, the study of past climates. Varied investigative techniques focus on physical, chemical, isotopic, and biological indicators, and they provide surprisingly complete histories of changes in time and space.

This report especially focuses on the last tens of millions of years. This interval has been characterized by slow cooling, leading from a largely ice-free world to ice-age cycling in response to orbital changes. Both the cooling trend and the ice-age cycling were punctuated occasionally by abrupt shifts. The last approximately 10,000 years have been a reduced-ice interglacial during the ice-age cycling, but they have experienced a variety of climate changes linked to changing volcanism, ocean currents, solar output, and—recently evident—human perturbation.

CHAPTER 3

Temperature and Precipitation History of the Arctic

Lead Authors: Gifford H. Miller*, University of Colorado, Boulder, CO; Julie Brigham-Grette*, University of Massachusetts, Amherst, MA

Contributing Authors: Lesleigh Anderson, U.S.Geological Survey, Denver, CO; Henning Bauch, GEOMAR, University of Kiel, Germany; Mary Anne Douglas, University of Alberta, Edmonton, Alberta, Canada; Mary E. Edwards, University of Southampton, United Kingdom; Scott Elias, Royal Holloway, University of London, United Kingdom; Bruce Finney, Idaho State University, Pocatello, ID; Svend Funder, University of Copenhagen, Denmark; Timothy Herbert, Brown University, Providence, RI; Larry Hinzman, University of Alaska, Fairbanks, AK; Darrell Kaufman, Northern Arizona University, Flagstaff, AZ; Glen MacDonald, University of California, Los Angeles, CA; Alan Robock, Rutgers University, Rutgers, NJ; Mark Serreze, University of Colorado, Boulder, CO; John Smol, Queen's University, Kingston, Ontario, Canada; Robert Spielhagen, GEOMAR, University of Kiel, Germany; Alexander P. Wolfe, University of Alberta, Edmonton, Alberta, Canada; Eric Wolff, British Antarctic Survey, Cambridge, United Kingdom

*SAP 1.2 Federal Advisory Committee Member

ABSTRACT

The Arctic has undergone dramatic changes in temperature and precipitation during the Cenozoic Era, the past 65 million years (m.y.) of Earth history. These past conditions are not perfect analogues for climates of the next century, primarily because of differences in forcings (orbital) and because some important feedbacks operate on longer than centennial timescales (e.g., ice sheets). Most of this report focuses on the past 3.5 m.y., for which boundary conditions, such as continental configurations and ocean gateways, were similar to those of the present; for older periods boundary conditions differed from modern ones. Studies of past climates help to reveal key processes of climate change such as response to elevated greenhouse-gas concentrations, and they provide insights to future climate behavior. At times both warmer and colder than recently for which comparisons can be made, Arctic surface air temperature changes exceeded global- or hemispheric-average temperature changes. This evaluation supports the concept of Arctic amplification, in which strong positive feedbacks—processes that amplify the effects of a change in the controls on global temperature—produce larger changes in temperature in the Arctic than elsewhere.

At the start of the Cenozoic, 65 million years ago (Ma), the planet was ice free; there was no sea ice in the Arctic Ocean, nor was there a Greenland nor an Antarctic ice sheet. Atmospheric CO_2 levels were notably elevated relative to those of the pre-industrial world (Berner and Kothavala, 2001; Jansen et al., 2007). General cooling through the Cenozoic is attributed mainly to a slow drawdown of greenhouse gases in the atmosphere through the weathering of silicic rocks that exceeded the release of stored carbon through volcanism and reprocessing (Berner and Kothavala, 2001). During the past 65 m.y., atmospheric CO_2 has decreased about 1,200 parts per million by volume (ppmv), or on average 1 ppmv for every 50 thousand years (k.y.). This rate is much more gradual than the rate of atmospheric CO_2 increase during the past 150 years of about 100 ppmv due to combustion of fossil fuel.

As the Arctic cooled, high-elevation mountain glaciers formed as did seasonal sea ice in the Arctic Ocean, but a detailed record of changes in the Arctic is available only for the last few million years. A global warm period that affected both cold and warm seasons in the middle Pliocene, about 3.5 Ma, is well represented in the Arctic; at that time extensive deciduous forests occupied lands that now support only polar desert and tundra. Global oceanic and atmospheric circulation was substantially different between 3 and 2.5 Ma than subsequently. The development of the first continental ice sheets over North America and Eurasia led to changes in the circulation of both the atmosphere and oceans. The onset of continental glaciation is most clearly defined by the first appearance of rock fragments in sediment cores from the central Atlantic Ocean about 2.6 Ma. These rock fragments, often referred to as ice-rafted debris (IRD), are too heavy to have blown or been washed into the central Atlantic; they must have been delivered by large icebergs emanating from continental ice sheets. The first appearance of IRD marks the onset of the Quaternary Period (2.6–0 Ma), generally equated with "ice-age" time, even though for a small fraction (about 10%) of the time the ice sheets were very likely to have been as small as or smaller than their present size. From about 2.7 to about 0.8 Ma, the ice sheets came and went about every 41 k.y., the same timing as cycles in the tilt of Earth's axis. Ice sheets grew when Earth's tilt was at a minimum, resulting in less seasonality (cooler summers, warmer winters), and they melted when tilt was at a maximum and seasonality was at its greatest (warmer summers and cooler winters). For the past 600 k.y., ice sheets have grown larger and ice-age times have been longer, lasting about 100 k.y.; those icy intervals have been separated by brief warm periods (interglaciations), when sea level and ice volumes were close to those at present. The duration of interglaciations ranges from about 10 k.y. to perhaps 40 k.y. The cause of the shift from 41 k.y. to 100 k.y. glacial cycles is still being debated. Most explanations center on the continued gradual planetary cooling that may have produced larger ice sheets that were more resistant to melting, or with removal of soft sedimentary cover over bedrock in glaciated regions that, once removed, increased the frictional coupling of the ice sheet to its bed, resulting in steeper ice-sheet profiles and thicker ice sheets, again more resistant to melting (e.g., Clark and Pollard, 1998; Raymo et al., 2006; Huybers, 2007; Bintanja and van de Wal, 2008).

The relatively warm planetary state during which human civilization developed is the most recent of the warm interglaciations, the Holocene (about 11.5–0 kiloannum (thousands of years ago (ka)). During the penultimate warm interval, about 130–120 ka, solar energy in summer in the northern high latitudes was greater than at any time in the current warm interval. As a consequence, the Arctic summer was about 5°C warmer than at present and almost all glaciers melted completely except for the Greenland Ice Sheet, and even it was reduced in size substantially from its present extent. With the increased ice melt, sea level was about 5 meters (m) higher than at present; the extra melt came from both Greenland and Antarctica as well as from small glaciers (Overpeck et al., 2006; Meier et al., 2007). Although sea ice is difficult to reconstruct, the evidence suggests that the central Arctic Ocean retained some permanent ice cover or was periodically ice free, even though the flow of warm Atlantic water into the Arctic Ocean was very likely to have been greater than during the present warm interval.

The Last Glacial Maximum (LGM) peaked about 21 ka when mean annual temperatures in parts of the Arctic were as much as 20°C lower than at present. Ice recession was well underway by 16 ka, and most of the Northern Hemisphere ice sheets had melted by 7 ka. Summertime solar energy rose steadily in the Arctic from 21 ka to a maximum (10% higher than at present) about 11 ka and has been decreasing since then, primarily in response to the precession of the equinoxes causing Earth's distance from the Sun during Northern Hemisphere summer to decrease from 21 to 11 ka and then to increase to the present. The extra energy received in early Holocene summers warmed summers throughout the Arctic about 1°–3°C above 20th century averages, enough to completely melt many small glaciers throughout the Arctic (although the Greenland Ice Sheet was only slightly smaller than present). Summer sea ice limits were substantially smaller than their 20th century average, and the flow of Atlantic water into the Arctic Ocean was substantially greater. As summer solar energy decreased in the second half of the Holocene, glaciers re-established or advanced, sea ice extended, and the flow of warm Atlantic water into the Arctic Ocean diminished. Late Holocene cooling reached its nadir during the Little Ice Age (about 1250–1850 AD), when sun-blocking volcanic eruptions and perhaps other causes added to the orbital cooling, allowing most Arctic glaciers to reach their maximum Holocene extent. During the warming of the past century and a half, glaciers have receded throughout the Arctic, terrestrial ecosystems have advanced northward, and perennial Arctic Ocean sea ice has diminished.

Paleoclimate reconstructions indicate that Arctic temperature changes typically have been larger than corresponding hemispheric or globally averaged changes. This behavior is observed with conditions both warmer and colder than recently, indicating that Arctic amplification is a pervasive feature of the climate system.

3.1 INTRODUCTION

Recent instrumental records show that during the last few decades, surface air temperatures throughout much of the far north have risen more rapidly than temperatures in lower latitudes and usually about twice as fast (Delworth and Knutson, 2000; Knutson et al., 2006). The remarkable reduction in *ARCTIC OCEAN** summer **sea ice** in 2007 (Figure 3.1) has outpaced the most recent predictions from available **climate models** (Stroeve et al., 2008), but it is in concert with widespread reductions in **glacier** length, increased borehole temperatures, increased coastal erosion, changes in vegetation and wildlife habitats, the northward migration of marine life, and degradation of **permafrost**. On the basis of the past century's trend of increasing **greenhouse gases**, climate models forecast continuing warming into the foreseeable future (Figure 3.2) and a continuing **amplification** in the Arctic of global changes (Serreze and Francis, 2006). As outlined by the Arctic Climate Impact Assessment (Arctic Climate Impact Assessment, 2005), the sensitivity of the Arctic to changed **forcing** is due to strong **positive feedbacks** in the Arctic climate system (see section 2.2). These **feedbacks** strongly amplify changes to the climate of the Arctic and also affect the global climate system.

The short time interval for which instrumental data are available in the Arctic is not sufficient to characterize natural variability, so a paleoclimatic perspective is required.

Because strong Arctic feedbacks act on **climate changes** caused by either nature or by humans, natural variability and human-caused changes are large in the Arctic, and separating them requires understanding and characterization of the natural variability component. The short time interval for which instrumental data are available in the Arctic is not sufficient to characterize that natural variability, so a paleoclimatic perspective is required.

This chapter focuses primarily on the history of temperature and precipitation in the Arctic. These topics are important in their own right, and they also set the stage for understanding the histories of the *GREENLAND ICE SHEET* and the *ARCTIC OCEAN* sea ice, which are described in Chapters 5 (History of the Greenland Ice Sheet)

*For bold terms, refer to Glossary; for italic terms, refer to Plate 1; for geologic ages, refer to Plate 2.

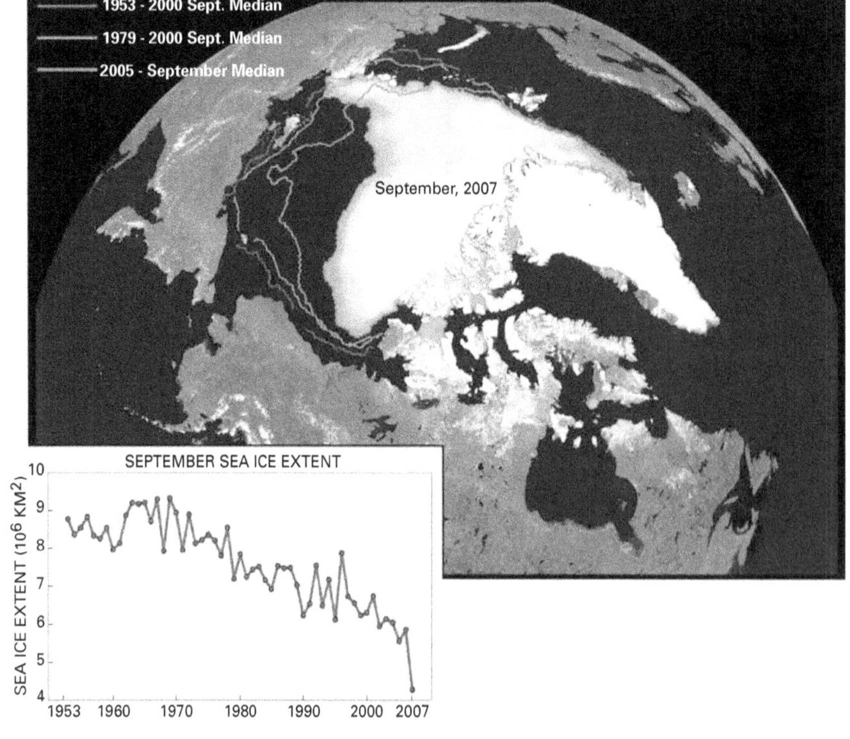

Figure 3.1. Median extent of sea ice in September 2007, compared with averaged intervals during recent decades. Red curve = 1953–2000; orange curve = 1979–2000; green curve = September 2005. Inset: sea ice extent time series plotted in square kilometers, shown from 1953–2007 in the graph below (Stroeve et al., 2008). The reduction in Arctic Ocean summer sea ice in 2007 was greater than that predicted by most recent climate models. [Copyright 2008 American Geophysical Union, reproduced by permission of American Geophysical Union.]

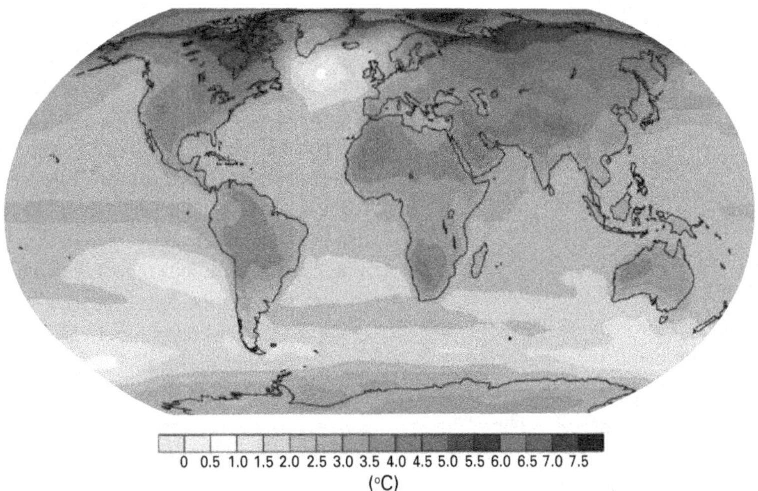

Figure 3.2. Projected surface temperature changes for the last decade of the 21st century (2090–2099) relative to the period 1980-1999. The map shows the Intergovernmental Panel on Climate Change multi-Atmosphere-Ocean coupled Global Climate Model average projection for the A1B (balanced emphasis on all energy resources) scenario. The most significant warming is projected to occur in the Arctic. (Intergovernmental Panel on Climate Change, 2007, Figure SPM6.)

and 6 (History of Arctic Sea Ice). Because of the great interest in rates of change, and because of some technical details in extracting rates of change from the broad history of temperature or precipitation, careful consideration of rates of change is deferred to Chapter 4 (Past Rates of Climate Change in the Arctic).

Before providing the history of temperature and precipitation in the Arctic, this chapter supplements the discussion in Chapter 2 (Paleoclimate Concepts) on forcings, feedbacks, and proxies by providing additional information on those aspects particularly relevant to the histories of temperature and precipitation in the Arctic. The climate history of the past 65 **m.y.** is then summarized; it focuses on temperature and precipitation changes that span the full range of the Arctic's natural climate variability and response under different forcings. The authors place special emphasis on relevant intervals in the past with a mean climate state warmer than the 20th century average. Where possible, causes of these changes are discussed. The forcings for past warm times, as understood, differ from the greenhouse gas forcing anticipated in the coming decades so that the **paleorecord** is an imperfect **analogue** for the future. But the paleorecord is clear that Arctic temperature changes typically have been larger than Northern Hemisphere or globally averaged changes.

The paleorecord is clear that Arctic temperature changes typically have been larger than Northern Hemisphere or globally averaged changes.

3.2 FEEDBACKS INFLUENCING ARCTIC TEMPERATURE AND PRECIPITATION

Changes in the seasonal and areal distribution of snow and ice will exert strong influences on the planetary energy balance.

The most commonly used measure of the climate is the mean surface air temperature (Figure 3.3), which is influenced by climate forcings and climate **feedbacks**. As discussed with references in Chapter 2, Paleoclimate Concepts, section 2.2, important forcings during the past several millennia have been changes in the distribution of solar radiation that resulted from features of Earth's orbit, volcanism, changes in atmospheric greenhouse-gas concentrations and, to a lesser extent, small variations in solar **irradiance**. On longer time scales (tens of millions of years), the long-term increase in the solar constant (a 30% increase in the past 4,600 m.y.) was important, and the redistribution of continental landmasses caused by plate motions also affected the planetary energy balance.

How much the temperature changes in response to a forcing of a given magnitude (or in response to the net magnitude of a set of forcings in combination) depends on the sum of all of the feedbacks. Feedbacks can act in days or less or endure for millions of years. The focus here is on faster feedbacks. For example, a warming may have many causes (such as brighter Sun, higher concentration of greenhouse gases in the atmosphere, less blocking of the Sun by volcanic emissions). Whatever the cause, warmer air moving over the ocean tends to entrain more water vapor, which itself is a greenhouse gas, so more water vapor in the atmosphere leads to a further rise in global mean surface temperature (Pierrehumbert et al., 2007). The discussion below focuses on those feedbacks especially linked to the Arctic. Several processes linked to ice-age cycling are included here, because of the dominant role of northern land in supporting ice-sheet growth, although ice-age processes (like some of the other processes discussed below) clearly extend well beyond the Arctic.

3.2.1 Ice-Albedo Feedback

Ice and snow present highly reflective surfaces. The albedo of a surface is defined as the reflectivity of that surface to the wavelengths of solar radiation. Fresh ice and snow have the highest albedo of any widespread surfaces on the planet (Figure 3.4), so it is apparent that changes in the seasonal and areal distribution of snow and ice will exert strong influences on the planetary

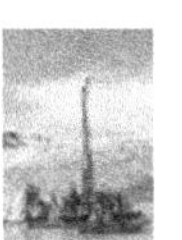

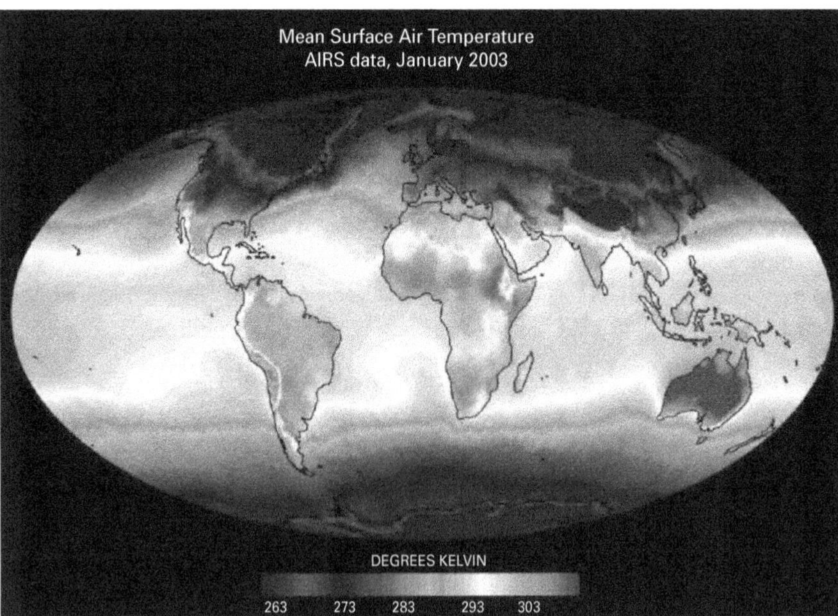

Figure 3.3. Global mean observed near-surface air temperatures for the month of January 2003, derived from the Atmospheric Infrared Sounder (AIRS) data. Contrast between equatorial and Arctic temperatures is greatest during the Northern Hemisphere winter. The transfer of heat from the tropics to the polar regions is a primary feature of Earth's climate system. (Color scale is in Kelvin degrees such that 0°C=273.15 Kelvin.) [Source: *http://daac.gsfc.nasa.gov/data/datapool/AIRS/index.html.*]

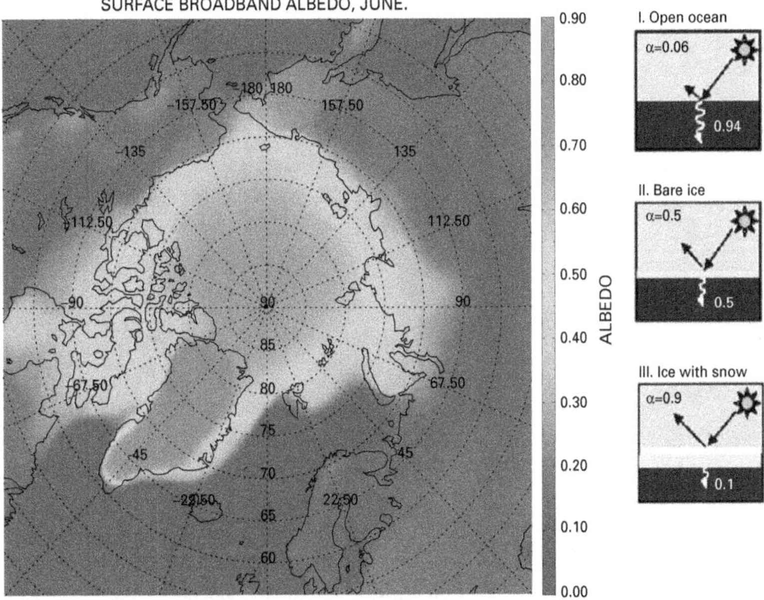

Figure 3.4. Albedo values in the Arctic. *A)* Advanced Very High Resolution Radiometer (AVHRR)-derived Arctic albedo values in June, 1982–2004 multi-year average, showing the strong contrast between snow- and ice-covered areas (green through red) and open water or land (blue). [Image courtesy of X. Wang, University of Wisconsin-Madison, CIMSS/NOAA.] *B)* Albedo feedbacks. Albedo is the fraction of incident sunlight that is reflected. Snow, ice, and glaciers have high albedo. Dark objects such as the open ocean, which absorbs some 93% of the Sun's energy, have low albedo (about 0.06). Bare ice has an albedo of 0.5; however, sea ice covered with snow has an albedo of nearly 90% [Source: *http://nsidc.org/seaice/processes/albedo.html.*]

energy balance (Peixoto and Oort, 1992). Open ocean, on the other hand, has a low albedo; it absorbs almost all solar energy when the Sun angle is high. Changes in albedo are most important in the Arctic summer, when solar radiation is at a maximum, whereas changes in the winter albedo have little influence on the energy balance because little solar radiation reaches the surface then. In general, warming reduces ice and snow whereas cooling allows them to extend, so the changes in ice and snow act as positive feedbacks to amplify climate changes (e.g., Lemke et al., 2007).

3.2.2 Ice-Insulation Feedback

In addition to its effects on albedo, sea ice also causes a positive insulation feedback, primarily in the wintertime. Ice effectively blocks heat transfer between relatively warm ocean (at or above the freezing point of seawater) and cold atmosphere (which, in the Arctic winter, averages −40°C (Chapman and Walsh, 2007). If sea ice is thinned by warming, then the ocean heats the overlying atmosphere in winter months, amplifying that warming.

Feedbacks involving snow insulation of the ground are also important, through their effects on vegetation and on permafrost temperature and its influence on storage or release of greenhouse gases, as described in the next subsections (e.g., Ling and Zhang, 2007).

3.2.3 Vegetation Feedback

A related terrestrial feedback involves changing vegetation. A warming climate can cause **tundra** to give way to shrub vegetation. However, the shrub vegetation has a lower albedo than tundra, and the shrubs thus cause further warming (Figure 3.5) (Chapin et al., 2005; Goetz et al., 2007). Interactions involving the **boreal** forest and deciduous forest can also be important. When, as a result of warming, deciduous forest replaces evergreen boreal forest, then winter surface albedo increases—an example of a **negative feedback** to the warming climate (Bonan et al., 1992; Rivers and Lynch, 2004). Alternatively, if warming allows evergreen boreal forest to advance northward replacing tundra or shrub vegetation, then the lower

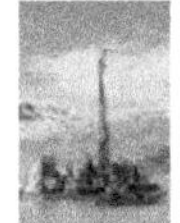

If sea ice is thinned by warming, then the ocean heats the overlying atmosphere in winter months, amplifying that warming.

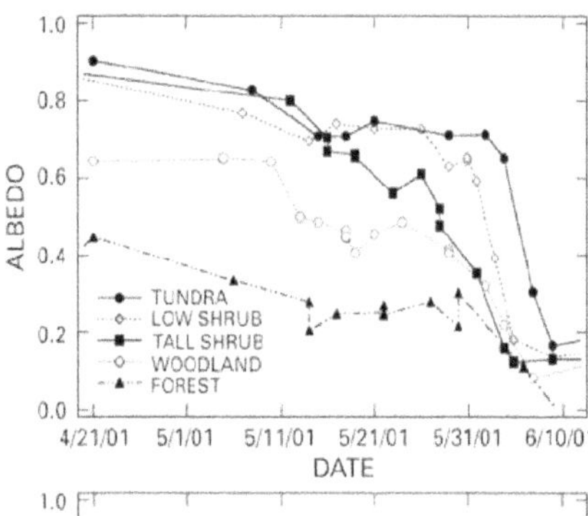

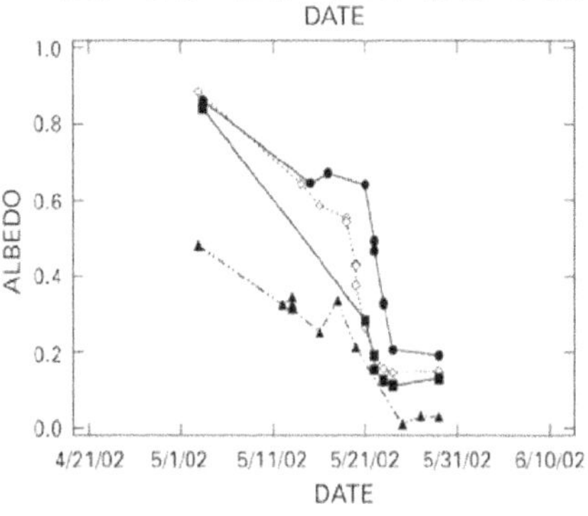

Figure 3.5. Changes in vegetation cover throughout the Arctic can influence albedo, as can altering the onset of snow melt in spring. A) Progression of the melt season in northern Alaska, May 2001 (top) and May 2002 (bottom), demonstrates how areas with exposed shrubs show earlier snow melt. B) Dark branches against reflective snow alter albedo (Sturm et al., 2005; photograph courtesy of Matt Sturm). [Copyright 2005 American Geophysical Union, reproduced by permission of American Geophysical Union.]

albedo of the boreal forest, especially in spring when snow cover may bury tundra, results in earlier warming and hence exerts a positive feedback on warming. This warming feedback would tend to be partially balanced for a period of time by the removal of atmospheric $\mathbf{CO_2}$ as stored carbon in forest ecosystems, which have more above-ground carbon than shrub ecosystems (Denman et al., 2007).

3.2.4 Permafrost Feedback

As permafrost thaws under a warmer summer climate, it is likely to release more greenhouse gases.

Additional but poorly understood feedbacks in the Arctic involve changes in the extent of permafrost and how changes in cloud cover interact both with permafrost and with the release of **carbon dioxide** and **methane** from the land surface. Feedbacks between permafrost and climate became widely recognized only in recent decades (building on the works of Kvenvolden, 1988, 1993; MacDonald, 1990; and Haeberli et al., 1993). As **permafrost** thaws under a warmer summer climate (Figure 3.6), it is likely to release more greenhouse gases such as carbon dioxide and methane from the decomposition of organic matter previously sequestered in permafrost and in widespread Arctic **yedoma** deposits (e.g., Vörösmarty, 2001; Thomas et al., 2002; Smith et al., 2004; Archer, 2007; Walter et al., 2007). Because carbon dioxide and methane are greenhouse gases, atmospheric temperature is likely to increase in turn, a positive feedback. Walter et al. (2007) suggest that methane bubbling from the thawing of newly formed thermokarst lakes across parts of the Arctic during deglaciation could account for as much as 33–87% of the increase in atmospheric methane measured in ice cores. Such a release would have contributed a strong and rapid positive feedback to warming during the last deglaciation, and it likely continues today (Walter et al., 2006).

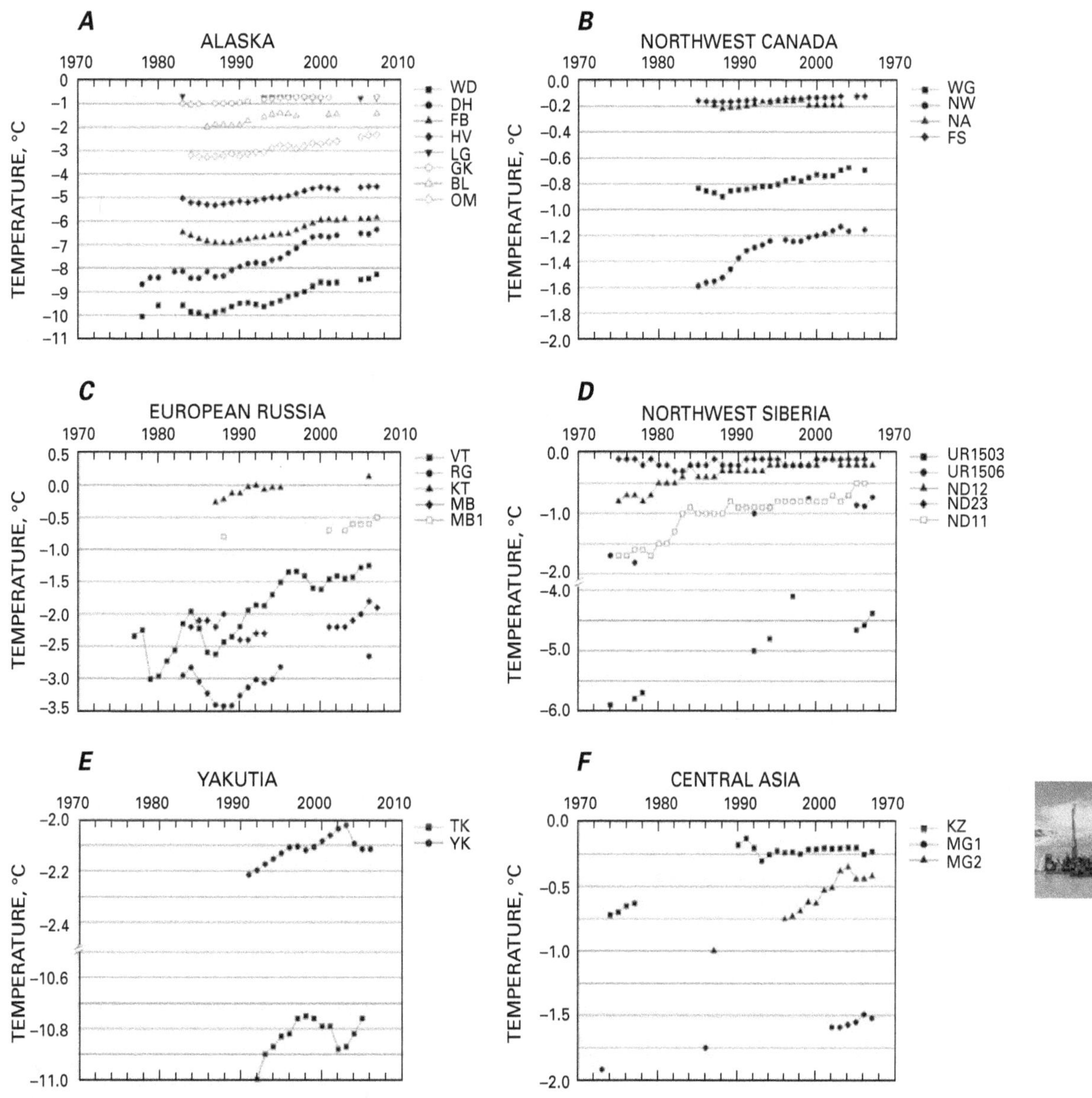

Figure 3.6. Warming trend in Arctic permafrost (permanently frozen ground), 1970–present. Local effects can modify this trend. *A*) Sites in Alaska: WD = West Dock; DH = Deadhorse; FB = Franklin Bluffs; HV = Happy Valley; LG = Livengood; GK = Gulkana; BL = Birch Lake; OM = Old Man. *B*) Sites in northwest Canada: WG = Wrigley; NW = Norman Wells; NA = Northern Alberta; FS = Fort Simpson. *C*) Sites in European Russia: VT = Vorkuta; RG = Rogovoi; KT = Karataikha; MB = Mys Bolvansky. *D*) Sites in northwest Siberia: UR = Urengoi; ND = Nadym. *E*) Sites in Yakutia: TK = Tiksi; YK = Yakutsk. *F*) Sites in central Asia: KZ = Kazakhstan; MG = Mongolia (Brown and Romanovsky, 2008). [Copyright 2008 Wiley Interscience. Reproduced by permission of Wiley Interscience.]

3.2.5 Freshwater Balance Feedback and Thermohaline Circulation

The *Arctic Ocean* is almost completely surrounded by continents (Figure 3.7). Because precipitation is low over the ice-covered ocean (Serreze et al., 2006), the freshwater input to the *Arctic Ocean* largely derives from the runoff from large rivers in Eurasia and North America and by the inflow of relatively low-salinity Pacific water through the *Bering Strait*. The *Yenisey*, *Ob*, and *Lena* are among the nine largest rivers on Earth, and there are several other large rivers, such as the *Mackenzie*, that feed into the *Arctic Ocean* (see Vörösmarty et al., 2008). The freshwater discharged by these rivers dilutes the saltiness of ocean surface waters, maintaining low salinities on the broad, shallow, and seasonally ice-free seas bordering the *Arctic Ocean*. The largest of these border the Eurasian continent, where they serve as the dominant area in the *Arctic Ocean* in which sea ice is produced (for some fundamentals on Arctic **sea ice**, see Barry et al., 1993). Sea ice forms along the Eurasian margin and then drifts toward *Fram Strait*; its transit time is 2–3 years in the current regime. In the *Amerasian* part of the *Arctic Ocean*, the clockwise-rotating Beaufort Gyre is the dominant ice-drift feature (see Figure 6.1).

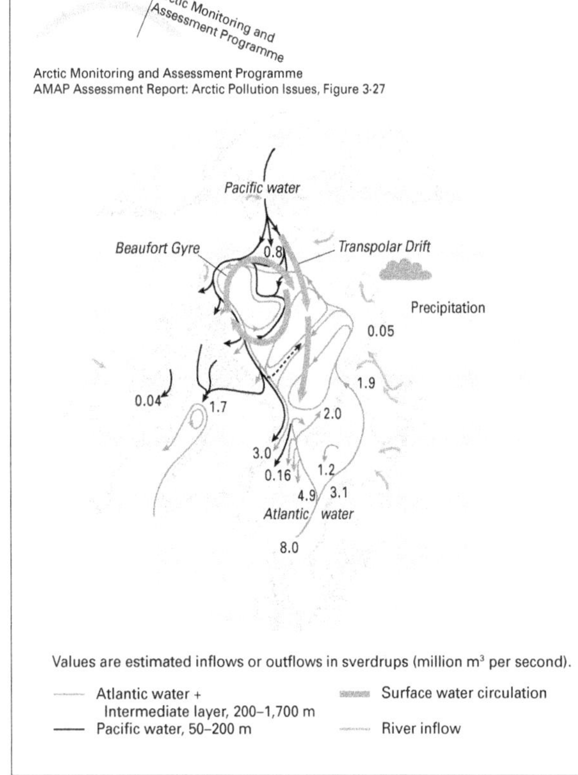

Figure 3.7. Inflows and outflows of water in the Arctic Ocean. Red lines = components and paths of the surface and Atlantic Water layer in the Arctic; black arrows = pathways of Pacific water inflow from 50–200 m depth; blue arrows = surface-water circulation; green arrows = major river inflow; red arrows = movements of density-driven Atlantic water and intermediate water masses into the Arctic (AMAP, 1998, Figure 3.27). [Reproduced by permission of Arctic Monitoring and Assessment Program.]

Surface currents transport low-salinity surface water (its upper 50 m) and sea ice (freshwater) out of the *Arctic Ocean* (e.g., Schlosser et al., 2000). Surface waters are primarily exported from the *Arctic Ocean* to the northern *North Atlantic* (*Nordic Seas*) through western *Fram Strait*, after which they follow the east coast of *Greenland* and exit the *Nordic Seas* into the *North Atlantic* through *Denmark Strait*. A smaller volume of surface water flows out through the inter-island channels of the *Canadian Arctic Archipelago*, and it eventually reaches the *North Atlantic* through the *Labrador Sea*. The low-saline outflow from the *Arctic Ocean* is compensated by a relatively warm inflow of saline Atlantic water through eastern *Fram Strait*. Despite its warmth, Atlantic water has sufficiently high salt content that its density is higher than the low-salinity surface waters. The inflowing relatively dense Atlantic water is forced to sink beneath the colder, but fresher, surface water upon entering the *Arctic Ocean*. North of *Svalbard*, Atlantic water spreads as a **boundary current** into the Arctic Basin and forms the Atlantic Water layer (Morison et al., 2000). The strong vertical gradients of salinity and temperature in the *Arctic Ocean* produce a relatively stable stratification. However, recent observations have shown that in some areas in the Eurasian part of the *Arctic Ocean*, the warm Atlantic layer mixes with the surface mixed layer (Rudels et al., 1996; Steele and Boyd, 1998; Schauer et al., 2002), thereby limiting sea ice formation and promoting vertical heat transfer to the Arctic atmosphere in winter. In recent decades circum-Arctic glaciers and **ice sheets** have been losing mass (more snow and ice melting in summer than accumulates as snow in winter) (Dowdeswell et al., 1997; Rignot and Thomas, 2002; Meier et al., 2007), and since the 1930s river runoff to the *Arctic Ocean* has been increasing (Peterson et al., 2002). Recent studies suggest that changes in river runoff strongly influence the stability of *Arctic Ocean*

stratification (Steele and Boyd, 1998; Martinson and Steele, 2001; Björk et al., 2002; Boyd et al., 2002; McLaughlin et al., 2002; Schlosser et al., 2002).

In the *North Atlantic*, primarily in the *Nordic Seas* and the *Labrador Sea*, wintertime cooling of the relatively warm and salty waters increases its density. The denser waters then sink and flow southward to participate in the global **thermohaline** circulation ("thermo" for temperature and "haline" for salt), the two components that determine density. This circulation system also is referred to as the meridional overturning circulation (MOC). Although the two terms are sometimes used interchangeably, the MOC is confined to the Atlantic Ocean where the phenomenon is quantified by using tracers that show surface waters sinking in the *Nordic* and *Labrador Seas*. The thermohaline circulation refers to a conceptual **model** of vertical ocean circulation that encompasses the global ocean and is driven by the fact that colder or saltier water sinks because it is denser than warmer or less salty water.

Continuing surface inflow from the south, which replaces the water sinking in the *Nordic* and *Labrador Seas* (MOC), promotes persistent open water rather than sea ice in these regions. In turn, this lack of sea ice promotes notably warmer conditions, especially in wintertime, over and near the *North Atlantic* and extending downwind across Europe and beyond (Seager et al., 2002). Salt rejected from sea ice growing nearby very likely contributes to the density of the adjacent sea water and to its sinking.

If the surface waters are made sufficiently less salty by an increase in freshwater from runoff of melting ice or from direct precipitation, then the rate of sinking of those surface waters will diminish or stop (e.g., Broecker et al., 1985). Results of numerical models indicate that if freshwater runoff into the *Arctic Ocean* and the *North Atlantic* increases as surface waters warm in the northern high latitudes, then the thermohaline circulation in the *North Atlantic* will weaken, with consequences for marine ecosystems and energy transport (e.g., Rahmstorf, 1996, 2002; Marotzke, 2000; Schmittner, 2005).

Reducing the rate of *North Atlantic* thermohaline circulation likely has global as well as regional effects (e.g., Obata, 2007). Oceanic overturning is an important mechanism for transferring atmospheric CO_2 to the deep ocean. Reducing the rate of deep convection in the ocean would allow a higher proportion of **anthropogenic** CO_2 to remain in the atmosphere. Similarly, a slowdown in thermo-haline circulation would reduce the turnover of nutrients from the deep ocean, with potential consequences across the Pacific Ocean.

3.2.6 Feedback During Glacial-Interglacial Cycles

The polar ice sheets currently cover approximately 14 million square kilometers (km^2), whereas at their **Quaternary** maxima, as recently as 21 **ka**, they covered approximately twice that area, including the modern sites of New York and Chicago. The growth and decay of the Quaternary ice sheets were paced by the orbital variations often called **Milankovitch forcings** (e.g., Imbrie et al., 1993) described in Chapter 2 (Paleoclimate Concepts). There is little doubt that the orbital forcings drove this **glacial-interglacial** cycling, but a remarkably rich and varied literature debates the detailed mechanisms (see, e.g., Roe and Allen, 1999).

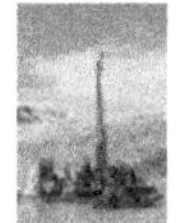

The generally accepted explanation of the glacial-interglacial cycling is that ice sheets grew when limited summer sunshine at high northern latitudes allowed survival of accumulated snow, and ice sheets shrank when abundant summer sunshine in the north melted the ice. The north is more important than the south because the Antarctic has remained ice covered during this cycling of the last million years and more, and there is no other high-latitude land in the south on which ice sheets could grow.

The large global cooling of the ice ages has been successfully explained only if the reduced greenhouse effect is included.

The increased reflectivity produced by expanded ice contributed to cooling. This effect is the ice-albedo feedback as described above, but with slower response controlled by the flow of the great ice sheets. Atmospheric dust was more abundant in the ice ages than in the intervening warm interglacials, and that additional ice-age dust contributed to cooling by blocking sunlight. The changes in Earth's orbit and ice-sheet growth led to complex changes in the ocean-atmosphere system that shifted carbon dioxide from the air to the ocean and reduced the atmospheric greenhouse effect. The carbon-dioxide changes lagged behind the orbital forcing, and thus carbon dioxide was clearly a feedback, but the large global cooling of the ice ages has been successfully explained only if the reduced greenhouse effect is included (Jansen et al., 2007). By analogy, overspending a credit card induces debt, which is made larger by interest payments on that debt. The interest payments clearly lag the debt in time and did not cause the debt, but they contribute to the size of the debt, and the debt cannot be explained quantitatively unless the interest payments are included.

Until sea ice forms, heat stored in the ocean's surface waters is transferred to the atmosphere, limiting the extreme cold Arctic air temperatures despite the lack of solar energy.

Abrupt climate changes have been associated with the ice-age cycles. The most prominent and best known of these are linked to jumps in the wintertime extent of sea ice in the NORTH ATLANTIC, which in turn were linked to changes in the large-scale circulation of the ocean (e.g., Alley, 2007), as described in the previous section. The associated temperature changes were very large around the NORTH ATLANTIC (as much as 10°C or more) but much smaller in remote regions, and they were in the opposite direction in the far south (northern cooling was accompanied by slight southern warming). Hence, the globally averaged temperature changes were small and were probably linked primarily to ice-albedo feedback and small changes in the strength of the greenhouse effect. As reviewed by Alley (2007), the large ice-age ice sheets seem to have both triggered these abrupt swings and created conditions under which triggering was easier. Although such events remain possible, they are less likely without the large ice sheet on CANADA.

3.2.7 Arctic Amplification

The positive feedbacks outlined above amplify the Arctic response to climate forcings. The ice-albedo feedback is potentially strong in the Arctic because it hosts so much snow and ice (see Serreze and Francis, 2006 for additional discussion); if conditions are too warm for snow or ice to form, no ice-albedo feedback can exist. Climate models initialized from modern or similar conditions and forced in various ways are in widespread agreement that global temperature trends are amplified in the Arctic and that the largest changes are over the ARCTIC OCEAN during the cold season (autumn through spring) (e.g., Manabe and Stouffer, 1980; Holland and Bitz, 2003; Meehl et al., 2007). Summer changes over the ARCTIC OCEAN are relatively damped, although summer changes over Arctic lands are likely to be substantial (Serreze and Francis, 2006). The strong wintertime changes over the ARCTIC OCEAN are linked to the insulating character of sea ice.

To understand **Arctic amplification**, think first of an unperturbed climate in balance on annual time scales. During summer, solar energy melts the sea ice cover. As the ice cover melts, areas of open water are exposed. The albedo of the open water is much lower than that of sea ice, so the open water gains heat. Because much of the solar energy goes into melting ice and warming the ocean, the surface air temperature does not rise much and, indeed, over the melting ice it stays fairly close to the freezing point. Through autumn and winter, when little or no solar energy is received, this ocean heat is released back to the atmosphere. Until sea ice forms, heat stored in the ocean's surface waters is transferred to the atmosphere, limiting the extreme cold Arctic air temperatures despite the lack of solar energy. The formation of sea ice itself further releases heat back to the atmosphere. Once the sea ice is formed, it insulates the atmosphere from the relatively warm ocean waters allow much colder surface air temperatures to develop.

However, if the climate warms (regardless of the forcing) then the summer melt season lengthens and intensifies, and more areas of low-albedo open water form in summer and absorb solar radiation. As more heat is gained in the upper ocean, more heat is released back to the atmosphere in autumn and winter; this additional heat is expressed as a rise in air temperature. Furthermore, because the ocean now contains more heat, the ice that forms in autumn and winter is thinner, and therefore less insulating than before. This thinner ice melts more easily in summer and produces even more low-albedo open water that absorbs solar radiation, meaning even larger releases of heat to the atmosphere in autumn and even thinner ice the next spring, and so on. The process can also work in reverse. An initial Arctic cooling melts less ice during the summer and creates less low-albedo open water. If less summer heat is gained in the ocean, then less heat is released back to the atmosphere in autumn and winter, and air temperatures fall further.

Although the albedo feedback over the ocean seems to dominate, an albedo feedback over land is much more direct. Under a warming climate, snow melts earlier in spring and thus low-albedo tundra, shrub, and forest cover is exposed earlier and fosters further spring warming. Similarly, later autumn snow cover will foster further autumn warming. More snow-free days produce a longer period of surface warming and imply warmer summers. Again, the process can work in reverse: initial cooling leads to more snow cover, fostering further cooling. Collectively, these processes result in stronger net positive feedbacks to forced temperature change (regardless of forcing mechanism) than is typical globally, thereby producing "Arctic amplification."

During longer time intervals, an ice sheet such as the ***Laurentide Ice Sheet*** on North America can grow, or an ice sheet such as that on *Greenland* can melt. This growth or melting in turn influences albedo, freshwater fluxes to the ocean, broad patterns of atmospheric circulation, greenhouse-gas storage or release in the ocean and on land, and more. The climate amplification due to processes such as ice-sheet growth and decay are considered "slow feedbacks" because they require millennia to take effect (e.g., Edwards et al., 2007).

3.3 PROXIES OF ARCTIC TEMPERATURE AND PRECIPITATION

Temperature and precipitation are especially important climate variables. **Climate change** is typically driven by changes in key forcing factors, which are then amplified or retarded by regional feedbacks that affect temperature and precipitation (section 4.2 and 3.2). Because feedbacks have strong regional variability, spatially variable responses to hemispherically symmetric forcing are common throughout the Arctic (e.g., Kaufman et al., 2004). Consequently, spatial patterns of temperature and precipitation must be reconstructed regionally.

Reconstructing temperature and precipitation in pre-industrial times requires reliable proxies (see section 2.3 for a general discussion of proxies) that can be used to derive qualitative or, preferably, quantitative estimates of past climates. To capture the expected spatial variability, **proxy** climate reconstructions must be spatially distributed and span a wide range of geological time. In general, the use of several proxies to reconstruct past climates provides the most robust evidence for past changes in temperature and precipitation.

3.3.1 Proxies for Reconstruction of Temperature

3.3.1A Vegetation and Pollen Records

In general, the use of several proxies to reconstruct past climates provides the most robust evidence for past changes in temperature and precipitation.

Estimates of past temperature from data that describe the distribution of vegetation (primarily fossil pollen assemblages but also plant macrofossils such as fruits and seeds) may be relative (warmer or colder) or quantitative (number of degrees of change). Most information pertains to the growing season, because plants are dormant in the winter and so are less influenced by climate than during the growing season (but see below). For example, evidence of boreal forest vegetation (the presence of one or more boreal tree species) would be more strongly associated with warmer growing seasons than would evidence of treeless tundra—and the general position of northern treeline today approximates the location of the July 10°C isotherm.

Indicator species are species with well-studied and relatively restricted modern climatic ranges. The appearance of these species in the fossil record indicates that a certain climate milestone was reached, such as exceeding a minimum summer temperature threshold for successful growth or a winter minimum temperature of freezing tolerance (Figure 3.8). This methodology was developed early in Scandinavia (Iversen, 1944); Matthews et al. (1990) used indicator species to constrain temperatures during the last **interglaciation** in northwest *CANADA*, and Ritchie et al. (1983) used indicator species to highlight early **Holocene** warmth in northwest *CANADA*. This technique has also been used extensively with fossil insect assemblages.

Methodologies for the numerical estimation of past temperatures from pollen assemblages follow one of two approaches. The first is the inverse-modeling approach, in which fossil data from one or more localities are used to provide temperature estimates for those localities (this approach also underlies the relative estimates of temperature described above). A modern "calibration set" of data (in this case, pollen assemblages) is related by equations to observed modern temperature, and the functions thus obtained are then applied to fossil data. This method has been developed and applied in Scandinavia (e.g., Seppä et al., 2004). A variant of the inverse approach is analogue analysis, in which a large modern dataset with assigned climate data forms the basis for comparison with fossil spectra. Good matches are derived statistically, and the resulting set of analogues provides an estimate of the past mean temperature and accompanying uncertainty (Anderson et al., 1989, 1991).

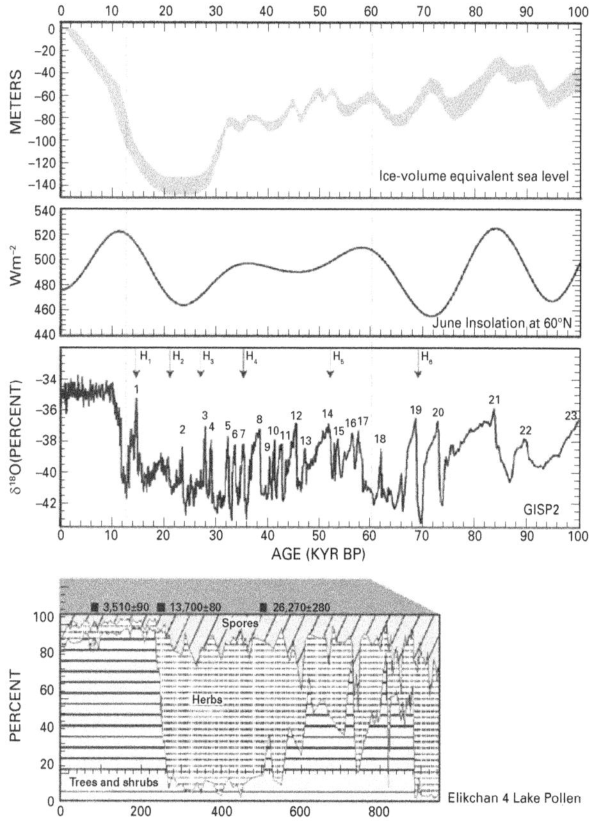

Figure 3.8 Upper three panels reflect changes as follows: top, sea level (Lambeck et al., 2002); middle, June insolation at 60°N (Berger and Loutre, 1991); bottom, air temperature over Greenland Ice Sheet (numbers refer to interstadials) based on the $\delta^{18}O$ record in ice (Grootes et al., 1993) during the past 100 ka (ages in calendar years). Arrows denote timing of Heinrich events. The lowermost panel shows changes in the percentages of tree and shrub pollen, herb pollen, and spores at Ilikchan 4 Lake in the Magadan region of Chukotka, Russia (Lozhkin and Anderson, 1996). Lake core x-axis is depth, not time; the base of the core dates to about 60 ka BP. Between approximately 55 ka and 72 ka, treeline recovered for short intervals to nearly Holocene conditions. These conditions could reflect warm interstadials that also appear in the Greenland ice core record about the same time.

Inverse modeling relies upon observed modern relationships. Some plant species were more abundant in the past than they are today, and the fossil pollen spectra they produced may have no recognizable modern counterpart—so-called "no-analogue" assemblages. Outside the envelope of modern observations, fossil pollen spectra, which are described in terms of pollen abundance, cannot be reliably related to past climate. This problem led to the adoption of a second approach to estimating past temperature (or other climate variable) called forward modeling. The pollen data are not used to develop numerical values but are used to test a "hypothesis" about the status of past temperature (a key ingredient of climate). The hypothesis may be a conceptual model of the status of past climate, but typically it is represented by a climate-model simulation for a given time in the past. The climate simulation drives a vegetation model that assigns vegetation cover on the basis of bioclimatic rules (such as the winter minimums or required warmth of summer growing temperatures mentioned above). The resultant map is compared with a map of past vegetation developed from the fossil data. The philosophy of this approach is described by Prentice and Webb (1998). Such data and models have been compared for the Arctic by Kaplan et al. (2003) and Wohlfahrt et al. (2004). The great advantage of this approach is that underlying the model simulation are hypothesized climatic mechanisms; those mechanisms allow not only the description but also an explanation of past climate changes.

3.3.1B Dendroclimatology

Seasonal differences in climate variables such as temperature and precipitation throughout many parts of the world, including the high latitudes, are known to produce annual rings that reflect distinct changes in the way trees grow and respond, year after year, to variations in the weather (Fritts, 1976). Alternating light and dark bands (couplets) of low-density early wood (spring and summer) and higher density late wood (summer to late summer) have been used for decades to reproduce long time series of regional climate change thought to directly influence the production of **meristematic** cells in the trees' vascular cambium, just below the bark. Cambial activity in many parts of the northern boreal forests can be short; late wood production very likely starts in late June and annual-ring width is complete by early August (e.g., Esper and Schweingruber, 2004). Fundamental to the use of tree rings is the fact that the average width of a tree ring couplet reflects some combination of environmental factors, largely temperature and precipitation, but it can also reflect local climatic variables such as wind stress, humidity and soil properties (see Bradley, 1999, for review). As a general guideline, growing season conditions favorable for the production of wide annual rings tend to be characterized by warmer than average summers with sufficient precipitation to maintain adequate soil moisture. Narrow tree rings occur during unusually cold or dry growing seasons.

As a general guideline, growing season conditions favorable for the production of wide annual rings tend to be characterized by warmer than average summers.

The extraction of a climate signal from ring width and wood density (**dendroclimatology**) relies on the identification and calibration of regional climate factors and on the ability to distinguish local climate influences from regional noise (Figure 3.9). How sites for tree sampling are selected is also important depending upon the climatological signal of interest. Trees in marginal growth sites, perhaps on drier substrates or near an ecological transition, are likely to be most sensitive to minor changes in temperature stress or moisture stress. On the other hand, trees in less-marginal sites likely reflect conditions of more widespread change. In the high latitudes, research is commonly focused on trees at both the latitude and elevation limits of tree growth or of the forest-tundra ecotone.

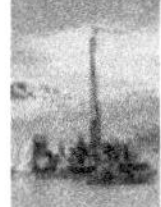

Figure 3.9. Annual tree rings composed of seasonal early and late wood are clear in this 64 year-old *Larix siberica* from western Siberia (Esper and Schweingruber, 2004). Initial growth was restricted; narrow rings average 0.035 mm/year, punctuated by one thicker ring (one single arrow). Later (two arrows), tree-ring width abruptly at least doubled for more than three years. Ring widths increased to 0.2 mm/year. [Photograph courtesy of Jan Esper, Swiss Federal Research Institute.]

Pencil-sized increment cores or sanded trunk cross sections are routinely used for stereomicroscopic examination and measurement (Figure 3.10*A*). A number of tree species are examined, most commonly varieties of the genera *Larix* (larch), *Pinus* (pine), and *Picea* (spruce). Raw ring-width time series are typically generated at a resolution of 0.01 mm along one or more radii of the tree, and these data are normalized for changes in ring width that reflect the natural increase in tree girth (a young tree produces wider rings). Ring widths for a number of trees are then averaged to produce a master curve for a particular site. The replication of many time series throughout a wide area at a particular site permits extraction of a climate-related signal and the elimination of anomalous ring biases caused by changes in competition or the ecology of any particular tree. Abrupt growth that caused a large change in ring width (Figure 3.10*B*) can be causally evaluated only on forest-site characteristics; that is, if the change isn't replicated in nearby trees, it's probably not related to climate.

The oxygen isotope composition of the calcareous shells of planktic foraminifers accurately records the oxygen isotope composition of ambient seawater, modulated by the temperature at which the organisms built their shells.

Dendroclimatology is statistically laborious, and a variety of approaches are used by the science community. Ring widths or ring density must first be calibrated by a response-function analysis in which tree growth and monthly climatic data are compared for the instrumental period. Once this is done, then cross-dated tree ring series reaching back millennia can be used as predictors of past change. Principal-components analysis, along with some form of multiple regression analysis, is commonly used to identify key variables. A comprehensive review of statistical treatments is beyond the scope of this report, but summaries can be found in Fritts (1976), Briffa and Cook (1990), Bradley (1999, his Chapter 10), and Luckman (2007).

3.3.1c Marine Isotopic Records

The oxygen isotope composition of the calcareous shells of planktic **foraminifers** accurately records the oxygen isotope composition of ambient seawater, modulated by the temperature at which the organisms built their shells (Epstein et al., 1953; Shackleton, 1967; Erez and Luz, 1982; Figure 3.11). (The term **$\delta^{18}O$** refers to the proportion of the heavy isotope, ^{18}O, relative to the lighter, more abundant isotope, ^{16}O.) However, the low horizontal and vertical temperature variability found in *Arctic Ocean* surface waters (less than –1°C) has little effect on the oxygen isotope composition of *N. pachyderma* (sin.) (maximum 0.2‰, according to Shackleton, 1974). Because meteoric waters, discharged into the ocean by precipitation and (indirectly) by river runoff, have considerably lower $\delta^{18}O$ values than do ocean waters, a reasonable correlation can be interpreted between salinity and the oxygen isotope composition of Arctic surface waters despite the complications of seasonal sea ice (Bauch et al., 1995; LeGrande and Schmidt, 2006). Accordingly, the spatial variability of surface-water salinity in the *Arctic Ocean* is recorded today by the $\delta^{18}O$ of planktic foraminifers (Spielhagen and Erlenkeuser, 1994; Bauch et al., 1997).

Figure 3.10. Typical tree ring samples. *A*) Increment cores taken from trees with a small-bore hollow drill. They can be easily stored and transported in plastic soda straws for analysis in the laboratory. *B*) Alternatively, cross sections or disks can be sanded for study. A cross section of *Larix decidua* root shows differing wood thickness within single rings, caused by exposure. [Photographs courtesy of Jan Esper and Holger Gärtner, respectively, Swiss Federal Research Institute.]

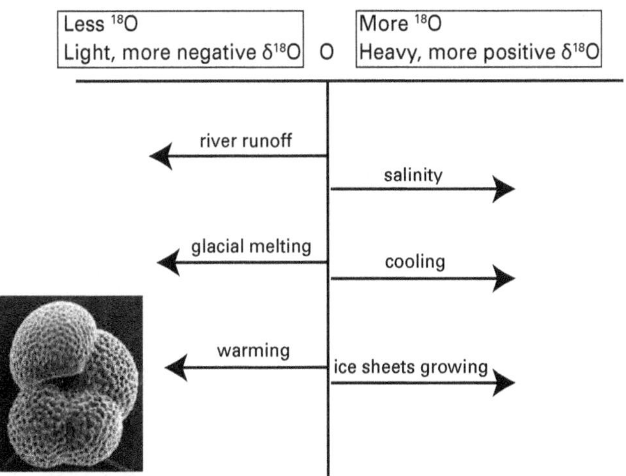

Figure 3.11. Fourteen microscopic marine plankton known as foraminifers (see inset) grow a shell of calcium carbonate ($CaCO_3$) that is in or near isotopic equilibrium with ambient sea water. The oxygen isotope ratio measured in these shells can be used to determine the temperature of the surrounding waters. (The oxygen-isotope ratio is expressed in $\delta^{18}O$ parts per mil (‰)$=10^3[(R_{sample}/R_{standard})-1]$, where $R_x=(^{18}O)/(^{16}O)$ is the ratio of isotopic composition of a sample compared with that of an established standard, such as ocean water.) However, factors other than temperature can influence the ratio of ^{18}O to ^{16}O. Warmer seasonal temperatures, glacial meltwater, and river runoff with depleted values all will produce a more negative (lighter) $\delta^{18}O$ ratio. On the other hand, cooler temperatures or higher salinity waters will drive the ratio up, making it heavier, or more positive. The growth of large continental ice sheets selectively removes the lighter isotope (^{16}O), leaving the ocean enriched in the heavier isotope (^{18}O).

The $\delta^{18}O$ values of planktic foraminifers in cores of ancient sediment from the deep *Arctic Ocean* vary considerably on millennial time scales (e.g., Aksu, 1985; Scott et al., 1989; Stein et al., 1994; Nørgaard-Pedersen et al., 1998, 2003, 2007a,b; Polyak et al., 2004; Spielhagen et al., 2004, 2005). The observed variability in foraminiferal $\delta^{18}O$ commonly exceeds the change in the isotopic composition of seawater that results merely from storing, on glacial-interglacial time scales, isotopically light freshwater in glacial ice sheets (about 1.0–1.2‰ $\delta^{18}O$) (Fairbanks, 1989; Adkins et al., 1997; Schrag et al., 2002). Changes with time in freshwater balance of the near-surface waters, and in the temperature of those waters, are both recorded in the $\delta^{18}O$ values of foraminifer shells. Moreover, in cases where independent evidence of a regional warming of surface waters is available (e.g., in the eastern *Fram Strait* during the **Last Glacial Maximum**; Nørgaard-Pedersen et al., 2003), this warming is thought to have been caused by a stronger influx of saline Atlantic Water. Because salinity influences $\delta^{18}O$ of foraminfer shells from the *Arctic Ocean* more than temperature does, it is difficult to reconstruct temperatures in the past on the basis of systematic variations in calcite $\delta^{18}O$ in *Arctic Ocean* sediment cores.

3.3.1D Lacustrine Isotopic Records

Isotopic records preserved in lake sediment provide important paleoclimatic information on landscape change and hydrology. Lakes are common in high-latitude landscapes, and sediment deposited continuously provides uninterrupted, high-resolution records of past climate (Figure 3.12).

Isotopic records preserved in lake sediment provide important paleoclimatic information on landscape change and hydrology.

Oxygen isotope ratios in precipitation reflect climate processes, especially temperature (see 3.3.1E). The oxygen isotope ratios of shells and other materials in lakes primarily reflect ratios of the lake water. The isotopic ratios in the lake water are dominantly controlled by the isotopic ratios in precipitation—unless evaporation from the lake is sufficiently rapid, compared with inflow of new water, to shift the isotopic ratios towards heavier values by preferentially

Figure 3.12. Lake El'gygytgyn in the Arctic Far East of Russia. Open and closed lake systems in the Arctic differ hydrologically according to the balance between inflow and outflow and the ratio of precipitation to evaporation. These parameters are the dominant influence on lake stable-isotopic chemistry and on the depositional character of the sediments and organic matter. Lake El'gygytgyn is annually open and flows to the Bering Sea during July and August, but the outlet closes by early September as lake level drops and storms move beach gravels that choke the outlet. [Photograph by J. Brigham-Grette.]

The most common way to deduce temperature from ice cores is through the isotopic content their water.

removing isotopically lighter water. Those lakes that have streams entering and leaving (open lakes) have isotopic ratios that are generally not affected much by evaporation, as do some lakes supplied only by water flow through the ground (closed lakes). These lakes allow isotopic ratios of shells and other materials in them to be used to reconstruct climate, especially temperature. However, some closed lakes are affected notably by evaporation, in which case the isotopic ratios of the lake are at least in part controlled by lake hydrology. Unless independent evidence of lake hydrology is available, quantitative interpretation of $\delta^{18}O$ is difficult. Consequently, $\delta^{18}O$ is normally combined with additional climate proxies to constrain other variables and strengthen interpretations. For example, in rare cases, ice core records that are located near lakes can provide an oxygen isotope record for direct comparison (Fisher et al., 2004; Anderson and Leng, 2004; Figure 3.13). Oxygen isotope ratios are relatively easy to measure on carbonate shells or other carbonate materials. Greater difficulty, which limits the accuracy (i.e., the time-resolution) of the records, is associated with analyses of oxygen isotopes in silica from **diatom** shells (Leng and Marshall, 2004) and in organic matter (Sauer et al., 2001; Anderson et al., 2001). Additional uncertainty arises with organic matter because its site of origin is unknown: although some of it grew in the lake, some was also washed in and is likely to have been stored on the landscape for an indeterminate time previously.

3.3.1e Ice Cores

The most common way to deduce temperature from ice cores (Figures 3.13 and 3.14) is through the isotopic content their water, i.e., the ratio of $H_2{}^{18}O$ to $H_2{}^{16}O$, or of HDO to H_2O (where **D** is deuterium, ^2H). The ratios are expressed as $\delta^{18}O$ and **δD** respectively, relative to standard mean ocean water (SMOW). Pioneering studies (Dansgaard, 1964) showed how $\delta^{18}O$ is related to climatic variables in modern precipitation. At high latitudes both $\delta^{18}O$ and δD are generally, with some caveats, considered to represent

CORE LOCATIONS

1. Meli Lake (Anderson et al., 2001)
2. Tangled Up Lake (Anderson et al., 2001)
3. Farewell Lake (Hu et al., 2001)
4. Grandfather Lake (Hu et al., 2003)
5. Prospector Col, Mt Logan (Fisher et al., 2004)
6. Jellybean Lake (Anderson et al., 2005)
7. Marcella Lake (Anderson et al., 2007)
8. Toronto Lake (Wolfe et al., 1996)
9. Agassiz Ice Cap (Fisher et al., 1995)
10. Summit Greenland
11. Lakes SS6 and Braya Sø (Anderson and Leng, 2004)
12. Lake Igelsjön (Hammarlund et al., 2003)
13. Lake Tibetus (Hammarlund et al., 2002)
14. Vuolep Allakasjaure (Rosqvist et al., 2004)
15. Lake 850 (Shemesh et al., 2001)
16. Chuna Lake (Jones et al., 2004)
17. Lakes Yarnyshnoe and Poteryanny Zub (Wolfe et al., 2000)
18. Middendorf Lake (Wolfe et al., 2000)
19. Derevanoi Lake (Wolfe et al., 2000)
20. Dolgoe Lake (Wolfe et al., 2000)

Ice Cores

Lake Cores

Figure 3.13. Locations of Arctic and sub-Arctic lakes (blue) and ice cores (green) whose oxygen isotope records have been used to reconstruct Holocene paleoclimate. [Map adapted from the Atlas of Canada, © 2002. Her Majesty the Queen in Right of Canada, Natural Resources Canada/Sa Majesté la Reine du chef du Canada, Resources naturelles Canada.]

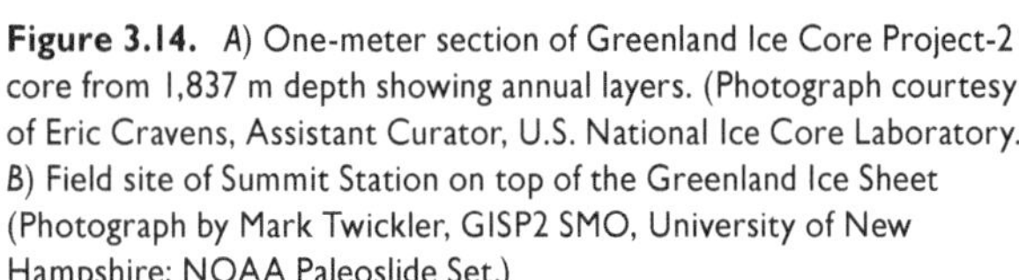

Figure 3.14. A) One-meter section of Greenland Ice Core Project-2 core from 1,837 m depth showing annual layers. (Photograph courtesy of Eric Cravens, Assistant Curator, U.S. National Ice Core Laboratory.) B) Field site of Summit Station on top of the Greenland Ice Sheet (Photograph by Mark Twickler, GISP2 SMO, University of New Hampshire; NOAA Paleoslide Set.)

the mean annual temperature at the core site, and the use of both measures together offers additional information about conditions at the source of the water vapor (e.g., Dansgaard et al., 1989). Recent work by Werner et al. (2000), however, demonstrates that changes in the seasonal cycle of precipitation over the ice sheets can affect measurements of ice-core temperature.

The center of the Greenland Ice Sheet has not finished warming from the ice age.

The underlying idea is that an air mass loses water vapor by condensation as it travels from a warm source to a cold (polar) site. This point is easily shown by the nearly linear relationship between precipitation and temperature over modern ice sheets (Figure 3.15). Water that contains the heavy isotopes has a lower vapor pressure, so the heavy isotope preferentially condenses into rain or snow, and the air mass becomes progressively depleted of the heavy isotope it moves to colder sites. It can easily be shown from spatial surveys (Johnsen et al., 1989) and, indeed, from modeling studies using models enabled with water isotopes (e.g., Hoffmann et al., 1998; Mathieu et al., 2002) that a good spatial relationship between temperature and water isotope ratio exists. The relationship is

$$\delta = aT + b$$

where T is mean annual surface temperature, and δ is annual mean $\delta^{18}O$ or δD value in precipitation in the polar regions, and the slope, a, has values typically around 0.6 for *Greenland* $\delta^{18}O$.

Temperature is not the only factor that can affect isotopic ratios. Changes in the season when snow falls, in the source of the water vapor, and other things are potentially important as well (Jouzel et al., 1997; Werner et al., 2000) (Figure 3.16). For this reason, it is common whenever possible to calibrate the isotopic ratios using additional **paleothermometers**. For short intervals, instrumental records of temperature can be compared with isotopic ratios (e.g., Shuman et al., 1995). The few comparisons that have been done (summarized in Jouzel et al., 1997) tend to show δ/T gradients that are slightly lower than the spatial gradient. Accurate reconstructions of past temperature, but with low time resolution, are obtained from the use of borehole thermometry. The center of the *Greenland Ice Sheet* has not finished warming from the ice age, and the remaining cold temperatures reveal how cold the ice age was (Cuffey et al., 1995; Johnsen et al., 1995). Additional paleothermometers are available that use a thermal **diffusion** effect. In this effect, gas isotopes are separated slightly when an abrupt temperature change at the surface creates a temperature difference between the surface and the region a few tens of meters down, where bubbles are pinched off from the interconnected pore spaces in old snow (called **firn**). The size of the gas-isotope shift reveals the size of an abrupt warming, and the number of years between the indicators of an abrupt change in the ice and in the bubbles trapped in ice reveals the temperature before the abrupt change—if the snowfall rate before the abrupt change is known (Severinghaus et al., 1998; Severinghaus and Brook, 1999; Huber et al., 2006). These methods show that the value of

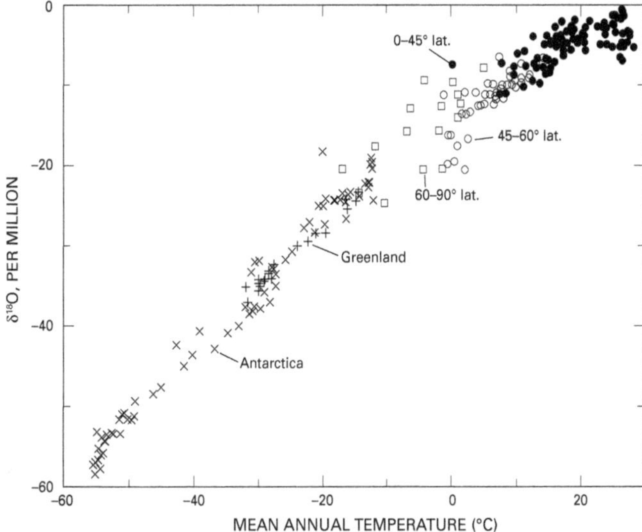

Figure 3.15. Relation between isotopic composition of precipitation and temperature in the parts of the world where ice sheets exist. Sources of data as follows: International Atomic Energy Agency (IAEA) network (Fricke and O'Neil, 1999); calculated as the means of summer and winter data of their Table 1 for all sites with complete data. Open squares = poleward of 60° latitude (but with no inland ice-sheet sites); open circles = 45°–60° latitude; filled circles = equatorward of 45° latitude; X = data from Greenland (Johnsen et al., 1989); + = data from Antarctica (Dahe et al., 1994). About 71% of Earth's surface area is equatorward of 45°, where dependence of $\delta^{18}O$ on temperature is weak to nonexistent. Only 16% of Earth's surface falls in the 45°–60° band, and only 13% is poleward of 60°. The linear array is clearly dominated by data from the ice sheets. (Source: Alley and Cuffey, 2001.) [Reproduced by permission of Mineralogical Society of America.]

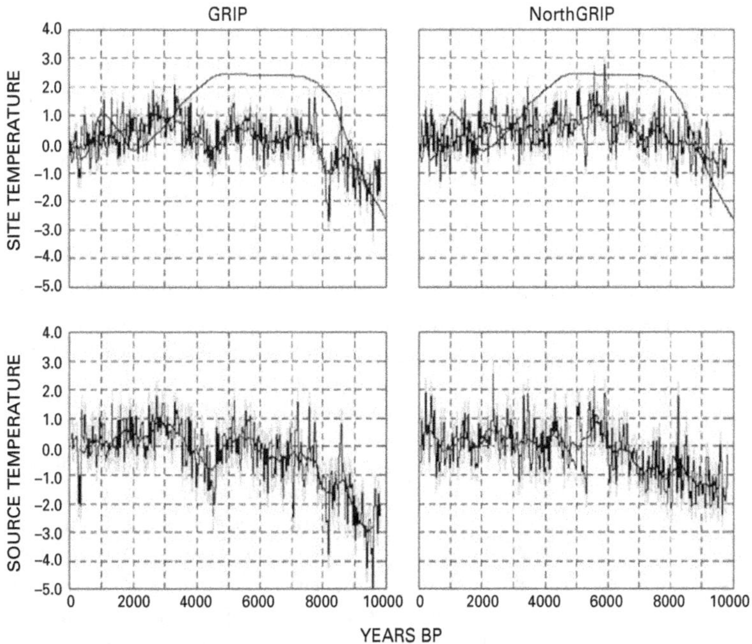

Figure 3.16. Paleotemperature estimates of site and source waters from on Greenland: GRIP and NorthGRIP, Masson-Delmotte et al., 2005). GRIP (left) and NorthGRIP (right) site (top) and source (bottom) temperatures derived from GRIP and NorthGRIP $\delta^{18}O$ and deuterium excess corrected for seawater $\delta^{18}O$ (until 6000 BP). Shaded lines in gray behind the black line = an estimate of uncertainties due to the tuning of the isotopic model and the analytical precision. Solid line (in part above zigzag line) = GRIP temperature derived from the borehole-temperature profile (Dahl-Jensen et al., 1998). [Copyright 2005; reproduced by permission of American Geophysical Union.]

the δ/T slope produced by many of the large changes recorded in *Greenland* ice cores was considerably less (typically by a factor of 2) than the spatial value, probably because of a relatively larger reduction in winter snowfall in colder times (Cuffey et al., 1995; Werner et al., 2000; Denton et al., 2005). The actual temperature changes were therefore larger than would be predicted by the standard calibration.

In summary, water isotopes in polar precipitation are a reliable proxy for mean annual air temperature, but for quantitative use, some means of calibrating them is required. They may be calibrated either against instrumental data by using an alternative estimate of temperature change, or through modeling, even for ice deposited during the Holocene (Schmidt et al., 2007).

3.3.1F Fossil Assemblages and Sea-Surface Temperatures

Different species live preferentially at different temperatures in the modern ocean. Modern observations can be used to learn the preferences of species. An inherent assumption is that species maintain their preferences through time. With that assumption, the mathematical expression of these preferences plus the history of where the various species lived in the past can then be used to interpret past temperatures (Imbrie and Kipp, 1971; CLIMAP, 1981). This line of reasoning is primarily applied to near-surface (planktic) species, and especially to foraminifers, diatoms, and **dinoflagellates**. The presence or absence and the relative abundance of species can be used. Such methods are now commonly supported by sea-surface temperature estimates using emerging biomarker techniques outlined below.

Organic biomarkers can usually be recovered from Arctic sediments that do not preserve carbonate or siliceous microfossils.

3.3.1G Biogeochemistry

Within the past decade, two new organic proxies have emerged that can be used to reconstruct past ocean surface temperature. Both measurements are based on quantifying the proportions of **biomarkers**—molecules produced by restricted groups of organisms—preserved in sediments. In the case of the "**$U^{k'}_{37}$ index**" (Brassell et al., 1986; Prahl et al., 1988), a few closely related species of **coccolithophorid algae** are entirely responsible for producing the 37-**carbon ketones** ("**alkenones**") used in the paleotemperature index, whereas **crenarcheota** (archea) produce the tetra-ether lipids that make up the **TEX_{86} index** (Wuchter et al., 2004). Although the specific function that the alkenones and glycerol dialkyl tetraethers serve for these organisms is unclear, the relationship of the biomarker $U^{k'}_{37}$ index to temperature has been confirmed experimentally in the laboratory (Prahl et al., 1988) and by extensive calibrations of modern surface sediments to overlying surface ocean temperatures (Muller et al., 1998; Conte et al., 2006; Wuchter et al., 2004).

Biomarker reconstructions have several advantages for reconstructing sea surface conditions in the Arctic. First, in contrast to $\delta^{18}O$ analyses of marine carbonates (outlined above), the confounding effects of salinity and ice volume do not compromise the utility of biomarkers as paleotemperature proxies (a brief discussion of caveats in the use of $U^{k'}_{37}$ is given below). Both the $U^{k'}_{37}$ and TEX_{86} proxies can be measured reproducibly to high precision (analytical errors correspond to about 0.1°C for $U^{k'}_{37}$ and 0.5°C for TEX_{86}), and sediment extractions and gas or liquid chromatographic detections can be automated for high sampling rates. The abundances of biomarkers also provide insights into the composition of past ecosystems, so that links between the physical oceanography of the high latitudes and carbon cycling can be assessed. And lastly, organic biomarkers can usually be recovered from Arctic sediments that do not preserve carbonate or siliceous microfossils. It should be noted, however, that the harsh conditions of the northern high latitudes mean that the organisms producing the alkenone and tetraethers possibly were excluded at certain times and places; thus, continuous records cannot be guaranteed.

The principal caveats in using biomarkers for paleotemperature reconstructions come from ecological and evolutionary considerations. Alkenones are produced by algae that are restricted to the region of abundant light (the photic zone), so paleotemperature estimates based on them apply to this layer, which approximates the sea surface temperature. In the vast majority of the ocean, the alkenone signal recorded by sediments closely correlates with mean annual sea-surface temperature (Muller et al., 1998; Conte et al., 2006; Figure 3.17). However, in the case of highly seasonal high-latitude oceans,

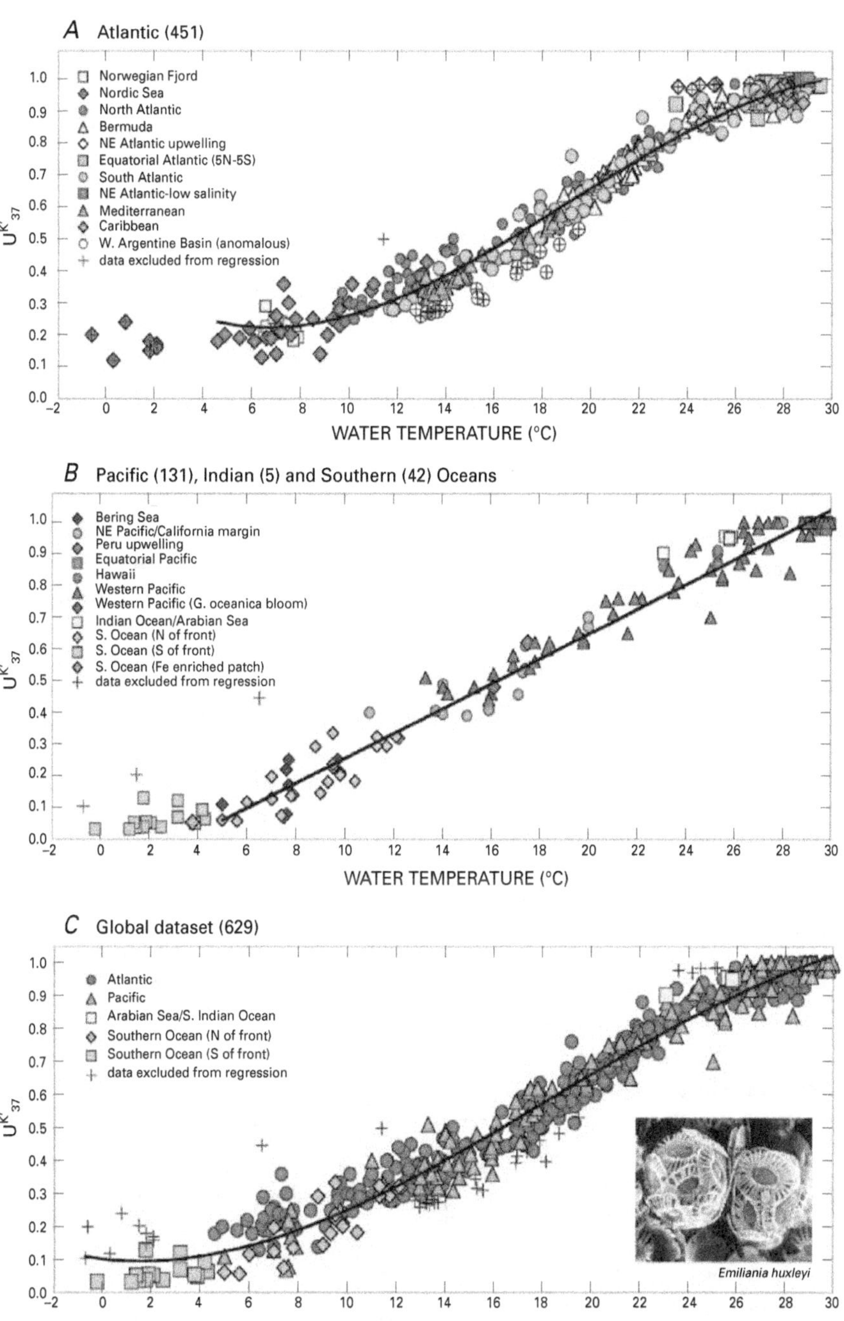

Figure 3.17 Biomarker alkenone. $U^{K'}_{37}$ versus measured water temperature for ocean surface mixed layer (0–30 m) samples. *A*) Atlantic region. Empirical third-order polynomial regression for samples collected in >4°C waters is $U^{K'}_{37} = -1.004 \times 10^{-4}T^3 + 5.744 \times 10^{-3}T^2 - 6.207 \times 10^{-2}T + 0.407$ ($r^2 = 0.98$, n = 413). (Outlier data from the southwest Atlantic margin and northeast Atlantic upwelling regime is excluded.) *B*) Pacific, Indian, and Southern Ocean regions: the empirical linear regression of Pacific samples is $U^{K'}_{37} = 0.0391T - 0.1364$ ($r^2 = 0.97$, n = 131). Pacific regression does not include the Indian and Southern Ocean data. *C*) Global data: the empirical third-order polynomial regression, excluding anomalous southwest Atlantic margin data, is $U^{K'}_{37} = -5.256 \times 10^{-5}T^3 + 2.884 \times 10^{-3}T^2 - 8.4933 \times 10^{-3}T + 9.898$ ($r^2 = 0.97$, n = 588). + = sample excluded from regressions (Conte et al, 2006). [Copyright 2006; reproduced by permission of American Geophysical Union.]

Use of these biomarker proxies is based on the assumption that the modern relation between organic proxies and temperature was the same in the past.

where coccolithophorid blooms typically occur during the summer months, the temperatures inferred from the alkenone $U^{k'}_{37}$ index may better approximate summer surface temperatures than mean annual sea-surface temperature. Furthermore, past changes in the season of production could bias long-term time series of past temperatures that are based on the $U^{k'}_{37}$ proxy. Depending on water column conditions, past production could have been highly focused toward a short summer or a more diffuse late spring–early fall productive season. A survey of modern surface sediments in the *North Atlantic* and *Nordic Seas* (Rosell-Melé et al., 1995) shows that at colder water temperatures the original unsaturation ratio as defined by Brassell et al. (1986) most reliably estimates surface water temperatures because it includes the $U^{k'}_{37}$:4 alkenone type, which is common in the *Nordic Seas* although it is rare or absent in most of the world ocean including the Antarctic. The Brassell et al. (1986) unsaturation ratio provides reliable surface water temperature estimates as low as 6°C, but errors increase at lower temperatures (Bendle and Rosell-Melé, 2004).

In contrast to the near-surface restriction of the algae producing the $U^{k'}_{37}$ proxy, the marine crenarcheota that produce the tetraether membrane lipids used in the TEX_{86} index can range widely through the water column. In situ analyses of particles suspended in the water column show that the **tetraether lipids** are most abundant in winter and spring months in many ocean provinces (Wuchter et al., 2005) and are present in large amounts below 100 m depth. However, it appears that the chemical basis for the TEX_{86} proxy is fixed by processes in the upper lighted (photic) zone, so that the sedimentary signal originates near the sea surface (Wuchter et al., 2005), just as for the $U^{k'}_{37}$ proxy. No studies have yet been conducted to assess how high-latitude seasonality affects the TEX_{86} proxy.

Diatom shells and remains of non-biting midge flies are among the biological indicators most commonly used to reconstruct ancient Arctic climate.

As for many other proxies, use of these biomarker proxies is based on the assumption that the modern relation between organic proxies and temperature was the same in the past. The two modern (and genetically closely related) species producing the alkenones in the $U^{k'}_{37}$ proxy can be traced back in time in a continuous lineage to the **Eocene** (about 50 **Ma**), and alkenone occurrences coincide with the fossil remains of the ancestral lineage in the same sediments (Marlowe et al., 1984). One might suppose that past evolutionary events in the broad group of algae that includes these species might have produced or eliminated other species that generated these chemicals but with a different relation to temperature. However, other such species would cause jumps in climate reconstructions at times of evolutionary events in the group, and no such jumps are observed. The TEX_{86} proxy can be applied to marine sediments 70–100 million years old. The working assumption is, therefore, that both organic proxies can be applied accurately to sediments containing the appropriate chemicals.

Because these biomarker proxies depend on changes in relative abundance of chemicals, it is important that natural processes after death of the producing organisms do not preferentially break down one chemical and thus change the ratio. Fortunately, the ratio appears to be stable (Prahl et al., 1989; Grice et al., 1998; Teece et al., 1998; Herbert, 2003; Schouten et al., 2004). An additional complication is that sediments can be moved around by ocean currents, so that the material sampled at one place might have been produced in another place under different climate conditions (Thomsen et al., 1998; Ohkouchi et al., 2002). Ordinarily, lengthy transport of biomarkers into a depositional site is rare and volumes are small compared with the supply from the productive ocean above, so that the proxy indeed records local climate. However, at some times and places, the Arctic has been comparatively unproductive, so that transport from other parts of the ocean, or from land in the case of the TEX_{86} proxy, likely was important (Weijers et al., 2006).

3.3.1h Biological Proxies in Lakes

Lakes and ponds are common in most Arctic regions and provide useful records of climate change (Smol and Cumming, 2000; Cohen, 2003; Schindler and Smol, 2006; Smol 2008). Many different biological climate proxies are preserved in Arctic lake and pond sediments (Pienitz et al., 2004). **Diatom** shells (Douglas et al., 2004) and remains of non-biting midge flies (chironomid head capsules; Bennike et al., 2004) are among the biological indicators most commonly used to reconstruct ancient Arctic climate (Figure 3.18). The approach generally

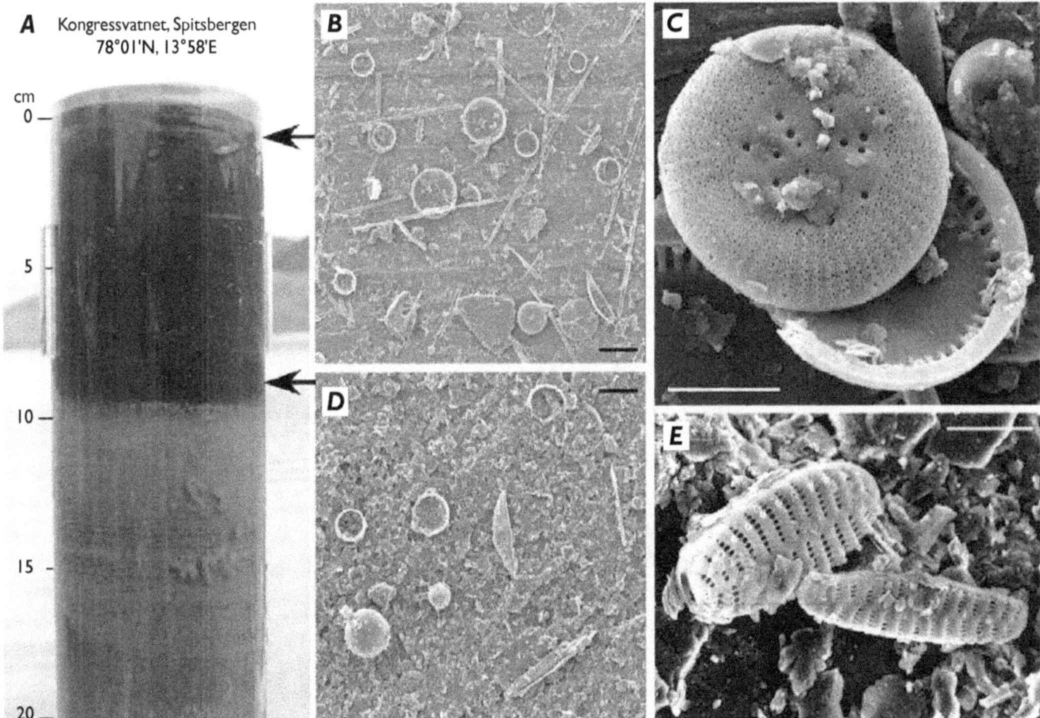

Figure 3.18. Diatom assemblages are determined by a variety of environmental conditions in Arctic lake systems. Changes from one assemblage to another reflect changes in environmental controls such as light, nutrient availability, water chemistry, or temperature. Shown here is the upper 20 cm of a sediment core from Kongressvatnet, a deep polar lake on western Spitsbergen, Svalbard. An abrupt transition (*A*) from laminated silts to organic sediments at ~9 cm dated to 1880 A.D. records the retreat of glaciers from the lake's drainage catchment since end of the Little Ice Age. Recent assemblages (*B*) are characterized by abundant and diverse free-floating diatoms such as *Cyclotella* spp. (*C*). In contrast, diatom assemblages were less diverse and primarily bottom-dwelling during the early 20th century with a greater abundance of the silica remains of chysophytes (flagellate golden algae) (*D*). Across the Arctic, small diatoms of the Family Fragilariaceae (*E*) are common in sediments pre-dating the 20th century, forming communities that, in many instances, indicate stable limnological conditions for several millennia.

used by those who study the history of lakes (paleolimnologists) is first to identify useful species—those that grow only within a distinct range of conditions. Then, the modern conditions preferred by these indicator species are determined, as are the conditions beyond which these indicator species cannot survive. (Typically used are surface sediment calibration sets or training sets to which are applied statistical approaches such as canonical correspondence analysis and weighted averaging regression and calibration; see Birks, 1998.) The resulting mathematical relations (or transfer functions such as those used in marine records) are then used to reconstruct the environmental variables of interest, on the basis of the distribution of indicator assemblages preserved in dated sediment cores (Smol, 2008). Where well-calibrated transfer functions are not available, such as for some parts of the Arctic, less-precise climate reconstructions are commonly based on the known ecological and life-history characteristics of the organisms.

Ideally, sedimentary characteristics would be linked directly to key climatic variables such as temperature (e.g., Pienitz and Smol, 1993; Joynt and Wolfe, 2001; Bigler and Hall, 2003; Bennike et al., 2004; Larocque and Hall, 2004; Wooler et al., 2004; Finney et al., 2004; other chapters in Pienitz et al., 2004; Barley et al., 2006; Weckström et al., 2006). However, lake sediments typically record conditions in the lake that are only indirectly related to climate (Douglas and Smol, 1999). For example, lake ecosystems are strongly influenced by the length of the ice-free versus the ice-covered season, by the Sun-blocking effect of any snow cover on ice (Figure 3.19) (e.g., Smol, 1988; Douglas et al., 1994; Sorvari and Korhola, 1998; Douglas and Smol, 1999; Sorvari et al., 2002; Rühland et al., 2003; Smol and Douglas, 2007a) and by the existence or absence of a seasonal layer of warm water near the lake surface that remains separate from colder waters beneath (Figure 3.20). Shells and other features in

Shells and other features in the lake sediment record the species living in the lake and conditions under which they grew.

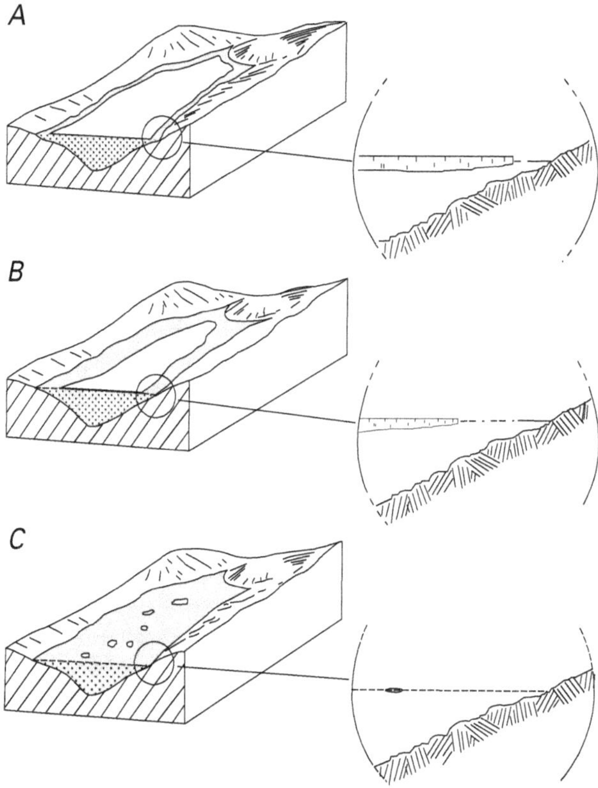

Figure 3.19. Changing ice and snow conditions on an Arctic lake during relatively *A*) cold, *B*) moderate, and *C*) warm conditions. During colder years, a permanent raft of ice may persist throughout the short summer, precluding the development of large populations of phytoplankton, and restricting much of the primary production to a shallow, open open-water moat. Many other physical, chemical and biological changes occur in lakes that are either directly or indirectly affected by snow and ice cover (see Table 1; Douglas and Smol, 1999). Modified from Smol (1988). [Reproduced by permission of E.Schweizerbart'sche Verlagsbuchhandlung OHG, *http://www.schweizerbart.de/*.]

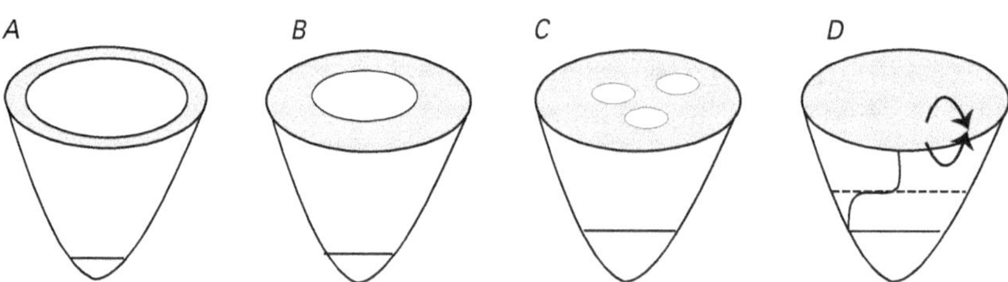

Figure 3.20. Lake ice melts as it continues to warm (*A–D*). Eventually, in deeper lakes (not ponds) thermal stratification (horizontal lines) may also occur (or be prolonged) during the summer months (*D*), further altering the limnological characteristics of the lake. Modified from Douglas (2007). [Reproduced by permission of The Paleontological Society Papers.]

the lake sediment record the species living in the lake and conditions under which they grew. These factors rather directly reflect the ice and snow cover and lake stratification and only indirectly reflect the atmospheric temperature and precipitation that control the lake conditions.

3.3.1i Insect Proxies

Insects are common and typically are preserved well in Arctic sediment. Because many insect types live only within narrow ranges of temperature or other environmental conditions, the remains of particular insects in old sediments provide useful information on past climate.

Calibrating the observed insect data to climate involves extensive modern and recent studies, together with careful statistical analyses. For example, fossil beetles are typically related to temperature using what is known as the Mutual Climatic Range method (Elias et al., 1999; Bray et al., 2006). This method quantitatively assesses the relation between the modern geographical ranges of selected beetle species and modern meteorological data. A "climate envelope" is determined, within which a species can thrive. When used with paleodata, the method allows for the reconstruction of several parameters such as mean temperatures of the warmest and coldest months of the year.

3.3.1j Sand Dunes

When plant roots anchor the soil, sand cannot blow around to make dunes. In the modern Arctic, and especially in *Alaska* (Figure 3.21) and *Russia*, sand dunes are forming and migrating in many places where dry, cold conditions restrict vegetation. During the last glacial interval and at some other times, dunes formed in places that now lack active dunes and indicate colder or drier conditions at those earlier times (Carter, 1981; Oswald et al., 1999; Beget, 2001; Mann et al., 2002). Some wind-blown mineral grains are deposited in lakes. The rate at which sand and silt are deposited in lakes increases as nearby vegetation is removed by cooling or drying, so analysis of the sand and silt in lake sediments provides additional information on the climate (e.g., Briner et al., 2006).

Sand dunes are forming and migrating in many places where dry, cold conditions restrict vegetation.

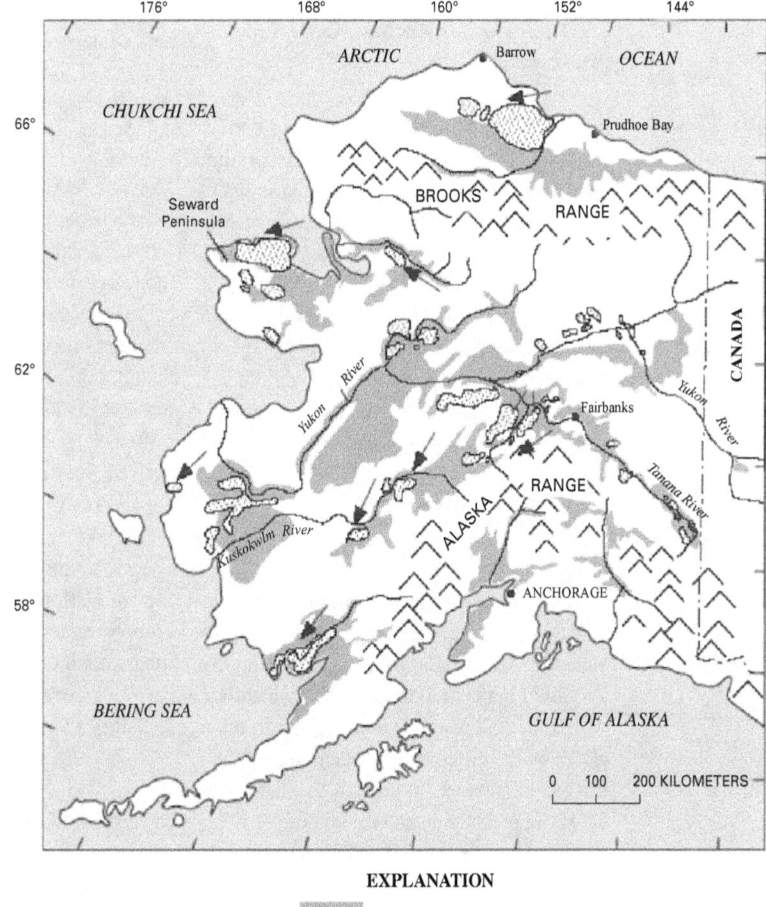

Figure 3.21. The form and distribution of wind-blown silt (loess), wind-blown sand (dunes), and other deposits of wind-blown sediment in Alaska have been used to infer both Holocene and last-glacial paleowind directions. (Compiled from multiple sources by Muhs and Budahn, 2006.)

3.3.2 Proxies for Reconstruction of Precipitation

In the case of sand dunes described above, separating the effects of changing temperature from those of changing precipitation is likely to be difficult, but additional indicators such as insect fossils in lake sediments very likely help by constraining the temperature. In general, precipitation is more difficult to estimate than is temperature, so reconstructions of changes in precipitation in the past are less common, and typically less quantitative, than are reconstructions of past temperature changes.

Different plants live in wet and dry places, so indications of past vegetation provide estimates of past wetness.

3.3.2a Vegetation-Derived Precipitation Estimates

Different plants live in wet and dry places, so indications of past vegetation provide estimates of past wetness. Plants do not respond primarily to rainfall but instead to moisture availability. Availability is primarily controlled in most places by the difference between precipitation and evaporation (or, in deserts, the difference between precipitation and the amount of water that could be evaporated if sufficient water were available)., although some soils carry water downward so efficiently that dryness occurs even without much evaporation.

Much modern tundra vegetation grows where precipitation exceeds evaporation. Plants such as Sphagnum (bog moss), cotton-grass (*Eriophorum*), and cloudberry (*Rubus chamaemorus*) indicate moist growing conditions. In contrast, grasses dominate dry tundra and polar semi-desert. Such differences are evident today (Oswald et al., 2003) and can be reconstructed from pollen and larger plant materials (macrofossils) in sediments. Some regions of *Alaska* and *Siberia* retain sand dunes that formed in the Last Glacial Maximum but are inactive today; typically, those regions are near areas that had grasses then but now have plants requiring greater moisture (Colinvaux, 1964; Ager and Brubaker, 1985; Lozhkin et al., 1993; Goetcheus and Birks, 2001; Zazula et al., 2003).

In Arctic regions, deep snow cover very likely allows the persistence of shrubs that would be killed if exposed during the harsh winter cold and wind. For example, dwarf willow can survive if snow depths exceed 50 cm (Kaplan et al., 2003). *Siberian* stone pine requires considerable winter snow to weigh down and bury its branches (Lozhkin et al, 2007). The presence of these species therefore indicates certain minimum levels of winter precipitation.

Moisture levels can also be estimated quantitatively from pollen assemblages by means of formal techniques such as inverse and forward modeling, following techniques also used to estimate past temperatures. Moisture-related transfer functions have been developed, in Scandinavia for example (Seppä and Hammarlund, 2000). Kaplan et al. (2003) compared pollen-derived vegetation with vegetation derived from model simulations for the present and key times in the past. The pollen data indicated that model simulations for the Last Glacial Maximum tended to be "too moist"—the simulations generated shrub-dominated **biomes** whereas the pollen data indicated drier tundra dominated by grass.

3.3.2b Lake-Level-Derived Precipitation Estimates

In addition to their other uses in **paleoclimatology** as described above, lakes act as natural rain gauges. If precipitation increases relative to evaporation, lakes tend to rise, so records of past lake levels provide information about the availability of moisture.

Most of the water reaching a lake first soaked into the ground and flowed through spaces as groundwater, before it either seeped directly into the lake or else came back to the surface in a stream that flowed into the lake. Smaller amounts of water fall directly on the lake or flow over the land surface to the lake without first soaking in (e.g., MacDonald et al., 2000b). Lakes lose water to streams ("overflow"), as outflow into groundwater, and by evaporation. If water supply to a lake increases, the lake level will rise and the lake will spread. This spread will increase water loss from the lake by increasing the area for evaporation, by increasing the area through which groundwater is leaving and the "push" (hydraulic head) causing that outflow, and perhaps by forming a new outgoing stream or increasing the size of an existing stream. Thus, the level of a lake adjusts in response to changes in the balance between precipitation and evaporation in the region feeding water to the lake (the catchment). Because either an increase in precipitation or a reduction in evaporation will cause a lake level to rise, an independent estimate of either precipitation or evaporation is required before one can estimate the other on the basis of a history of lake levels (Barber and Finney, 2000).

Former lake levels can be identified by deposits such as the fossil shoreline they leave (Figure 3.22); sometimes these deposits are preserved under water and can be recognized in sonar surveys or other data, and these deposits can usually be dated. Furthermore, the sediments of the lake very likely retain a signature of lake-level fluctuations: coarse-grained material generally lies near the shore and finer grained materials offshore (Digerfeldt, 1988), and these too can be identified, sampled, and dated (Abbott et al., 2000).

For a given lake, modern values of the major inputs and outputs can be obtained empirically, and a model can then be constructed that simulates lake-level changes in response to changing precipitation and evaporation. Allowable pairs of precipitation and evaporation can then be estimated for any past lake level. Particularly in cases where precipitation is the primary control of water depth, it is possible to model lake level responses to past changes in precipitation (e.g., Vassiljev, 1998; Vassiljev et al., 1998). For two lakes in interior *Alaska*, this technique suggested that precipitation about 12,500 years ago was as much as 50% lower than modern (Barber and Finney, 2000).

Biological groups living within lakes also leave fossil assemblages that can be interpreted in terms of lake level by comparing them with modern assemblages. In all cases, factors other than water depth (e.g., conductivity and salinity) likely influence the assemblages

For a given lake, modern values of the major inputs and outputs can be obtained empirically, and a model can then be constructed that simulates lake-level changes in response to changing precipitation and evaporation.

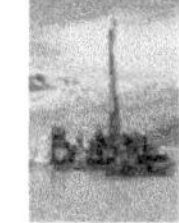

Figure 3.22. Unnamed, hydrologically closed lake in the Yukon Flats Wildlife Refuge, Alaska. Concentric rings of vegetation developed progressively inward as water level fell, owing to a negative change in the lake's overall water balance. Historic Landsat imagery and air photographs indicate that these shorelines formed within the last 40 years or so. (Photograph by Lesleigh Anderson.)

(MacDonald et al., 2000b), but these factors are themselves likely to be indirectly related to water depth. Aquatic plants, which are represented by pollen and macrofossils, tend to dominate from nearshore to moderate depths, and shifts in the abundance of pollen or seeds in one of more sediment profiles can indicate relative water-level changes (Hannon and Gaillard, 1997; Edwards et al., 2000). Diatom and chironomid (midge) assemblages may also be related quantitatively to lake depth by means of inverse modeling and the transfer functions used to reconstruct past lake levels (Korhola et al., 2000; Ilyashuk et al., 2005).

During the past 65 m.y. (the Cenozoic), the Arctic has experienced a greater change in temperature, vegetation, and ocean surface characteristics than has any other Northern Hemisphere latitudinal band.

The great variety of lakes, and the corresponding range of sedimentary indicators, requires that field scientists be broadly knowledgeable in selecting which lakes to study and which techniques to use in reconstructions. For some important case studies, see Hannon and Gaillard, 1997; Abbott et al., (2000), Edwards et al., (2000), Korhola et al., 2000; Pienitz et al., (2000), Anderson et al., (2005), and Ilyashuk et al., 2005).

3.3.2c Ice-Core-Derived Precipitation Estimates

Ice cores provide a direct way of recording the net accumulation rate at sites with permanent ice. The initial thickness of an annual layer in an ice core (after mathematically accounting for the amount of air trapped in the ice) is the annual accumulation. Most ice cores are drilled in cold regions that produce little meltwater or runoff. Furthermore, **sublimation** or condensation and snow drift generally account for little accumulation, so that accumulation is not too different from the precipitation (e.g., Box et al., 2006). The thickness of layers deeper in the core must be corrected for the thinning produced as the ice sheet spreads and thins under its own weight, but for most samples this correction can be made with much accuracy by using simple ice flow models (e.g., Alley et al., 1993; Cuffey and Clow, 1997).

Ice cores provide a direct way of recording the net accumulation rate at sites with permanent ice.

The annual-layer thickness can be recorded using any component that varies regularly with a defined seasonal cycle. Suitable components include visible layering (e.g. Figure 3.14*A*), which responds to changes in snow density or impurities (Alley et al., 1997), the seasonal cycle of water isotopes (Vinther et al., 2006), and seasonal cycles in different chemical species (e.g., Rasmussen et al., 2006). Using more than one component gives extra security to the combined output of counted years and layer thicknesses.

Although the correction for strain (layer thinning) increases the uncertainty in estimates of absolute precipitation rate deeper in ice cores, estimates of changes in relative accumulation rate over subsections of the record along an ice core can be considered reliable (e.g., Kapsner et al., 1995). Because the accumulation rate combines with the temperature to control the rate at which snow is transformed to ice, and because the isotopic composition of the trapped air (Sowers et al., 1989) and the number of trapped bubbles in a sample (Spencer et al., 2006) record the results of that transformation, then accumulation rates can also be estimated from measurements of these parameters plus independent estimation of past temperature using techniques described above.

3.4 ARCTIC CLIMATE DURING THE PAST 65 M.Y.

During the past 65 m.y. (the **Cenozoic**), the Arctic has experienced a greater change in temperature, vegetation, and ocean surface characteristics than has any other Northern Hemisphere latitudinal band (e.g., Sewall and Sloan, 2001; Bice et al., 2006; and see results presented below). Because of the magnitude of climate change in the Arctic, a broad review of the evolution of climate during the Cenozoic offers insights into the positive feedbacks within the Arctic system that result in global impacts. Evidence from which the Cenozoic history of climate in the Arctic is reconstructed is presented below, focusing especially on warm times as identified by climate and environmental proxies outlined in section 3.3.

3.4.1 Early Cenozoic and Pliocene Warm Times

Records of the $\delta^{18}O$ composition of bottom-dwelling foraminifers from the global ocean document a long-term cooling of the deep sea during the past 70 m.y. (Chapter 2, Figure 2.8; Zachos et al., 2001) and the development of large

Northern Hemisphere continental ice sheets at 2.6–2.9 Ma (Duk-Rodkin et al., 2004). As discussed below and in Chapter 4 (Past Rates of Climate Change in the Arctic), Arctic climate history is broadly consistent with the global data reported by Zachos et al. (2001): general cooling and increase in ice was punctuated by short-lived and longer lived reversals, by variations in cooling rate, and by additional features related to growth and shrinkage of ice once the ice was well established. A detailed Arctic Ocean record that is equivalent to the global results of Zachos et al. (2001) is not yet available, and because the Arctic Ocean is geographically somewhat isolated from the world ocean (e.g., Jakobsson and Macnab, 2006), the possibility exists that some differences would be found. Emerging **paleoclimate** reconstructions from the Arctic Ocean derived from recently recovered sediment cores on the Lomonosov Ridge (Backman et al., 2006; Moran et al., 2006) shed new light on the Cenozoic evolution of the Arctic Basin, but the data have yet to be fully integrated with the evidence from terrestrial records or with the sketchy records from elsewhere in the Arctic Ocean (see Chapter 6, History of Arctic Sea Ice).

Data clearly show warm Arctic conditions during the **Cretaceous** and early Cenozoic. For example, late Cretaceous (70 Ma) Arctic Ocean temperatures of 15°C (compared to near-freezing temperatures today) are indicated by TEX_{86}-based estimates (Jenkyns et al., 2004). The same indicator shows that peak Arctic Ocean temperatures near the North Pole rose from about 18°C to more than 23°C during the short-lived (multi-**millenial**) **Paleocene**-Eocene Thermal Maximum about 55 **Ma** (Figure 3.23) (Moran et al., 2006; also see Sluijs et al., 2006, 2008). This rise was synchronous with warming on nearby land from a previous temperature of about 17°C to peak temperature during the event of about 25°C (Weijers et al., 2007). By about 50 Ma, Arctic Ocean temperatures were about 10°C and relatively fresh surface waters were dominated by aquatic ferns (Brinkhuis et al., 2006). Restricted connections to the world ocean allowed the fern-dominated interval to persist for about 800,000 years; return of more-vigorous interchange between the Arctic and North Atlantic oceans was accompanied by a warming in the central Arctic Ocean of about 3°C (Brinkhuis et al., 2006). On Arctic

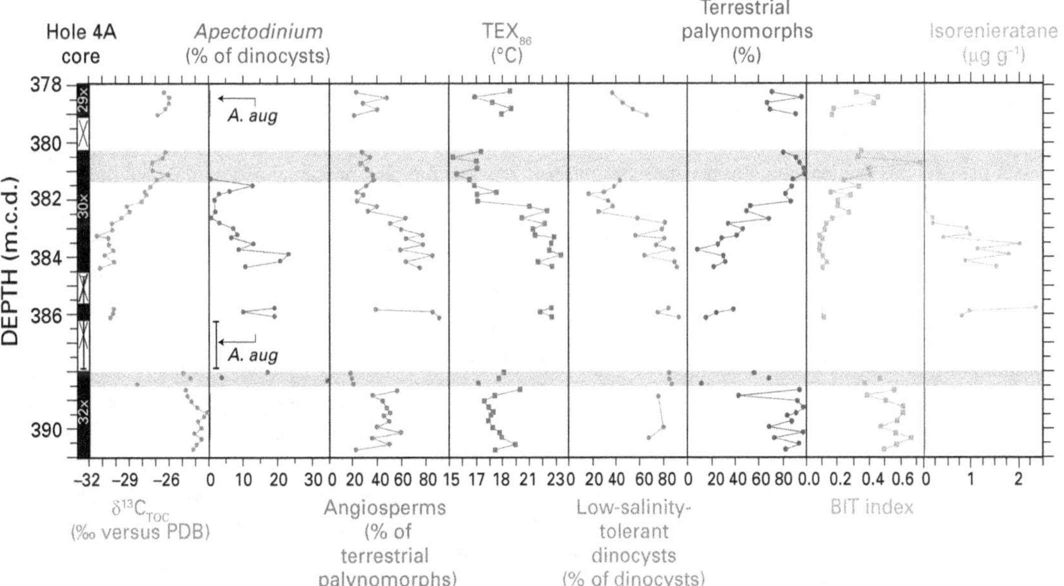

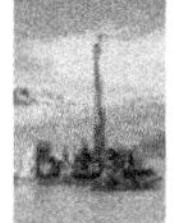

Figure 3.23. Recovered sections and palynological and geochemical results across the Paleocene-Eocene Thermal Maximum about 55 Ma; IODP Hole 302-4A (87°52.00'N.; 136°10.64'E.; 1,288 m water depth, in the central Arctic Ocean basin). Mean annual surface-water temperatures (as indicated in the TEX_{86} column) are estimated to have reached 23°C, similar to water in the tropics today. (Error bars that represent Core 31X show the uncertainty of its stratigraphic position. Orange bars = intervals affected by drilling disturbance.) Stable carbon isotopes are expressed relative to the PeeDee Belemnite standard. Dinocysts tolerant of low salinity comprise *Senegalinium* spp., *Cerodinium* spp., and *Polysphaeridium* spp., whereas *Membranosphaera* spp., *Spiniferites ramosus* complex, and *Areoligera-Glaphyrocysta* cpx. represent typical marine species. Arrows and *A. aug* (second column) indicate the first and last occurrences of dinocyst *Apectodinium augustum*—a diagnostic indicator of Paleocene-Eocene Thermal Maximum warm conditions (Sluijs et al., 2006.) [Reprinted by permission of Macmillan Publishers Ltd.]

lands during the Eocene (55–34 Ma), forests of *Metasequoia* dominated a landscape characterized by organic-rich floodplains and wetlands quite different from the modern tundra (McKenna, 1980; Francis, 1988; Williams et al., 2003).

Terrestrial evidence shows that warm conditions persisted into the early **Miocene** (23–16 Ma), when the central *Canadian Arctic* Islands were covered in mixed conifer-hardwood forests similar to those of southern Maritime *Canada* and New England today (Whitlock and Dawson, 1990). *Metasequoia* was still present although less abundant than in the Eocene. Still younger, deposits known as the Beaufort Formation and tentatively dated to about 8–3 Ma (and thus within Miocene to **Pliocene** times) record an extensive riverside forest of pine, birch, and spruce, which lived throughout the *Canadian Arctic Archipelago* before geologic processes formed many of the channels that now divide the islands.

About 3 Ma, forests covered large regions near the Arctic Ocean that are currently polar deserts.

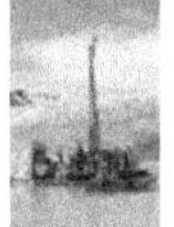

The relatively warm climates of the early Cenozoic gradually gave way to colder times of the Quaternary ice age, which was marked by cyclic growth and shrinkage of continental ice sheets. By the Pliocene (5–1.8 Ma), continental configurations were similar to those of the present, and most Pliocene plant and animal species were similar to those that remain today. A well-documented warm period in the middle Pliocene (about 3 Ma), just before the planet transitioned into the Quaternary ice age, supported forests that covered large regions near the *Arctic Ocean* that are currently polar deserts. Fossils of *Arctica islandica* (a marine bivalve that does not live near seasonal sea ice) in marine deposits as young as 3.2 Ma on *Meighen Island* at 80°N., likely record the peak Pliocene mean warmth of the ocean (Fyles et al., 1991). As compared with recent conditions, warmer conditions then are widely indicated (Dowsett et al., 1994). At a site on *Ellesmere Island*, application of a novel technique for paleoclimatic reconstruction based on ring-width and isotopic measurements of wood suggests mean-annual temperatures 14°C warmer than recently (Ballantyne et al., 2006). Additional data from records of beetles and plants indicate mid-Pliocene conditions as much as 10°C warmer than recently for mean summer conditions, and even larger wintertime warming to a maximum of 15°C or more (Elias and Matthews, 2002).

Many lines of proxy evidence show that atmospheric CO_2 was higher in the warm Cretaceous than it was recently.

Much attention has been focused on learning the causes of the slow, bumpy slide from Cretaceous hothouse temperatures to the recent ice age. As discussed below, changes in greenhouse-gas concentrations appear to have played the dominant role, and linked changes in continental positions, in sea level, and in oceanic circulation also contributed.

Based on general circulation models of climate, Barron et al. (1993) found that continental position had little effect on temperature difference between Cretaceous and modern temperatures (also see Poulsen et al., 1999 and references therein). Years later, Donnadieu et al. (2006), using more sophisticated climate modeling, found that continental motions and their effects on atmospheric and oceanic circulation modified global average temperature by almost 4°C from Early to Late Cretaceous; this result does not compare directly with modern conditions, but it does suggest that continental motions can notably affect climate. However, despite much effort, modeling does not indicate that the motion of continents by itself can explain either the long-term cooling trend from the Cretaceous to the ice age or the "wiggles" within that cooling.

The direct paleoclimatic data provide one interesting perspective on the role of oceanic circulation in the warmth of the later Eocene. When the *Arctic Ocean* was filled with water ferns living in "brackish" water (less salty than normal marine water) in an ocean that was ice free or nearly so, the oceanic currents reaching the near-surface *Arctic Ocean* must have been greatly weakened relative to today for the fresh water to persist. Thus, heat transport by oceanic currents cannot explain the *Arctic Ocean* warmth of that time. The resumption of stronger currents and normal salinity was accompanied by a warming of about 3°C (Brinkhuis et al., 2006), important but not dominant in the temperature difference between then and now.

As discussed in section 2.2.4, the atmospheric CO_2 concentration has changed during tens of millions of years in response to many processes, and especially to those processes linked to **plate tectonics** and perhaps also to biological evolution. Many lines of proxy evidence (see Royer, 2006) show that atmospheric CO_2 was higher in the warm Cretaceous than it was recently, and that it subsequently fell in parallel with the cooling (Figure 3.24). Furthermore, models find that the changing CO_2 concentration is sufficient to explain much of the cooling (e.g., Bice et al., 2006; Donnadieu et al., 2006).

A persistent difficulty is that models driven by changes in greenhouse gases (mostly CO_2) tend to underestimate Arctic warmth (e.g., Sloan and Barron, 1992). Many possible explanations have been offered for this situation: underestimation of CO_2 levels (Shellito et al., 2003; Bice et al., 2006); an enhanced greenhouse effect from polar stratospheric clouds during warm times (Sloan and Pollard, 1998; Kirk-Davidoff et al., 2002); changed planetary **obliquity** (Sewall and Sloan, 2004); reduced biological productivity that provided fewer cloud-condensation nuclei and thus fewer reflective clouds (Kump and Pollard, 2008); and greater heat transport by tropical cyclones (Korty et al., 2008). Several of these mechanisms use feedbacks not normally represented in climate models that serve to amplify warming in the Arctic. Consideration of the literature cited above and of additional materials points to some combination of stronger greenhouse-gas forcing (see Alley (2003) for a review) and to stronger long-term feedbacks than typically are included in models, rather than to large change in Earth's orbit, although that cannot be excluded.

It is thought that greenhouse gases were the primary control on Arctic temperature changes because the warmth of the Paleocene-Eocene Thermal Maximum took place in the absence of any ice—and therefore the absence of any ice-albedo or snow-albedo feedbacks. As described above (see Sluijs et al., 2008 for an extensively referenced summary of the event together with new data pertaining to the Arctic), this thermal maximum was achieved by a rapid (within a few centuries or less), widespread warming coincident with a large increase in atmospheric greenhouse-gas concentrations from a biological source (whether from sea-floor methane, living biomass, soils, or other sources remains debated). Following the thermal maximum, the anomalous warmth decayed more slowly and the extra greenhouse gases dissipated for tens of thousands of years, to roughly 100,000 years ago. The event in the Arctic seems to have been positioned within a longer interval of restricted oceanic circulation into the *Arctic Ocean* (Sluijs et al., 2008), and it was too fast for any notable effect of plate

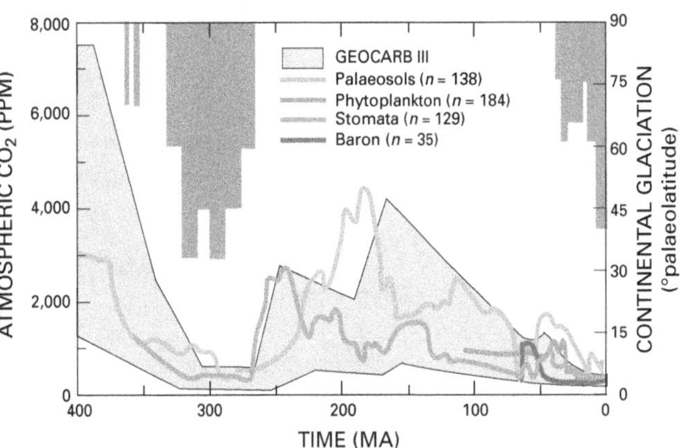

Figure 3.24. Atmospheric CO_2 and continental glaciation 400 Ma to present. Vertical blue bars = timing and palaeolatitudinal extent of ice sheets (after Crowley, 1998). Plotted CO_2 records represent five-point running averages from each of four major proxies (see Royer, 2006 for details of compilation). Also plotted are the plausible ranges of CO_2 derived from the geochemical carbon cycle model GEOCARB III (Berner and Kothavala, 2001). All data adjusted to the Gradstein et al. (2004) time scale. Continental ice sheets grow extensively when CO_2 is low. (After Jansen, 2007, in Intergovernmental Panel on Climate Change, 2007, Figure 6.1.)

tectonics or evolving life. The reconstructed CO_2 change thus is strongly implicated in the warming (e.g., Zachos et al., 2008).

Taken very broadly, the Arctic changes parallel the global ones during the Cenozoic, except that changes in the Arctic were larger than globally averaged ones (e.g., Sluijs et al., 2008). The global changes parallel changing atmospheric carbon-dioxide concentrations, and changing CO_2 is the likely cause of most of the temperature change (e.g., Royer, 2006; Royer et al., 2007).

Changing CO_2 is the likely cause of most of the temperature change.

The well-documented warmth of the Pliocene is not fully explained. This interval is recent enough that continental positions were substantially the same as today. As reviewed by Jansen et al. (2007), many reconstructions show notable Arctic warmth but little low-latitude change; however, recent work suggests the possibility of low-latitude warmth as well (Haywood et al., 2005). Reconstructions of Pliocene atmospheric CO_2 concentration (reviewed by Royer, 2006) generally agree with each other within the considerable uncertainties, but they allow values above, similar to, or even below the typical levels just before major human influence. Data remain equivocal on whether the ocean transported more heat during Pliocene warmth (reviewed by Jansen et al., 2007). The high-latitude warmth thus is likely to have originated primarily from changes in greenhouse-gas concentrations in the atmosphere, or from changes in oceanic or atmospheric circulation, or from some combination, perhaps with a slight possibility that other processes also contributed.

3.4.2 The Early Quaternary: Ice-Age Warm Times

A major reorganization of the climate system occurred between 3.0 and 2.5 Ma. As a result, the first continental ice sheets developed in the North American and *Eurasian Arctic* and marked the onset of the Quaternary Ice Ages (Raymo, 1994). For the first 1.5–2.0 Ma, ice age cycles appeared at a 41-**k.y.** interval, and the climate oscillated between glacial and interglacial states (Figure 3.25). A prominent but apparently short-lived interglacial (warm interval) about 2.4 Ma is recorded especially well in the *Kap København* Formation, a 100-m-thick sequence of estuarine sediments that covered an extensive lowland area near the northern tip of *Greenland* (Funder et al., 2001).

The rich and well-preserved fossil fauna and flora in the *Kap København* Formation (Figure 3.26) record warming from cold conditions into an interglacial and then subsequent cooling during 10,000–20,000 years. During the peak warmth, forest trees reached the *Arctic Ocean* coast, 1,000 kilometers (km) north of the northernmost trees today. Based on this warmth, Funder et al. (2001) suggested that the *Greenland Ice Sheet* must have been reduced to local **ice caps** in mountain areas (Figure 3.26*A*) (see Chapter 5, History of the Greenland Ice Sheet). Although finely resolved time records are not available throughout the *Arctic Ocean* at that time, by analogy with present faunas along the Russian coast, the coastal zone would have been ice-free for 2 to 3 months in summer. Today this coast of *Greenland* experiences year-round sea ice, and models of diminishing sea ice in a warming world generally indicate long-term persistence of summertime sea ice off these shores (e.g., Holland et al., 2006). Thus, the reduced sea ice off northern *Greenland* during deposition of the *Kap København* Formation suggests a widespread warm time in which Arctic sea ice was much diminished.

During *Kap København* times, precipitation was higher and temperatures were warmer than at the peak of the current interglacial about 7 ka, and the temperature difference was larger during winter than during summer. Higher temperatures during deposition of the *Kap København* were not caused by notably greater solar **insolation**, owing to the relative repeatability

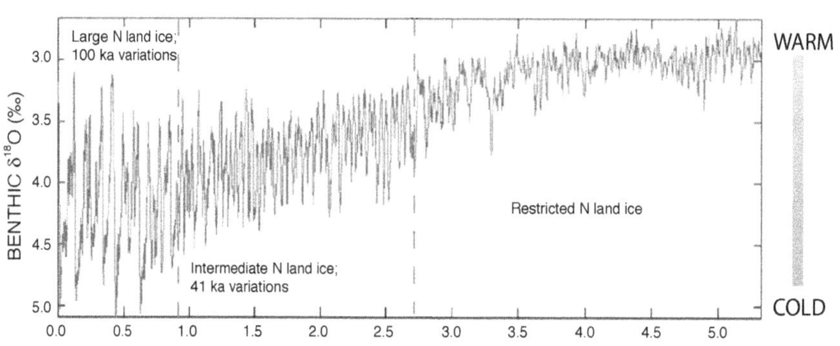

Figure 3.25. The average isotope composition ($\delta^{18}O$) of bottom-dwelling foraminifers in a globally distributed set of 57 sediment cores that record the last 5.3 m.y. (modified from Lisiecki and Raymo, 2005). The $\delta^{18}O$ is controlled primarily by global ice volume and deep-ocean temperature, with less ice or warmer temperatures (or both) upward in the core. The influence of Milankovitch frequencies of Earth's orbital variation are present throughout, but glaciation increased about 2.7 Ma concurrently with establishment of a strong 41-k.y. variability linked to Earth's obliquity (changes in tilt of Earth's spin axis), and the additional increase in glaciation about 1.2–0.7 Ma parallels a shift to stronger 100-k.y. variability. Dashed lines are used because the changes seem to have been gradual. The general trend toward higher $\delta^{18}O$ that runs through this series reflects the long-term drift toward a colder Earth that began in the early Cenozoic (see Figure 2.8). *http://lorraine-lisiecki.com/stack.html.*

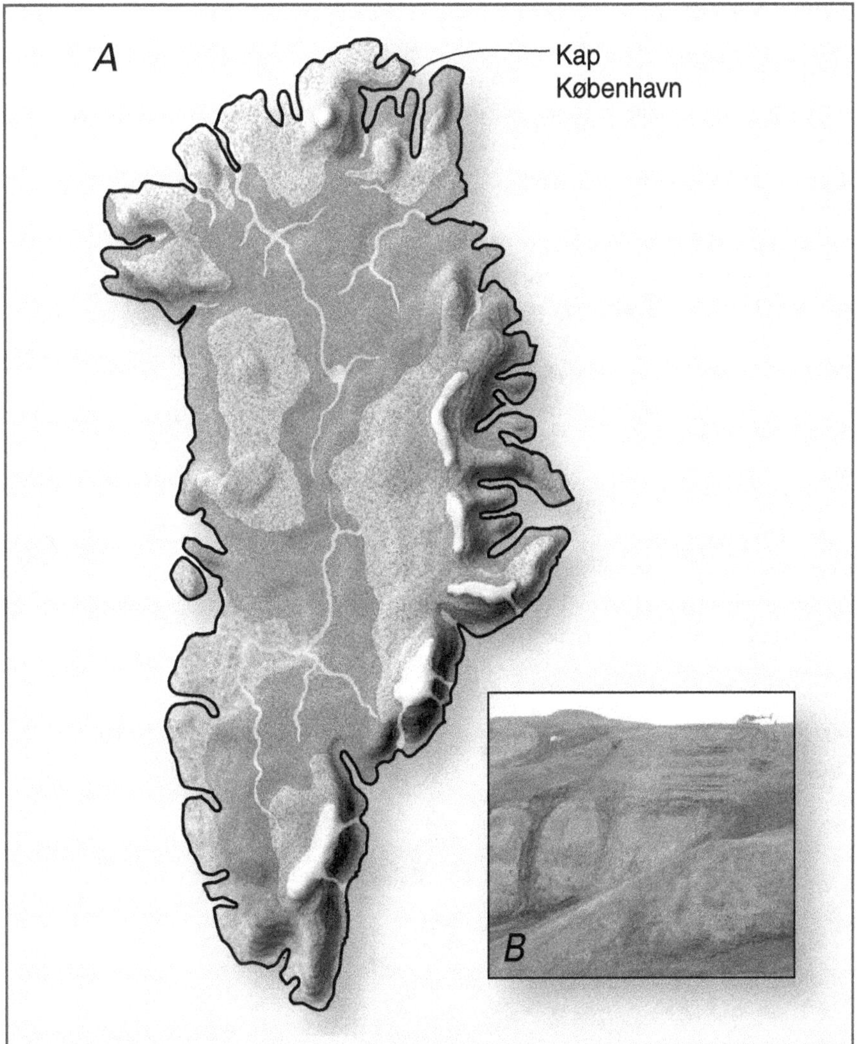

Figure 3.26. A) Greenland without ice for the last time? Dark green = boreal forest; light green = deciduous forest; brown = tundra and alpine heaths; white = ice caps. The north-south temperature gradient is constructed from a comparison of North Greenland and northwest European temperatures, using standard lapse rate; distribution of precipitation assumed to retain the Holocene pattern. Topographical base, from model by Letreguilly et al. (1991) of Greenland's sub-ice topography after isostatic recovery. B) Upper part of the Kap København Formation, North Greenland. The sand was deposited in an estuary about 2.4 Ma; it contains abundant well-preserved leaves, seeds, twigs, and insect remains. (Figure and photograph by S.V. Funder.)

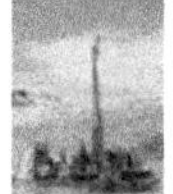

of the Milankovitch variations during millions of years (e.g., Berger et al., 1992). As discussed above, uncertainties in estimation of atmospheric CO_2 concentration, ocean heat transport, and perhaps other factors at the time of the *KAP KØBENHAVN* Formation are sufficiently large to preclude strong conclusions about the causes of the unusual warmth.

Potentially correlative records of warm interglacial conditions are found in deposits on coastal plains along the northern and western shores of *ALASKA*. High sea levels during interglaciations repeatedly flooded the *BERING STRAIT*, and they rapidly modified the configuration of the coastlines, altered regional **continentality** (isolation from the moderating influence of the sea), and reinvigorated the exchange of water masses between the North Pacific, *ARCTIC*, and *NORTH ATLANTIC OCEANS*. Since the first submergence of the *BERING STRAIT* about 5.5–5 Ma (Marincovich and Gladenkov, 2001), this marine gateway has allowed relatively warm

High sea levels during interglaciations repeatedly flooded the Bering Strait, and they rapidly modified the configuration of the coastlines.

Pacific water from as far south as northern Japan to reach as far north as the *Beaufort Sea* (Brigham-Grette and Carter, 1992). The *Gubik Formation* of northern *Alaska* records at least three warm high sea stands in the early Quaternary (Figure 3.27). During the Colvillian transgression, about 2.7 Ma, the *Alaskan Coastal Plain* supported open boreal forest or spruce-birch woodland with scattered pine and rare fir and hemlock (Nelson and Carter, 1991). Warm marine conditions are confirmed by the general character of the **ostracode** fauna, which includes *Pterygocythereis vannieuwenhuisei* (Brouwers, 1987), an extinct species of a genus whose modern northern limit is the *Norwegian Sea* and which, in the northwestern Atlantic Ocean, is not found north of the southern cold-temperate zone (Brouwers, 1987). Despite the high sea level and relative warmth indicated by the Colvillian transgression, erratics (rocks not of local origin) in Colvillian deposits southwest of *Barrow, Alaska*, indicate that glaciers then terminated in the *Arctic Ocean* and produced icebergs large enough to reach northwest *Alaska* at that time.

Subsequently, the Bigbendian transgression (about 2.5 Ma) was also warm, as indicated by rich molluscan faunas such as the gastropod *Littorina squalida* and the bivalve *Clinocardium californiense* (Carter et al., 1986). The modern northern limit of both of these mollusk species is well to the south (Norton Sound, *Alaska*). The presence of sea otter bones suggests that the limit of seasonal ice on the *Beaufort Sea* was restricted during the Bigbendian interval to positions north of the Colville River and thus well north of typical 20th-century positions (Carter et al., 1986); modern sea otters cannot tolerate severe seasonal sea-ice conditions (Schneider and Faro, 1975).

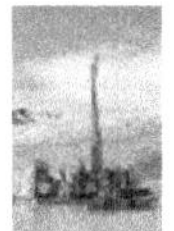

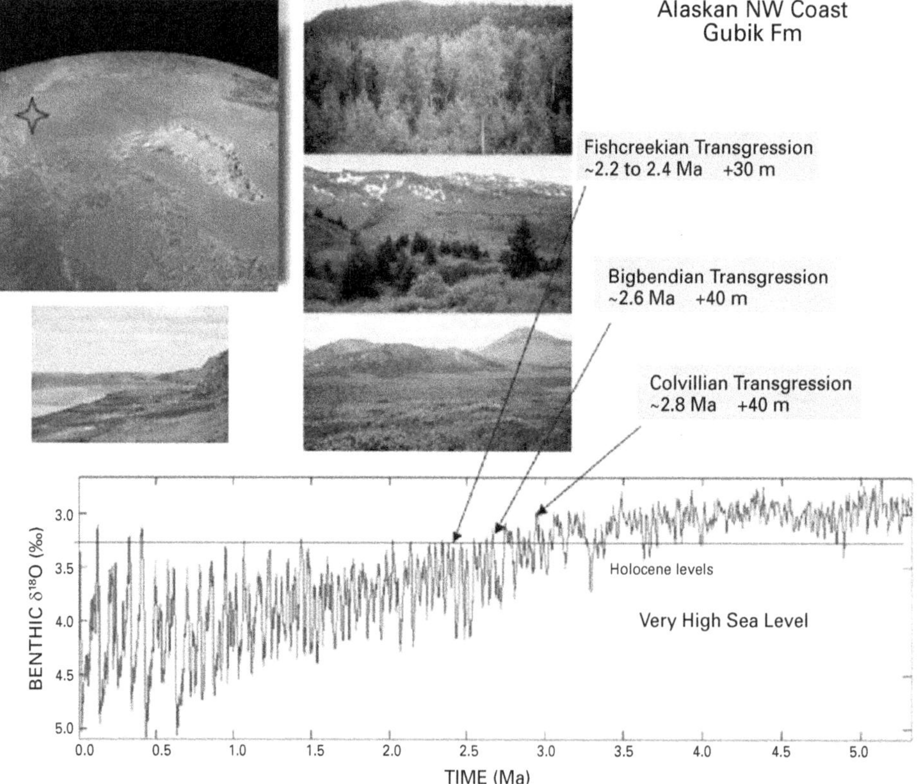

Figure 3.27. The largely marine Gubik Formation, North Slope of Alaska, contains three superposed lower units that record relative sea level as high as +30 to +40 m. Pollen in these deposits suggests that borderland vegetation at each of these times was less forested; boreal forests or spruce-birch woodlands at 2.7 Ma gave way to larch and spruce forests at about 2.6 Ma and to open tundra by about 2.4 Ma (see photographs by Robert Nelson, Colby College, who analyzed the pollen; oldest at top). Isotopic reference time series of Lisecki and Raymo (2005) suggests best as assignments for these sea level events (Brigham-Grette and Carter, 1992). [Reprinted by permission of Arctic Institute of North America.]

The youngest of these early Quaternary events of high sea level is the Fishcreekian transgression (about 2.1–2.4 Ma), suggested to be the same age as the *Kap København* Formation on *Greenland* (Brigham-Grette and Carter, 1992). However, age control is not complete, and Brigham (1985) and Goodfriend et al. (1996) suggested that the Fishcreekian could be as young as 1.4 Ma. This deposit contains several mollusk species that currently are found only to the south. Moreover, sea otter remains and the intertidal gastropod *Littorina squalida* at Fish Creek suggest that perennial sea ice was absent or severely restricted during the Fishcreekian transgression (Carter et al., 1986). Correlative deposits rich in mollusk species that currently live only well to the south are reported from the coastal plain at *Nome, Alaska* (Kaufman and Brigham-Grette, 1993).

The available data clearly indicate episodes of relatively warm conditions that correlate with high sea levels and reduced sea ice in the early Quaternary. The high sea levels suggest melting of land ice (see Chapter 5, History of the Greenland Ice Sheet). Thus the correlation of warmth with diminished ice on land and at sea (see Chapter 6, History of Arctic Sea Ice)—indicated by recent instrumental observations, model results, and data from other time intervals—is also found for this time interval. Improved time resolution of histories of forcing and response will be required to assess the causes of the changes, but estimates of forcings indicate that they were relatively moderate and thus that the strong Arctic amplification of climate change was active in these early Quaternary events.

3.4.3 The Mid-Pleistocene Transition: 41-k.y. and 100-k.y. Worlds

Since the late Pliocene, the cyclical waxing and waning of continental ice sheets have dominated global climate variability. The variations in sunshine caused by features of Earth's orbit have been very important in these ice-sheet changes, as described in Chapter 2 (Paleoclimate Concepts).

After the onset of **glaciation** in North America about 2.7 Ma (Raymo, 1994), ice grew and shrank as Earth's obliquity (tilt) varied in its 41 k.y. cycle. But between 1.2 and 0.7 Ma, the variations in ice volume became larger and slower, and an approximately 100-k.y. period has dominated especially during the last 700 k.y. or so (Figure 3.25). Although Earth's **eccentricity** varies with an approximately 100-k.y. period, this variation does not cause as much change in sunshine in the key regions of ice growth as did the faster cycles, so the reasons for the dominant 100-k.y. period in ice volume remain obscure. Roe and Allen (1999) assessed six different explanations of this behavior and found that all fit the data rather well. The record is still too short to allow the data to demonstrate the superiority of any one model.

The available data clearly indicate episodes of relatively warm conditions that correlate with high sea levels and reduced sea ice in the early Quaternary.

Models for the 100-k.y. variability commonly assign a major role to the ice sheets themselves and especially to the *Laurentide Ice Sheet* on North America, which dominated the total global change in ice volume (e.g., Marchant and Denton, 1996). For example, Marshall and Clark (2002) modeled the growth and shrinkage of the *Laurentide Ice Sheet* and found that during growth the ice was frozen to the **bed** beneath and unable to move rapidly. After many tens of thousands of years, ice had thickened sufficiently that it trapped Earth's heat and thawed the bed, which allowed faster flow. Faster flow of the ice sheet lowered the upper surface, which allowed warming and melting (see Chapter 5, History of the Greenland Ice Sheet). Behavior such as that described could cause the main variations of ice volume to be slower than the main variations in sunshine caused by Earth's orbital features, and the slow-flowing ice might partly ignore the faster variations in sunshine until the shift to faster flow allowed a faster response. Note that this explanation remains a hypothesis, and other possibilities exist. Alternative hypotheses require interactions in the Southern Ocean between the ocean and sea ice and between the ocean and the atmosphere (Gildor et al., 2002). For example, Toggweiler (2008) suggested that because of the close connection between the southern westerly winds and meridional overturning circulation in the Southern Ocean, shifts in wind fields very likely control the exchange of CO_2 between the ocean and the atmosphere. Carbon models support the notion that weathering and the burial of carbonate can be perturbed in ways that alter deep ocean carbon storage and that result in 100-k.y. CO_2 cycles (Toggweiler, 2008). Others have suggested that 100-k.y. cycles and CO_2 might be controlled by variability in obliquity cycles (i.e., two or three 41-k.y. cycles

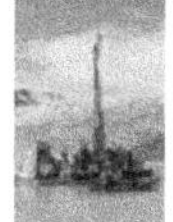

The mid-Pleistocene transition is very likely to be at least in part related to the continuation of the gradual global cooling that began in the early Cenozoic.

(Huybers, 2006) or by variable **precession** cycles (altering the 19-k.y. and 23-k.y. cycles (Raymo, 1997)). Ruddimann (2006) recently furthered these ideas but suggested that since 900 ka, CO_2-amplified ice growth continued at the 41-k.y. intervals but that polar cooling dampened ice ablation. His CO_2-feedback hypothesis suggests a mechanism that combines the control of 100-k.y. cycles with precession cycles (19 k.y. and 23 k.y.) and with tilt cycles (41 k.y.). The cause of the switch in the length of climate cycles from about 41 k.y. to about 100 k.y., known as the mid-**Pleistocene** transition, also remains obscure. This transition is of particular interest because it does not seem to have been caused by any major change in Earth's orbital behavior, and so the transition likely reflects a fundamental threshold within the climate system.

The mid-Pleistocene transition is very likely to be at least in part related to the continuation of the gradual global cooling that began in the early Cenozoic, as described above (Raymo et al., 1997, 2006; Ruddiman, 2003). If, for example, the 100-k.y. cycle requires that the *Laurentide Ice Sheet* grow sufficiently large and thick to trap enough of Earth's internal heat that thaws the ice-sheet bed (Marshall and Clark, 2002), then long-term cooling may have reached the threshold at which the ice sheet became large enough.

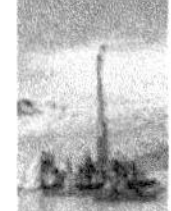

The globally averaged temperature change during one of the large 100-k.y. ice-age cycles was about 5°–6°C.

However, such a cooling model does not explain the key observation (Clark et al., 2006) that the ice sheets of the last 700 k.y. configured a larger volume (Clark et al., 2006) into a smaller area (Boellstorff, 1987; Balco et al., 2005a,b) than was true of earlier ice sheets. Clark and Pollard (1998) used this observation to argue that the early *Laurentide Ice Sheet* must have been substantially lower in elevation than in the late Pleistocene, possibly by as much as 1 km. Clark and Pollard (1998) suggested that the tens of millions of warm years back to the Cretaceous and earlier had produced thick soils and broken-up rocks below the soil. When glaciations began, the ice advanced over these water-saturated soils, which deformed easily. Just as grease on a griddle allows batter poured on top to spread easily into a wide, thin pancake, deformation of the soils beneath the growing ice (Alley, 1991) would have produced an extensive ice sheet that did not contain a large volume of ice. As successive ice ages swept the loose materials to the edges of the ice sheet, and as rivers removed most of the materials to the sea, hard bedrock was exposed in the central region. And, just as the bumps and friction of an ungreased waffle iron slow spreading of the batter to give a thicker, not-as-wide breakfast than on a greased griddle, the hard, bumpy bedrock produced an ice sheet that did not spread as far but which contained more ice.

Other hypotheses also exist for these changes. A complete explanation of the onset of extensive glaciation on North America and Eurasia as well as *Greenland* about 2.8 Ma, or of the transition from 41-k.y. to 100-k.y. ice age cycles, remains the object of ongoing investigations.

3.4.4 A Link Between Ice Volume, Atmospheric Temperature, and Greenhouse Gases

The globally averaged temperature change during one of the large 100-k.y. ice-age cycles was about 5°–6°C (Jansen et al., 2007). The larger changes were measured in the Arctic and close to the ice sheets, such as a change of 21°–23°C atop the *Greenland Ice Sheet* (Cuffey et al., 1995). The total change in sunshine reaching the planet during these cycles was near zero, and the orbital features served primarily to move sunshine from north to south and back, or from equator to poles and back, depending on the cycle considered (see Chapter 2, Paleoclimate Concepts).

As discussed by Jansen et al. (2007), and in section 3.2.6 above, many factors probably contributed to the large temperature change despite very small global change in total sunshine. Cooling produced growth of reflective ice that reduced the amount of sunshine absorbed by the planet. Complex changes especially in the ocean reduced atmospheric carbon dioxide, and both oceanic and terrestrial changes reduced atmospheric methane and **nitrous oxide**, all of which are greenhouse gases; the changes in carbon dioxide were most important. Various changes produced additional dust that blocked sunshine from reaching the planet (e.g., Mahowald et al., 2006). Cooling caused regions formerly forested to give way to grasslands or tundra that also reflected more sunshine. While Earth's orbit features drove the ice-age cycles, these feedbacks are required to provide quantitatively accurate explanations of the changes.

The relation between climate and carbon dioxide has been relatively constant for at least 650,000 years (Siegenthaler et al., 2005), and the growth and shrinkage of ice, cooling and warming of the globe, and other changes have repeated along similar although not identical paths. However, some of the small differences between successive cycles are of interest, as discussed next.

3.4.5 Marine Isotope Stage 11: A Long Interglaciation

Following the mid-Pleistocene transition, the growth and decay of ice sheets followed a 100-k.y. cycle: brief, warm interglaciations lasted from 10 to about 40 ka, after which ice progressively extended to a maximum limit, and then the icy interval terminated rapidly by the transition into the next warm interglaciation (e.g., Kellogg, 1977; Ruddiman et al., 1986; Jansen et al., 1988; Bauch and Erlenkeuser, 2003; Henrich and Baumann, 1994). As discussed above, this 100-k.y. cycle is unlikely to be linked to the 100-k.y. variation of the eccentricity, or out-of-roundness, of Earth's orbit about the Sun, because there is so little change in solar isolation reaching the Earth due to this effect.

The eccentricity exhibits an additional cycle of just greater than 400,000 years, such that the orbit goes from almost round to more eccentric to almost round in about 100,000 years, but the maximum eccentricity reached in this 100,000-year cycle increases and decreases within a 400,000-year cycle (Berger and Loutre, 1991; Loutre, 2003). When the orbit is almost round, there is little effect from Earth's precession, which determines whether Earth is closer to the Sun or farther from the Sun during a particular season such as northern summer. About 400,000 years ago, during **Marine Isotope Stage (MIS)** 11, the 400,000-year cycle caused a nearly round orbit to persist. The interglacial of MIS 11 lasted longer then previous or subsequent interglacials (see Droxler et al., 2003, and references therein; Kandiano and Bauch, 2007; Jouzel et al., 2007), perhaps because the summer sunshine (insolation) at high northern latitudes did not become low enough at the end of the first 10,000 years of the interglacial to allow ice growth at high northern latitudes—because the persistently nearly round orbit (i.e., of low eccentricity) prevented adequate cooling during northern summer (Figure 3.28).

The relation between climate and carbon dioxide has been relatively constant for at least 650,000 years.

As discussed in Chapter 5 (History of the Greenland Ice Sheet), indications of Arctic and subarctic temperatures at this time versus more-recent interglacials are inconsistent (also see Stanton-Frazee et al., 1999; Bauch et al., 2000; Droxler and Farrell, 2000; Helmke and Bauch, 2003). Sea level seems to have been higher at this time than at any time since, and data from *Greenland* are consistent with notable shrinkage or loss of the ice sheet accompanying the notable warmth, although the age of this shrinkage is not constrained well enough to be sure that the warm time recorded was indeed MIS 11 (Chapter 5).

3.4.6 Marine Isotope Stage 5e: The Last Interglaciation

The warmest millennia of at least the past 250,000 years occurred during MIS 5, and especially during the warmest part of that interglaciation, MIS 5e (e.g., McManus et al., 1994; Fronval and Jansen, 1997; Bauch et al., 1999; Kukla, 2000). At that time global ice volumes were smaller than they are today, and Earth's orbital parameters aligned to produce a strong positive anomaly in solar radiation during summer throughout the Northern Hemisphere (Berger and Loutre, 1991). Between 130 and 127 ka, the average solar radiation during the key summer months (May, June, and July)

Between 130 and 127 ka, the average solar radiation during the key summer months (May, June, and July) was about 11% greater than solar radiation at present throughout the Northern Hemisphere.

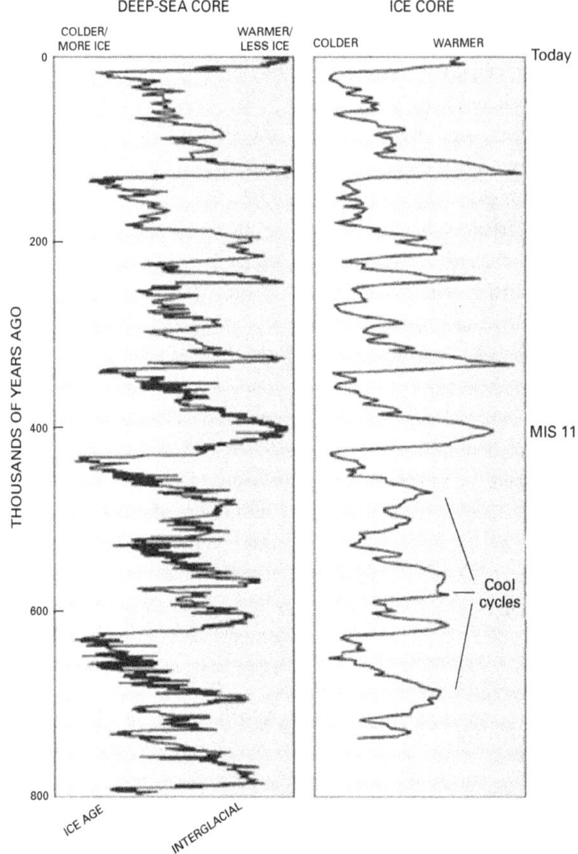

Figure 3.28. Glacial cycles of the past 800 k.y. derived from marine-sediment and ice cores (McManus, 2004). The history of deep-ocean temperatures and global ice volume inferred from $\delta^{18}O$ measured in bottom-dwelling foraminifera shells preserved in Atlantic Ocean sediments. Air temperatures over Antarctica inferred from the ratio of deuterium to hydrogen in ice from central Antarctica (EPICA, 2004). Marine Isotope Stage 11 (MIS 11) is an interglacial whose orbital parameters were similar to those of the Holocene, yet it lasted about twice as long as most interglacials. Note the smaller magnitude and less-pronounced interglacial warmth of the glacial cycles that preceded MIS 11. Interglaciations older than MIS 11 were less warm than subsequent interglaciations. [Copyright 2004, reprinted by permission of Macmillan Publishers Ltd.]

Marine Isotope Stage 5e demonstrates the strength of positive feedbacks on Arctic warming.

was about 11% greater than solar radiation at present throughout the Northern Hemisphere, and a slightly greater anomaly, 13%, has been measured over the Arctic. Greater solar energy in summer, melting of the large Northern Hemisphere ice sheets, and intensification of the North Atlantic Drift (Chapman et al., 2000; Bauch and Kandiano, 2007) combined to reduce Arctic Ocean sea ice, to allow expansion of boreal forest to the Arctic Ocean shore throughout large regions, to reduce permafrost, and to melt almost all glaciers in the Northern Hemisphere (CAPE Last Interglacial Project Members, 2006).

High solar radiation in summer during MIS 5e, amplified by key boundary-condition feedbacks (especially sea ice, seasonal snow cover, and atmospheric water vapor; see above), collectively produced summer temperature anomalies 4°–5°C above present over most Arctic lands, substantially above the average Northern Hemisphere summer temperature anomaly (0°–2°C above present; CLIMAP Project Members, 1984; Bauch and Erlenkeuser, 2003). MIS 5e demonstrates the strength of positive feedbacks on Arctic warming (CAPE Last Interglacial Project Members, 2006; Otto-Bleisner et al., 2006).

3.4.6a Terrestrial Records of Marine Isotope Stage 5e

At high northern latitudes, summer temperatures exert the dominant control on glacier mass balance, unless they are accompanied by strong changes in precipitation (e.g., Oerlemans, 2001; Denton et al., 2005; Koerner, 2005). Summer temperature is also the most effective predictor of most biological processes, although seasonality and the availability of moisture very likely also influence some biological parameters such as dominance by evergreen or by deciduous vegetation (Kaplan et al., 2003). For these reasons, most studies of conditions during MIS 5e have focused on reconstructing summer temperatures. Terrestrial MIS 5e climate, especially, has been reconstructed from diagnostic assemblages of biotic proxies preserved in lake, peat, river, and shallow marine **archives** and from isotopic changes preserved in ice cores and carbonate deposits in lakes. Estimated winter and summer temperatures, and hence seasonality, are well constrained for Europe but are poorly known for most other Arctic regions; likewise, precipitation reconstructions are limited to qualitative estimates in most cases where they are available, and they are not available for most regions.

During MIS 5e, all sectors of the Arctic had summers that were warmer than at present, but the magnitude of warming differed from one place to another (Figure 3.29) (CAPE Last Interglacial Project Members, 2006). Positive summer temperature anomalies were largest around the Atlantic sector, where summer warming was typically 4°–6°C. This anomaly extended into *Siberia*, but it decreased from *Siberia* westward to the European sector (0°–2°C), and eastward toward *Beringia* (2°–4°C). The *Arctic coast of Alaska* had sea-surface temperatures 3°C above recent values and considerably less summer sea ice than recently, but much of interior *Alaska* had smaller anomalies (0°–2°C) that probably extended into western *Canada*. In contrast, northeastern *Canada* and parts of *Greenland* had summer temperature anomalies of about 5°C and perhaps more (see Chapter 5 for a discussion of *Greenland*).

Precipitation and winter temperatures are more difficult to reconstruct for MIS 5e than are summer temperatures. In northeastern Europe, the latter part of MIS 5e was characterized by a marked increase in winter temperatures. A large positive winter temperature anomaly also occurred in *Russia* and western *Siberia*, although the timing is not as well constrained (Troitsky, 1964; Gudina et al., 1983; Funder et al., 2002). Qualitative precipitation estimates for most other sectors indicate wetter conditions than in the Holocene.

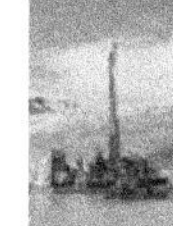

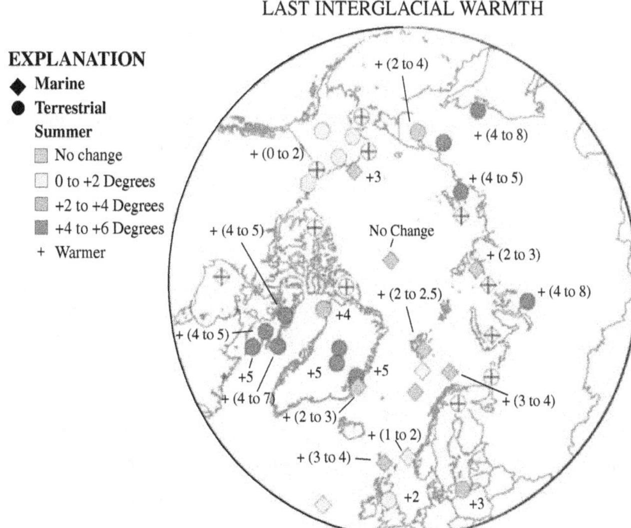

Figure 3.29. Polar projection showing regional maximum last interglacial summer temperature anomalies relative to present summer temperatures; derived from paleotemperature proxies (see Tables 1 and 2 in CAPE Last Interglacial Project Members, 2006). Circles = terrestrial sites; squares = marine sites. [Copyright 2006, reprinted by permission of Elsevier.]

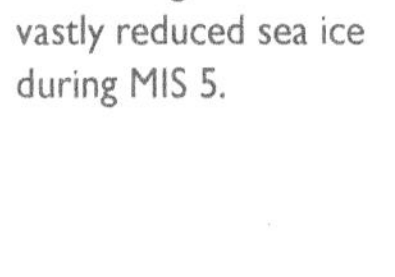

Landforms and fossils from the western Arctic and Bering Strait indicate vastly reduced sea ice during MIS 5.

3.4.6b Marine Records of Marine Isotope Stage 5e

Low sedimentation rates in the central Arctic Ocean and the rare preservation of carbonate fossils limit the number of sites at which MIS 5e can be reliably identified in sediment cores. MIS 5e sediments from the central Arctic Ocean usually contain high concentrations of planktonic (surface-dwelling) foraminifers and **coccoliths**, which indicate a reduction in summer sea-ice coverage that permitted increased biological productivity (Gard, 1993; Spielhagen et al., 1997, 2004; Jakobsson et al., 2000; Backman et al., 2004; Polyak et al., 2004; Nørgaard-Pedersen et al., 2007a,b). However, occasional dissolution of carbonate fossils complicates the interpretation of microfossil concentrations. Also, marine sediments from MIS 5a, slightly younger and cooler than MIS 5e, sometimes have higher microfossil concentrations than do MIS 5e sediments (Gard, 1986, 1987).

Arctic Ocean sediment cores recently recovered from the Lomonosov Ridge, north of Greenland, have revived the discussion of MIS 5e conditions in the Arctic Ocean. Unusually high concentrations of a subpolar foraminifer species, one which usually dwells in waters with temperatures well above freezing, were found in MIS 5e zones and interpreted to indicate warm interglacial conditions and much reduced sea-ice cover in the interior Arctic Ocean (Nørgaard-Pedersen et al., 2007a,b). Interpretation of these and other microfossils is complicated by the strong vertical stratification in the Arctic Ocean; today, warm Atlantic water (temperatures greater than 1°C) is in most areas isolated from the atmosphere by a relatively thin layer of cold (less than 1°C) fresher water; this cold water limits the transfer of heat to the atmosphere. It is not always possible to determine whether warm-water foraminifers found in marine sediment from the Arctic Ocean lived in warm waters that remained isolated from the atmosphere below the cold surface layer, or whether the warm Atlantic water had displaced the cold surface layer and was interacting with the atmosphere and affecting its energy balance.

Landforms and fossils from the western Arctic and Bering Strait indicate vastly reduced sea ice during MIS 5e (Figure 3.30). The winter sea-ice limit is estimated to have been as much as 800 km farther north than its average 20th-century position, and summer sea ice was likely to have been much reduced relative to present (Brigham-Grette and Hopkins, 1995). These reconstructions are consistent with the northward migration of treeline by hundreds of kilometers throughout much of Alaska and nearby Chukotka and with the elimination of tundra from Chukotka to the Arctic Ocean coast (Lozhkin and Anderson, 1995).

Sufficient data are not yet available to allow unambiguous reconstruction of MIS 5e conditions in the central Arctic Ocean. Key uncertainties are related to the extent and duration of Arctic Ocean sea ice. The vertical structure of the upper 500 m of the water column is also climatically important but poorly known, in particular whether the strong vertical stratification characteristic of the modern regime persisted throughout MIS 5e, or whether reduced sea ice and changes in the hydrologic cycle and winds destabilized this stratification and allowed Atlantic water to reside at the surface in larger areas of the Arctic Ocean.

3.4.7 Marine Isotope Stage 3 Warm Intervals

The temperature and precipitation history of MIS 3 (about 70–30 ka) is difficult to reconstruct because of the paucity of continuous records and the difficulty in providing a secure time frame. The $\delta^{18}O$ record of temperature change over the Greenland Ice Sheet and other ice-core data show that the North Atlantic region experienced repeated episodes of rapid, high-magnitude climate change, that temperatures rapidly increased by as much as 15°C (reviewed by Alley, 2007 and references therein), and that each warm period lasted several hundred to a few thousand years. These brief climate excursions are found not only in the Greenland Ice Sheet but are also recorded in cave sediments in China (Wang et al., 2001; Dykoski, et al., 2005) and in high-resolution marine records off California (Behl and Kennett, 1996), and in the Caribbean Sea's Cariaco Basin (Hughen et al., 1996.), the Arabian Sea (Schulz et al., 1998) and the Sea of Okhotsk (Nürnberg and Tiedmann, 2004), among many other sites. The ice-core records from Greenland contain indications of climate change in many regions on the same time scale (for example, the methane trapped in ice-core bubbles was in part produced in tropi-

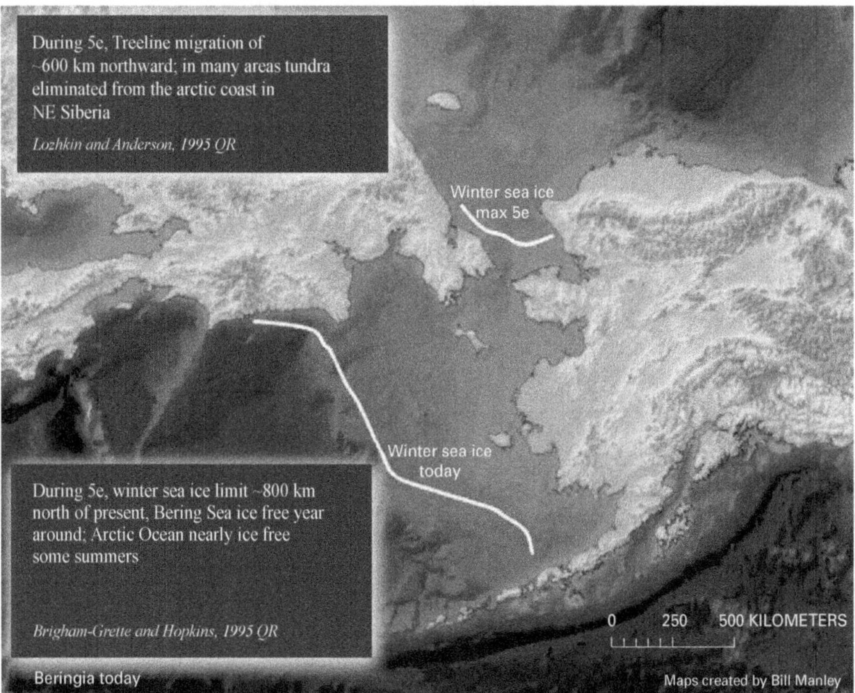

Figure 3.30. Winter sea-ice limit during MIS 5e and at present. Fossiliferous paleoshorelines and marine sediments were used by Brigham-Grette and Hopkins (1995) to evaluate the seasonality of coastal sea ice on both sides of the Bering Strait during the last interglaciation. Winter sea limit is estimated to have been north of the narrowest section of the strait, 800 km north of modern limits. Pollen data derived from Last Interglacial lake sediments suggest that tundra was nearly eliminated from the Russian coast at this time (Lozhkin and Anderson, 1995). In Chukokta during the warm interglaciation, additional open water favored some taxa tolerant of deeper winter snows. (Base map by William Manley, *http://instaar.colorado.edu/QGISL/*.)

cal wetlands and was essentially all produced beyond the *Greenland Ice Sheet*; Severinghaus et al., 1998). These ice-core records demonstrate clearly that the climate-change events were synchronous throughout widespread areas, and that the ages of events from many regions agree within the stated uncertainties. These events were thus hemispheric to global in nature (see review by Alley, 2007) and are considered a sign of large-scale coupling between the ocean and the atmosphere (Bard, 2002). The cause of these events is still debated. However, Broecker and Hemming (2001) and Bard (2002) among others suggested that they were likely the result of major and abrupt reorganizations of the ocean's thermohaline circulation, probably related to ice sheet instabilities that introduced large quantities of fresh water into the *North Atlantic* (Alley, 2007). Such large and abrupt **oscillations**, which were linked to changes in *North Atlantic* surface conditions and probably to the large-scale oceanic circulation, persisted into the Holocene (MIS 1); the youngest was only about 8.2 **ka** (Alley and Ágústdóttir, 2005). However, it appears that the abrupt 8.2 ka cooling was linked to an ice-age cause, a catastrophic flood from a very large lake that had been dammed by the melting *Laurentide Ice Sheet*.

Within MIS 3, land ice was somewhat reduced compared with the colder times of MIS 2 and MIS 4, but Arctic temperatures generally were much lower and ice more extensive than in MIS 1 (with certain exceptions). Sea level was lower at that time, the coastline was well offshore in many places, and the increased continentality very likely contributed to warmer summer temperatures that presumably were offset by colder winter temperatures.

These ice-core records demonstrate clearly that the climate-change events were synchronous throughout widespread areas, and that the ages of events from many regions agree within the stated uncertainties.

For example, on the *New Siberian Islands* in the *East Siberian Sea*, Andreev et al. (2001) documented the existence of graminoid-rich tundra thought to have covered wide areas of the emergent shelf while summer temperatures were perhaps as much as 2°C warmer than during the

20th century. At Elikchan 4 Lake in the upper Kolyma drainage, the sediment record contains at least three intervals (especially one about 38 ka) when summer temperatures and treeline reached late Holocene conditions (Anderson and Lozhkin, 2001). Insect faunas nearby in the lower Kolyma are thought to have thrived in summers that were 1°–4.5°C warmer than recently for similar intervals of MIS 3 (Alfimov et al., 2003). In general, variable paleoenvironmental conditions were typical of the traditional Karaginskii-MIS 3 period throughout Arctic *Russia*; however, stratigraphic confusion within the limits of radiocarbon-dating precludes the widespread correlation of events.

Relative warmth during MIS 3 appears to have been strongest in eastern *Beringia*; some evidence suggests that between 45 and 33 ka temperatures were only 1°–2°C lower than at present (Elias, 2007). The warmest interval in interior *Alaska* is known as the Fox Thermal Event, about 40–35 ka, which was marked by spruce forest tundra (Anderson and Lozhkin, 2001). Yet in the Yukon forests were most dense a little earlier, about 43–39 ka. In general (Anderson and Lozhkin, 2001), the warmest interstadial interval in all of *Beringia* possibly was 44–35 ka; it is well represented in proxies from interior sites and little or no vegetation response in areas closest to *Bering Strait*. Climatic conditions in eastern *Beringia* appear to have been harsher than modern conditions for all of MIS 3. In contrast,MIS 3 climates of western *Beringia* achieved modern or near modern conditions during several intervals. Moreover, although the transition from MIS 3 to MIS 2 was clearly marked by a transition from warm-moist to cold-dry conditions in western *Beringia*, this transition is absent or subtle in all but a few records in *Alaska* (Anderson and Lozhkin, 2001).

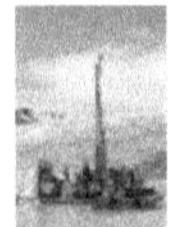

3.4.8 Marine Isotope Stage 2, the Last Glacial Maximum (30 to 15 ka)

The Last Glacial Maximum (**LGM**)was particularly cold both in the Arctic and globally. During peak cooling of the LGM, planetary temperatures were about 5°–6°C lower than at present (Farrera et al., 1999; Braconnot et al., 2007; Jansen et al., 2007), whereas mean annual temperatures in central *Greenland* were depressed more than 20°C (Cuffey et al., 1995; Dahl-Jensen et al., 1998)

3.4.9 Marine Isotope Stage 1, the Holocene: The Present Interglaciation

In the face of rising solar energy in summer that was tied to orbital features and to rising greenhouse gases, Northern Hemisphere ice sheets began to recede from near their largest extent shortly after 20 ka, and the rate of recession noticeably increased after about 16 ka (see, e.g., Alley et al., 2002 for the timing of various events during the deglaciation). Most coastlines became ice-free before 12 ka, and ice continued to melt rapidly as summer insolation reached a peak (about 9% above modern insolation) about 11 ka. The transition from MIS 2 to MIS 1, which marks the start of the Holocene interglaciation, is commonly placed at the abrupt termination of the cold event called the **Younger Dryas**; that termination recently was estimated at about 11.7 ka (Rasmussen et al., 2006).

A wide variety of evidence from terrestrial and marine archives indicates that peak Arctic summertime warmth was achieved during the early Holocene, when most regions of the Arctic experienced sustained temperatures that exceeded observed 20th century values. This period of peak warmth, which is geographically variable in its timing, is generally referred to as the Holocene Thermal Maximum. The ultimate driver of the warming was orbital forcing, which produced increased summer solar radiation across the Northern Hemisphere. At 70°N., insolation in June now is near a local minimum (the maximum was recorded about 11–12 ka). June insolation about 4 ka was about 15 W/m^2 larger than recently, and June insolation at the Holocene peak was about 45 W/m^2 larger than recently, for a total change of about 10% (Figure 3.31; Berger and Loutre, 1991). Winter (January) insolation about 11 ka was only slightly lower than today, in large part because there is almost zero insolation that far north in January.

By 6 ka, sea level and ice volumes were close to those observed more recently, and climate forcings such as atmospheric carbon-dioxide concentration differed little from pre-industrial conditions (e.g., Jansen et al., 2007). (The exception is that far-northern summer insolation steadily decreased throughout the Holocene.) High-resolution (decades to centuries) archives containing many climate proxies are available

During peak cooling of the LGM, planetary temperatures were about 5°–6°C lower than at present, whereas mean annual temperatures in central Greenland were depressed more than 20°C.

Most coastlines became ice-free before 12 ka, and ice continued to melt rapidly as summer insolation reached a peak (about 9% above modern insolation) about 11 ka.

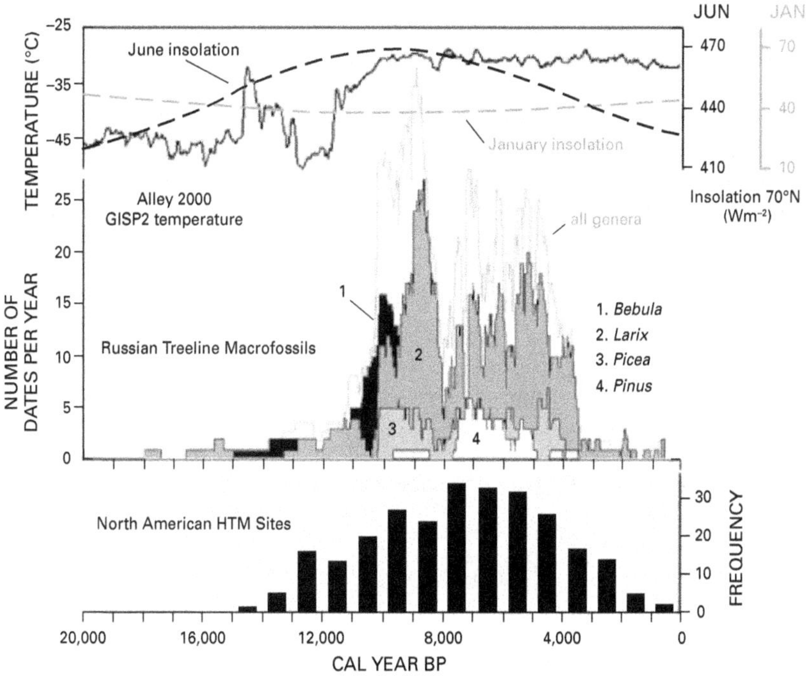

Figure 3.31. The Arctic Holocene Thermal Maximum. Items compared, top to bottom: seasonal insolation patterns at 70°N. (Berger and Loutre, 1991) and reconstructed Greenland air temperature from the GISP2 drilling project (Alley, 2000); age distribution of radiocarbon-dated fossil remains of various tree genera from north of present treeline (MacDonald et al., 2007) and the frequency of Western Arctic sites that experienced Holocene Thermal Maximum conditions (Kaufman et al., 2004). [Reprinted with permission of Philosophical Transactions of the Royal Society.]

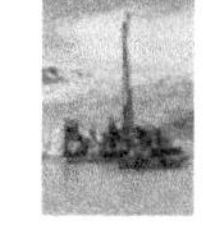

for most of the Holocene throughout the Arctic. Consequently, the mid- to late-Holocene record allows evaluation of the range of natural climate variability and of the magnitude of climate change in response to relatively small changes in forcings.

3.4.9A The Holocene Thermal Maximum

Many of the Arctic paleoenvironmental records for the Holocene Thermal Maximum (HTM) appear to have recorded primarily summertime conditions. Many different proxies have been exploited to derive these reconstructions by use of biological indicators such as pollen, diatoms, **chironomids**, dinoflagellate cysts, and other microfossils; elemental and isotopic geochemical indexes from lacustrine sediments, marine sediments, and ice cores; borehole temperatures; and age distributions of radiocarbon-dated tree stumps north of (or above) current treeline, marine mollusks, and whale bones (Kaufman et al., 2004).

A recent synthesis of 140 Arctic paleoclimatic and paleoenvironmental records extending from *Beringia* westward to *Iceland* (Kaufman et al., 2004) outlines the nature of the Holocene Thermal Maximum in the western Arctic (Figure 3.32). Fully 85% of the sites included in the synthesis contained evidence of a Holocene thermal maximum. Its average duration extended from 2,100 years in *Beringia* to 3,500 years in *Greenland*. The interval 10–4 ka contains the greatest number of sites recording Holocene Thermal Maximum conditions and the greatest spatial extent of those conditions in the western Arctic (Figure 3.32*B*). In the western Arctic the timing of this thermal maximum begins and ends along a strong geographic gradient (Figure 3.32*C*). The thermal maximum began first in *Beringia*, where warmer-than-present summer conditions became established at 14–13 ka. Intermediate ages for its initiation (10–8 ka) are apparent in the *Canadian Arctic* Islands and in central *Greenland*. The Holocene Thermal Maximum on *Iceland* occurred a bit later,

Fully 85% of the sites included in the synthesis contained evidence of a Holocene thermal maximum.

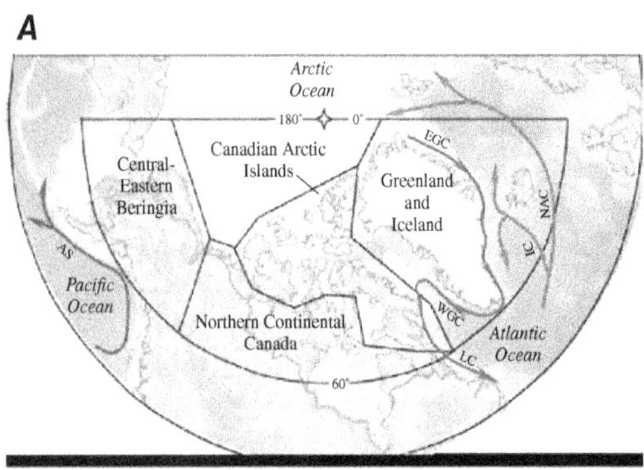

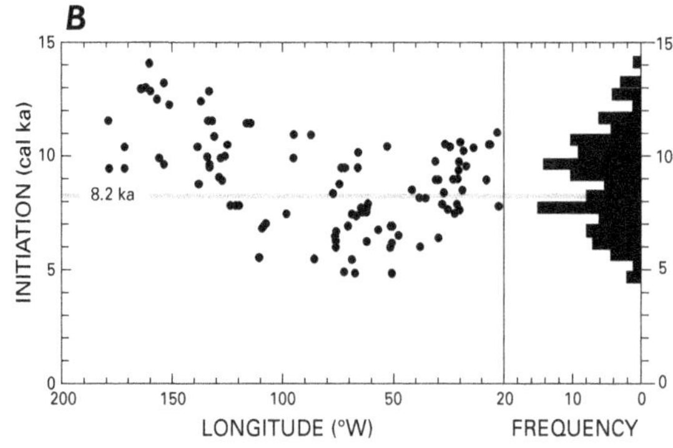

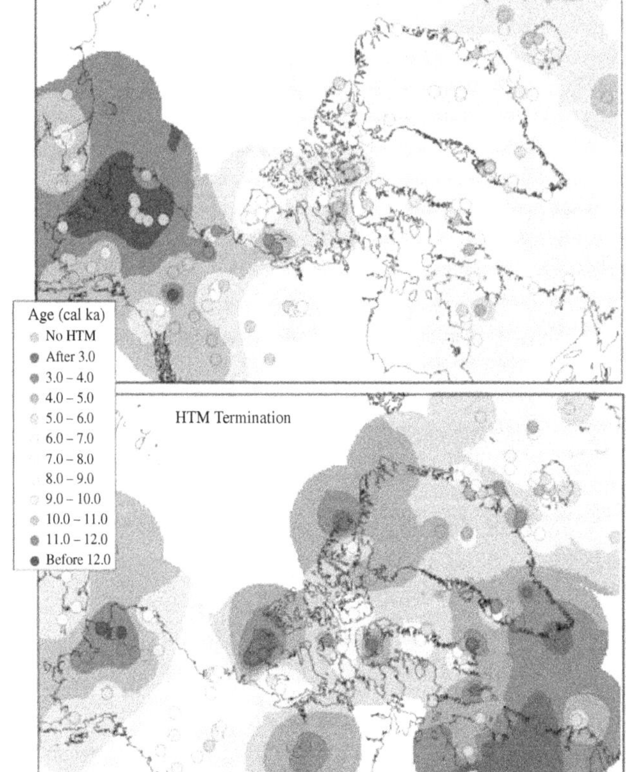

Figure 3.32. The timing of initiation and termination of the Holocene Thermal Maximum in the western Arctic (Kaufman et al., 2004). *A*) Regions reviewed in Kaufman et al. (2004). *B*) Initiation of the Holocene Thermal Maximum in the western Arctic. Longitudinal distribution (left) and frequency distribution (right). *C*) Spatial-temporal pattern of the Holocene Thermal Maximum in the western Arctic. Upper panel = initiation; lower panel = termination. Dot colors bracket ages of the Holocene Thermal Maximum; ages contoured using the same color scheme. Gray dots = equivocal evidence for the Holocene Thermal Maximum. [Copyright 2004, reprinted with permission of Elsevier.]

8–6 ka. The onset on *SVALBARD* was earlier, by 10.8 ka (Svendsen and Mangerud, 1997). The latest general onset (7–4 ka) of Holocene Thermal Maximum conditions affected the continental portions of central and eastern *CANADA* experienced. Similarly, the earliest termination of the Holocene Thermal Maximum occurred in *BERINGIA*, although most regions registered summer cooling by 5 ka. Much of the pattern of the onset of the Holocene Thermal Maximum can be explained at least in part by proximity to cold winds blowing off the melting *LAURENTIDE ICE SHEET* in *CANADA*, which depressed temperatures nearby until the ice melted back. **Milankovitch cycling** has also been suggested to explain the spatial variability of the Holocene Thermal Maximum (Maximova and Romanovsky, 1988).

Records for sea-ice conditions in the *ARCTIC OCEAN* and adjacent channels have been developed by radiocarbon-dating indicators including the remains of open-water proxies such as whales and walrus, warm-water marine mollusks, and changes in the microfauna preserved in marine sediments. These reconstructions, presented in more detail in Chapter 6 (History of Arctic Sea Ice), parallel the terrestrial record for the most part. The data demonstrate that an increased mass of warm Atlantic water moved into the *ARCTIC OCEAN* beginning about 11.5 ka. It peaked about 8–5 ka which, coupled with increased summer insolation, decreased the area of perennial sea-ice cover during the early Holocene. Decreased sea-ice cover in the western Arctic during the early Holocene also may be indicated by changes in concentrations of sodium from sea salt in the Penny Ice Cap (eastern *CANADIAN ARCTIC*; Fisher et al., 1998) and the *GREENLAND ICE SHEET* (Mayewski et al., 1997). In most regions, perennial sea ice increased in the late Holocene, although it has been suggested that sea ice declined in the *CHUKCHI SEA* (de Vernal et al., 2005), possibly in response to changing rates of Atlantic water inflow in *FRAM STRAIT*.

As summer temperatures increased through the early Holocene, in North America treeline expanded northward into regions formerly mantled by tundra, although the northward extent appears to have been limited to perhaps a few tens of kilometers beyond its recent position (Seppä et al., 2003; Gajeswski and MacDonald, 2004). In contrast, treeline advanced much farther across the *EURASIAN ARCTIC*. Tree macrofossils (Kremenetski et al., 1998; MacDonald et al., 2000a,b; 2007) collected at or beyond the current treeline indicate that tree genera such as birch (*Betula*) and larch (*Larix*) advanced beyond the modern limits of treeline across most of northern Eurasia between 11 and 10 ka (Figures 5.33 and 5.34). Spruce (*Picea*) advanced slightly later than the other two genera. Interestingly, pine (*Pinus*), which now forms the conifer treeline in *FENNOSCANDIA* and the *KOLA PENINSULA*, does not appear to have established appreciable forest cover at or beyond the present treeline in those regions at the far west of Europe until around 7 ka (MacDonald et al. 2000a). However, quantitative reconstructions of temperature from the *KOLA PENINSULA* and adjacent Fennoscandia suggest that summer temperatures were warmer than modern temperatures by 9 ka (Seppä and Birks, 2001, 2002; Hammarlund et al., 2002; Solovieva et al., 2005), and the development of extensive pine cover at and north of the present treeline appears to have been delayed relative to this warming. In the *TAIMYR PENINSULA* of *SIBERIA* and across nearby regions, the most northerly limit reached by trees during the Holocene was more than 200 km north of the current treeline. The treeline appears to have begun its retreat across northern Eurasia about 4 ka. The timing of the Holocene Thermal Maximum in the *EURASIAN ARCTIC* overlaps the widest expression of the Holocene Thermal Maximum in the western Arctic (Figure 3.33), but it differs in two respects. The timing of onset and termination in Eurasia show much less variability than in North America, and the magnitude of the treeline expansion and retreat is far greater in the *EURASIAN ARCTIC*. Fossil pollen and other indicators of

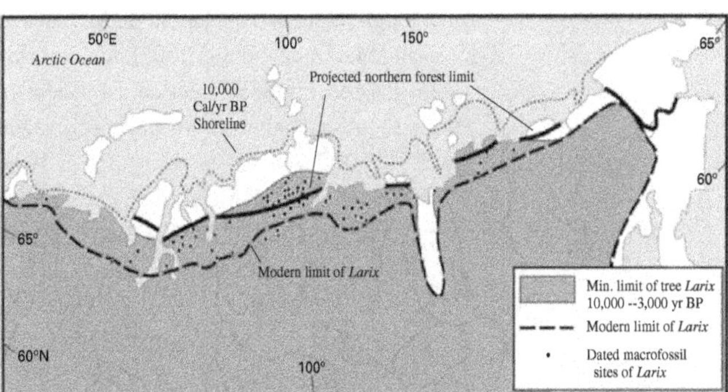

Figure 3.33 The northward extension of larch (*Larix*) treeline across the Eurasian Arctic. Treeline today compared with treeline during the Holocene Thermal Maximum and with anticipated northern forest limits (Arctic Climate Impact Assessment, 2005) due to climate warming (MacDonald et al., 2007). [Reprinted with permission of Philosophical Transactions of the Royal Society.]

Areas south of the Arctic Circle appear to have experienced deep thawing (100–200 m depth) from the early Holocene until about 4–3 ka.

vegetation or temperature from the northern Eurasian margin also support the contention of a prolonged warming and northern extension of treeline during the early through middle Holocene (see for example Hyvärinen, 1976; Seppä, 1996; Clayden et al., 1997; Velichko et al., 1997; Kaakinen and Eronen, 2000; Pisaric et al., 2001; Seppä and Birks, 2001, 2002; Gervais et al., 2002; Hammarlund et al., 2002; Solovieva et al., 2005).

Changes in landforms suggest that during the early to middle Holocene, permafrost in *Siberia* degraded. A synthesis of Russian data by Astakhov (1995) indicates that melting permafrost was apparent north of the Arctic Circle only in the European North, not in *Siberia*. In the *Siberian* North, permafrost partially thawed only very locally, and thawing was almost entirely confined to areas under thermokarst lakes that actively formed there during the early through middle Holocene. Areas south of the Arctic Circle appear to have experienced deep thawing (100–200 m depth) from the early Holocene until about 4–3 ka, when cooler summer conditions led permafrost to develop again. The deep thawing and subsequent renewal of surface permafrost in these regions produced an extensive thawed layer sandwiched between shallow (20–80 m deep) more recently frozen ground and deeper Pleistocene permafrost throughout much of northwestern *Siberia*.

Summer cooling during the second half of the Holocene led to the expansion of mountain glaciers and ice caps around the Arctic.

Quantitative estimates of the Holocene Thermal Maximum summer temperature anomaly along the northern margins of Eurasia and adjacent islands typically range from 1° to 3°C. The geographic position of northern treeline across Eurasia is largely controlled by summer temperature and the length of the growing season (MacDonald et al., 2007), and in some areas the magnitude of treeline displacement there suggests a summer warming equivalent of 2.5°–7.0°C (see for example Birks, 1991; Wohlfarth et al., 1995; MacDonald et al., 2000a; Seppä and Birks, 2001, 2002; Hammarlund et al., 2002; Solovieva et al., 2005). Sea-surface temperature anomalies during the Holocene Thermal Maximum were as much as 4°–5°C higher than during the late Holocene for the eastern *North Atlantic Sector* and adjacent *Arctic Ocean* (Salvigsen, 1992; Koç et al., 1993). Anomalies in summer temperature in the western Arctic during the Holocene Thermal Maximum ranged from 0.5° to 3°C (mean, 1.65°C). The largest anomalies were in the *North Atlantic Sector* (Kerwin et al., 1999; Kaufman et al., 2004; Flowers et al., 2008).

3.4.9b Neoglaciation

Many climate proxies are available to characterize the overall pattern of Late Holocene climate change. Following the Holocene Thermal Maximum, most proxy summer temperature records from the Arctic indicate an overall cooling trend through the late Holocene. Cooling is first recognized between 6 and 3 ka, depending on the threshold for change of each particular proxy. Records that exhibit a shift by 6–5 ka typically reflect intensified summer cooling about 3 ka (Figure 3.34).

Summer cooling during the second half of the Holocene led to the expansion of mountain glaciers and ice caps around the Arctic. The term "Neoglaciation" is widely applied to this episode of glacier growth, and in some cases re-formation, following the maximum glacial retreat during the Holocene Thermal Maximum (Porter and Denton, 1967). The former extent of glaciers is inferred from dated **moraines** and proglacial sediments deposited in lakes and marine settings. For example, ice-rafted detritus (Andrews et al., 1997) and the glacial geologic record (Funder, 1989) indicate that **outlet glaciers** of the *Greenland Ice Sheet* advanced during 6–4 ka (see Chapter 5, History of the Greenland Ice Sheet). Multiproxy records from 10 glaciers or glaciated areas in *Norway* show evidence for increased activity by 5 ka (Nesje et al., 2001; Nesje et al., 2008). Major advances of outlet glaciers of northern Icelandic ice caps begin by 5 ka (Stötter et al., 1999; Geirsdottir et al., in press). In the *European Arctic*, glaciers expanded on *Franz Josef Land* (Lubinski et al., 1999) and *Svalbard* (Svendsen and Mangerud, 1997) by 4 ka, although sustained growth primarily began around 3 ka. An early Neoglacial advance of mountain glaciers is registered in *Alaska*, most prominently in the *Brooks Range*, the highest-latitude mountains in the state (Ellis and Calkin, 1984; Calkin, 1988). In southwest *Alaska*, mountain glaciers in the Ahklun Mountains did not reform until about 3 ka (Levy et al., 2003). Neoglacial advances began in *Arctic Canada* by 5 ka (Miller et al., 2005).

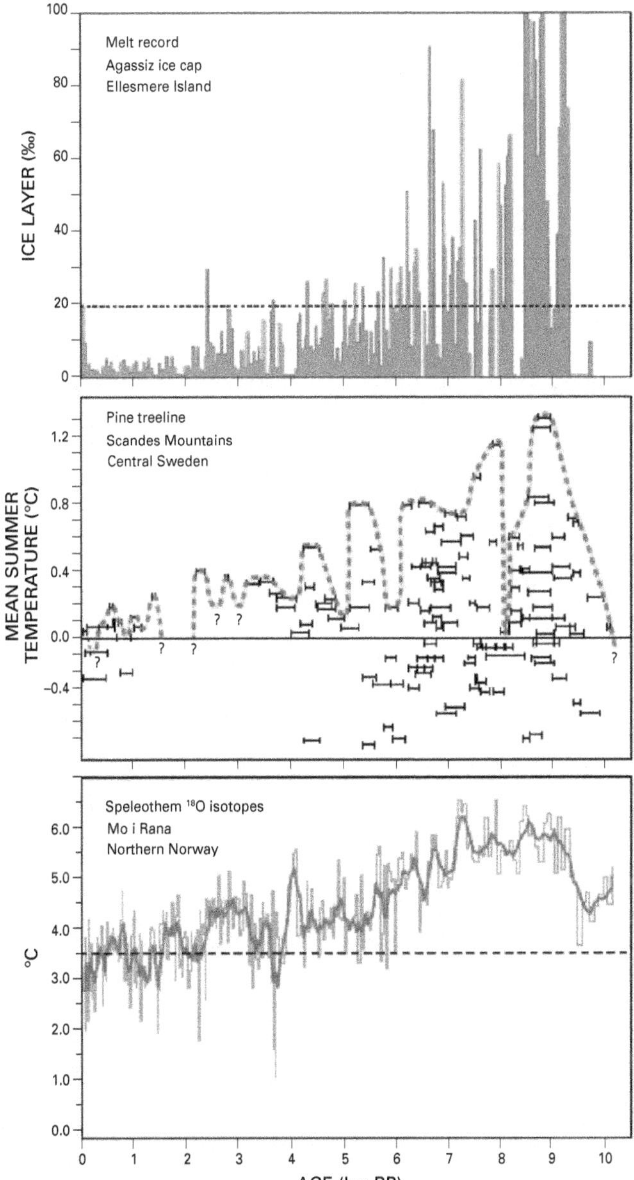

Figure 3.34. Arctic temperature reconstructions. Upper panel: Holocene summer melting on the Agassiz Ice Cap, northern Ellesmere Island, Canada. "Melt" indicates the fraction of each core section that contains evidence of melting (from Koerner and Fisher, 1990). Middle panel: Estimated summer temperature anomalies in central Sweden. Black bars = elevation of ^{14}C-dated sub-fossil pine wood samples (*Pinus sylvestris* L.) in the Scandes Mountains, central Sweden, relative to temperatures at the modern pine limit in the region. Dashed line = upper limit of pine growth is indicated by the dashed line. Changes in temperature estimated by assuming a lapse rate of 6°C km^{-1} (from Dahl and Nesje, 1996, on the basis of samples collected by L. Kullman and by G. and J. Lundqvist). Lower panel: Paleotemperature reconstruction from oxygen isotopes in calcite sampled along the growth axis of a stalagmite from a cave at Mo i Rana, northern Norway. Growth ceased around AD 1750 (from Lauritzen, 1996; Lauritzen and Lundberg, 1998). Figure from Bradley (2000). [Copyright 2000, reproduced with permission of Elsevier.]

The most consistent records of an Arctic Medieval Climate Anomaly come from the North Atlantic sector of the Arctic.

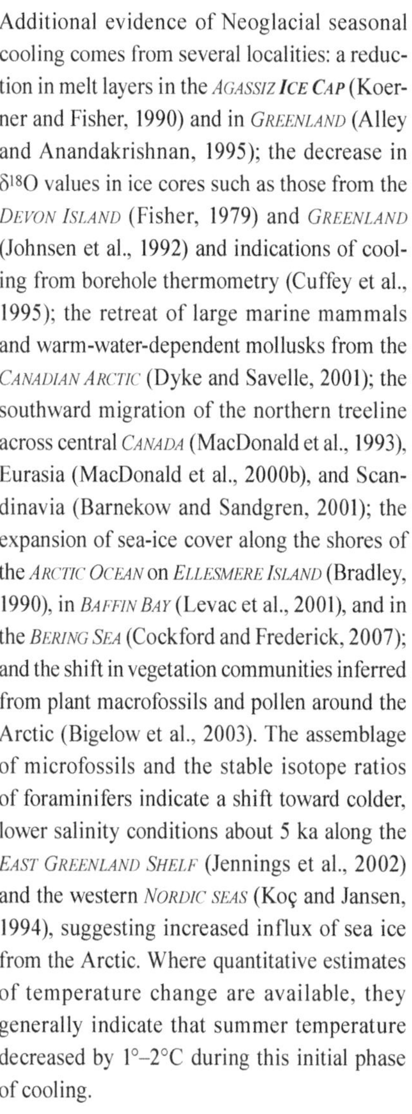

Additional evidence of Neoglacial seasonal cooling comes from several localities: a reduction in melt layers in the *Agassiz Ice Cap* (Koerner and Fisher, 1990) and in *Greenland* (Alley and Anandakrishnan, 1995); the decrease in $\delta^{18}O$ values in ice cores such as those from the *Devon Island* (Fisher, 1979) and *Greenland* (Johnsen et al., 1992) and indications of cooling from borehole thermometry (Cuffey et al., 1995); the retreat of large marine mammals and warm-water-dependent mollusks from the *Canadian Arctic* (Dyke and Savelle, 2001); the southward migration of the northern treeline across central *Canada* (MacDonald et al., 1993), Eurasia (MacDonald et al., 2000b), and Scandinavia (Barnekow and Sandgren, 2001); the expansion of sea-ice cover along the shores of the *Arctic Ocean* on *Ellesmere Island* (Bradley, 1990), in *Baffin Bay* (Levac et al., 2001), and in the *Bering Sea* (Cockford and Frederick, 2007); and the shift in vegetation communities inferred from plant macrofossils and pollen around the Arctic (Bigelow et al., 2003). The assemblage of microfossils and the stable isotope ratios of foraminifers indicate a shift toward colder, lower salinity conditions about 5 ka along the *East Greenland Shelf* (Jennings et al., 2002) and the western *Nordic Seas* (Koç and Jansen, 1994), suggesting increased influx of sea ice from the Arctic. Where quantitative estimates of temperature change are available, they generally indicate that summer temperature decreased by 1°–2°C during this initial phase of cooling.

The general pattern of an early- to middle-Holocene Thermal Maximum followed by Neoglacial cooling forms a multi-millennial trend that, in most places, culminated in the 19th century. Superposed on the long-term cooling trend were many centennial-scale warmer and colder summer intervals, which are expressed to a varying extent and are interpreted with various levels of confidence in different proxy records. In northern Scandinavia, evidence for notable late Holocene cold intervals before the 16th century includes narrow tree rings (Grudd et al., 2002), lowered treeline (Eronen et al., 2002), and major glacier advances (Karlén, 1988) between 2.6 and 2.0 ka. An extended analysis of these many centennial-scale warmer and colder intervals in *Russia* was published by Velichko and Nechaev (2005).

Warmth during the Medieval interval is generally ascribed to lack of explosive volcanoes that produce particles that block the Sun and perhaps to greater brightness of the Sun.

3.4.9c The Medieval Climate Anomaly (MCA)

Probably the most oft-cited warm interval of the late Holocene is the **Medieval Climate Anomaly** (MCA), earlier referred to as the Medieval Warm Period (MWP). The anomaly was recognized on the basis of several lines of evidence in Western Europe, but the term is commonly applied to other regions to refer to any of the relatively warm intervals of various magnitudes and at various times between about 950 and 1200 AD (Lamb, 1977) (Figure 3.35). In the Arctic, evidence for climate variability, such as relative warmth, during this interval is based on glacier extents, marine sediments, **speleothems**, ice cores, borehole temperatures, tree rings, and archaeology. The most consistent records of an Arctic Medieval Climate Anomaly come from the *North Atlantic sector* of the Arctic. The summit of *Greenland* (Dahl-Jensen et al., 1998), western *Greenland* (Crowley and Lowery, 2000), Swedish Lapland (Grudd et al., 2002), northern *Siberia* (Naurzbaev et al., 2002), and *Arctic Canada* (Anderson et al., 2008) were all relatively warm around 1000 AD. During Medieval time, Inuit populations moved out of *Alaska* into the eastern *Canadian Arctic* and hunted whale from skin boats in regions perennially ice-covered in the 20th century (McGhee, 2004).

The evidence for Medieval warmth throughout the rest of the Arctic is less clear. However, some indications of Medieval warmth include the general retreat of glaciers in southeastern *Alaska* (Reyes et al., 2006; Wiles et al., 2008) and the wider tree rings in some high-latitude tree-ring records from Asia and North America (D'Arrigo et al., 2006). However D'Arrigo et al. (2006) emphasized the uncertainties involved in estimating Medieval Climate Anomaly warmth relative to that of the 20th century, owing in part to the sparse geographic distribution of proxy data as well as to the less coherent variability of tree growth temperature estimates for this anomaly. Hughes and Diaz (1994) argued that the Arctic as a whole was not anomalously warm throughout Medieval time (also see Bradley et al., 2003b, and National Research Council, 2006). Warmth during the Medieval interval is generally ascribed to lack of explosive volcanoes that produce particles that block the Sun and perhaps to greater brightness of the Sun (Crowley, 2000; Goosse et al., 2005; also see Jansen et al., 2007). Warming around the *North Atlantic* and adjacent regions may have been linked to changes in oceanic circulation as well (Broecker, 2001).

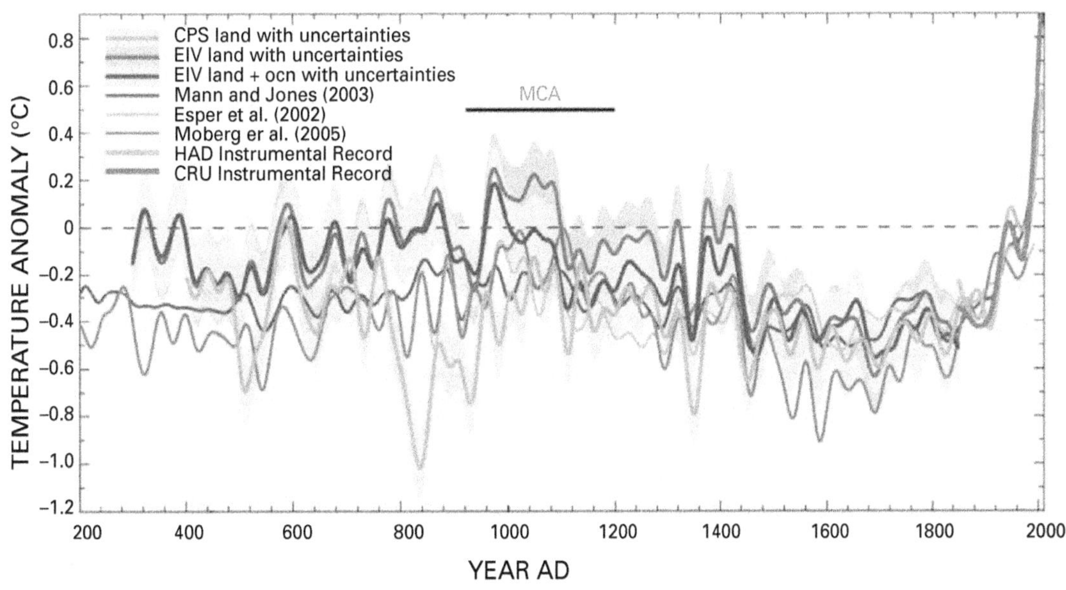

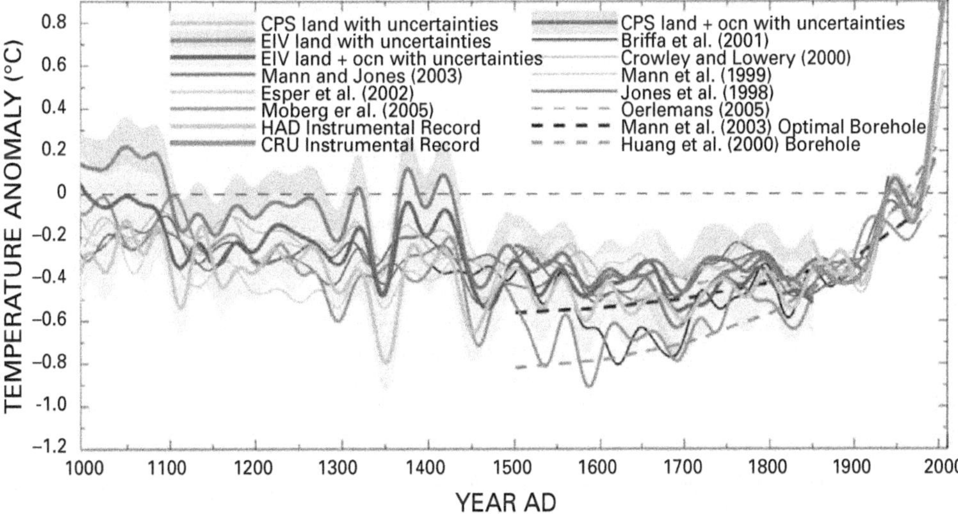

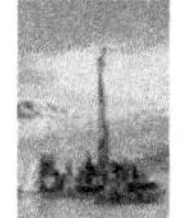

Figure 3.35. Updated composite proxy-data reconstruction of Northern Hemisphere temperatures for most of the last 2,000 years, compared with other published reconstructions. Estimated confidence limits, 95%. All series have been smoothed with a 40-year lowpass filter. MCA = Medieval Climate Anomaly, about 950–1200 AD. The array of reconstructions demonstrates that the warming documented by instrumental data during the past few decades exceeds that of any warm interval of the past 2,000 years, including that estimated for the MCA. (Figure from Mann et al., 2008.) CPS = composite plus scale methodology; CRU = East Anglia Climate Research unit, a source of instrumental data; EIV = error-in-variables; HAD = Hadley Climate Center. [Reprinted by permission of National Academy of Sciences.]

3.4.9D Climate of the Past Millennium and the Little Ice Age

The trend toward colder summers after about 1250 AD coincides with the onset of the Little Ice Age.

Given the importance of understanding climate in the most recent past and the richness of the available evidence, intensive scientific effort has resulted in numerous temperature reconstructions for the past millennium (Jones, et al., 1998; Mann et al., 1998; Briffa et al., 2001; Esper et al., 2002; Crowley et al., 2003; Mann and Jones, 2003; Moberg et al., 2005; National Research Council, 2006; Jansen et al., 2007), and especially the last 500 years (Bradley and Jones, 1992; Overpeck et al., 1997). Most of these reconstructions are based on annually resolved proxy records, primarily from tree rings, and they attempt to extract a record of air-temperature change over large regions or entire hemispheres. Data from *Greenland* ice cores and a few annually laminated lake sediment records are typically included in these compilations, but few other records of quantitative temperature changes spanning the last millennium are available from the Arctic. In general, the temperature records are broadly similar: they show modest summer warmth during Medieval times, a variable, but cooling climate from about 1250 to 1850 AD, followed by warming as shown by both paleoclimate proxies and the instrumental record. Less is known about changes in precipitation, which is spatially and temporally more variable than temperature.

The trend toward colder summers after about 1250 AD coincides with the onset of the **Little Ice Age**, which persisted until about 1850 AD, although the timing and magnitude of specific cold intervals were different in different places. **Proxy** climate records, both glacial and non-glacial from around the Arctic and for the Northern Hemisphere as a whole, show that the coldest interval of the Holocene was sustained sometime between about 1500 and 1900 AD (Bradley et al., 2003a). Recent evidence from the *Canadian Arctic* indicates that, following their recession in Medieval times, **glacier**s and ice sheets began to expand again between 1250 and 1300 AD. Expansion was further amplified about 1450 AD (Anderson et al., 2008).

Icelandic written records indicate that the duration and extent of sea ice in the Nordic Seas were high during the Little Ice Age.

Glacier mass balances throughout most of the Northern Hemisphere during the Holocene are closely correlated with summer temperature (Koerner, 2005), and the widespread evidence of glacier re-advances across the Arctic during the Little Ice Age is consistent with estimates of summer cooling that are based on tree rings. The climate history of the Little Ice Age has been extensively studied in natural and historical archives, and it is well documented in Europe and North America (Grove, 1988). Historical evidence from the Arctic is relatively sparse, but it generally agrees with historical records from northwest Europe (Grove, 1988). Icelandic written records indicate that the duration and extent of sea ice in the *Nordic Seas* were high during the Little Ice Age (Ogilvie and Jónsson, 2001).

The average temperature of the Northern Hemisphere during the Little Ice Age was less than 1°C lower than in the late 20th century (Bradley and Jones, 1992; Hughes and Diaz, 1994; Crowley and Lowery, 2000), but regional temperature anomalies varied. Little Ice Age cooling appears to have been stronger in the Atlantic sector of the Arctic than in the Pacific (Kaufman et al., 2004), perhaps because ocean circulation promoted the development of sea ice in the *North Atlantic*, which further amplified Little Ice Age cooling there (Broecker, 2001; Miller et al., 2005).

The Little Ice Age also shows evidence of multi-decadal climatic variability, such as widespread warming during the middle through late 18th century (e.g., Cronin et al., 2003). Although the initiation of the Little Ice Age and the structure of climate fluctuations during this multi-centennial interval vary around the Arctic, most records show warming beginning in the late 19th century (Overpeck et al., 1997). The end of the Little Ice Age was apparently more uniform both spatially and temporally than its initiation (Overpeck et al., 1997).

The climate change that led to the Little Ice Age is manifested in proxy records other than those that reflect temperature. For example, it was associated with a positive shift in transport of dust and other chemicals to the summit of *Greenland* (O'Brien et al., 1995), perhaps related to deepening of the Icelandic low-pressure system (Meeker and Mayewski, 2002). According to modeling studies, the negative phase (see *http://www.ldeo.columbia.edu/res/pi/**NAO**/*) of the **North Atlantic Oscillation** could have been amplified during the Little Ice Age (Shindell et al., 2001) whereas, in the North Pacific, the Aleutian low was significantly weakened during the Little Ice Age (Fisher et al., 2004; Anderson et al., 2005).

Seasonal cooling into the Little Ice Age resulted from the orbital changes as described above, together with increased explosive volcanism and probably also decreased solar luminosity as recorded by sunspot numbers as far back as 1600 AD (Renssen et al., 2005; Ammann et al., 2007; Jansen et al., 2007).

3.4.10 Placing 20th Century Warming in the Arctic in a Millennial Perspective

Much scientific effort has been devoted to learning how 20th-century and 21st-century warmth compares with warmth during earlier times (e.g., National Research Council, 2006; Jansen et al., 2007). Owing to the orbital changes affecting midsummer sunshine (a drop in June insolation of about 1 W/m^2 at 75°N. and 2 W/m^2 at 90°N. during the last 1,000 years; Berger and Loutre, 1991), additional forcing was needed in the 20th century to give the same summertime temperatures as achieved in the Medieval Warm Period.

After it evaluated globally or even hemispherically averaged temperatures, the National Research Council (2006) found that "Presently available proxy evidence indicates that temperatures at many, but not all, individual locations were higher during the past 25 years than during any period of comparable length since A.D. 900" (p. 3). Greater uncertainties for hemispheric or global reconstructions were identified in assessing older comparisons. As reviewed next, some similar results are available for the Arctic.

Thin, cold ice caps in the eastern *Canadian Arctic* preserve intact—but frozen—vegetation beneath them that was killed by the expanding ice. As these ice caps melt, they expose this dead vegetation, which can be dated by radiocarbon with a precision of a few decades. A recent compilation of more than 50 radiocarbon dates on dead vegetation emerging from beneath thin ice caps on northern *Baffin Island* shows that some ice caps formed more than 1,600 years ago and persisted through Medieval times before melting early in the 21st century (Anderson et al., 2008).

Fifty radiocarbon dates on dead vegetation emerging from beneath thin ice caps on northern *Baffin Island* shows that some ice caps formed more than 1,600 years ago.

Records of the melting from ice caps offer another view by which 20th century warmth can be placed in a millennial perspective. The most detailed record comes from the *Agassiz Ice Cap* in the Canadian High Arctic, for which the percentage of summer melting of each season's snowfall is reconstructed for the past 10 k.y. (Fisher and Koerner, 2003). The percent of melt follows the general trend of decreasing summer insolation from orbital changes, but some brief departures are substantial. Of particular note is the significant increase in melt percentage during the past century; current percentages are greater than any other melt intensity since at least 1,700 years ago, and melting is greater than any in sustained interval since 4–5 ka.

As reviewed by Smol and Douglas (2007b), changes in lake sediments record climatic and other changes in the lakes. Extensive changes especially in the post-1850 interval are most easily interpreted in terms of warming above the Medieval warmth on *Ellesmere Island* and probably in other regions, although other explanations cannot be excluded (also see Douglas et al., 1994). D'Arrigo et al. (2006) show tree-ring evidence from a few North American and Eurasian records that imply that summers were cooler in the Medieval Warm Period than in the late 20th century, although the statistical confidence is weak. Tree-ring and treeline studies in western *Siberia* (Esper and Schweingruber, 2004) and *Alaska* (Jacoby and D'Arrigo, 1995) suggest that warming since 1970 is has been optimal for tree growth and follows a circumpolar trend. Hantemirov and Shiyatov (2002) records from the Russian *Yamal Penisula*, well north of the Arctic Circle, show that summer temperatures of recent decades are the most favorable for tree growth within the past 4 millennia.

Whole-Arctic reconstructions are not yet available to allow confident comparison of late 20th century warmth with Medieval temperatures, nor has the work been done to correct for the orbital influence and thus to allow accurate comparison of the remaining forcings.

For all times considered, Arctic temperatures differ from recent values more than do hemispheric or global temperatures.

During the warm times (HTM, LIG) summer insolation anomalies north of 60°N. typically were only 10–20% greater than the anomalies for corresponding times averaged across the Northern Hemisphere as a whole.

3.5 ARCTIC AMPLIFICATION

The scientific understanding of climate processes shows that Arctic climate exhibits strong positive feedbacks (Serreze and Francis, 2006; Serreze et al., 2007a). As outlined in section 3.2, these feedbacks especially depend on the interactions of snow and ice with sunlight, the ocean, and the land surface (including its vegetation). For example, higher temperature tends to melt reflective ice and snow, more solar heat is then absorbed, and absorption of that heat promotes further warming (ice-albedo feedback). Furthermore, higher temperature tends to allow dark shrubs to replace low-growing tundra that is easily covered by snow, intensifying the snow-albedo feedback. Also, higher temperature tends to melt sea ice that insulates the cold wintertime air from the warmer ocean beneath, further warming the air (ice-insulation feedback). Similarly strong negative feedbacks are not known to stabilize Arctic climate, so physical understanding indicates that climate changes should be amplified in the Arctic as compared with lower latitude sites (reviewed by Serreze and Francis, 2006).

The following subsections summarize data from times for which meaningful comparison can be made between Arctic and more-widespread (hemispheric or global) temperatures to assess whether the paleoclimatic record is consistent with this physical expectation of Arctic amplification. For all times considered, Arctic temperatures differ from recent values more than do hemispheric or global temperatures. Because the paleoclimatic temperature changes had different forcings with different implications for feedbacks, as described next, no single number or function can be produced from the paleoclimatic data for the strength of Arctic amplification. Furthermore, the paleoclimatic data summarized here are from climate states that were relatively stable over millennia or longer, and they do not provide detailed information on the path by which Arctic and widespread regions reached those climate states (Serreze and Francis, 2006). (For rates of change in the Arctic, see chapter 4.) Nonetheless, the paleoclimatic data show that Arctic climate has been strongly variable, confirming physical understanding.

3.5.1 Forcings That Enhance Arctic Amplification

For three of the time intervals considered below (the Holocene Thermal Maximum or HTM, about 6 ka, the Last Glacial Maximum or LGM, about 21 ka, and MIS 5e, also known as the last interglaciation or LIG, about 130–120 ka), the climate changes were primarily forced by regular variations in Earth's orbital parameters, with some effect from changes in greenhouse gases and in ice-sheet size affecting the LGM and LIG. The anomalies of mean annual incoming solar radiation (insolation) averaged across the whole planet were less than 0.4% for all times considered, with the orbital changes serving primarily to shift sunlight around on the planet seasonally or geographically. However, during the three considered intervals the summer insolation forcing was relatively uniform throughout the Northern Hemisphere. During the warm times (HTM, LIG) summer insolation anomalies north of 60°N. typically were only 10–20% greater than the anomalies for corresponding times averaged across the Northern Hemisphere as a whole. For example, at the peak of the last interglaciation (130–125 ka), the Arctic (60°–90°N.) summer (May-June-July) insolation was 12.7% above present, while the Northern Hemisphere insolation was 11.4% above present (Berger and Loutre, 1991). At the same time, the Southern Hemisphere summer (November, December, January) insolation at 60°S. was 6% less than present. In contrast, greenhouse gas forcings are globally uniform because of the short mixing time for the global atmosphere (<2 years) and long residence times of most greenhouse gases (10 to 100 years).

As discussed below in section 3.5.3, the summertime Northern Hemisphere orbital forcing of the HTM, LGM and LIG linked to Earth's precession and axial tilt was largely offset by opposing-sign wintertime anomalies in the low-latitude Northern Hemisphere, yielding smaller mean-annual forcing at low latitudes than at high latitudes. The seasonally varying part of the Arctic snow and ice can respond notably to a summertime insolation anomaly, providing much of the total ice-albedo feedback quickly. However, the main feedback at low latitudes is through water vapor, which depends greatly on the temperature of the oceanic mixed layer, and which requires a few years to adjust; therefore, the low-latitude response to a seasonal

forcing is suppressed relative to the response to a mean-annual forcing (Rind, 2006). Thus, one would expect a larger feedback-amplified temperature change in the Arctic than in lower latitudes in response to seasonal forcing than to mean-annual forcing, all else being equal.

3.5.2 Comparing Arctic Temperature Anomalies With Hemispheric or Global Anomalies

To assess the geographic differences in the climate response to relatively uniform hemispheric changes in seasonal insolation forcing, Arctic summer temperature anomalies can be compared with the Northern Hemisphere average summer temperature anomalies for the HTM and LIG because of the similar forcing in the Arctic and Northern Hemisphere. During the Pliocene (and during earlier warm times discussed below), warmth persisted much longer than the cycle time of insolation changes resulting from Earth's orbital features (about 20 k.y. and about 40 k.y.). Consequently, Arctic anomalies for these times are compared with global temperature anomalies.

A difficulty is that for some of the younger times, global and Arctic estimates of temperature anomalies are available but hemispheric estimates are not. Although global estimates clearly include hemispheric data, those data have not been summarized in anomaly maps or hemispheric anomaly estimates that are published in the refereed scientific literature. To obtain hemispheric estimates here, note (as described in more detail below) that climate models driven by the known forcings show considerable fidelity in reproducing the global anomalies indicated by the data for the relevant times, and that hemispheric anomalies can be assessed within these models. The hemispheric anomalies so produced are consistent with the available paleoclimate data, and so they are used here.

The Palaeoclimate Modeling Intercomparison Project (PMIP2; Harrison et al., 2002, and see *http://pmip2.lsce.ipsl.fr/*) coordinates an international effort to compare paleoclimate simulations produced by a range of climate models, and to compare these climate model simulations with data-based **paleoclimate reconstructions** for the Holocene thermal maximum (6 ka) and for the Last Glacial Maximum (LGM; 21 ka). A comparison of simulations for 6 and 21 ka by the project is reported by Braconnot et al. (2007).

As part of this Palaeoclimate Modeling Intercomparison Project effort, Harrison et al. (1998) compared global (mostly Northern Hemisphere) vegetation patterns simulated by using the output of 10 different climate model simulations for 6 ka. The model simulations closely agreed with the vegetation reconstructed from paleoclimate

The general agreement between data and models provides confidence that climate-model simulations of past times may be compared with paleoclimate-based reconstructions of summer temperatures for the Arctic.

records. Similar comparisons on a regional basis for the Northern Hemisphere north of 55°N. (Kaplan et al., 2003), the Arctic (**CAPE Project** Members, 2001), Europe (Brewer et al., 2007), and North America (Bartlein et al., 1998) also showed close matches between paleoclimate data and models for the HTM. Comparison of models and data for the Last Glacial Maximum (Bartlein et al., 1998; Kaplan et al., 2003), and Last Interglaciation (CAPE Last Interglacial Project Members, 2006; Otto-Bliesner et al., 2006) reached similar conclusions. (Also see Pollard and Thompson, 1997; Farrera et al., 1999; Pinot et al., 1999; Kageyama et al., 2001.) Paleoclimate data correspond reasonably well with model simulations of the Holocene Thermal Maximum, Last Interglaciation warmth, and Last Glacial Maximum cold. Climate models generally underestimate the magnitude of change, and there are significant variations between different models, although the PMIP2 experiments, which include an interactive ocean, more closely match the paleodata than did the original PMIP experiments (Masson-Delmotte et al., 2006; Braconnot et al., 2007). The general agreement between data and models provides confidence that climate-model simulations of past times may be compared with paleoclimate-based reconstructions of summer temperatures for the Arctic. Clearly, however, additional data and additional analyses of existing as well as new data would improve confidence in the results and reduce uncertainties. Notably, current understanding of paleo-sea ice conditions in the ARCTIC OCEAN is clearly insufficient and requires both new data generation and a development of better ice proxies for paleorecords.

Intervals when the Arctic was warmer-than-present in the recent past as reconstructed from proxy data and independently supported by climate model experiments remain imperfect analogues for future greenhouse gas warming because the forcings are different. This point was stressed in Chapter 9 section 9.6.2 on page 724 of the 2007 **Intergovernmental Panel on Climate Change** Fourth Assessment WG1 Report:

> *"As with analyses of the instrumental record discussed in Section 9.6.2, some studies using palaeoclimatic data have also estimated PDFs [climate sensitivity probability density function] for ECS [equilibrium climate sensitivity] by varying model parameters. Inferences about ECS made through direct comparisons between radiative forcing and climate response, without using climate models, show large uncertainties since climate feedbacks, and thus sensitivity, may be different for different climatic background states and for different seasonal characteristics of forcing (e.g., Montoya et al., 2000). Thus, sensitivity to forcing during these periods cannot be directly compared to that for atmospheric CO_2 doubling."*

Nevertheless, paleoclimate reconstructions provide essential examples of how the planetary climate system responds to a range of forcings and constrain the relative importance of a range of strong **climate feedback mechanisms**. In an Arctic context, paleoclimatic reconstructions allow a measure of the effectiveness of Arctic amplification across a range of different climate forcings and **boundary conditions**, and they serve as targets for climate model experiments.

3.5.3 Arctic Amplification Case Studies

We evaluate paleoclimate estimates of Arctic summer temperature anomalies relative to recent summer temperature anomalies, and the appropriate Northern Hemisphere or global summer temperature anomalies, together with their uncertainties, for the following time periods: the Holocene Thermal Maximum (HTM; about 8 ka), the Last Glacial Maximum (LGM; about 21 ka), the last interglaciation (LIG; 130–120 ka) and the middle Pliocene (MP; about 3.5–3.0 Ma); we also provide some discussion of older changes.

Holocene Thermal Maximum (HTM):

Arctic summer ΔT=1.7±0.8°C; Northern Hemisphere summer ΔT=0.5±0.3°C; Global ΔT=0°±0.5°C.

A recent summary of summer temperature anomalies in the western Arctic (Kaufman et al., 2004) built on earlier summaries (Kerwin et al., 1999; CAPE Project Members, 2001) is consistent with more-recent reconstructions (Kaplan and Wolfe, 2006; Flowers et al., 2007). Although the Kaufman et al. (2004) summary considered only the western half of the Arctic, the earlier summaries by Kerwin et al., (1999) and CAPE Project Members (2001) indicated that similar anomalies characterized the eastern Arctic, and all syntheses report the largest anomalies in the *North Atlantic sector*. Although few data were available for these reconstructions from the central *Arctic Ocean*, the circumpolar dataset provides an adequate reflection of air temperatures over the *Arctic Ocean* as well.

Climate models suggest that the average planetary anomaly was concentrated over the Northern Hemisphere. Braconnot et al. (2007) summarized the simulations from 10 different climate model contributions to the PMIP2 project that compared simulated summer temperatures at 6 ka with recent temperatures. The global average summer temperature anomaly at 6 ka was 0°±0.5°C, whereas the Northern Hemisphere anomaly was 0.5°±0.3°C. These patterns are similar to patterns in model results described by Hewitt and Mitchell (1998) and Kitoh and Murakami (2002) for 6 ka, and a global simulation for 9 ka (Renssen et al., 2006). All simulate little difference in summer temperature outside the Arctic when those temperatures are compared with pre-industrial temperatures.

The primary forcing responsible for the warmer Northern Hemisphere summers was orbital, from the combined effects of Earth's tilt and precession producing positive summer insolation anomalies across the Northern Hemisphere. The positive summer anomalies were mostly offset by winter negative anomalies on an annual basis, although overall there was a small net positive annual insolation anomaly across the Northern Hemisphere. However, because the net annual insolation anomaly was small, water vapor feedback, a strong feedback with hemispheric impacts, was not an important contributor to the global or hemispheric temperature fields (Rind, 2006). In contrast, *Arctic Ocean* sea ice, which is strongly influenced by summer insolation, was substantially reduced (e.g., Häggblom, 1982; Koç et al., 1993; Dyke et al., 1996, 1997; Duplessy et al., 2001; Vavrus and Harrison, 2003; Bennike, 2004; Braconnot et al., 2007; England et al., 2008), helping to produce an Arctic amplification of temperature.

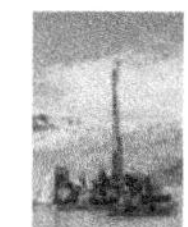

Last Glacial Maximum (LGM):

Arctic annual ΔT=18°±7°C; global and Northern Hemisphere annual ΔT=–5°±1°C.

Quantitative estimates of annual or seasonal temperature reductions during the peak of the Last Glacial Maximum are rare for the Arctic. Ice-core borehole temperatures, which offer the most compelling evidence (Cuffey et al., 1995; Dahl-Jensen et al., 1998), reflect mean annual temperatures. *Greenland* paleotemperature estimates cannot be partitioned by season, and it is not yet possible to estimate how much lower summer temperatures were during the LGM from the *Greenland* proxies. There are few terrestrial sites in the Arctic that were ice-free during the LGM and that contain secure paleotemperature proxies for summer (or winter) temperatures. One such site in *Beringia* suggests summer temperatures were about 20°C lower during the LGM (Elias et al., 1996). Because of the limited datasets for temperature reduction in the Arctic during the Last Glacial Maximum, a large uncertainty is specified.

The global-average temperature decrease during peak glaciations, based on paleoclimate proxy data, was 5°–6°C, with little difference between the Northern and Southern Hemispheres (Farrera et al., 1999; Braconnot et al., 2007). A similar temperature anomaly is derived independently from climate-model simulations (Otto-Bliesner et al., 2007).

The forcings responsible for LGM cold are primarily minimum summer insolation at high northern latitudes and reduction in greenhouse gases (both CO_2 and $\mathbf{CH_4}$). Ice-sheet growth in response to a combination of orbital and greenhouse forcings began at the end of the last interglaciation (about 120 ka) in northern North America and Eurasia. Slow feedbacks associated with ice-sheet growth (ice-elevation, ice-height, and glacial-isostasy) (Rind, 1987) coupled with fast feedbacks related to albedo and planetary water vapor, amplified the cooling, which reached a maximum during the LGM.

Last Interglaciation (LIG):

Arctic summer ΔT=5°±1°C; global and Northern Hemisphere summer ΔT=1°±1°C.

A recent summary of all available quantitative reconstructions of summer-temperature anomalies in the Arctic during peak Last Interglaciation warmth shows a spatial pattern generally similar to that shown by Holocene Thermal Maximum reconstructions. The largest anomalies are in the *NORTH ATLANTIC SECTOR* and the smallest anomalies are in the North Pacific sector, but those small anomalies are substantially larger (5°±1°C) than they were during the Holocene Thermal Maximum (CAPE Last Interglacial Project Members, 2006). A similar pattern of Last Interglaciation summer-temperature anomalies is apparent in climate model simulations (Otto-Bliesner et al., 2006). Global and Northern Hemisphere summer-temperature anomalies are derived from summaries in CLIMAP Project Members (1984), Crowley (1990), Montoya et al. (2000), and Bauch and Erlenkeuser (2003).

The primary forcings responsible for LIG Northern Hemisphere summer warmth include the unusual alignment of precession and obliquity terms in Earth's orbit such that Earth was closest to the Sun in Northern Hemisphere summer when tilt was at a maximum. This produced strong positive summer insolation anomalies across the Northern Hemisphere. These anomalies were accompanied by a slight increase in greenhouse gases relative to Holocene levels.

Middle Pliocene:

Arctic annual ΔT=12°±3°C; global annual ΔT=4°±2°C.

Widespread forests throughout the Arctic in the middle Pliocene offer a glimpse of a notably warm time in the Arctic, which had essentially modern continental configurations and connections between the *ARCTIC OCEAN* and the global ocean. Reconstructed Arctic temperature anomalies are available from several sites that show much warmth and probably no summer sea ice in the *ARCTIC OCEAN*. These sites include the *CANADIAN ARCTIC ARCHIPELAGO* (Dowsett et al., 1994; Elias and Matthews, 2002; Ballantyne et al., 2006), *ICELAND* (Buchardt and Símonarson, 2003), and the North Pacific (Heusser and Morley, 1996). A global summary of middle Pliocene biomes by Salzmann et al. (2008) concluded that Arctic mean-annual-temperature anomalies were in excess of 10°C; some sites indicate temperature anomalies of as much as 15°C. Estimates of global sea-surface temperature anomalies are from Dowsett (2007).

Global reconstructions of mid-Pliocene temperature anomalies from proxy data and general circulation models show an average warming of 4° ± 2°C , with greater warming over land than oceans, but with substantial warming of sea surface temperatures even at low to middle latitudes (Budyko et al., 1985; Chandler et al., 1994a, Raymo et al., 1996; Sloan et al., 1996; Dowsett et al., 1999; Haywood and Valdes, 2004; Jiang et al., 2005; Haywood and Valdes, 2006; Salzmann et al., 2008).

The forcing of the warmth of the middle Pliocene remains unclear. Orbital oscillations have continued throughout Earth history, but the Pliocene warmth persisted long enough to cross many orbital oscillations, which thus cannot have been responsible for the warmth. The most likely explanation is an elevated level of CO_2 (estimated as 360-400 ppmv by Jansen et al., 2007, p. 442, but with notable uncertainties as shown by their Figure 6.1), coupled with smaller *GREENLAND* and Antarctic ice sheets (Haywood and Valdes, 2004; Jansen et al., 2007); a possible role for altered oceanic circulation is also considered in some studies (Chandler et al. 1994b; also Jansen et al., 2007 and references therein).

This trend of larger Arctic anomalies was already well established during the greater warmth of the early Cenozoic peak warming and of the Cretaceous before that. Somewhat greater uncertainty is attached to these more ancient times in which continental configuration was somewhat different. Barron et al. (1995) estimated global-average temperatures about 6°C warmer in the Cretaceous than recently. As reviewed by Alley (2003) (also see Bice et al., 2006), subsequent work suggests upward revision of tropical sea-surface temperatures by as much as a few degrees. The Cretaceous peak warmth seems to have been somewhat higher than early Cenozoic values, or perhaps similar (Zachos et al., 2001). The early Cenozoic (late Paleocene; section 3.4.1) temperature records probably mostly recorded summertime conditions of about 18°C in the *ARCTIC OCEAN* and about 17°C on adjacent Arctic lands, followed during the short-lived (multi-millennial) Paleocene-Eocene Thermal Maximum by warming to about 23°C in the summer *ARCTIC OCEAN* and 25°C on adjacent lands (Moran et al., 2006; Sluijs et al.; 2006, 2008; Weijers et al., 2007). No evidence of wintertime ice exists, and temperatures very likely remained higher than during the mid-Pliocene. Hence, changes in the Arctic were much larger than the globally averaged change.

The Cretaceous and early-Cenozoic warmth was apparently forced by increased greenhouse-gas concentrations (e.g., Jansen et al., 2007; Donnadieu et al., 2006; Royer et al., 2007).

Case Study Summary

Based on these case studies, Arctic amplification appears to have operated at times when the Northern Hemisphere was both colder and warmer than recent levels, and in response to different forcings. This result is fully consistent with expectations based on physical understanding and models (e.g., Serreze and Francis, 2006). The magnitude of Arctic amplification probably depends on the extent to which slow versus fast feedbacks are engaged, and whether hemispherically uniform feedbacks (water vapor, greenhouse gas) are triggered.

3.6 SUMMARY

3.6.1 Major Features of Arctic Climate in the Past 65 m.y.

Section 3.4 summarized some of the extensive evidence for changes in Arctic temperatures, and to a lesser extent in Arctic precipitation, during the last 65 m.y. To some degree it also discussed "attribution"—the best scientific understanding of the causes of the climate changes. In this subsection, a brief synopsis is provided; for citations, the reader is referred to the extensive discussion just above.

At the start of the Cenozoic, 65 Ma, the Arctic was much warmer year around than it was recently; forests grew on all land regions and no perennial sea ice or *Greenland Ice Sheet* existed. Gradual but bumpy cooling has dominated most of the last 65 million years, and falling atmospheric CO_2 concentration apparently is the most important contributor to the cooling—although possible changing continental positions and their effect on atmospheric or oceanic circulation may also contribute. One especially prominent "bump," the Paleocene-Eocene Thermal Maximum about 55 Ma, warmed the *Arctic Ocean* more than 5°C and the Arctic landmass about 8°C, probably in a few centuries to a millennium or so, followed by cooling for about 100 k.y. Warming from release of much CO_2 (possibly initially as seafloor methane that was then oxidized to CO_2) is the most likely explanation. In the middle Pliocene (about 3 Ma) a modest warming was sufficient to allow deciduous trees on Arctic land that at present supports only High Arctic polar-desert vegetation; whether this warming originated from changes to circulation, CO_2, or some other cause remains unclear.

About 2.7 Ma, the cooling reached the threshold beyond which extensive continental ice sheets developed in the North American and *Eurasian Arctic*, and it marked the onset of the Quaternary Ice Age. Initially, the growth and shrinkage of the ice ages were directly controlled by changes in northern sunshine caused by features of Earth's orbit (the 41-k.y. cycle of sunshine that is tied to the obliquity (tilt) of Earth's axis is especially prominent). More recently, a 100-k.y. cycle has become more prominent, perhaps because the ice sheets became large enough that their behavior became important. Short, warm interglacials (usually lasting about 10,000 years, although the one about 440,000 years ago lasted longer) have alternated with longer glacial intervals. Recent work suggests that, in the absence of human influence, the current interglacial would continue for a few tens of thousands of years before the start of a new ice age (Berger and Loutre, 2002). Although driven by the orbital cycles, the large temperature differences between glacials and interglacials, and the globally synchronous response, reflect the effects of strong positive feedbacks, such as changes in atmospheric CO_2 and other greenhouse gases and in the areal extent of reflective snow and ice.

Arctic amplification appears to have operated at times when the Northern Hemisphere was both colder and warmer than recent levels, and in response to different forcings.

Interactions among the various orbital cycles have caused small differences between successive interglacials. More summer sunshine was received in the Arctic during the interglacial of about 130–120 ka than has been received in the current interglacial. Thus, summer temperatures in many places were about 4°–6°C warmer than recently, and these higher temperatures reduced ice on *Greenland* (Chapter 5, History of the Greenland Ice Sheet), raised sea level, and melted widespread small glaciers and ice caps.

Recent work suggests that, in the absence of human influence, the current interglacial would continue for a few tens of thousands of years before the start of a new ice age.

The cooling into and warming out of the most recent glacial were punctuated by numerous abrupt climate changes, and conditions persisted for millennia between jumps that were completed in years to decades. These events were very pronounced around the *North Atlantic*, but they had a much smaller effect on temperature elsewhere in the Arctic. Temperature changes

The warming from the Little Ice Age began for largely natural reasons, but it appears to have been accelerated by human contributions and especially by increasing CO_2 concentrations in the atmosphere.

extended to equatorial regions and caused a see-saw response in the far south (i.e., mean annual warming in the south when the north cooled). Large changes in extent of sea ice in the *North Atlantic* were probably responsible, linked to changes in regional to global patterns of ocean circulation; freshening of the *North Atlantic* favored expansion of sea-ice.

These abrupt temperature changes also were a feature of the current interglacial, the Holocene, but they ended as the *Laurentide Ice Sheet* on *Canada* melted away. Arctic temperatures in the Holocene broadly responded to orbital changes, and temperatures warmed during the middle Holocene when there was more summer sunshine. Warming generally led to northward migration of vegetation and to shrinkage of ice on land and sea. Smaller oscillations in climate during the Holocene, including the so-called Medieval Warm Period and the Little Ice Age, were linked to variations in the sun-blocking effect of particles from explosive volcanoes and perhaps to small variations in solar output, or in ocean circulation, or other factors. The warming from the Little Ice Age began for largely natural reasons, but it appears to have been accelerated by human contributions and especially by increasing CO_2 concentrations in the atmosphere (Jansen, 2007).

3.6.2 Arctic Amplification

Based on paleoclimate data, Arctic amplification is a pervasive feature of Earth's climate system over a range of forcing scenarios. It must be remembered that no particular past interval offers a perfect analogue for conditions in the coming decades as greenhouse gases increase, and that the paleoclimatic record often does not allow reconstruction of the spatially distributed rate of change in response to rapid forcings. Nonetheless, larger changes in the Arctic than hemispherically or globally are expected based on physical understanding and models, and this expectation is confirmed by the available paleoclimatic data.

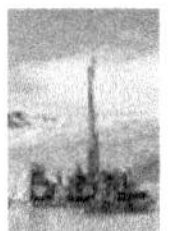

Past Rates of Climate Change in the Arctic

Lead Authors: James W.C. White, University of Colorado, Boulder, CO; Richard B. Alley*, Pennsylvania State University, University Park, PA

Contributing Authors: Anne Jennings, University of Colorado, Boulder, CO; Sigfus Johnsen, University of Copenhagen, Denmark; Gifford H. Miller*, University of Colorado, Boulder, CO; Steven Nerem, University of Colorado, Boulder, CO

*SAP 1.2 Federal Advisory Committee Member

ABSTRACT

Climate has changed on numerous time scales for various reasons and has always done so. In general, longer lived changes are somewhat larger but much slower than shorter lived changes. Processes linked with continental drift have affected atmospheric and oceanic currents and the composition of the atmosphere for tens of millions of years; in the Arctic, a global cooling trend has altered conditions near sea level from ice-free year-round to icy. Within the icy times, variations in Arctic sunshine over tens of thousands of years in response to features of Earth's orbit caused regular cycles of warming and cooling that were roughly half the size of the continental-drift-linked changes. This "glacial-interglacial" cycling has been amplified by colder times bringing reduced greenhouse gases and greater reflection of sunlight, especially from more-extended ice. This glacial-interglacial cycling has been punctuated by sharp-onset, sharp-end (in some instances less than 10 years) millennial oscillations, which near the North Atlantic were roughly half as large as the glacial-interglacial cycles but which were much smaller Arctic-wide and beyond. The current warm period of the glacial-interglacial cycle has been influenced by cooling events from single volcanic eruptions, slower but longer lasting changes from random fluctuations in frequency of volcanic eruptions and from weak solar variability, and perhaps by other classes of events. Human-forced climate changes appear similar, in terms of both size and duration, to the fastest natural changes of the past, but projected changes in the future may become anomalous compared to natural changes.

4.1 INTRODUCTION

Climate change*, as opposed to change in the weather (the distinction is defined below), occurs on all time scales, ranging from several years to billions of years. The rate of change, typically measured in degrees Celsius (°C) per unit of time (years, decades, centuries, or millennia, for example, if **climate** is being considered) is a key determinant of the effect of the change on living things such as plants and animals; collections and webs of living things, such as ecosystems; and humans and human societies. Consider, for example, a 10°C change in annual average temperature, roughly the equivalent to going from Birmingham, Alabama, to Bangor, Maine. If such a change took place during thousands of years, as happens when the Earth's orbit varies and portions of the planet receive more or less energy from the Sun, ecosystems and aspects of the environment, such as sea level, would change, but the slow change would allow time for human societies to adapt. A 10°C change that appears in 50 years or less, however, is fundamentally different (National Research Council, 2002). Ecosystems would be able to complete only very limited adaptation because trees, for example, typically are unable to spread that fast by seed dispersal. Human adaptation would be limited as well, and widespread challenges would face agriculture, industry, and public utilities in response to changing patterns of precipitation, severe weather, and other events. Such abrupt climate changes on regional scales are well documented in the **paleoclimate record** (National Research Council, 2002; Alley et al., 2003). This rate of change is about 100 times as fast as the warming of the last century.

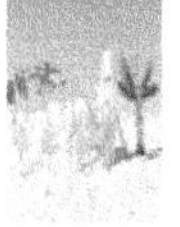

Not all parts of the climate system can change this rapidly. Global temperature change is slowed by the heat capacity of the oceans, for example (e.g., Hegerl et al., 2007). Local changes, particularly in continental interiors or where sea-ice changes modify the interaction between ocean and atmosphere, can be faster and larger. Changes in atmospheric circulation are potentially faster than changes in ocean circulation, owing to the difference in mass and thus inertia of these two circulating systems. This difference, in turn, influences important climate properties that depend on oceanic or atmospheric circulation. The concentration of **carbon dioxide** in the atmosphere, for example, depends in part on ocean circulation, and thus it does not naturally vary rapidly (e.g., Monnin et al., 2001). **Methane** concentration in the atmosphere, on the other hand, has increased by more than 50% within decades (Severinghaus et al., 1998), as this gas is more dependent on the distribution of wetlands, which in turn depend on atmospheric circulation to bring rains.

In the following pages we examine past rates of environmental change observed in Arctic paleoclimatic records. We begin with some basic definitions and clarification of concepts. Climate change can be evaluated absolutely, using numerical values such as those for temperature or rainfall, or it can be evaluated relative to the effects they produce (National Research Council, 2002). Different groups often have differing views on what constitutes "important." Hence, we begin with a common vocabulary.

*For bold terms, refer to Glossary; for italic terms, refer to Plate 1; for geologic ages, refer to Plate 2.

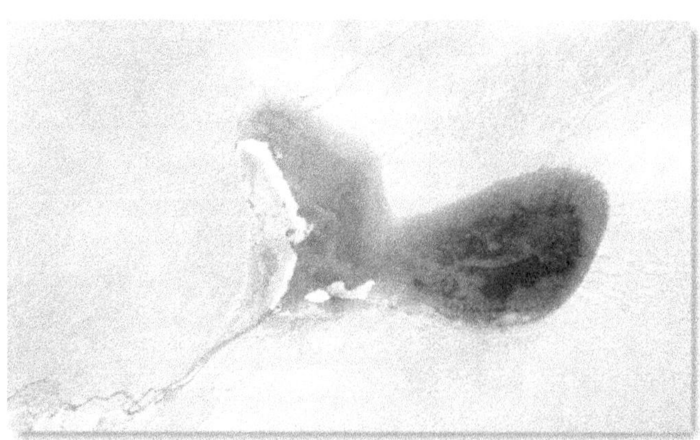

4.2 VARIABILITY VERSUS CHANGE; DEFINITIONS AND CLARIFICATION OF USAGE

Climate scientists and weather forecasters are familiar with opposite sides of very common questions. Does this hot day (or month, or year) prove that global warming is occurring? or does this cold day (or month, or year) prove that global warming is not occurring? Does global warming mean that tomorrow (or next month, or next year) will be hot? or does the latest argument against global warming mean that tomorrow (or next month, or next year) will be cold? Has the climate changed? When will we know that the climate has changed? To people accustomed to seven-day weather forecasts, in which the forecast beyond the first few days is not very accurate, the answers are often not very satisfying. The next sections briefly discuss some of the issues involved.

4.2.1 Weather Versus Climate

The globally averaged temperature difference between an ice age and an **interglacial** is about 5°–6°C (Cuffey and Brook, 2000; Jansen et al., 2007). The 12-hour temperature change between peak daytime and minimum nighttime temperatures at a given place, or the 24-hour change, or the seasonal change, may be much larger than that **glacial**-interglacial change (e.g., Trenberth et al., 2007). In assessing the "importance" of a climate change, it is generally accepted that a single change has greater effect on ecosystems and economies, and thus is more "important," if that change is less expected, arrives more rapidly, and stays longer (National Research Council, 2002). In addition, a step change that then persists for millennia might become less important than similar-sized changes that occurred repeatedly in opposite directions at random times.

Historically, climate has been taken as a running average of weather conditions at a place or throughout a region. The average is taken for a long enough time interval to largely remove fluctuations caused by "weather." Thirty years is often used for averaging.

Weather, to most observers, implies day-to-day occurrences, which are predictable for only about two weeks. Looking further ahead than that is limited by the chaotic nature of the atmospheric system; that is, by the sensitivity of the system to initial conditions (e.g., Lorenz, 1963; Le Treut et al., 2007), as described next. All thermometers have uncertainties, even if only a fraction of a degree, and all measurements by thermometers are taken at particular places and not in between. All temperature estimates at and between thermometers are thus subject to some uncertainty. A weather-forecasting **model** can correctly be started from a range of possible starting conditions that differ by an amount equal to or less than the measurement uncertainties. For short times of hours or even days, the different starting conditions provided by the modern observational system typically have little effect on the prediction of future weather; vary the starting data within the known uncertainties, and the output of the model will not be affected much out in time for a day or two. However, if the model is run for times beyond a few days to perhaps a couple of weeks, the different starting conditions produce very different weather forecasts. The forecasts are "bounded"—they do not produce blizzards in the tropics or tropical temperatures in the Arctic wintertime, for example; and they do produce "forecasts" recognizably possible for all regions covered—but the forecasts differ greatly in the details of where and when convective thunderstorms or frontal systems occur and how much precipitation will be produced during what time period. To many observers, "weather" refers to those features of Earth's coupled atmosphere-ocean system that are theoretically predictable to two weeks or so but not beyond.

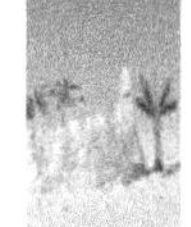

To many observers, "weather" refers to those features of Earth's coupled atmosphere-ocean system that are theoretically predictable to two weeks or so but not beyond.

For many climatologists, however, somewhat longer term events are often lumped under the general heading of "weather." The year-to-year temperature variability in global average temperature associated with the El Niño–La Niña phenomenon may be a few tenths of a degree Celsius (e.g., Trenberth et al., 2002), and similar or slightly larger variability can be caused by volcanic eruptions (e.g., Yang and Schlesinger, 2002). The influences of such phenomena are short lived compared with a 30-year average, but they are long lived compared with the two-week interval described just above. Volcanic eruptions may someday prove to be predictable beyond two weeks (U.S. Geological Survey scientists successfully predicted one of the

Mt. St. Helens eruptions more than two weeks in advance (Tilling et al., 1990)), and the effects following an eruption certainly are predictable for longer times. El Niños are predictable beyond two weeks. However, if one is interested in the climatic conditions at a particular place, a proper estimate would include the average behavior of volcanoes and El Niños, but it would not be influenced by the accident that the starting and ending points of the 30-year averaging period happened to sample a higher or lower number of these events than would be found in an average 30-year period.

As recorded in Greenland ice cores, local temperatures fell many degrees Celsius within a few decades about 13 ka. The temperature remained low for more than a millennium, and then it jumped up about 10°C in about a decade, and it has remained substantially elevated since.

The issues of the length of time considered and the starting time chosen are illustrated in Figure 4.1. Annual temperatures for the continental

CONTINENTAL US ANNUAL TEMPERATURES

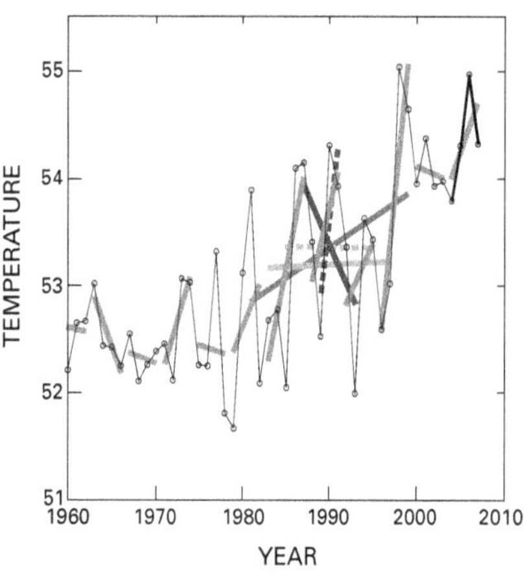

Figure 4.1. "Weather" versus "climate," in annual temperatures for the continental United States, 1960–2007. Red lines = trends for 4-year segments that show how the time period affects whether the trend appears to depict warming, cooling, or no change. Various lines show averages of different number of years, all centered on 1990: Dark blue dash = 3 years; dark blue = 7 years; light blue dash = 11 years; light blue = 15 years; and green = 19 years. The perceived trend can be warming, cooling, or no change depending on the length of time considered. Climate is normally taken as a 30-year average; all 30-year-long intervals (1960–1989 through 1978–2007) warmed significantly (greater than 95% confidence), whereas only 1 of the 45 possible trend-lines (17 are shown) has a slope that is markedly different from zero with more than 95% confidence. Thus, a climate-scale interpretation of these data indicates warming, whereas shorter-term ("weather") interpretations lead to variable but insignificant trends. Data from United States Historical Climatology Network, *http://www.ncdc.noaa.gov/oa/climate/research/cag3/cag3.html* (Easterling et al., 1996).

United States since 1960 are shown. The variability shown is linked to El Niño, volcanic eruptions, and other factors. If we use a 4-year window to illustrate the issue, it is apparent that for any given 4-year period, the temperature can appear to warm, to cool, or to stay flat. Also shown are the 3-, 7-, 11-, 15-, and 19-year linear trends centered on 1990. Depending on the number of years chosen, the trend can be strongly warming to strongly cooling. The warm El Niño years of 1987 and 1988, and the cooling trend in 1992 and 1993 caused by the eruption of Mt. Pinatubo, affect our perception of the time trend, or climate. Notice that of the 45 four-year regression lines possible between 1960 and 2007 (17 are shown in Figure 4.1) only one meets the usual statistical criterion of having a slope different from zero with at least 95% confidence. Climate is often considered as a 30-year average, and all 30-year regression lines that can be placed on Figure 4.1 (years 1960–1989, 1961–1990, …, 1978–2007) have a positive slope (warming) with greater than 95% confidence. Thus, all of the short-time-interval lines shown on Figure 4.1 are part of a warming climate over a 30 year interval but clearly reflect weather as well.

4.2.2 Style of Change

In some situations a 30-year climatology appears inappropriate. As recorded in GREENLAND ice cores, local temperatures fell many degrees Celsius within a few decades about 13 **ka** during the **Younger Dryas** time, a larger change than the **interannual variability**. The temperature remained low for more than a millennium, and then it jumped up about 10°C in about a decade, and it has remained substantially elevated since (Clow, 1997; Severinghaus et al., 1998; Cuffey and Alley, 2000). It is difficult to imagine any observer choosing the temperature average of a 30-year period that included that 10°C jump and then arguing that this average was a useful representation of the climate. The jump is perhaps the best-known and most-representative example of abrupt climate change (National Research Council, 2002; Alley et al., 2003), and the change is ascribed to what is now known colloquially as a "tipping point." Tipping points occur when a slow process reaches a threshold that "tips" the climate system into a new mode of operation (e.g., Alley, 2007).

Tipping points occur when a slow process reaches a threshold that "tips" the climate system into a new mode of operation

Analogy to a canoe tipping over suddenly in response to the slowly increasing lean of a paddler is appropriate.

Tipping behavior is readily described sufficiently long after the event, although it is much less evident that a climate scientist could have predicted the event just before it occurred, or that a scientist experiencing the event could have stated with confidence that conditions had tipped. Research on this topic is advancing, and quantitative statements can be made about detection of events, but timely detection may remain difficult (Keller and McInerney, 2007).

4.2.3 How to Talk About Rates of Change

The term "abrupt climate change" has been defined with some authority in the report of the National Research Council (2002). However, many additional terms such as "tipping point" remain colloquial, although arguably they can be related to well-accepted definitions. For the purposes of this report, preference will be given to common English words whenever possible, with explanations of what is meant, without relying on new definitions of words or on poorly defined words.

4.2.4 Spatial Characteristics of Change

The Younger Dryas cold event, introduced above in section 4.2.2, led to prominent cooling around the *North Atlantic*, weaker cooling around much of the Northern Hemisphere, and weak warming in the far south; uncertainty remains about changes in many places, and the globally averaged effect probably was minor (reviewed by Alley, 2007). The most commonly cited records of the Younger Dryas are those that show large signals. Informal discussions by many investigators with people outside our field indicate that the strong local signals are at least occasionally misinterpreted as global signals. It is essential to recognize the geographic as well as time limitations of climate events and their paleoclimatic records.

Further complicating this discussion is the possibility that an event may start in one region and then require some climatically notable time interval to propagate to other regions. Limited data supported by our basic understanding of how climate processes work suggest that the Younger Dryas cold event began and ended in the north, that the response was delayed by decades or longer in the far south, and that it was transmitted there through the ocean (Steig and Alley, 2003; Stocker and Johnsen, 2003). Cross-dating climate records around the world to the precision and accuracy needed to confirm that relative timing is a daunting task. The mere act of relating records from different areas then becomes difficult; an understanding of the processes involved is almost certainly required to support the interpretation.

It is essential to recognize the geographic as well as time limitations of climate events and their paleoclimatic records.

4.3 ISSUES CONCERNING RECONSTRUCTION OF RATES OF CHANGE FROM PALEOCLIMATIC INDICATORS

In an ideal world, a chapter on rates of change would not be needed. If climate records were available from all places and all times, with accurate and precise dates, then rate of change would be immediately evident from inspection of those records. However, as suggested in the previous section, such a simple interpretation is seldom possible.

Consider a hypothetical example. A group of tree trunks, bulldozed by a **glacier** and incorporated into glacial sediments, is now exposed at a coastal site. Many trees were killed at approximately the same time. The patterns of thick and thin rings, dense and less-dense wood, and isotopic variation of the wood layers contain climatic information (e.g., White et al., 1994). The climatic fluctuations that controlled the tree-ring characteristics can be dated precisely relative to each other—for example, this isotopic event occurred 7 years after that one. However, the precise age of the start and end of that climate record may not be available.

If climate records were available from all places and all times, with accurate and precise dates, then rate of change would be immediately evident from inspection of those records.

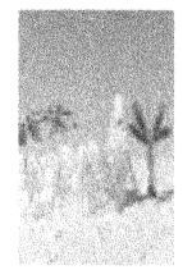

If much additional wood of various ages is available nearby, and if a large effort is expended, it may be possible to use the patterns of thick and thin rings and other features to match overlapping trees of different ages and thus to tie the record to still-living trees and provide a continuous record absolutely dated to the nearest year. If this is not possible, but the trees grew within the time span for which radiocarbon can be used, it may be possible to learn the age of the record to within a few decades or centuries, but no better. If the record is older than can be dated using radiocarbon, and other dating techniques are not available, even larger errors may be attached to estimates of the time interval occupied by the record.

Uncertainties are always associated with reconstructed climate changes (were the thick and thin rings controlled primarily by temperature changes or by moisture changes? for example), but once temperatures or rainfall amounts are estimated for each year, calculation of the rate of change from year to year will involve no additional error because each year is accurately identified. However, learning the spatial pattern of climate change may not be possible, because it will not be possible to relate the events recorded by the tree rings to events in records from other places with their own dating difficulties.

In Arctic and subarctic marine sediments, radiocarbon dating remains the standard technique for obtaining well-dated records during the last 40,000 to 50,000 years.

Sometimes, however, it is possible to learn the spatial pattern of the climate change and to learn how the rate of change at one place compared with the rate of change elsewhere. Volcanic eruptions are discrete events, and major eruptions typically are short lived (hours to days), so that the layer produced by a single eruption in various lake and marine sediments and glaciers is almost exactly the same age in all. If the same pattern of volcanic fallout is found in many cores of lake or ocean sediment or ice, then it is possible to compare the rate of change at those different sites. The uncertainties in knowing the time interval between two volcanic layers may be small or large, but whatever the time interval is, it will be the same in all cores containing those two layers.

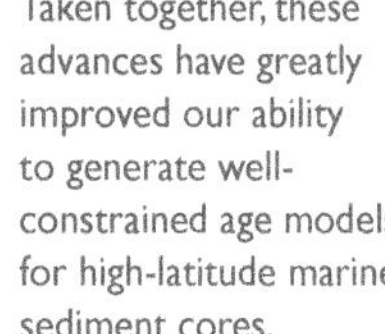

Taken together, these advances have greatly improved our ability to generate well-constrained age models for high-latitude marine sediment cores.

These and additional considerations motivate the additional discussion of rates of climate change provided here.

4.3.1 Measurement of Rates of Change in Marine Records

In Arctic and subarctic marine sediments, radiocarbon dating remains the standard technique for obtaining well-dated records during the last 40,000 to 50,000 years. Radiocarbon dating is relatively inexpensive, procedures are well developed, and materials that can be dated usually are more common than is true for other techniques. Radiocarbon dating is now conventionally calibrated against other techniques such as tree-ring or uranium-series-disequilibrium techniques, which are more accurate but less widely applicable. The calibration continues to improve (e.g., Stuiver et al., 1998; Hughen et al., 2000, 2004). Instruments also improve. In particular, the **accelerator mass spectrometer** (AMS) radiocarbon analysis allows dating of milligram quantities of **foraminifers**, mollusks, and other biogenic materials. A single seed or tiny shell can be dated, and this analysis of smaller samples than was possible with previous techniques in turn allows finer time resolution in a single core. Taken together, these advances have greatly improved our ability to generate well-constrained age models for high-latitude marine sediment cores. In addition, coring systems such as the Calypso corer have been deployed in the Arctic to recover much longer (10–60 m) sediment cores. This corer allows sampling of relatively long time intervals even in sites where sediment has accumulated rapidly. Sites with faster sediment accumulation allow easier "reading" of the history of short-lived events, so higher resolution paleoenvironmental records can now be generated from high-latitude continental-margin and deep-sea sites. Where dates can be obtained from many levels in a core, it is feasible to evaluate centennial and even multidecadal variability from these **archives** (e.g., Ellison et al., 2006; Stoner et al., 2007).

However, in the Arctic, particularly along eastern margins of oceans where cold polar and Arctic water masses influence the environment, little carbonate that can be dated by radiocarbon techniques is produced, and much of the carbonate produced commonly dissolves after the producing organism dies. In addition, the carbon used in growing the shells is commonly "old" (that is, the carbon entered the ocean some decades or centuries before being used by the creature in growing its shell; the date obtained is approximately the time when the carbon entered the ocean, and it must be corrected for the time interval between the carbon entering the ocean and being incorporated into the shell). This marine reservoir correction is often more uncertain in the Arctic than elsewhere (e.g., Björck et al., 2003) in part because of the strong but time-varying effect of **sea ice**, which blocks exchange between atmosphere and ocean. This uncertainty continues to hamper development of highly constrained chronologies. Some important regions, such as near the eastern side of *Baffin Island*, have received little study since radiocarbon dating by accelerator mass spectrometry was introduced, so the chronology and **Holocene** climate evolution of this important margin are still poorly known.

As researchers attempt to develop centennial to multidecadal climate records from marine cores and to correlate between records at submillennial resolution, the limits of the dating method are often reached, hampering our ability to determine whether high-frequency variability is synchronous or asynchronous between sites. Resource limitations generally restrict radiocarbon dating to samples no closer together than about 500-year intervals. In marine areas with rapid biological production where sufficient biogenic carbonate is available to obtain highly accurate dates, the instrumental error on individual radiocarbon dates may be as small as ±20 years. But, in many Arctic archives, it is not possible to obtain enough carbonate material to achieve that accuracy, and many dates are obtained with standard deviations (one sigma) errors of ±80 years to a couple of centuries.

A new approach that uses a combination of paleomagnetic secular variation (PSV) records and radiocarbon dating has improved relative correlation and chronology well above the accuracy that each of these methods can achieve on its own (Stoner et al., 2007). Earth's magnetic field varies in strength and direction with time, and the field affects the magnetization of sediments deposited. Gross features in the field (reversals of direction) have been used for decades in the interpretation of geologic history, but much shorter lived, smaller features are now being used that allow correlation among different records by matching the features.

This technique was applied to two high-accumulation-rate Holocene cores from shelf basins on opposite sides of the *Denmark Strait*. The large number of tie points between cores provided by the paleomagnetic secular-variation records and by numerous radiocarbon dates allowed matching of these cores at the centennial scale (Stoner et al., 2007). In addition, the study has supported development of a well-dated Holocene paleomagnetic secular-variation record for this region (Figure 4.2), which can be used to aid in the dating of nearby lacustrine cores and for synchronization of marine and terrestrial

Many dates are obtained with standard deviations (one sigma) errors of ±80 years to a couple of centuries.

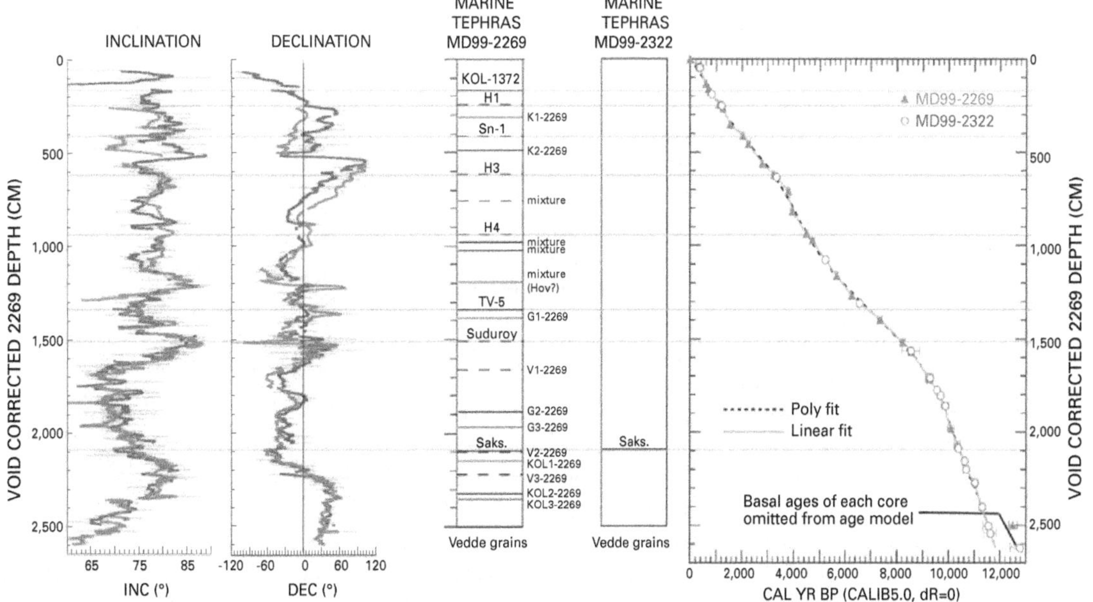

Figure 4.2. Paleomagnetic secular variations records (left), tephrochronology records (right), and calibrated radiocarbon ages for cores MD99-2269 and -2322 (center) provide a template for Holocene stratigraphy of the Denmark Straits region (after Stoner et al., 2007, and Kirstjansdottir et al., 2007). Solid lines = tephra horizons in core 2269. [Copyright 2007, reproduced by permission of American Geophysical Union.]

records. Traditionally, volcanic layers such as the Saksunarvatn **tephra** have been used as time markers for correlation, but they can be used only at the times of major eruptions and not between, whereas the new magnetic technique is continuous. The technique was tested by its ability to independently achieve the same correlations as the volcanic layer, and it functioned very well.

The Vedde Ash, a widely dispersed explosive Icelandic tephra, provides a 12,000-year-old constant-time horizon (an isochron) during the Younger Dryas cold period

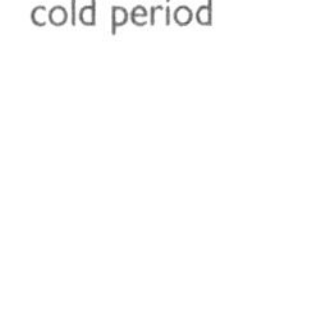

As noted above, tephra layers are an important source of chronological control in Arctic marine sediments. Explosive volcanic eruptions from Icelandic and Alaskan volcanoes have deposited widespread, geochemically distinct, tephra layers, each of which marks a unique time. Where the geochemistry of these events is documented, they provide **isochrones** that can be used to date and synchronize paleoclimate archives (e.g., marine, lacustrine, and ice-cores) and to evaluate leads and lags in the climate system. Where radiocarbon dates can be obtained at the same depth in a core as tephra layers, deviations of calibrated ages from the known age of a tephra can be used to determine the marine-reservoir age at that location and time (Eiriksson et al., 2004; Kristjansdottir, 2005; Jennings et al., 2006). An example is the Vedde Ash, a widely dispersed explosive Icelandic tephra that provides a 12,000-year-old constant-time horizon (an isochron) during the Younger Dryas cold period, when marine reservoir ages are poorly constrained and very different from today's. On the *North Iceland Shelf*, changes in the marine reservoir age are associated with shifts in the Arctic and polar fronts, which have important climatic implications (Eiriksson et al., 2004; Kristjansdottir, 2005). As many as 22 tephra layers have been identified in Holocene marine cores off north *Iceland* (Kristjansdottir et al., 2007). Eiriksson et al. (2004) recovered 10 known-age tephra layers of Holocene age. Some of the Icelandic tephras have wide geographic distributions either because they were ejected by very large explosive eruptions or because tephra particles were transported on sea ice whereas, nearer to their source, the tephra layers are more numerous and locally distributed. Transport on sea ice may spread the deposition time of a layer to months or years, but the layer will still remain a very short-interval time marker.

Most lake records span the time since deglaciation, typically the past 10,000 to 15,000 years.

4.3.2 Measurement of Rates of Change in Terrestrial Records

Terrestrial archives across the Arctic have been tapped to evaluate changes in the climate system in prehistoric times, with particular emphasis on changes in summer temperature, although moisture balance has been addressed in some studies. With sufficient age control, environmental proxies extracted from these archives can be used to evaluate rates of change. Archives that accumulate sediment in a regular and continuous pattern have the highest potential for reconstructing rates of change. The most promising archives are lake sediments and tree rings, both of which add material incrementally over time. Long-lived trees reach only to the fringes of the Arctic, so most reconstructions rely on climate proxies preserved in the sediments that accumulate in lake basins. Trees do extend to relatively high latitudes in *Alaska* and portions of the *Eurasian Arctic*, where they contribute high-resolution, usually annually resolved, paleoclimate records of the past several centuries, but they rarely exceed 400 years duration (Overpeck et al., 1997). The steady accumulation of calcium carbonate precipitates in caves may also provide a continuous paleoenvironmental record (Lauritzen and Lundberg, 2004), although these archives are relatively rare in the Arctic. This overview focuses on how well we can reconstruct times of rapid change in terrestrial sediment archives from the Arctic, focusing on changes that occurred on time scales of decades to centuries during the past 150,000 years or so, the late Quaternary.

Much of the terrestrial Arctic was covered by continental **ice sheets** during the **Last Glacial Maximum** (until about 15 ka), and large areas outside the ice sheet margins were too cold for lake sediment to accumulate. Consequently, most lake records span the time since deglaciation, typically the past 10,000 to 15,000 years. In a few Arctic regions, longer, continuous lacustrine records more than 100,000 years long have been recovered, and these rare records provide essential information about past environments and about rates of change in the more distant past (e.g., Lozhkin and Anderson, 1995; Brubaker et al., 2005; Hu et al., 2006; Brigham-Grette et al., 2007). In addition to these continuous records, discontinuous lake-sediment archives are found in formerly glaciated regions. These sites provide continuous records spanning several millennia through past warm times. In special settings, usually where the over-riding ice was very cold, slow-moving, and relatively thin, lake basins have preserved past sediment accumulations intact, despite subsequent over-riding by ice sheets during **glacial** periods (Miller et al., 1999; Briner et al., 2007).

The rarity of terrestrial archives that span the last **glaciation** hampers our ability to evaluate how rapid, high-magnitude changes seen in ice-core records (**Dansgaard-Oeschger**, or **D-O** events) and marine sediment cores (**Heinrich**, or H events) are manifested in the terrestrial arctic environment.

4.3.2A Climate Indicators and Ages

Deciphering rates of change from lake sediment, or any other geological archive, requires a reliable environmental **proxy** and a secure geochronology.

Climate and Environmental Proxies: Most high-latitude biological proxies record peak or average summer air temperatures. The most commonly employed paleoenvironmental proxies are biological remains, particularly pollen grains and the siliceous cell walls (frustules) of microscopic, unicellular algae called **diatoms**, which preserve well and are very abundant in lake sediment. In a summary of the timing and magnitude of peak summer warmth during the Holocene across the North American Arctic, Kaufman et al. (2004) noted that most records rely on pollen and plant macrofossils to infer growing-season temperature of terrestrial vegetation. Diatom assemblages primarily reflect changes in water chemistry, which also carries a strong environmental signal. More recently, biological proxies have expanded to include larval head capsules of non-biting midges (**chironomids**) that are well preserved in lake sediment. The distribution of the larval stages of chironomid taxa exhibit a strong summer-temperature dependence in the modern environment (Walker et al., 1997), which allows fossil assemblages to be interpreted in terms of past summer temperatures.

In addition to biological proxies that provide information about past environmental conditions, a wide range of physical and geochemical tracers also provide information about past environments. Biogenic silica (mostly produced by diatoms), organic carbon (mostly derived from the decay of aquatic organisms), and the isotopes of carbon and nitrogen in the organic carbon residues can be readily measured on small volumes of sediment, allowing the generation of closely spaced data—a key requirement for detecting rapid environmental change. Some lakes have sufficiently high levels of calcium and carbonate ions that calcium carbonate precipitates in the sediment. The isotopes of carbon and oxygen extracted from calcium carbonate deposits in lake sediment offer proxies of past temperatures and precipitation, and they have been used to reconstruct times of rapid climate change at high latitudes (e.g., Hu et al., 1999b).

Deciphering rates of change from lake sediment, or any other geological archive, requires a reliable environmental proxy and a secure geochronology.

Promising new developments in molecular **biomarkers** (Hu et al., 1999a; Sauer et al., 2001; Huang et al., 2004; D'Andrea and Huang, 2005) offer the potential of a wide suite of new climate proxies that might be measured at relatively high resolution as instrumentation becomes increasingly automated.

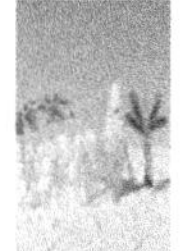

Dating Lake Sediment: In addition to the extraction of paleoenvironmental proxies at sufficient resolution to identify rapid environmental changes in the past, a secure geochronology also must be developed for the sedimentary archive. Methods for developing a secure depth-age relationship generally falls into one of three categories: direct dating, identification of key stratigraphic markers dated independently at other sites, and dating by correlation with an established record elsewhere. Much similarity exists between the techniques applied in lakes and in marine environments, although some differences do exist.

In addition to biological proxies that provide information about past environmental conditions, a wide range of physical and geochemical tracers also provide information about past environments.

Direct Dating: The strengths and weaknesses of various dating methods applied to Arctic terrestrial archives have been reviewed recently (Abbott and Stafford, 1996; Oswald et al., 2005; Wolfe et al., 2005). Radiocarbon is the primary dating method for archives dating from the past 15,000 years and sometimes beyond, although

For young sediment (20th century), the best dating methods are ^{210}Pb (age range of about 100–150 years) and identification of the atmospheric nuclear testing spike of the early 1960s, usually either with peak abundances of ^{137}Cs, $^{239,240}Pu$ or ^{241}Am.

conditions endemic to the Arctic (and described next) commonly prevent application of the technique back as far as 40,000 to 50,000 years, the limit achieved elsewhere. The primary challenge to accuracy of radiocarbon dates in Arctic lakes is the low primary productivity of both terrestrial and aquatic vegetation throughout most of the Arctic, coupled with the low rate at which organic matter decomposes on land. These two factors work together so that dissolved organic carbon incorporated into lake sediment contains a considerable proportion of material that grew on land, was stored on land for long times, and was then washed into the lake. The carbon in this terrestrial in-wash is much older than the sediment in which it is deposited, and it produces dissolved-organic-carbon ages that are anomalously old by centuries to millennia (Wolfe et al., 2005). Dissolved organic carbon contains many compounds, including humic acids; these acids tend to have the lowest reservoir ages among the compounds and so are most often targeted when no other options are available.

The large and variable reservoir age of dissolved organic carbon has led most researchers to avoid it for dating, and instead they concentrate on sufficiently large, identifiable organic remains such as seeds, shells, leaves, or other materials, typically called macrofossils. Macrofossils of things living on land, such as land plants, almost always yield accurate radiocarbon ages because the carbon in the plant was fully and recently exchanged (equilibrated) with the atmosphere. Similarly, aquatic plants are equilibrated with the carbon in the lake water, which for most lakes is equilibrated with the atmosphere. However, some lakes contain sufficient calcium carbonate, which typically contains old carbon not equilibrated with the atmosphere, such that the ^{14}C activity of the lake water is not in equilibrium with the atmosphere, a fundamental assumption for accurate radiocarbon dating. In these settings, known as hard-water lakes, macrofossils of terrestrial origin are targeted for dating. In lakes without this hard-water effect, either terrestrial or aquatic macrofossils may be targeted. Although macrofossil dates have been shown to be more reliable than bulk-carbon dates in Arctic lakes, in many instances terrestrial macrofossils washed into lake basins are derived from stored reservoirs (older rocks or sediments) in the landscape and have radiocarbon ages hundreds of years older than the deposition of the enclosing lake sediments.

Tephra layers provide time-synchronous marker horizons that can be used to constrain the geochronology of lacustrine sediment records.

For young sediment (20th century), the best dating methods are **^{210}Pb** (age range of about 100–150 years) and identification of the atmospheric nuclear testing spike of the early 1960s, usually either with peak abundances of **^{137}Cs**, **^{239,240}Pu** or **^{241}Am**. These methods usually provide high-precision age control for sediments deposited within the past century.

Some lakes preserve annual laminations, owing to strong seasonality in either biological or physical parameters. If laminations can be shown to be annual, chronologies can be derived by counting the number of annual laminations, or varves (Francus et al., 2002; Hughen et al., 1996; Snowball et al., 2002).

For late **Quaternary** sediments beyond the range of radiocarbon dating, dating methods include optically stimulated luminescence (OSL) dating, amino acid racemization (AAR) dating, cosmogenic radionuclide (CRN) dating, uranium-series disequilibrium (U-series) dating and, for volcanic sediment, potassium-argon or argon-argon (K-Ar or $^{40/39}$Ar) dating (e.g., Bradley, 1999; Cronin, 1999). With the exception of U-series dating, none of these methods has the precision to accurately date the timing of rapid changes directly. But these methods are capable of defining the time range of a sediment package and, if reasonable assumptions can be made about sedimentation rates, then the rate at which measured proxies changed can be derived within reasonable uncertainties. U-series dating has stringent depositional-system requirements that must be met to be applicable. For the terrestrial realm, calcium carbonate accumulations precipitated in a regular fashion in caves (flowstones, stalagmites, stalactites) offer the optimal materials. In these instances, high-precision ages can be derived for the entire Late Quaternary time period.

STRATIGRAPHIC MARKERS: As noted in the previous subsection, the Arctic includes major centers of volcanism in the *NORTH ATLANTIC* (*ICELAND*) and the North Pacific (*ALASKA* and Kamchatka) sectors. Explosive volcanism from both regions can produce large volumes of source- and time-diagnostic **tephra** distributed extensively across the Arctic. These tephra layers provide time-synchronous marker horizons that can be used to constrain the geochronology of lacustrine sediment records. The tephra layers can also serve to precisely synchronize records derived from lacustrine, marine, and ice-sheet archives, thereby allowing a better assessment of leads and lags in the climate system and the phasing of abrupt changes identified in different archives. Most tephras have diagnostic geochemical signatures that allow them to be securely identified with a source and, with modest age constraints, to a given eruptive event. If that event is well dated in regions near the source, such tephras then become dating tools in a technique known as tephrachronology.

As indicated in section 4.3.1, systematic centennial to millennial changes in Earth's magnetic field (paleomagnetic secular variation) (Figure. 4.2) have been used to correlate between several high-latitude lacustrine sedimentary archives and between marine and lacustrine records in the same region (Snowball et al., 2007; Stoner et al., 2007). Lacustrine records of paleomagnetic secular variation calibrated with varved sediments have been used for dating in Scandinavia (Saarinen, 1999; Ojala and Tiljander, 2003; Snowball and Sandgren, 2004)]. Recent work on marine sediments suggests that paleomagnetic secular variation can provide a useful means of correlating marine and terrestrial records.

"WIGGLE MATCHING": In some instances, very high resolution down-core analytical profiles from sedimentary archives with only moderate age constraints can be conclusively correlated with a well-dated high-resolution record at a distant locality, such as *GREENLAND* ice core records, with little uncertainty. Although the best examples of such correlations are not from the Arctic (e.g., Hughen et al., 2004a), this method remains a potential tool for providing age control for Arctic lake sediment records.

4.3.2B Potential for Reconstructing Rates of Environmental Change in the Terrestrial Arctic

A goal of paleoclimate research is to understand rapid changes on human time scales of decades to centuries. The major challenges in meeting this goal for the Arctic include uncertainties in the time scales of terrestrial archives and in the interpretation of various environmental proxies. Although uncertainties are widespread in both aspects, neither presents a fundamental impediment to the primary goal, quantifying rates of change.

PRECISION VERSUS ACCURACY: Many Arctic lake archives are dated with high precision, but with greater uncertainty in their accuracy. One can say, for example, that a particular climate change recorded in a section of core occurred within a 500-year interval with little uncertainty, but the exact age of the start and end of that 500-year interval are much less certain. This uncertainty is due to systematic errors in the proportion of old carbon incorporated into the humic acid fraction of the dissolved organic carbon used to date the lake sediment. Although this fraction, or "reservoir age," varies through the Holocene, changes in the reservoir age occur relatively slowly.

A goal of paleoclimate research is to understand rapid changes on human time scales of decades to centuries.

Figure 4.3 shows a segment of a sediment core from the eastern *CANADIAN ARCTIC*, for which six humic acid dates define an age-depth relation with an uncertainty of only ±65 years, but the humic acid ages are systematically 500–600 years too old. In this situation, rates of change for decades to centuries can be calculated with confidence, although determining whether a rapid change at this site correlated with a rapid change elsewhere is much less certain owing to the large uncertainty in the accuracy of the humic acid dates.

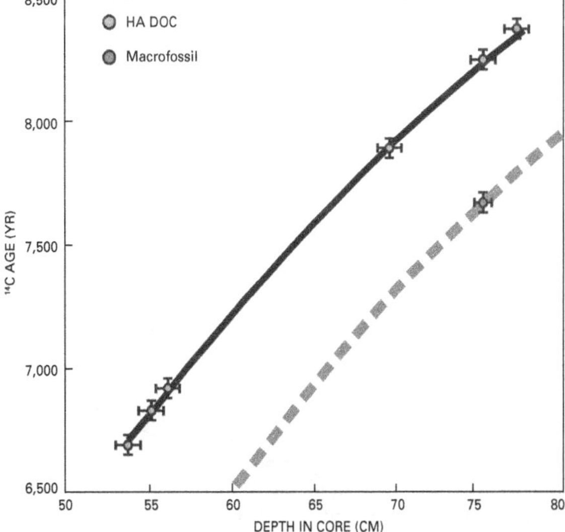

Figure 4.3. Precision versus accuracy in radio-carbon dates. Blue circle = accelerated mass spectrometry (AMS) ^{14}C date on the humic acid (HA) fraction of the total dissolved organic carbon (DOC) extracted from a sediment core from the eastern Canadian Arctic. Red circle = AMS ^{14}C date on macrofossil of aquatic moss from 75.6 cm, the same stratigraphic depth as a HA-DOC date. Dashed line is the best estimate of the age-depth model for the core. Samples taken 1–2 cm apart for HA-DOC dates show a systematic down-core trend suggesting that the precision is within the uncertainty of the measurements (±40 to ±80 years), whereas the discrepancy between macrofossil and HA-DOC dates from the same stratigraphic depth demonstrates an uncertainty in the accuracy of the HA-DOC ages of nearly 600 years. Data from Miller et al. (1999).

Ice-core records have figured especially prominently in the discussion of rates of change during the time interval for which such records are available.

Figure 4.4 similarly provides an example of rapid change in an environmental proxy in an Arctic lake sediment core, for which the rate of change can be estimated with certainty, but the timing of the change is less certain.

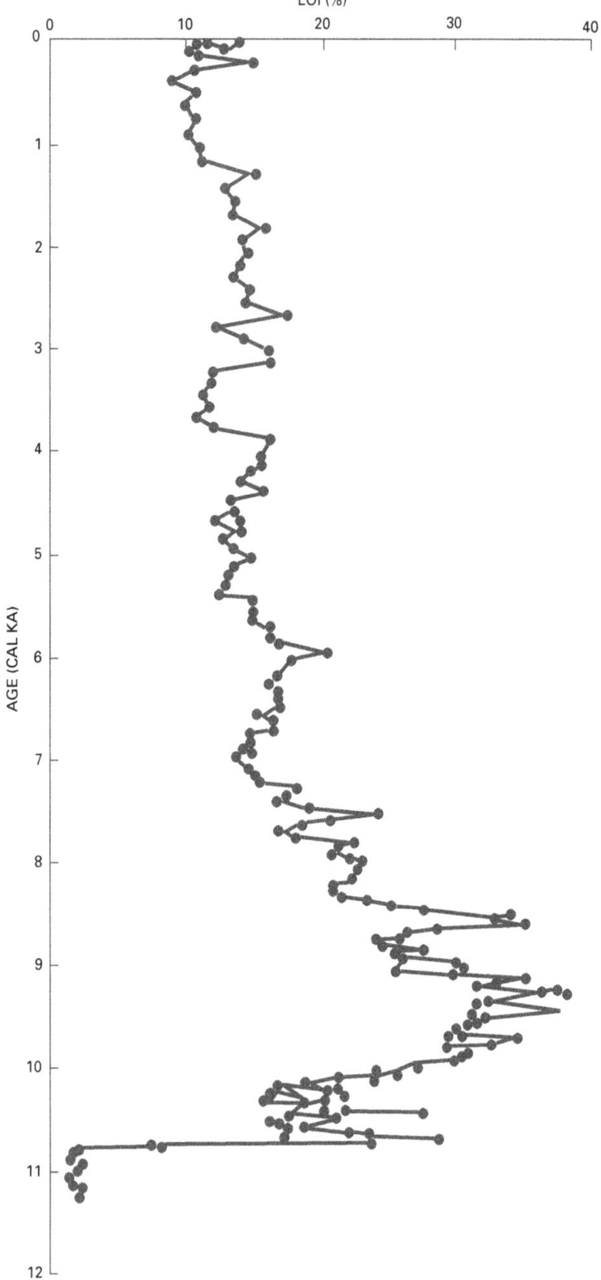

Figure 4.4. Down-core changes in organic carbon (measured as loss-on-ignition (LOI)) in a lake sediment core from the eastern Canadian Arctic. At the base of the record, organic carbon increased sharply from about 2% to greater than 20% in less than 100 years, but the age of the rapid change has an uncertainty of 500 years. Data are from Briner et al. (2006). [Copyright 2006, reproduced by permission of Elsevier.]

4.3.3 Measurement of Rates of Change in Ice-Core Records

Ice-core records have figured especially prominently in the discussion of rates of change during the time interval for which such records are available. One special advantage of ice cores is that they collect climate indicators from many different regions. In central *Greenland*, for example, the dust trapped in ice cores has been isotopically and chemically fingerprinted: it comes from central Asia (Biscaye et al., 1997), the methane has widespread sources in Arctic and in low latitudes (e.g., Harder et al., 2007), and the snowfall rate and temperature are primarily local indicators (see review by Alley, 2000). This aspect of ice-core records allows one to learn whether climate in widespread regions changed at the same time or different times and to obtain much better time resolution than is available by comparing individual records and accounting for the associated uncertainties in their dating.

Ice cores also exhibit very high time resolution. In many *Greenland* cores, individual years are recognized so that sub-annual dating is possible. Some care is needed in the interpretation. For example, the template for the history of temperature change in an ice core is typically the stable-isotope composition of the ice. (The calibration of this template to actual temperature is achieved in various ways, as discussed in Chapter 5, History of the Greenland Ice Sheet, but the major changes in the isotopic ratios correlate with major changes in temperature with very high confidence, as discussed there.) However, owing to post-depositional processes such as **diffusion** in **firn** and ice (Johnsen, 1977; Whillans and Grootes, 1985; Cuffey and Steig, 1998; Johnsen et al., 2000), the resolution of the isotope records does decrease with increasing age and depth. Initially the decrease is due to processes in the porous firn, and later it is due to more rapid diffusion in the warmer ice close to the bottom of the ice sheet. The isotopic resolution may reveal individual storms shortly after deposition but be smeared into several years in ice tens of thousands of years old. Normally in *Greenland*, accumulation rates of less than about 0.2 m/yr of ice are insufficient to preserve annual cycles for more than a few decades; higher accumulation rates allow the annual layers to survive the transformation of low-density snow to high-density ice, and the cycles then survive for millennia before being gradually smoothed.

Records of dust concentration appear to be almost unaffected by smoothing processes, but some chemical constituents seem to be somewhat mobile and thus to have their records smoothed over a few years in older samples (Steffensen et al., 1997; Steffensen and Dahl-Jensen, 1997). Unfortunately, despite important recent progress (Rempel and Wettlaufer, 2003), the processes of chemical diffusion are not as well understood as are isotopic ratios, so confident modeling of the chemical diffusion is not possible and the degree of smoothing is not as well quantified as one would like. Persistence of relatively sharp steps in old ice that is still in normal stratigraphic order demonstrates that the diffusion is not extensive. The high-resolution features of the dust and chemistry records have been used to date the glacial part of the ***GISP2*** core by using mainly annual cycles of dust (Meese et al., 1997) and the ***NGRIP*** core by using annual layers in different ionic constituents together with the visible dust layers (cloudy bands; Figure 4.5) back to 42 **ka** (Andersen et al., 2006, Svensson et al., 2006). Figure 4.5 shows the visible cloudy bands in a 72 ka section of the *NGRIP* core. The cloudy bands are generally assumed to be due to tiny gas bubbles that form on dust particles as the core is brought to surface. During storage of core in the laboratory, these bands fade somewhat. However, the very sharp nature of the bands when the core is recovered suggests that diffusive smoothing has not been important, and that high-time-resolution data are preserved.

4.4 CLASSES OF CHANGES AND THEIR RATES

The day-to-night and summer-to-winter changes are typically larger—but have less persistent effect on the climate—than long-lived features such as ice ages. This observation suggests that it is wise to separate rates of change on the basis of persistence. As discussed in Chapter 2, Paleoclimate Concepts, section 2.2 on **forcings**, effects from the aging of the Sun can be discounted on "short" time scales of 100 **m.y.** or less, but many other forcings must be considered. Several are discussed below. For the last ice-age cycle, special reliance is placed on *Greenland* ice-core records because of their high time resolution and confident paleothermometery. But *Greenland* is only a small part of the whole Arctic, and this limitation should be borne in mind.

4.4.1 Tectonic Time Scales

As also discussed in section 2.2 on forcings, drifting continents and related slow shifts in global biogeochemical cycling, together with evolving life forms, can have profound local and global effects on climate during tens of millions of years. If a continent moves from equator to pole, the climate of that continent will change greatly. In addition, by affecting ocean currents, ability to grow ice sheets, cloud patterns, and more, the moving continent may have an effect on global and regional climates as well, although this effect will in general be much more subtle than the effect on the continent's own climate (e.g., Donnadieu et al., 2006).

Drifting continents and related slow shifts in global biogeochemical cycling, together with evolving life forms, can have profound local and global effects on climate during tens of millions of years.

Figure 4.5. A linescan image of NGRIP ice core interval 2528.35–2530.0 m depth. Gray layers, annual cloudy bands; annual layers are about 1.5 cm thick. Age of this interval is about 72 ka, which corresponds with Greenland Interstadial 19 (Svensson et al., 2005). [Copyright 2005, reproduced by permission of American Geophysical Union.]

Within the last tens of millions of years, the primary direct effect of drifting continents on the Arctic probably has been to modify the degree to which the *Arctic Ocean* connects with the lower latitudes, by altering the "gateways" between land masses. The *Arctic Ocean*, primarily surrounded by land masses, has persisted throughout that time (Moran et al., 2006). Much attention has been directed to the possibility that the warmth of the Arctic during certain times, such as the **Eocene** (which began about 50 **Ma**), was linked to increased transport of ocean heat as compared with other, colder times. However, both models and data indicate that this possibility appears unlikely (e.g., Bice et al., 2000). The late Eocene *Arctic Ocean* appears to have supported a dense growth of pond weed (Azola), which is understood to grow in brackish waters (those notably fresher than full marine salinity) (Moran et al., 2006). A more-vigorous ocean circulation then would have introduced fully marine waters and would have transported the pond weed away. A great range of studies indicates that larger atmospheric carbon-dioxide concentrations during that earlier time were important in causing the warmth (Royer et al., 2007; Vandermark et al., 2007; and Tarduno et al., 1998).

The Arctic of about 50 Ma appears to have been ice free, at least near sea level, and thus minimum wintertime temperatures must have been above freezing.

The Arctic of about 50 **Ma** appears to have been ice free, at least near sea level, and thus minimum wintertime temperatures must have been above freezing. Section 6.3.1 includes some indications of temperatures in that time, with perhaps 20°C a useful benchmark for Arctic-wide average annual temperature. Recent values are closer to –15°C, which would indicate a cooling of roughly 35°C within about 50 m.y. The implied rate is then in the neighborhood of 0.7°C/million years or 0.0000007°C/yr. One could pick time intervals during which little or no change occurred, and intervals within the last 50 m.y. during which the rate of change was somewhat larger; a rough "**tectonic**" value of about 1°C/million years or less may be useful.

4.4.2 Orbital Time Scales

As described in section 2.2 on forcings, features of Earth's orbit cause very small changes in globally averaged incoming solar radiation (**insolation**) but large changes (more than 10%) in local sunshine. These orbital changes serve primarily to move sunshine from north to south and back or from poles to equator and back, depending on which of the orbital features is considered. The leading interpretation (e.g., Imbrie et al., 1993) is that ice sheets grow and the world enters an ice age when reduced summer sunshine at high northern latitudes allows survival of snow without melting; ice sheets melt, and the world exits an ice age, when greater summer sunshine at high northern latitudes melts snow there. Because the globally averaged forcing is nearly zero but the globally averaged response is large (e.g., Jansen et al., 2007), the Earth system must have strong amplifying processes (**feedbacks**). Changes in greenhouse-gas concentrations (especially carbon dioxide), how much of the Sun's energy is reflected (ice-albedo feedback, plus some changes in vegetation), and blocking of the Sun by dust are prominent in interpretations, and all appear to be required to explain the size and pattern of the reconstructed changes (Jansen et al., 2007).

The leading interpretation (e.g., Imbrie et al., 1993) is that ice sheets grow and the world enters an ice age when reduced summer sunshine at high northern latitudes allows survival of snow without melting.

The globally averaged change from ice-age to interglacial is typically estimated as 5°–6°C (e.g., Jansen et al., 2007). Changes in the Arctic clearly were larger. In central *Greenland*, typical glacial and interglacial temperatures differed by about 15°C, and the maximum warming from the most-recent ice age was about 23°C (Cuffey et al., 1995). Very large changes occurred where ice sheets grew during the ice age and melted during the subsequent warming, related to the cooling effect of the higher elevation of the ice sheets, but the elevation change is not the same as a climatic effect.

In central *Greenland*, the coldest time of the ice age was about 24 ka although, as discussed in Chapter 5, History of the Greenland Ice Sheet, some records place the extreme value of the most recent ice age slightly more recently. Kaufman et al. (2004) analyzed the timing of the peak warmth of the Holocene throughout broad regions of the Arctic; near the melting ice sheet on North America, peak warmth was delayed until most of the ice was gone, whereas far from the ice sheet peak warmth was reached before 8 ka, in some regions by a few millennia.

A useful order-of-magnitude estimate may be that the temperature change associated with the end of the ice age was about 15°C in about 15 thousand years (**k.y.**) or about 1°C/k.y.) or 0.001°C/yr, and peak rates were perhaps twice that. The ice-age cycle of the last few hundred thousand years is often described as consisting of about 90 k.y. of cooling followed by about 10 k.y. of warming, or something similar, implying faster warming than cooling (see Chapter 5, Figure 5.9). Thus, rates notably slower than 1°–2°C/k.y. are clearly observed at times.

Kaufman et al. (2004) indicated that the warmest times of the current or Holocene interglacial (**MIS** 1) in the western-hemisphere part of the Arctic were, for average land, 1.6±0.8°C above mean 20th-century values. Warmth peaked before 12 ka in western *Alaska* but after 3 ka in some places near Hudson Bay; a typical value is near 7–8 ka. Thus, the orbital signal during the Holocene has been less than or equal to approximately 0.2°C/k.y., or 0.0002°C/yr.

4.4.3 Millenial or Abrupt Climate Changes

Exceptional attention has been focused on the abrupt climate changes recorded in *Greenland* ice-cores and in many other records from the most recent ice age and earlier (see National Research Council, 2002; Alley et al., 2003; Alley, 2007).

The more recent of these changes has been well known for decades from many studies primarily in Europe that worked with lake and bog sediments and the **moraines** left by retreating ice sheets. However, most research focused on the slower ice-age cycles, which were easier to study in paleoclimatic archives.

The first deep ice core through the *Greenland Ice Sheet*, at *Camp Century* in 1966, produced a δ^{18}**O** isotope profile that showed unexpectedly rapid and strong climatic shifts through the entire last glacial period (Dansgaard et al., 1969, 1971; Johnsen et al., 1972). The fastest observed sharp transitions from cold to warm seemed to have been on the time scale of centuries, clearly much faster than **Milankovitch time scales**.

These results did not stimulate much additional research immediately; the record lay close to the glacier **bed**, and it may be that many investigators suspected that the records had been altered by ice-flow processes. There were, however, data from quite different archives pointing to the same possibility of large and rapid climate change. For example, the Grand Pile pollen profile (Woillard, 1978, 1979) showed that the last interglacial (MIS 5) ended rapidly during an interval estimated at 150±75 yrs, comparable to the *Camp Century* findings. The Grand Pile pollen data also pointed to many sharp warming events during the last ice age.

The next deep core in *Greenland* at the *Dye-3* radar station was drilled by the United States, Danish, and Swiss members of the *Greenland Ice Sheet* Program (Dansgaard et al., 1982). The violent climatic changes, as Willi Dansgaard termed them, matched the often-ignored *Camp Century* results. The cause for these strong climatic **oscillations** had already been hinted at by Ruddiman and Glover (1975) and Ruddiman and McIntyre (1981), who studied oceanic evidence for the large climatic oscillations involving strong warming into the **Bølling** interval, cooling into the Younger Dryas, and warming into the **Preboreal**. They assigned the cause for these strong climatic anomalies to **thermohaline circulation** changes combined with strong zonal winds partly driving the surface currents in the *North Atlantic*; these forces drove sharp north-south shifts of the polar front. In light of the ice core data, the oscillations around the Younger Dryas were part of a long row of similar events, which Dansgaard et al. (1984) and Oeschger et al. (1984) likewise assigned to circulation changes in the *North Atlantic*. Broecker et al. (1985) argued for bi-stable *North Atlantic* circulation as the cause for the *Greenland* climatic jumps.

The first deep ice core through the Greenland Ice Sheet, at Camp Century in 1966, produced a δ^{18}O isotope profile that showed unexpectedly rapid and strong climatic shifts through the entire last glacial period.

The results of the *Dye-3* core went a long way toward settling the issue of the existence of abrupt climate change. Further results from year-by-year ice sampling during the Younger Dryas warming from this same core pushed the definition of "abrupt" from the century time scale to

In the Holocene period, the approximately 160-year-long cold event about 8.2 ka, which produced 4°–5°C cooling in Greenland (Leuenberger et al., 1999), began in less than 20 years, and perhaps much less.

the decadal and nearly annual scale (Dansgaard et al., 1989). Alley et al. (1993) suggested the possibility that much of an abrupt change was completed in a single year for at least one climatic variable (snow accumulation at the *GISP2* site).

In addition to the *GISP2*, *GRIP*, and *Dye-3* cores, ice core evidence has been strengthened by new deep ice cores at Siple Dome in West Antarctica and *North-GRIP* in northern *Greenland*. New high-resolution measurement techniques have provided subannual resolution for several parameters, and these data have been used for the *North-GRIP* core to provide absolute dating, the GICC05 chronology, back to 60 ka (Svensson et al., 2005; Rasmussen et al., 2006; Vinther et al., 2006). The *GISP2* and *GRIP* ice cores have also been synchronized with the *North-GRIP* core through MIS 2 (Rasmussen et al., 2006).

The temperature shifts into the warm intervals in the millennial climate changes, which are called **interstadials** (Johnsen et al., 1992; Dansgaard et al., 1993), have been found to vary from 10° to 16°C on the basis of borehole thermometry (Cuffey et al., 1995; Johnsen et al., 1995; Jouzel et al., 1997) and of studies of the isotopic effect of thermal **firn** diffusion on gas isotopes (Severinghaus et al., 1998; Lang et al., 1999; Leuenberger et al., 1999; Landais et al., 2004; Huber et al., 2006).

The *North-GRIP* core, at the time of this report the most recent of the *Greenland* deep cores and the one on which the most effort was expended in counting annual layers, shows that typically the rapid warmings into interstadials are recorded as increases in only 20 years in the 20-year averages of isotopic values during MIS 2 and MIS 3; this information indicates temperature changes of 0.5°C/yr or faster.

In the Holocene period, the approximately 160-year-long cold event about 8.2 ka, which produced 4°–5°C cooling in *Greenland* (Leuenberger et al., 1999), began in less than 20 years, and perhaps much less. The cooling is believed to have been caused by the emptying of Lake Agassiz (reviewed by Alley and Ágústsdóttir, 2005), and the rapid transitions found bear witness to the dynamic nature of the *North Atlantic* circulation in jumping to a new mode.

The Younger Dryas and the 8.2 ka cold event (section 6.3.5A) are well known in Europe and in Arctic regions, but they appear to have been much weaker or absent in other Arctic regions (see reviews by Alley and Ágústsdóttir (2005) and Alley (2007); note that strong signals of these events are found in some but not all lower-latitude regions). The signal of the Younger Dryas did extend across the Arctic to *Alaska* (see Peteet, 1995a,b; Hajdas et al., 1998). Lake sediment records from the eastern *Canadian Arctic* contain evidence for both excursions (Miller et al., 2005).

The 8.2 ka event is recorded at two sites as a notable readvance of cirque glaciers and **outlet glaciers** of local **ice caps** at 8,200±100 years (Miller et al., 2005). In some lakes not dominated by runoff of meltwater from glaciers, a reduction in primary productivity is apparent at the same time. These records suggest that colder summers during the event without a dramatic reduction in precipitation produced positive mass balances and glacier re-advances. For most local glaciers, this readvance was the last important one before they receded behind their **Little Ice Age** margins. Organic carbon accumulation in a West *Greenland* lake sediment record suggests a decrease in biotic productivity synchronous with the negative $\delta^{18}O$ excursion in the *GRIP* ice core (Willemse and Törnqvist, 1999).

Few Arctic lakes contain records that extend through Younger Dryas time. And despite the strong signal indicative of rapid, dramatic Younger Dryas cooling in *Greenland* ice cores, no definitive records document or refute accompanying glacier expansion or cold around the edge of the *Greenland Ice Sheet* (Funder and Hansen, 1996; Björck et al., 2002) (discussed in Chapter 5, History of the *Greenland Ice Sheet*), near *Svalbard* (Svendson and Mangerud, 1992), or in *Arctic Canada* (Miller et al., 2005). These observations are consistent with the joint observations that the events primarily occurred in wintertime, whereas most paleoclimatic indicators are more sensitive to summertime conditions. Moreover, the events manifested primarily in the *North Atlantic* and surroundings, and their amplitude was reduced away from the *North Atlantic* (Denton et al., 2005; Alley, 2007; also see Björck et al., 2002). This means in turn that the rate of climate change associated with these events, although truly spectacular in the *North Atlantic*, was much smaller elsewhere (poorly constrained, but perhaps only one-tenth as large in many parts of the Arctic, and a region of zero temperature

change somewhere on the planet separated the northern regions of cooling from the southern regions of weak warming). The globally averaged signal in temperature change was weak, although in some regions rainfall seems to have changed very markedly (e.g., Cai et al., 2008).

4.4.4 Higher Frequency Events Especially in the Holocene

The Holocene record, although showing greatly muted fluctuations in temperature as compared with earlier times, is not entirely without variations. As noted above, a slow variation during the Holocene is linked with orbital forcing and decay of the great ice sheets. Riding on the back of this variation are oscillations of roughly 1°C or less, at various temporal spacings. Great effort has been expended in determining what is signal versus noise in these records, because the signals are so small, and issues of whether events are broadly synchronous or not become important.

A few rather straightforward conclusions can be stated with some confidence. Ice-core records from *Greenland* show the forcing and response of individual volcanic eruptions. A large explosive eruption caused a cooling of roughly 1°C in *Greenland*, and the cooling and then warming each lasted roughly 1 year (Grootes and Stuiver, 1997; Stuiver et al., 1997), although a cool "tail" lasted longer. Thus, the temperature changes associated with volcanic eruptions are strong, 1°C/year, but not sustained. Because volcanic eruptions are essentially random in time, accidental clustering in time can influence longer term trends **stochastically**.

The possible role of solar variability in Holocene changes (and in older changes; e.g., Braun et al., 2005) is of considerable interest. Ice-core records are prominent in reconstruction of solar forcing (e.g., Bard et al., 2007; Muscheler et al., 2007). Identification of climate variability correlated with solar variability then allows assessment of the solar influence and the rates of change caused by the solar variability.

The Holocene record, although showing greatly muted fluctuations in temperature as compared with earlier times, is not entirely without variations.

Much study has focused on the role of the Sun in the oscillations within the interval from the so-called **Medieval Climate Anomaly** through the Little Ice Age and the subsequent warming to recent conditions. The reader is especially referred to Hegerl et al. (2007). In *Greenland*, the Little Ice Age–Medieval Climate Anomaly oscillation had an amplitude of roughly 1°C. Attribution exercises show that much of this amplitude can be explained by volcanic forcing in response to the changing frequency of large eruptions (Hegerl et al., 2007). In addition, some of this temperature change might reflect oceanic changes (Broecker, 2000; Renssen et al., 2006), but some fraction is probably attributable to solar forcing (Hegerl et al., 2007). Human influences on the environment were measurable at this time and, thus, changes in land cover and small changes to **greenhouse gases** such as methane may have also played a role. Although the time from Medieval Climate Anomaly to Little Ice Age to recent warmth is about 1 millennium, there are warmings and coolings in that interval that suggest that the changes involved are probably closer to 1°C/century; some fraction of that change is attributable to solar forcing and some to volcanic and perhaps to oceanic processes. Because recent studies tend to indicate

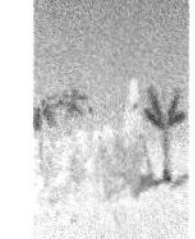

Rates of temperature changes associated with volcanic eruptions are strong, 1°C/year, but not sustained.

greater importance for volcanic forcing than for solar forcing (Hegerl et al., 2007), changes of 0.3°C/century may be a reasonable estimate of an upper limit for the solar forcing observed (but with notable uncertainty). Weak variations of the ice-core isotopic ratios that correlate with the sunspot cycles and other inferred solar periodicities similarly indicate a weak solar influence (Grootes and Stuiver, 1997; Stuiver et al., 1997). Whether a weak solar influence acting on millennial time scales is evident in poorly quantified paleoclimatic indicators (Bond et al., 2001) remains a hotly debated topic. The ability to explain the Medieval Climate Anomaly–Little Ice Age oscillation without appeal to such a periodicity and the evidently very small role of any solar forcing in those events largely exclude a major role for such millennial oscillations in the Holocene.

The warming from the Little Ice Age extends into the instrumental record.

The warming from the Little Ice Age extends into the instrumental record, generally consistent with the considerations above. In the instrumental data (Parker et al., 1994; also see Delworth and Knutson, 2000), the Arctic sections, particularly the *North Atlantic sector*, show warming of roughly 1°C in the first half of the 20th century (and with peak warming rates of twice that average). The warming likely arose from some combination of volcanic, solar, and human (McConnell et al., 2007) forcing, and perhaps some oceanic forcing. The warming was followed by weak cooling and then a similar warming in the latter 20th century (roughly 1°C per 30 years) primarily attributable to human forcing with little and perhaps opposing natural forcing (Hegerl et al., 2007).

As noted in section 2.2 on forcings (see above; also see Bard and Delaguye, 2008), the lack of correlation between indicators of climate and indicators of past magnetic-field strength, or between indicators of climate and indicators of in-fall rate of extraterrestrial materials, means that any role of these possible forcings must be minor and perhaps truly zero.

4.5 SUMMARY

The discussion in the previous section produced estimates of peak rates of climate change associated with different causes. These estimates are plotted in a summary fashion in Figure 4.6. As one goes to longer times, the total size of changes increases, but the rate of change decreases. Such behavior is unsurprising; a sprinter changes position very rapidly but does not sustain the rate, so that in a few hours the marathon runner covers more ground. To illustrate this concept, regression lines were added through the tectonic, ice-age, volcano, volcanoes, and solar points; abrupt climate changes and human-caused changes were omitted from this regression because of difficulty in estimating an Arctic-wide value.

The local effects of the abrupt climate changes in the *North Atlantic* are clearly anomalous compared with the general trend of the regression lines, and changes were both large and rapid. These events have commanded much scientific attention for precisely this reason. However, globally averaged, these events are unimpressive: they fall well below the regression lines, thus demonstrating clearly the difference between global and regional behavior. An Arctic-wide assessment of abrupt climate changes would yield rates of change that would plot closer to the regression lines than do either the local *Greenland* or global values.

Thus far, human influence does not stand out relative to other, natural causes of climate change. However, the projected changes can easily rise above those trends, especially if human influence continues for more than a hundred years and rises above the **IPCC** "mid-range" A1B scenario. No generally accepted way exists to formally assess the effects or importance of size versus rate of climate change, so no strong conclusions should be drawn from the observations here.

The data clearly show that strong natural variability has been characteristic of the Arctic at all time scales considered.

The data clearly show that strong natural variability has been characteristic of the Arctic at all time scales considered. The data suggest the twin hypotheses that the human influence on rate and size of climate change thus far does not stand out strongly from other causes of climate change, but that projected human changes in the future may do so.

The report here relied much more heavily on ice-core data from *Greenland* than is ideal in assessing Arctic-wide changes. Great opportunities exist for generation and synthesis of other data sets to improve and extend the results here, using the techniques described in this chapter. If widely applied, such research could remove the over-reliance on *Greenland* data.

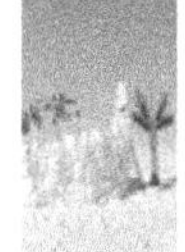

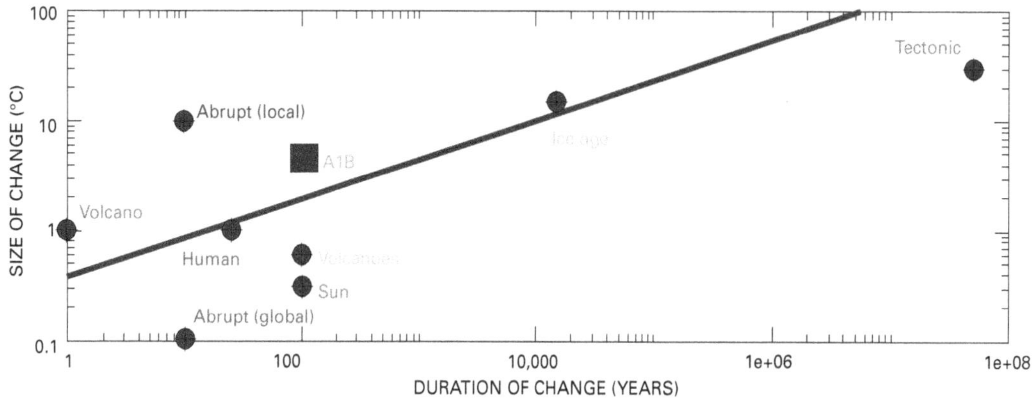

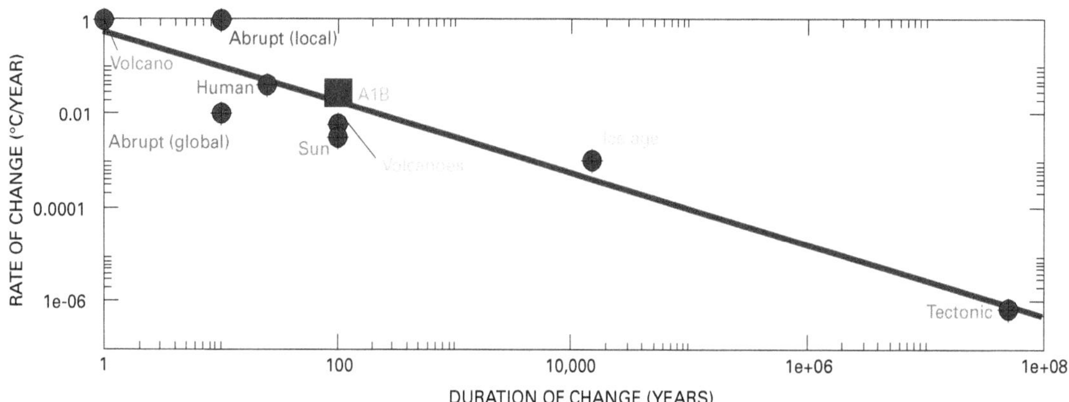

Figure 4.6. Summary of estimated peak rates of change and sizes of changes associated with various classes of cause. Error bars are not provided because of difficulty of quantifying them, but high precision is not implied. Both panels have logarithmic scales on both axes (log-log plots) to allow the huge range of behavior to be shown in a single figure. The natural changes during the Little Ice Age–Medieval Climate Anomaly have been somewhat arbitrarily partitioned as 0.6°C for changes in volcanic-eruption frequency (labeled "volcanoes" to differentiate from the effects of a single eruption, labeled "volcano"), and 0.3°C for solar forcing to provide an upper limit on solar causes; a larger volcanic role and smaller solar role would be easy to defend (Hegerl et al., 2007), but a larger solar role is precluded by available data and interpretations. The abrupt climate changes are shown for local Greenland values and for a poorly constrained global estimate of 0.1°C. These numbers are intended to represent the Arctic as a whole, but much Greenland ice-core data have been used in determinations. The instrumental record has been used to assess human effects (see Delworth and Knutson, 2000 and Hegerl et al., 2007). The "human" contribution may have been overestimated and natural fluctuations may have contributed to the late-20th-century change, but one also cannot exclude the possibility that the "human" contribution was larger than shown here and that natural variability offset some of the change. The ability of climate models to explain widespread changes in climate primarily on the basis of human forcing, and the evidence that there is little natural forcing during the latter 20th century (Hegerl et al., 2007), motivate the plot as shown. Also included for scaling is the projection for the next century (from 1980–1999 to 2080–2099 means) for the IPCC SRES A1B emissions scenario (one often termed "middle of the road") scaled from Figure 10.7 of Meehl et al. (2007); see also Chapman and Walsh (2007). This scenario is shown as the black square labeled A1B; a different symbol shows the fundamental difference of this scenario-based projection from data-based interpretations for the other results on the figure. Human changes could be smaller or larger than shown as A1B, and they may continue to possibly much larger values further into the future. There is no guarantee that human disturbance will end before the end of the 21st century, as plotted here. The regression lines pass through tectonic, ice-age, solar, volcano, and volcanoes; they are included solely to guide the eye and not to imply mechanisms.

History of the Greenland Ice Sheet

Lead Author: Richard B. Alley*, Pennsylvania State University, University Park, PA

Contributing Authors: John T. Andrews, University of Colorado, Boulder, CO; Garry K.C. Clarke, University of British Columbia, Vancouver, British Columbia, Canada; Kurt M. Cuffey, University of California, Berkeley, CA; Svend Funder, University of Copenhagen, Denmark; Shawn J. Marshall, University of Calgary, Alberta, Canada; Jerry X. Mitrovica, University of Toronto, Ontario, Canada; Daniel R. Muhs, U.S. Geological Survey, Denver, CO; Bette Otto-Bliesner, National Center for Atmospheric Research, Boulder, CO

*SAP 1.2 Federal Advisory Committee Member

ABSTRACT

The Greenland Ice Sheet is expected to shrink or disappear with warming, a conclusion based on a survey of paleoclimatic and related information. Recent observations show that the Greenland Ice Sheet has melted more in years with warmer summers. Mass loss by melting is therefore expected to increase with warming. But whether the ice sheet shrinks or grows, and at what pace, depend also on snowfall and iceberg production. The Arctic is a complicated system. Reconstructions of past climate and ice sheet configuration (the "paleo-record") are valuable sources of information that complement process-based models. The paleo-record shows that the Greenland Ice Sheet consistently lost mass when the climate warmed and grew when the climate cooled. Such changes have occurred even at times of slow or zero sea-level change, so changing sea level cannot have been the cause of at least some of the ice-sheet changes. In contrast, there are no documented major ice-sheet changes that occurred independent of temperature changes. Moreover, snowfall has increased when the climate warmed, but the ice sheet lost mass nonetheless; increased accumulation in the ice sheet's center has not been sufficient to counteract increased melting and flow near the edges. Most documented forcings of change, and the changes to the ice sheet themselves, spanned periods of several thousand years, but limited data also show rapid response to rapid forcings. In particular, regions near the ice margin have responded within decades. However, major changes of central regions of the ice sheet are thought to require centuries to millennia. The paleo-record does not yet strongly constrain how rapidly a major shrinkage or nearly complete loss of the ice sheet could occur. The evidence suggests nearly total loss may result from warming of more than a few degrees above mean 20th century values, but this threshold is poorly defined (perhaps as little as 2°C or more than 7°C). Paleoclimatic records are sufficiently sketchy that the ice sheet may have grown temporarily in response to warming, or changes may have been induced by factors other than temperature, without having been recorded.

5.1 THE GREENLAND ICE SHEET

5.1.1 Overview

The Greenland Ice Sheet contains by far the largest volume of any present-day Northern Hemisphere ice mass.

The *Greenland Ice Sheet** (Figure 5.1) contains by far the largest volume of any present-day Northern Hemisphere ice mass. The **ice sheet** is approximately 1.7 million square kilometers (km^2) in area, extending as much as 2,200 km north to south. The maximum ice thickness is 3,367 m, its average thickness is 1,600 m (Thomas et al., 2001), and its volume is 2.9 million km^3 (Bamber et al., 2001). Some of the bedrock beneath this ice has been depressed below sea level by the weight of the ice, and a little of this bedrock would remain below sea level following removal of the ice and rebound of the bedrock (Bamber et al., 2001). However, most of the ice that rests on bedrock is above sea level and so would contribute to sea-level rise if it were melted: if the entire ice sheet melted, it is estimated that sea-level would rise about 7.3 m (Lemke et al., 2007).

More-recent trends are toward warming temperatures, increasing snowfall, and more rapidly increasing meltwater runoff.

The ice sheet consists primarily of old snow that has been squeezed to ice under the weight of new snow that accumulates every year. Snow accumulation on the upper surface tends to increase ice-sheet size. Ice sheets lose mass primarily by melting in low-elevation regions, and by forming icebergs that break off the ice margins (**calving**) and drift away to melt elsewhere. **Sublimation**, snowdrift (Box et al., 2006), and melting or freezing at the **bed** beneath the ice are minor terms in the budget, although melting beneath floating extensions called ice shelves before icebergs break off may be important (see section 5.1.2, below).

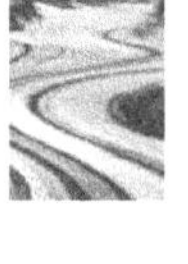

Estimates of net snow accumulation on the *Greenland Ice Sheet* have been presented by Hanna et al. (2005) and Box et al. (2006), among others. Hanna et al. (2005) found for 1961–1990 (an interval of moderately stable conditions before more-recent warming) that surface snow accumulation (precipitation minus evaporation) was about 573 gigatons per year (Gt/yr) and that 280 Gt/yr of meltwater left the ice sheet. The difference of 293 Gt/yr is similar to the estimated iceberg **calving** flux within broad uncertainties (Reeh, 1985; Bigg, 1999; Reeh et al., 1999). (For reference, return of 360 Gt of ice to the ocean would raise global sea level by 1 millimeter (mm); Lemke et al., 2007.) More-recent trends are toward warming temperatures, increasing snowfall, and more rapidly increasing meltwater runoff (Hanna et al., 2005; Box et al., 2006). Large **interannual variability** causes the statistical significance of many of these trends to be relatively low, but the independent trends exhibit internal consistency (e.g., warming is expected to increase both melting and snowfall, on the basis of modeling experiments and simple physical arguments, and both trends are observed in independent studies (Hanna et al., 2005; Box et al., 2006)).

Increased iceberg calving has also been observed in response to faster flow of many **outlet glaciers** and shrinkage or loss of ice shelves (see section 5.1.2, below, for discussion of the parts of an ice sheet) (e.g., Rignot and Kanagaratnam, 2006; Alley et al., 2005). The Intergovernmental Panel on Climate Change (**IPCC**; Lemke et al., 2007) found that "Assessment of the data and techniques suggests a mass balance of the *Greenland Ice Sheet* of between +25 and –60 Gt (–0.07 to 0.17 mm) **SLE** (**sea level equivalent**) per year from 1961–2003 and –50 to –100 Gt (0.14 to 0.28 mm SLE) per year from 1993–2003, with even larger losses in 2005." Updates are provided by Alley et al. (2007b) (Figure 5.2) and by Cazenave (2006). Rapid changes have been occurring in the ice sheet, and in the ability to observe the ice sheet, so additional updates are virtually certain to be produced.

The long-term importance of these trends is uncertain—short-lived **oscillation** or harbinger of further shrinkage? This uncertainty motivates some of the interest in the history of the ice sheet.

*For bold terms, refer to Glossary; for italic terms, refer to Plate 1; for geologic ages, refer to Plate 2.

Figure 5.1. Satellite image (SeaWiFS) of the Greenland Ice Sheet and surroundings, from July 15, 2000 (*http://www.gsfc.nasa.gov/gsfc/earth/pictures/earthpic.htm*).

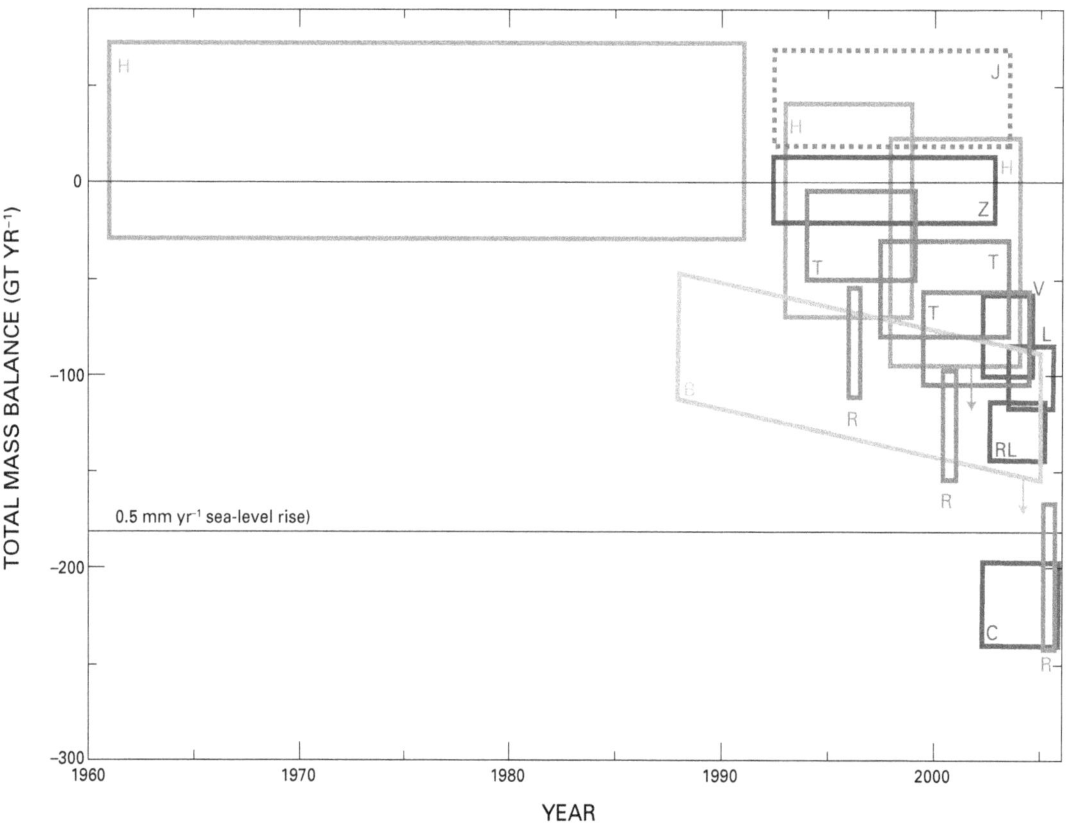

Figure 5.2. Recently published estimates of the mass balance of the Greenland Ice Sheet through time (modified from Alley et al., 2007b). A total mass balance of 0 indicates neither growth nor shrinkage, and –180 Gt yr^{-1} indicates ice-sheet shrinkage contributing to sea-level rise of 0.5 mm/yr, as indicated. Each box extends from the beginning to the end of the time interval covered by the estimate; the upper and lower lines indicate the uncertainties in the estimates. A given color is associated with a particular technique, and the different letters identify different studies. Two estimates have arrows attached, because those authors indicated that the change is probably larger than shown. The dotted box in the upper right is a frequently cited study that applies only to the central part of the ice sheet, which is thickening, and misses the faster thinning in the margins. [Reprinted from Annals of Glaciology by permission of International Glaciological Society.]

5.1.2 Ice-Sheet Behavior

Where delivery of snow or ice (typically as snowfall) exceeds removal (typically by meltwater runoff), a pile of ice develops. Such a pile that notably deforms and flows is called a **glacier**, **ice cap**, or ice sheet. (For a more comprehensive overview, see Paterson, 1994; Hughes, 1998; Van der Veen, 1999; or Hooke, 2005, among well-known texts.) Use of these terms is often ambiguous. "Glacier" most typically refers to a relatively small mass in which flow is directed down one side of a mountain, whereas "ice cap" refers to a small mass with flow diverging from a central dome or ridge, and "ice sheet" to a very large ice cap of continental or subcontinental scale. A faster moving "jet" of ice flanked by slower flowing parts of an ice sheet or ice cap may be referred to as an **ice stream**, but also as an outlet glacier or simply glacier (especially if the configuration of the underlying bedrock is important in delineating the faster moving parts), complicating terminology. Thus, the prominent *Jakobshavn Glacier* (Jakobshavn Isbrae, or Jakobshavn ice stream) is part of the ice sheet on *Greenland*, flowing in a deep bedrock trough but with slower moving ice flanking the faster-moving ice near the surface.

A glacier or ice sheet spreads under its own weight, deforming internally. The deformation rate increases with the cube of the **driving stress**, which is proportional to the ice thickness and to the surface slope of the ice. Ice may also move by sliding across the interface between the bottom of the ice and what lies beneath it, i.e., its substrate. Ice motion is typically slow or zero where the ice is frozen to the substrate, but is faster where the ice-substrate interface is close to the melting point. Ice motion can also take place through the deformation of subglacial sediments. This mechanism is important only where subglacial sediments are present and thawed. The contribution of these basal processes ranges from essentially zero to almost all of the total ice motion. Except for floating ice shelves (see below in this section), *Greenland*'s ice generally does not exhibit the gross dominance by basal processes seen in some West Antarctic ice streams.

Most glaciers and ice sheets tend toward a steady configuration. Snow accumulation in higher, colder regions supplies mass, which flows to lower, warmer regions where mass is lost by melting and runoff of the meltwater or by calving of icebergs that drift away to melt elsewhere.

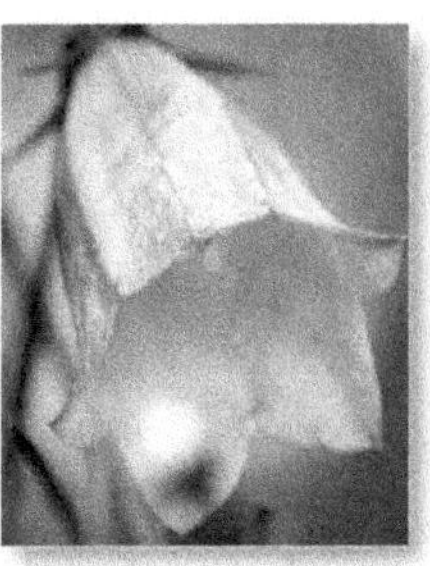

Some ice masses tend to an oscillating condition, marked by ice buildup during a period of slow flow, and then a short-lived surge of rapid ice flow; however, under steady climatic conditions, these oscillations repeat with some regularity and without huge changes in the average size across cycles.

Accelerations in ice flow, whether as part of a surging cycle, or in response to long-term ice-sheet evolution or climatically forced change, may occur through several mechanisms. These mechanisms include thawing of a formerly frozen bed, increase in meltwater reaching the bed causing increased lubrication (Joughin et al., 1996; Zwally et al., 2002; Parizek and Alley, 2004), and changes in meltwater drainage causing retention of water at the base of the glacier, which increases lubrication (Kamb et al., 1985). Ice-flow slowdown can similarly be induced by reversing these causes.

Recently, attention has been focused on changes in ice shelves. Where ice flows into a bordering water body, icebergs may calve from grounded (non-floating) ice. Alternatively, the flowing ice may remain attached to the glacier or ice sheet as it flows into the ice-marginal body of water. The attached ice floats on the water and calves from the end of the floating extension, which is called an **ice shelf**. Ice shelves frequently run aground on local high spots in the bed of the water body on which they float. Ice shelves that occupy embayments or fjords may rub against the rocky or icy sides, and friction from this restrains, or "buttresses," ice flow. Loss of this buttressing through shrinkage or loss of an ice shelf then allows faster flow of the ice feeding the ice shelf (Payne et al., 2004; Dupont and Alley, 2005, 2006).

Although numerous scientific papers have addressed the effects of changing lubrication or loss of ice-shelf buttressing affecting ice flow, comprehensive ice-flow **models** generally have not incorporated these processes. These comprehensive models also failed to accurately project recent ice-flow accelerations in *Greenland* and in some parts of the Antarctic ice sheet (Alley et al., 2005; Lemke et al., 2007; Bamber et al., 2007). This issue was cited by IPCC

(2007), which provided sea-level projections "excluding future rapid dynamical changes in ice flow" (Table SPM3, WG1) and noted that this exclusion prevented "a best estimate or an upper bound for sea level rise" (p. SPM 15). A paleoclimatic perspective can help inform our understanding of these issues.

Fast-flowing marginal regions can be affected within years, whereas the full response of the slow-flowing central regions to a step-change at the coast requires a few millennia.

As noted above in this section, when subjected to a **step forcing** (e.g., a rapid warming that moves temperatures from one sustained level to another), an ice sheet typically responds by evolving to a new steady state (Paterson, 1994). For example, an increase in accumulation rate thickens the ice sheet. The thicker ice sheet discharges mass faster and, if the ice margin does not move as the ice sheet thickens, the ice sheet becomes on average steeper, which also speeds ice discharge. These changes eventually cause the ice sheet to approach a new configuration—a new steady state—that is in balance with the new forcing. For central regions of cold ice sheets, the time required to complete most of the response to a step change in rate of accumulation (i.e., the response time) is proportional to the ice thickness divided by the accumulation rate. These characteristic times are a few thousands of years (millennia) for the modern *Greenland Ice Sheet* and a few times longer for the ice-age ice sheet (e.g., Alley and Whillans, 1984; Cuffey and Clow, 1997).

A change in the position of the ice-margin will steepen or flatten the mean slope of the ice sheet, speeding or slowing flow. The edge of the ice-sheet will respond first. This response, in turn, will cause a wave of adjustment that propagates toward the ice-sheet center. Fast-flowing marginal regions can be affected within years, whereas the full response of the slow-flowing central regions to a step-change at the coast requires a few millennia.

Warmer ice deforms more rapidly than colder ice. In inland regions, ice sheet response to temperature change is somewhat similar to response to accumulation-rate change, with cooling causing slower deformation, which favors thickening hence higher ice flux through the increased thickness (and perhaps with increasing surface slope also speeding flow), re-establishing equilibrium. However, because most of the deformation occurs in deep ice, and a surface-temperature change requires many millennia to penetrate to that deep ice to affect deformation, most of the response is delayed for a few millennia or longer while the temperature change penetrates to the deep layers, and then the response requires a few more millennia. The calculation is not simple, because the motion of the ice carries its temperature along with it. If melting of the upper surface of an ice sheet develops over a region in which the bottom of the ice is frozen to the substrate, thawing of that basal interface may be caused by penetration of surface meltwater to the bed if water-filled crevasses develop at the surface. The actual penetration of the water-filled crevasse is likely to occur in much less than a single year, perhaps in only a few minutes, rather than over centuries to millennia (Alley et al., 2005).

Numerous ice-sheet models (e.g., Huybrechts, 2002) demonstrate the relative insensitivity of inland ice thickness to many environmental parameters. This insensitivity has allowed reasonably accurate ice-sheet reconstructions using computational models that assume **perfectly plastic** ice behavior and a fixed **yield strength** (Reeh, 1984; the only piece of information needed in these reconstructions of inland-ice configuration is the footprint of the ice sheet; one need not specify accumulation rate hence **mass flux**, for example). This insensitivity can be understood from basic physics.

As noted above in this section, the stress that drives ice deformation increases linearly with ice thickness and with the surface slope, and the rate of ice deformation increases with the cube of this stress. Velocity from deformation is obtained by integrating the deformation rate through thickness, and ice flux is the depth-averaged velocity multiplied by thickness. Therefore, for ice frozen to the bed, the ice flux increases with the cube of the surface slope and the fifth power of the thickness. (Ice flux in an ice sheet with a thawed bed would retain strong dependence on surface slope and thickness, but with different numerical values.) If the ice-marginal position is fixed (say, because the ice has advanced to the edge of the continental shelf and cannot advance farther across the very deep water), then the typical surface slope of the ice sheet is also proportional to the ice thickness (divided by the fixed half-width), giving an eighth-power dependence of ice flux on inland thickness. Although an eighth-power dependence is not truly perfectly plastic, it does serve to greatly limit inland-thickness changes—doubling the inland thickness would increase ice flux 256-fold. Because of this insensitivity of the inland thickness to many controlling parameters, changes in ice-sheet volume are controlled more by changes in the areal extent of the ice sheet than by changes in the thickness in central regions (Reeh, 1984; Paterson, 1994).

Such simple mechanistic scalings of ice sheet behaviors can be useful in a pragmatic sense, and they have been used to interpret ice-sheet behavior in the past. However, in modern usage, our physical understanding of ice sheet behaviors is implemented in fully coupled three-dimensional (or reduced-dimensional) **ice-dynamical models** (e.g., Huybrechts, 2002; Parizek and Alley, 2004; Clarke et al., 2005), which help researchers assimilate and understand relevant data.

5.2 PALEOCLIMATIC INDICATORS BEARING ON ICE-SHEET HISTORY

The basis for paleoclimatic reconstruction is discussed in Cronin (1999) and Bradley (1999), among other sources. Here, additional attention is focused on those indicators that help in reconstruction of the history of the ice sheet. Marine indicators are discussed first, followed by terrestrial **archives**.

The Greenland Ice Sheet has at many times in the past been more extensive than it is now, and much of that extension occupied regions that now are below sea level.

5.2.1 Marine Indicators

As discussed in section 5.3 below, the *Greenland Ice Sheet* has at many times in the past been more extensive than it is now, and much of that extension occupied regions that now are below sea level. Furthermore, iceberg-rafted debris and meltwater from the ice sheet can leave records in marine settings related to the extent of the ice sheet and its flux of ice. Marine sediments also preserve important indicators of temperature and of other conditions that may have affected the ice sheet.

Research cruises to the marine shelf and slope margins of West and East *Greenland* dedicated to understanding changes over the times most relevant to the *Greenland Ice Sheet*'s history have been undertaken only in the last 10 to 20 years. Initially, attention was focused along the *East Greenland Shelf* (Marienfeld, 1992b; Mienert et al., 1992; Dowdeswell et al., 1994a), but in the last few years several cruises have extended to the West *Greenland* margin as well (Lloyd, 2006; Moros et al., 2006). Research on adjacent deep-sea basins, such as *Baffin Bay* or *Fram Basin* off North *Greenland*, is more complicated because the late **Quaternary** (less than 450 thousand years old (**ka**)) sediments contain inputs from several adjacent ice sheets (Aksu, 1985; Hiscott et al., 1989; Andrews et al., 1998a; Dyke et al., 2002). (We use calendar years rather than radiocarbon years unless indicated; conversions include those of Stuiver et al., 1998 and Fairbanks et al., 2005; all ages specified as "ka" or "**Ma**" are in years before present, where "present" is conventionally taken as the year 1950.) Regardless, only a few geographic areas on the *Greenland* shelf have been investigated. In terms of time, the majority of marine cores from the *Greenland*

The use of datable volcanic ashes offers the possibility of linking records from around Greenland.

shelf span the retreat from the last ice age (less than 15 ka). The use of datable volcanic ashes (tephras—a recognizable **tephra** or ash layer from a single eruption is commonly found throughout broad regions and has the same age in all cores) from Icelandic sources offers the possibility of linking records from around *Greenland* from the time of the layer known as Ash Zone II (about 54 ka) to the present (with appropriate cautions; Jennings et al., 2002a).

The sea-floor around *Greenland* is relatively shallow above "sills" formed during the **rifting** that opened the modern oceans. Such **sills** connect *Greenland* to *Iceland* through *Denmark Strait* and to *Baffin Island* through *Davis Strait*. These 500–600-m-deep sills separate sedimentary records of ice sheet histories into "northern" and "southern" components. Even farther north, sediments shed from north *Greenland* are transported especially into the *Fram Basin* of the *Arctic Ocean* (Darby et al., 2002).

The circulation of the ocean around *Greenland* today transports debris-bearing icebergs from the ice sheet. This circulation occurs largely in a clockwise pattern: cold, fresh waters exit the *Arctic Ocean* through *Fram Strait* and flow southward along the East *Greenland* margin as the East *Greenland* Current (Hopkins, 1991). These waters turn north after rounding the southern tip of *Greenland*. In the vicinity of *Denmark Strait*, warmer water from the Atlantic (modified Atlantic Water from the Irminger Current) turns and flows parallel to the East *Greenland* Current. This surface current is called the West *Greenland* Current once it has rounded the southern tip of *Greenland*. On the *East Greenland Shelf*, this modified Atlantic Water becomes an "intermediate-depth" water mass (reaching to the deeper parts of the continental shelf, but not to the depths of the ocean beyond the continental shelf), which moves along the deeper topographic troughs on the continental shelf and penetrates into the margins of the calving *Kangerdlugssuaq* ice stream (Jennings and Weiner, 1994; Syvitski et al., 1996). *Baffin Bay* contains three water masses: Arctic Water in the upper 100–300 meters (m) in all areas, West *Greenland* Intermediate Water (modified Atlantic Water) between 300–800 m, and Deep *Baffin Bay* Water throughout the Bay at depths greater than 1200 m (Tang et al., 2004).

Large quantities of coarse rock material found far from land indicate transport in ice.

Some of the interest in the *Greenland Ice Sheet* is linked to the possibility that meltwater could greatly influence the formation of deep water in the *North Atlantic*. Furthermore, changes in **deep-water formation** in the past are linked to **climate changes** that affected the ice sheet (e.g., Alley, 2007). The major deep-water flow is directed southward through and south of Denmark Strait (McCave and Tucholke, 1986). The sediment deposit known as the *Eirik Drift* off southwest *Greenland* is a product of this flow (Stoner et al., 1995). Convection in the *Labrador Sea* forms an upper component of this *North Atlantic* Deep Water.

Evidence from marine cores and seismic data has been used to reconstruct variations in the *Greenland Ice Sheet* during the last **glacial** cycle (and, occasionally, into older times). Four types of evidence apply:

1. ice-rafted debris and indications of changes in sediment sources;
2. glacial deposition onto **trough-mouth fans**;
3. stable-isotope and biotic data that indicate intervals when meltwater was released from the ice sheet; and
4. geophysical data that indicate sea-floor erosion and deposition.

Each is discussed briefly in section 5.2.1, below.

5.2.1a Ice-Rafted Debris And Its Provenance

Coarse-grained rock material (such as sand and pebbles) cannot be carried far from a continent by wind or current, so the presence of such material in marine cores is of great interest. Small amounts might be delivered in tree roots or attached to uprooted kelp holdfasts (Gilbert, 1990; Smith and Bayliss-Smith, 1998), and rarely a meteorite might be identified, but large quantities of coarse rock material found far from land indicate transport in ice, and so this material is called ice-rafted debris (IRD). Both **sea ice** and icebergs can carry coarse material, complicating interpretations. However, iceberg-rafted debris usually includes some number of grains larger than 2 mm in size and consistent with the grain-size distribution of glacially transported materials, whereas the sediment entrained in sea ice is typically finer (Lisitzin, 2002). In order to link the *Greenland Ice Sheet* with ice-rafted debris described in marine cores, we must be able to link that debris to specific bedrock sites (i.e., identify its **provenance** or site of origin). However, such studies are only in their infancy. Proxies for sediment source include **radiogenic isotopes** (such as εNd; Grousset et al., 2001; Farmer et al., 2003), **biomarkers** that can be linked to different outcrops of dolomite (Parnell et al., 2007), magnetic properties of sediment (Stoner et al., 1995), and quantitative mineralogical assessment of sediment composition (Andrews, 2008).

5.2.1b Trough Mouth Fans

The sediments in trough-mouth fans contain histories of sediment sources that may include ice sheets. Sediment is commonly transferred across the continental shelf along large troughs that form major depositional features called trough-mouth fans (TMF) where the troughs widen and flatten at the continental rise (Vorren and Laberg, 1997; O'Cofaigh et al., 2003). Along the East *Greenland* margin, trough-mouth fans exist off *Scoresby Sund* (Dowdeswell et al., 1997), the *Kangerdlugssuaq* Trough (Stein, 1996), and the *Angmagssalik* (also spelled *Angamassalik*) Trough (St. John and Krissek, 2002). Along the West *Greenland* margin, the most conspicuous such fan is a massive body off *Disko Bay* associated with erosion by *Jakobshavn Glacier* and other outlet glaciers in that region. During periods when the ice sheet reached the **shelf break**, glacial sediments were shed downslope as debris flows (producing coarse, poorly sorted deposits containing large grains in a fine-grained matrix), whereas periods when the ice sheet was well back from the shelf break are marked by sediments containing materials typical of open-marine environments, such as shells of **foraminifers**, and typical terrestrial materials including ice-rafted debris.

5.2.1c Foraminifers and Stable-Isotopic Ratios of Shells

Foraminifers—mostly marine, single-celled planktonic animals, commonly with chalky shells—are widely distributed in sediments, and shells of surface-dwelling (planktic) and bottom-dwelling (benthic) species are commonly found. The particular species present and the chemical and isotopic characteristics of the chalky shells reflect environmental conditions. Variations in the ratios of the stable isotopes of oxygen, ^{18}O to ^{16}O (**$\delta^{18}O$**) are especially widely used. These ratios respond to changes in the global ice volume. Water containing the lighter isotope (^{16}O) evaporates from the ocean more readily, and ice sheets are ultimately composed of that evaporated water, so during times when the ice sheets are larger, the ocean is isotopically heavier. This effect is well known, and it can be corrected for with considerable confidence if the age of a sample is known. Temperature also affects $\delta^{18}O$; warmer air temperatures favor incorporation of the lighter isotope into the shell. Near ice sheets, the abrupt appearance of light isotopes is most commonly associated with meltwater that delivered isotopically light and fresh water (Jones and Keigwin, 1988; Andrews et al., 1994). Around the *Greenland Ice Sheet*, most such records are from near-surface planktic foraminifers of the species *N. pachyderma sinistral* (Fillon and Duplessy, 1980; van Kreveld et al., 2000; Hagen and Hald, 2002), although there are some data from benthic foraminifers (Andrews et al., 1998a; Jennings et al., 2006).

Useful records of many time intervals have been assembled from terrestrial indicators.

5.2.1d Seismic and Geophysical Data

Several major shelf troughs and trough-mouth fans have been studied by seismic investigations. Most are high-resolution studies of the sediments nearest the sea floor (seismostratigraphy; O'Cofaigh et al., 2003), although some data on deeper strata are available (airgun profiles; see Stein, 1996; Wilken and Mienert, 2006). Sonar reveals the shape of the upper surface of the sediment, and features such as the tracks left by drifting icebergs that plowed through the sediment (Dowdeswell et al., 1994b; Dowdeswell et al., 1996; Syvitski et al., 2001) and the streamlining of the sediment surface caused by **glaciation**.

5.2.2 Terrestrial Indicators

Land-based records, like their marine equivalents, can reveal the history of changes in areal extent of ice and of the climate conditions that existed around the ice sheet. Terrestrial records are typically more discontinuous in space and time than are marine records, because net erosion (which removes sediments containing climatic records) is dominant on land whereas net deposition is dominant in most marine settings. Nonetheless, useful records of many time intervals have been assembled from terrestrial indicators. Here, common indicators are briefly described. This treatment is representative rather than comprehensive. Furthermore, the great wealth of indicators, and the interwoven nature of their interpretation, preclude any simple subdivision.

5.2.2a Geomorphic Indicators

The land surface itself records the action of ice and thus provides information on ice-sheet history. Glacial deposits known as moraines are especially instructive, but others are also important.

Moraines are composed of sediment deposited around glaciers from material carried on, in, or under the moving ice (e.g., Sugden and John, 1976). A preserved moraine may mark either the maximum extent reached by ice during some advance or a still-stand during retreat. Normally, older moraines are destroyed by ice readvance, although remnants of moraines overrun by a subsequent advance are occasionally preserved and identifiable, especially if the ice that readvanced was frozen to its bed and thus nearly or completely stationary where the ice met the moraine. Because most older moraines are reworked by subsequent advances, most existing moraines record only the time of the most recent glacial maximum and pauses or subsidiary readvances during retreat.

Glacier extent can usually be used as a proxy for temperature.

The limiting ages of moraines can be estimated from radiocarbon (carbon-14) dating of carbon-bearing materials incorporated into a moraine (the moraine must be younger than those materials) or deposited in lakes that formed on or behind moraines following ice retreat (the moraine must be older than those materials). Increasingly, moraines are dated by measurement of beryllium-10 or other isotopes produced in boulders by cosmic rays (e.g., Gosse and Phillips, 2001). Cosmic rays penetrate only about 1 m into rock. Thus, boulders that are quarried from beneath the ice following erosion of about 1 m or more of overlying material, or large boulders that fell onto the ice and rolled over during transport, typically start with no cosmogenic nuclides in their upper surfaces but accumulate those nuclides proportional to exposure time. Corrections for loss of nuclides by boulder erosion, for inheritance of nuclides from before deposition, and other factors may be nontrivial but potentially reveal further information. Additional techniques of dating can sometimes be used, including historical records and the increase with time of the size of lichen colonies (e.g., Locke et al., 1979; Geirsdottir et al., 2000), soil development, and breakdown of rocks (clast weathering).

Related information on glacial behavior and ages is also available from the land surface. For ages of events, a boulder need not be in a moraine to be dated using cosmogenic isotopes, and surfaces **striated** and polished by glacial action can be dated similarly. Glacial retreat often reveals wood or other organic material that died when it was overrun during an advance and that can also be dated using radiocarbon techniques.

In moraines produced by small glaciers, the highest elevation to which a moraine extends is commonly close to the **equilibrium-line altitude** at the time when the moraine formed. (The equilibrium-line altitude is the altitude above which net snow accumulation occurred and below which mass loss occurred—mass moved into the glacier above that elevation and out below that elevation, controlling the deposition of rock material.) Glaciation produces identifiable landforms, especially if the ice was thawed at the base and thus slid freely across its substrate, so contrasts in the appearance of landforms can be used to map the limits of glaciation (or of wet-based glaciation) where moraines are not available.

Glaciers respond to many environmental factors, but for most glaciers the balance between snow accumulation and melting is the major control on glacier size. Furthermore, with notable exceptions, melting is usually affected more by temperature than is accumulation. The equilibrium vapor pressure (the ability of warmer air to hold more moisture) increases roughly 7% per °C. For a variety of glaciers that balance snow accumulation by melting, the increase in melting is approximately 35% (±10%) per °C (e.g., Oerlemans, 1994, 2001; Denton et al., 2005). Thus, glacier extent can usually be used as a proxy for temperature (duration and warmth of the melt season), primarily summertime temperature.

5.2.2B Biological Indicators and Related Features

Living things are sensitive to climate. The species found in a tropical rain forest differ from those found on the tundra. By comparing modern species from different places that have different climates, or by looking at changes in species at one place for the short interval of the instrumental record, the relation with climate can be estimated. Assuming that this relation has not changed with time, longer records of climate then can be estimated from occurrence of the remains of different species in older sediments (e.g., Schofield et al., 2007). These climate records then can be tied, to some degree, to the state of the ice sheet.

Lake sediments are especially valuable as sources of biotic indicators, because sedimentation (and thus the record) is continuous and the ecosystems in and around lakes tend to be rich (e.g., Bjorck et al., 2002; Andresen et al., 2004; Ljung and Bjorck, 2004). Pollen (e.g., Ljung and Bjorck, 2004; Schofield et al., 2007), microfossils, and macrofossils (such as **chironomids**, also called midge flies (Brodersen and Bennike, 2003)) are all used to great advantage in reconstructing past climates. The isotopic composition of shells or of inorganic precipitates in lakes records some combination of temperature and of the isotopic composition of the water. Physical aspects of lake sediments, including those linked to biological processes (e.g., loss on ignition, which primarily measures the relative abundance of organic matter in the sediment) are also related to climate. In places where the weight of the ice previously depressed the land below sea level and subsequent rebound raised the land back above sea level and formed lakes (see section 5.2.2C, below), the time of onset of lacustrine conditions and the modern height of the lake together provide key information on ice-sheet history (e.g., Bennike et al., 2002).

Raised marine deposits in *Greenland* and surroundings provide an additional and important source of biological indicators of climate change. Many marine deposits now reside above sea level, because of the interplay of changing sea level, geological processes of uplift and subsidence, and isostatic response (ice-sheet growth depressing the land and subsequent ice-sheet shrinkage allowing rebound, with a lagged response; see section 5.2.2C, below). Biological materials within those deposits, and especially shells, can be dated by radiocarbon or uranium-thorium techniques (see section 5.2.2D, below). Those dates then help fill in the history of relative sea level that can be used to infer ice-sheet loading histories and to reconstruct climates on the basis of the species present (e.g., Dyke et al., 1996).

5.2.2C Glacial Isostatic Adjustment and Relative Sea-Level Indicators Near the Ice Sheet

Within the geological literature, sea level is generally defined as the elevation of the sea surface relative to some adjacent geological feature. (This convention contrasts with the concept of an absolute sea level whose position (the sea surface) is measured relative to some absolute datum, such as the center of Earth.) This definition of sea level is consistent with geological markers of past sea-level change (such as ancient shorelines, shells, and driftwood), which reflect changes in the absolute height of either the sea surface or the geological feature (i.e., an ancient shoreline can be exposed because the surface of the ocean dropped, or land uplifted, or a net combination of land and ocean height changes). During the time periods considered in this report, the dominant processes responsible for such changes, at least on a global scale, have been the mass transfer between ice reservoirs and oceans associated with the ice-age cycles, and the deformational response of Earth to this transfer of mass. This deformational response is formally termed **glacial isostatic adjustment**.

The growth and shrinkage of ice have generally been sufficiently slow that glacial isostatic adjustment of the solid Earth is characterized by both immediate **elastic** and slow viscous (i.e., flow) effects. As an example, if a large ice sheet were to form instantly and then persist for more than a few thousand years, the land would respond by nearly instantaneous elastic sinking, followed by slow subsidence toward isostatic equilibrium as deep, hot rock moved outward from beneath the ice sheet. Roughly speaking, the final depression would be about 30% of the thickness of the ice. Thus the ancient ***Laurentide Ice Sheet***, which covered most of *Canada* and the northeastern United States and whose peak thickness was 3–4 km, produced a crustal depression of about 1 km. (For comparison,

The ancient Laurentide Ice Sheet, which covered most of Canada and the northeastern United States and whose peak thickness was 3–4 km, produced a crustal depression of about 1 km.

that ice sheet contained enough water to make a layer about 70 m thick across the world oceans, much less than the local deformation beneath the ice.) Outside the depressed region covered by ice, land is gradually pushed upward to form a **peripheral bulge**. As the ice subsequently melts, the central region of depression rebounds, and relative sea level will fall for thousands of years beyond the end of the melting phase. For example, at sites in Hudson Bay, sea-level continues to fall on the order of 1 centimeter per year (cm/yr) despite the disappearance of most of the *Laurentide Ice Sheet* some 8,000 years ago. Moreover, the loss of ice cover allows the peripheral bulge to subside, leading to a sea-level rise in such areas (e.g., along the east coast of the United States) that also continues to the present (but involving slower rates of change than for the regions that were beneath the central part of the former ice sheet). As one considers sites farther away from the high-latitude ice cover, in the so-called "**far field**," the sea-level change is dominated during deglaciation by the addition of meltwater into the global oceans. However, in periods of stable ice cover, for example during much of the present interglacial, changes in sea level continue as a consequence of the ongoing gravitational and deformational effects of glacial isostatic adjustment. As an example, glacial isostatic adjustment is responsible for a fall in sea level in parts of the equatorial Pacific of about 3 m during the last 5,000 years and for the associated exposure of corals and ancient shoreline features of this age (Mitrovica and Peltier, 1991; Dickinson, 2001; Mitrovica and Milne, 2002). We will return to this point in section 5.2.2D, below.

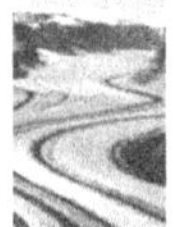

Nearby (**near field**) relative sea-level changes, where the term "relative" denotes the height of an ancient marker relative to the present-day level of the sea, have commonly been used to constrain models of the geometry of ice complexes, particularly since the **Last Glacial Maximum** (which peaked at about 24 **ka** in *Greenland*) (e.g., Lambeck et al., 1998; Peltier, 2004). Fleming and Lambeck (2004) compared a set of about 600 relative sea-level data points from sites in *Greenland*; all but the southeast coast and the west coast near *Melville Bugt* (Bay) were represented. Numerical models of glacial isotatic adjustment constrained the history of the *Greenland Ice Sheet* after the Last Glacial Maximum. The Fleming and Lambeck (2004) data set comprised primarily fossil mollusk shells that lived at or below the sea surface but that now are exposed above sea level; because of the unknown depth at which the mollusks lived, they provide a limiting value on sea level. However, Fleming and Lambeck (2004) also included observations on the transition of modern lakes from formerly marine conditions, and constraints associated with the present (sub-sea) location of initially terrestrial archaeological sites (see also Weidick, 1996; Kuijpers et al., 1999). Tarasov and Peltier (2002, 2003) analyzed their own compilation of local sea-level records by coupling glacial isostatic adjustment and climatological models; from this information they inferred ice history into the last interglacial.

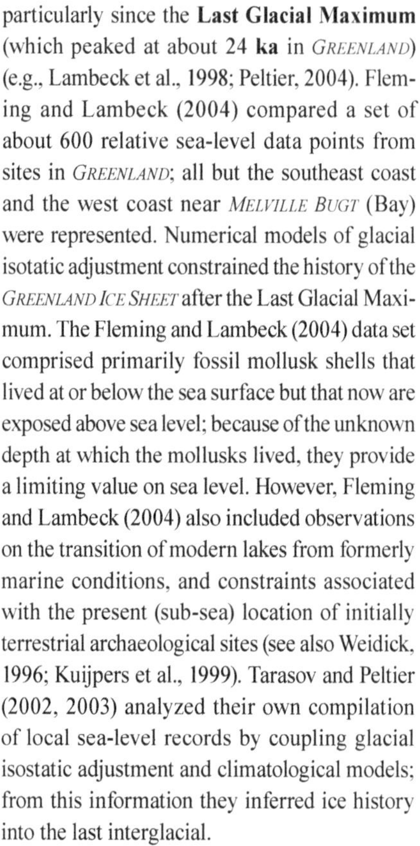

Like all glacial isostatic adjustment models, these studies are hampered by uncertainty about the **viscoelastic structure** of Earth (Mitrovica, 1996), which is generally prescribed by the thickness of the elastic plate and the radial profile of viscosity within the underlying mantle, and this uncertainty has implications for the robustness of the inferred ice history. In addition, the analysis of sea-level records in *Greenland* is complicated by signals from at least two other distant sources:

1. the adjustment of the peripheral bulge associated with the (de)glaciation of the larger North American *Laurentide Ice Sheet*, because this bulge extends into *Greenland* (e.g., Fleming and Lambeck, 2004); and

2. the net addition of meltwater from contemporaneous melting (or, in times of glaciation, growth) of all other global ice reservoirs.

Therefore, some constraints on the volume and extent of the *Laurentide Ice Sheet*, and the volume of more-distant ice sheets and glaciers, are required for the analysis of sea-level data from *Greenland*.

5.2.2D Far-Field Indicators of Relative Sea-Level High-Stands

Past changes in the volume of the *Greenland Ice Sheet* are recorded in far-field sea level. All other sources of sea-level change, as well as the change due to glacial isostatic adjustment, are also recorded in far-field sea-level records, so a single history of sea level provides information related to ice-volume change (and to other factors such as thermal expansion and contraction of ocean water) but no information on the relative contribution of individual sources.

The record of past sea level can be reconstructed in many ways. An especially powerful method of reconstruction uses the record of marine deposits or emergent coral reefs that are now found above sea level on geologically relatively stable coasts and islands (that is, in regions not markedly affected by processes linked to **plate tectonics**). Such records are high-water marks (or "bathtub rings") of past high sea levels. Coastal landforms and deposits provide powerful and independent records of sea-level history compared with the often-cited deep-sea oxygen-isotope record of glacial and interglacial periods. For recording sea-level history, coastal landforms have two advantages as compared with the deep-sea oxygen-isotope record:

1. if corals are present, they can be dated directly; and
2. estimates of ancient sea level may—depending on the geological setting—be possible.

Coastal landforms record high stands of the sea when coral-reefs grew as fast as sea level rose (Figure 5.3*A*) or when a stable sea-level high stand eroded marine terraces into bedrock (Figure 5.3*B*). Thus, emergent marine deposits,

The record of past sea level can be reconstructed in many ways. An especially powerful method of reconstruction uses the record of marine deposits or emergent coral reefs that are now found above sea level.

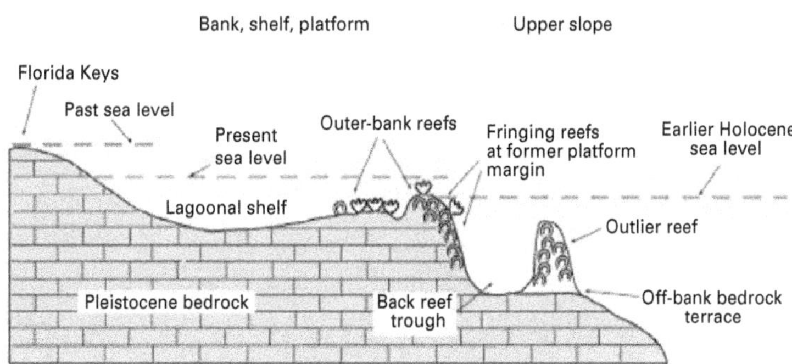

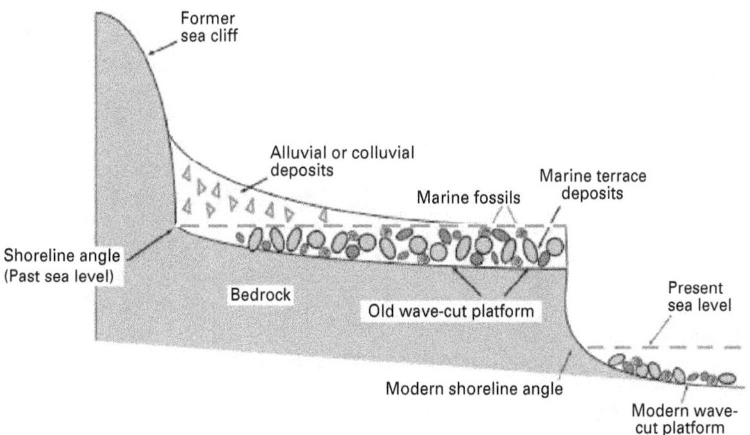

Figure 5.3. Cross sections showing idealized geomorphic and stratigraphic expression of *A*) coastal landforms and deposits found on low-wave-energy carbonate coasts of Florida and the Bahamas, and *B*) high-wave-energy rocky coasts of Oregon and California. (Vertical elevations are greatly exaggerated.)

either reefs or terraces, on geologically active, rising coastlines record interglacial periods (Figure 5.4). On a geologically stable or slowly sinking coast, reefs will emerge only from sea-level stands that were higher than at present (Figure 5.4). Past sea levels can thus be determined from stable coastlines, or even rising coastlines if one can make reasoned models of uplift rates. Geologic records of high sea-level stands on geologically relatively stable coasts are especially useful. Although valuable geologic records are found on rising coasts, estimates of past sea level derived from such coasts depend on assumptions about the rate of **tectonic** uplift, and therefore they embody more uncertainty.

The direct dating of emergent marine deposits is possible because uranium (U) is dissolved in ocean water but thorium (Th) and protactinium (Pa) are not. Certain marine organisms, particularly corals, co-precipitate U directly from seawater during growth. All three of the naturally occurring isotopes of uranium—^{238}U and ^{235}U (both primordial parents) and ^{234}U (a decay product of ^{238}U)—are therefore incorporated into living corals. ^{238}U decays to ^{234}U, which in turn decays to ^{230}Th. The parent isotope ^{235}U decays to ^{231}Pa. Thus, activity ratios of $^{230}Th/^{234}U$, $^{238}U/^{234}U$, and $^{231}Pa/^{235}U$ can provide three independent clocks for dating the same fossil coral (e.g., Edwards et al., 1997). Since the 1980s, most workers have employed thermal ionization mass spectrometry (TIMS) to measure U-series nuclides; this method has increased precision, requires only small samples, and can extend the useful time period for dating back to at least about 500,000 years.

The coastlines where the most reliable records of past high sea levels can be found are in the tropics and subtropics, where ocean temperatures are warm enough that coral-reefs grow. Within this broad equatorial region, the ideal coastlines for studies of past high sea levels are those that are distant from boundaries of tectonic plates. Such coastlines lie near geologically relatively quiescent continental margins or as islands well within the interiors of large tectonic plates. Even in such locations, however,

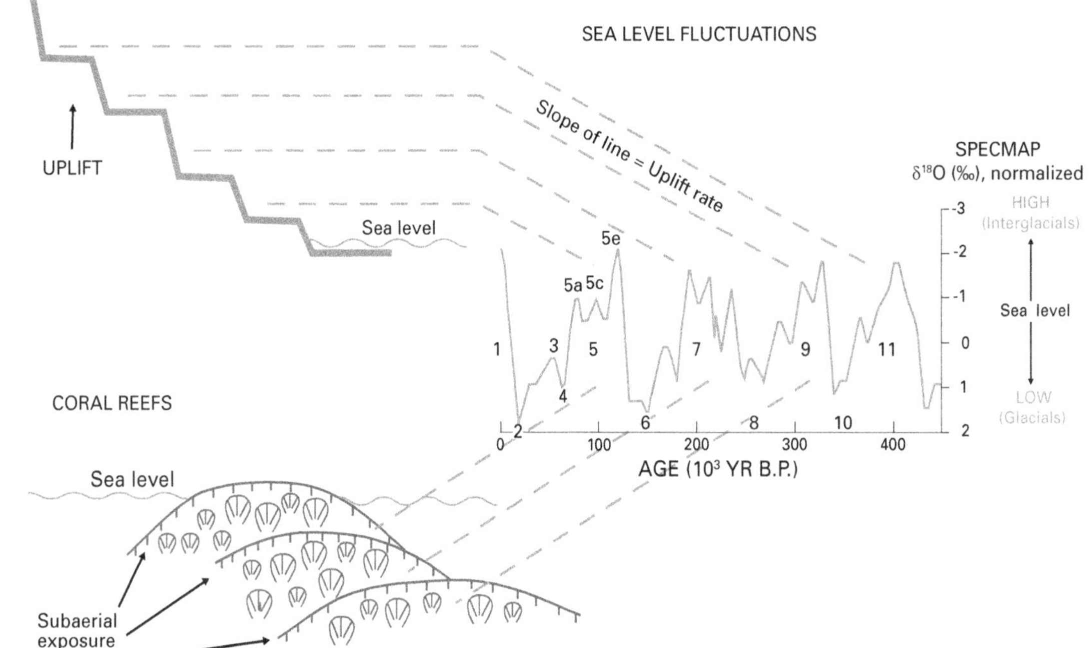

Figure 5.4. Relation of oxygen isotope records in foraminifers of deep-sea sediments to emergent reef or wave-cut terraces on an uplifting coastline (upper) and a tectonically stable or slowly subsiding coastline (lower). Emergent marine deposits record interglacial periods. Oxygen isotope data shown are from the SPECMAP record (Imbrie et al., 1984). Redrawn from Muhs et al. (2004).

interpreting past sea levels can include much uncertainty. We highlight two major reasons for this uncertainty.

First, many islands well within the crustal tectonic plate that underlies the Pacific Ocean, for example, are part of **hot-spot volcanic chains**. (A major source of internal heat, called a hot spot, leads to a volcano on the overriding tectonic plate; as the plate drifts laterally, the slower moving hot spot becomes positioned below a different part of the plate, and a new volcano is formed as the previously active volcano becomes extinct. Eventually, a chain of volcanoes is produced, such as the Hawaiian-Emperor seamount chain.) As a volcano grows in elevation, its weight isostatically depresses the land it sits on in the same way that the weight of an ice sheet does, and the cold upper elastic layer of the Earth flexes to form a broad ring-shaped ridge around the low caused by the volcano. Oahu, in the Hawaiian Island chain, is a good example of an island that is apparently experiencing slow uplift, and an associated local sea-level fall, due to volcanic loading on the "Big Island" of Hawaii (Muhs and Szabo, 1994).

Second, the existence of a sea-level highstand of a given age in a stable geologic setting does not necessarily imply that ice volumes were lower at that time relative to the present day, even if the highstand is dated to a previous interglacial. As discussed above, glacial isostatic adjustment, because it involves slow viscous flow of rock, produces global-scale changes in sea-level even during periods when ice volumes are stable. As an example, for the last 5,000 years (long after the end of the last glacial interval), ocean water has moved away from the equatorial regions and toward the former **Pleistocene** ice complexes to fill the voids left by the subsidence of the peripheral bulge regions produced by the ice sheets. As a result, sea level has fallen (and continues to fall) about 0.5 mm/yr in those far-field equatorial regions (Mitrovica and Peltier, 1991; Mitrovica and Milne, 2002). This process, known as equatorial ocean siphoning, has developed so-called 3-meter beaches and exposed coral reefs that have been dated to the end of the last deglaciation and that are endemic to the equatorial Pacific (e.g., Dickinson, 2001). Thus, the interpretation of such apparent highstands requires correction for glacial isostatic adjustments such that the residual record reflects true changes in ice volume.

5.2.2e Geodetic Indicators

Geodetic data are yielding both local and regional constraints on recent changes in the mass of ice-sheets. As an example, land-based measurements of changes in gravity and crustal motions, estimated by using the geographic positioning system (GPS), are being used to monitor deformation (associated with changes in the distribution of mass) at the periphery of the *Greenland Ice Sheet* (e.g., Kahn et al., 2007). A drawback of these techniques is that few sites have been monitored because of the difficulty of establishing high-quality GPS sites. In contrast, data from the Gravity Recovery and Climate Experiment (GRACE) satellite mission are revealing trends in gravity across the polar ice sheets (at a spatial resolution of about 400 km) from which estimates of both regional and integrated mass flux are being obtained (e.g., Velicogna and Wahr, 2006). A general problem in all attempts to infer recent ice sheet balance, whether from land-based or satellite gravity, GPS, or even altimeter measurements of ice height (e.g., Johannessen et al., 2005; Thomas et al., 2006), is that a measurements must be corrected for the continuing influence of glacial isostatic adjustments. As discussed above (section 5.2.2c), this correction involves uncertainty associated with both the ice sheet history and the viscoelastic structure of Earth.

Rapid melting of different ice sheets will have substantially different signatures, or fingerprints, in the spatial pattern of sea-level change.

Accurate glacial isostatic adjustment corrections are also central to regional estimates of ice-sheet mass balance. For the last century global sea-level change has been inferred principally by analyzing records from widely distributed tide gauges (simple sea-level monitoring devices). Most residual rates (those corrected for glacial isostatic adjustment) of tide gauges yield an average 20th century sea-level rise in the range 1.5–2.0 mm/yr (Douglas, 1997) (for additional information on recent trends in sea level, see Solomon et al., 2007).

Furthermore, geographic trends in the residual rates may constrain the sources of the meltwater. In particular, Mitrovica et al. (2001) and Plag and Juttner (2001) have demonstrated that the rapid melting of different ice sheets will have substantially different signatures, or fingerprints, in the spatial pattern of sea-level change. These patterns are linked to the gravitational effects of the lost ice (sea level is raised near an ice sheet because of the gravitational attraction of the ice mass for the adjacent ocean water) and to the elastic (as opposed to **viscoelastic**) **deformation** of Earth driven by the rapid unloading. Some ambiguity in determining the source of meltwater arises because of uncertainty in both the original correction for glacial isostatic adjustment and in the correction for the poorly known signature of ocean thermal expansion, as well as from the non-uniform distribution of tide gauge sites.

Ice cores preserve information about many climatic variables that affected the ice sheet, and about how the ice sheet responded to changes in those variables.

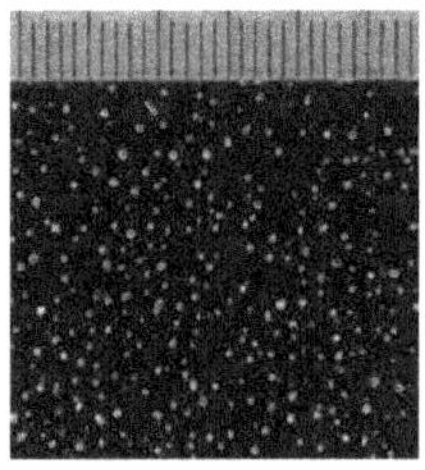

Temperature histories derived from ice cores are especially accurate.

Other geodetic indicators related to Earth's rotational state also constrain estimates of recent changes in the mass of ice-sheets (Munk, 2002; Mitrovica et al., 2006). Earth's rotation is affected by any redistribution of mass on or inside the planet. Transfer of mass from the poles to the equator slows the planet's rotation (like a spinning ice skater extending her arms to slow her rotation). Moreover, any transfer of mass that is not symmetric about the poles causes "wobble," or true polar wander (that is, the position of the north rotation pole moves relative to the surface of the planet). True polar wander for the last century has been estimated using both astronomical and satellite geodetic data. In contrast, changes in the rotation rate (or, as geodesists say, length of day), have been determined for the last few decades by using satellite measurements and for the last few millennia by using observations of eclipses recorded by ancient cultures. Specifically, the timing of ancient eclipses recorded by these cultures differs from the timing one would expect by simply projecting the Earth-Moon-Sun system back in time using the modern rotation rate of Earth. The discrepancy indicates a gradual slowing of Earth's rate of rotation (Munk, 2002). The difference in the rotation-rate history during the last few millennia (after correcting for slowing of Earth's rotation associated with the "drag" of the tides) as compared with the rotation rate of last few decades provides a measure of any anomalous recent melting of polar ice reservoirs. (This difference does not uniquely constrain the individual sources of the meltwater because all sources will be about equally efficient, for a given mass loss rate, at driving these changes in rotation.) True polar wander, after correction for glacial isostatic adjustment, serves as an important complement to this rotation-rate analysis because it does give some information about the source of the meltwater. As an example, melting from the Antarctic, because it is located at the pole, generates very little true polar wander, whereas melting from the *GREENLAND ICE SHEET*, whose center of mass lies about 15 degrees off Earth's rotation axis, is capable of driving substantial true polar wander (Munk, 2002; Mitrovica et al., 2006).

5.2.2F ICE CORES

Ice cores preserve information about many climatic variables that affected the ice sheet and about how the ice sheet responded to changes in those variables.

Temperature histories derived from ice cores are especially accurate. Several indicators are used, as described next, such as the isotope ratios of accumulated snow, ice-sheet temperature profiles (using borehole thermometry), and various techniques based on use of gas-isotope indicators. Agreement among these different indicators increases confidence in the results.

Let us first consider isotopic ratios of the oxygen and hydrogen in accumulated snow (e.g., Jouzel et al., 1997). The ocean contains both normal and "heavy" water: roughly one molecule in 500 incorporates at least one extra neutron in the nucleus of an oxygen or hydrogen atom. The lighter molecules evaporate more easily, and the heavier molecules condense (and thus precipitates) more easily. As water that evaporated from the ocean is carried by an air mass inland over an ice sheet, the heavy molecules preferentially rain or snow out. The colder the air mass, the more vapor is removed, the more depleted of the heavy molecules is the remaining vapor, and the lighter the isotopic ratios in the next rain or snow. Hence, the isotopic composition of precipitation is linked to temperature of the air mass and, over polar ice sheets, the temperature of the air mass is typically linked to the surface temperature. Oxygen- and hydrogen-isotope ratios are both studied, and they help locate the source of precipitation, track the changing isotopic composition of the moving air mass ("path effects"), and indicate the ice-sheet temperature as well. Because site temperature is most important for this review, one species is sufficient. Results will be discussed here as $\delta^{18}O$, the difference between the ^{18}O:^{16}O ratio of a sample and of standard mean ocean water, normalized by the ratio of the standard and expressed not as percent (%) but as per mil (‰) (percent is parts per hundred, and per mil is parts per thousand).

Although linked to site temperature, $\delta^{18}O$ can be affected by many factors (Jouzel et al., 1997; Alley and Cuffey, 2001), such as change in the ratio of summertime to wintertime precipitation. Hence, additional means of determining past temperatures are required. One of the most reliable is based on the physical temperature of the ice. Just as a frozen turkey takes a long time in a hot oven to warm in the middle, intermediate depths of the central *Greenland Ice Sheet* are colder than ice above or below. Surface ice temperatures equilibrate with air temperature, and basal ice receives some warmth from Earth's heat flow, but the center of the ice sheet has not finished warming from the ice-age cold. If ice flow is understood well at a site, the modern profile of the physical temperature of the ice with increasing depth provides a low-time-resolution history of the surface temperature with increasing time. Joint interpretation of the isotopic ratios and temperatures measured in boreholes (Cuffey et al., 1995; Cuffey and Clow, 1997), or independent interpretation of the borehole temperatures and then comparison with the isotopic ratios (Dahl-Jensen et al., 1998), helps to outline the history of surface air temperature. Furthermore, the relation between isotopic ratio and temperature (α ‰ per °C) becomes a useful paleoclimatic indicator, and changes in this ratio α with time can be used to test hypotheses about the overall changes in seasonality of snowfall and other factors.

The isotopic composition of gases trapped in bubbles in the ice sheet provides an additional indicator of temperature. New-fallen snow contains many interconnected air spaces. Snow turns to ice without melting in central regions of cold ice sheets through solid-state mechanisms that operate more rapidly under higher temperature or higher pressure. Snow in an ice sheet usually transforms to ice within the top few tens of meters. The intermediate material is called **firn**, and the transformation is complete when bubbles are isolated so that the air spaces are no longer interconnected to the surface. Wind moving over the ice sheet typically mixes gases in the pore spaces of the firn only in the uppermost few meters or less. **Diffusion** mixes the gases deeper than this. Gases are slightly separated by gravity (Sowers et al., 1992), with the air trapped in bubbles slightly isotopically heavier than in the free atmosphere, proportional to the thickness of the air column in which diffusion dominates.

If a sudden temperature change occurs at the surface, the resulting temperature change of the firn beneath requires typically about 100 years to penetrate to the depth of bubble trapping. However, when a temperature gradient is applied across gases in diffusive equilibrium, the gases are separated by thermal fractionation as well as by gravity, with the heavier gases moving thermally to the colder end (Severinghaus et al., 1998). Equilibrium of gases is obtained in a few years, far faster than the time for heat flow to remove the temperature gradient across the firn. Within a few years after an abrupt temperature change at the surface, newly forming bubbles will begin to trap air with very slight (but easily measured) anomalies in gas-isotope compositions, and this trapping of slightly anomalous air will continue for a century or so. Because different gases have different sensitivities to temperature gradients and to gravity, measuring isotopic ratios of several gases (such as argon and nitrogen) allows researchers to determine the temperature difference that existed vertically in the firn at the time of bubble trapping and to determine the thickness of firn in which wind was not mixing the gas (Severinghaus et al., 1998). If the surface temperature changed very quickly, the magnitude of the temperature difference across the firn will peak at the magnitude of the surface-temperature change; for a slower change, the temperature difference

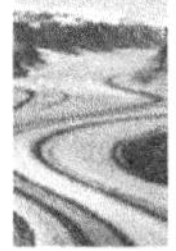

Past ice-accumulation rates are most readily obtained by measuring the thickness of annual layers in ice cores corrected for ice-flow thinning.

Ages in ice cores are most commonly estimated by counting annual layers.

across the firn will always be less than the total temperature change at the surface. If the climate was relatively steady before an abrupt temperature change, such that the depth-density profile of the firn came into balance with the temperature and the accumulation rate, and if the accumulation rate is known independently (see below), then the number of years or amount of ice between the gas-phase and ice-phase indications of abrupt change provides information on the mean temperature before the abrupt change (Severinghaus et al., 1998). With so many independent thermometers, highly confident paleothermometry is possible.

Ice cores can provide information on climatic indicators other than temperature. Past ice-accumulation rates are most readily obtained by measuring the thickness of annual layers in ice cores corrected for ice-flow thinning (e.g., Alley et al., 1993). In other methods, the thickness of firn can be approximated by measurements of gas-isotope fractionation or of the number and density of bubbles (Spencer et al., 2006); these measurements combined with temperature estimates constrain accumulation rates as well. Aerosols (very small liquid and solid particles) of all types fall with snow and during intervals when snow is not falling, and are incorporated into the ice sheet; with knowledge of the accumulation rate (hence dilution of the aerosols), time histories of atmospheric loading of those aerosols can be estimated (e.g., Alley et al., 1995a). Dust and volcanic fallout (e.g., Zielinski et al., 1994) help constrain the cooling effects of aerosols (particles) blocking the Sun. Cosmogenic isotopes (beryllium-10 is most commonly measured) reflect cosmic-ray bombardment of the atmosphere, which is modulated by the strength of Earth's magnetic field and by solar activity (e.g., Finkel and Nishiizumi, 1997). The observed correlation in paleoclimatic records between indicators of climate and indicators of solar activity (Stuiver et al., 1997; Muscheler et al., 2005; Bard and Frank, 2006)—and the lack of correlation with indicators of magnetic-field strength (Finkel and Nishiizumi, 1997; Muscheler et al., 2005)—help researchers understand climate changes.

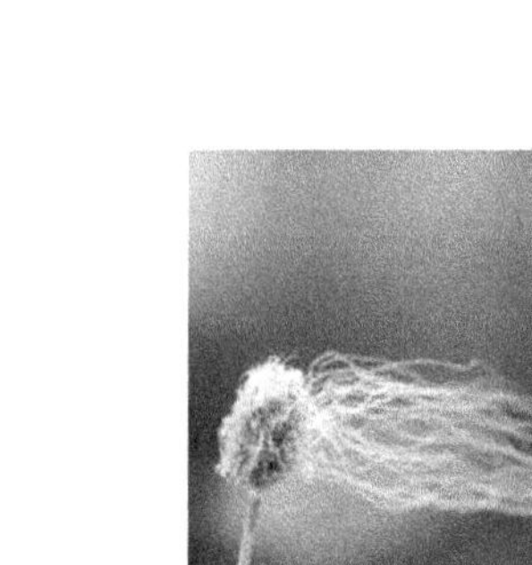

Ages in ice cores are most commonly estimated by counting annual layers (e.g., Alley et al., 1993; Andersen et al., 2006) and by correlation with other records (Blunier and Brook, 2001). Several indicators of atmospheric composition from GREENLAND ice cores that were matched with similar (but longer) records from Antarctica (Suwa et al., 2006) showed that old ice exists in central GREENLAND (Suwa et al., 2006; Chappellaz et al., 1997) at depths where flow processes have mixed the layers (Alley et al., 1997). In regions of continuous and unmixed layers, other features in ice cores, such as chemically distinctive ash from particular volcanic eruptions, can be correlated with independently dated records (e.g., Finkel and Nishiizumi, 1997; Zielinski et al., 1994). Flow models also can be used to aid in dating.

The past elevation of ice-sheets is indicated by the total gas content of the ice (Raynaud et al., 1997) at a given depth and age. As noted above in this section, bubbles are pinched off (pore close-off) from interconnected air spaces in the firn a few tens of meters down. The density of the ice at this pore close-off is nearly constant, with a small and fairly well known correction for climatic conditions. Because air pressure varies with elevation and elevation varies with ice thickness, the total number of trapped molecules of gas per unit volume of ice is correlated with ice-sheet thickness. Small elevation changes cannot be detected (because of additional fluctuations in total gas content that are likely linked to changing layering in the firn that affects trapped bubbles), but elevation changes of greater than 500 m are detectable with confidence (Raynaud et al., 1997).

Additional information on ice-sheet changes comes from the current distribution of isochronous surfaces (surfaces that have the same age throughout) in the ice sheet. An explosive volcanic eruption will deposit an acidic ash layer of a single age on the surface of the ice sheet, and that layer can be identified after burial by using radar (Whillans, 1976). Ages of reflectors can be determined at ice-core sites (e.g., Eisen et al., 2004), and the layers can then be mapped throughout broad areas (Jacobel and Welch, 2005). A model can be used to predict the current distribution of isochronous surfaces (as well as some other properties, such as temperature) for any hypothesis that combines the history of climatic forcing (primarily accumulation rate affecting burial and temperature) and ice-sheet flow (primarily changes in surface elevation and extent) (e.g., Clarke et al., 2005). Optimal histories can be estimated in this way.

5.3 HISTORY OF THE GREENLAND ICE SHEET

5.3.1 Ice-Sheet Onset and Early Fluctuations

Prior to 65 million years ago (Ma), dinosaurs lived on a high-CO_2, warm world that usually lacked permanent ice at sea level. The high latitudes were warm; Tarduno et al. (1998) provided a minimum estimate of the mean-annual temperature during this time of over 14°C at 71°N based on occurrence of crocodile-like champsosaurs (also see Markwick, 1998; Vandermark et al., 2007). Sluijs et al. (2006) showed that the ocean surface warmed near the North Pole from about 18°C to peak temperatures of 23°C during the short-lived **Paleocene-Eocene Thermal Maximum** about 55 Ma. Such warm temperatures preclude permanent ice near sea level and, indeed, no evidence of such ice has been found (Moran et al., 2006).

Prior to 65 million years ago (Ma), dinosaurs lived on a high-CO_2, warm world that usually lacked permanent ice at sea level.

Cooling following the Paleocene-Eocene Thermal Maximum may have allowed ice to reach sea level fairly quickly; sand and coarser materials found in a core from the *Arctic Ocean* sea floor and dated at about 46 **Ma** (Moran et al., 2006; St. John, 2008) are most easily (but not with absolute certainty) interpreted as indicating ice rafting linked to glaciers. Ice-rafted debris likely traceable at least in part to glaciers rather than to sea ice is found in a core recovered from about 75°N latitude in the *Norwegian-Greenland Sea* off East *Greenland*; the core is dated between about 38 and 30 Ma (late Eocene into **Oligocene** time). Certain characteristics of this debris point to an East *Greenland* source and exclude *Svalbard*, the next-nearest land mass (Eldrett et al., 2007). It is not known whether this ice-rafted debris represents isolated mountain glaciers or more-extensive ice-sheet cover.

The central *Arctic Ocean* sediment core of Moran et al. (2006) shows a highly condensed record that suggests erosion or little deposition across this interval of ice rafting off *Greenland* studied by Eldrett et al. (2007; see previous paragraph) and until about 16 Ma. Ice-rafted debris, interpreted as representing iceberg as well as sea-ice transport, was actively delivered to the open-ocean site studied by Moran et al. (2006) at 16 Ma, and volumes increased about 14 Ma and again about 3.2 Ma (also see Shackleton et al., 1984; Thiede et al., 1998; Kleiven et al., 2002). St. John and Krissek (2002) suggested onset of sea-level glaciation in southeastern *Greenland* at about 7.3 Ma, on the basis of ice-rafted debris near *Greenland* in the *Irminger Basin*. Because of its geographical pattern, the increase in ice-rafted debris about 3.2 Ma is thought to have had sources in *Greenland*, Scandinavia, and the North American landmass (*Laurentide Ice Sheet*). However, tying the debris to particular source rocks (e.g., Hemming et al., 2002) has not been possible. Additionally, no direct evidence shows whether this debris was supplied to the ocean by an extensive ice sheet or by vigorous glaciers that drained coastal mountains in the absence of ice from *Greenland*'s central lowlands. Despite the lack of conclusive evidence, *Greenland* seems to have supported at least some glaciation since at least 38 Ma; glaciation left more records after about 14 Ma (middle **Miocene**). Thus, as Earth cooled from the "hothouse" conditions extant during the time of dinosaurs, ice sheets began to form on *Greenland*.

Thus, as Earth cooled from the "hothouse" conditions extant during the time of dinosaurs, ice sheets began to form on Greenland.

Following the establishment of ice in *Greenland*, a notable warm interval about 2.4 million years (**m.y.**) ago is recorded by the *Kap København* Formation of North *Greenland* (Funder et al., 2001). This formation is a 100-m-thick unit of sand, silt, and clay deposited primarily in shallow marine conditions. Fossil biota in the deposit switch from Arctic to subarctic to **boreal** assemblages during the depositional interval. The unit was deposited rapidly, perhaps in 20,000 years or less. Funder et al. (2001) postulated complete deglaciation of *Greenland* at this time, primarily on the basis of the great summertime warmth indicated at this far-northern site, although clearly there is no comprehensive record of the whole ice sheet.

5.3.2 The Most Recent Million Years

Fragmented records on land combined with lack of unequivocal indicators in the ocean complicate ice-sheet reconstructions. Nonetheless, many additional indications of ice-sheet change are available between the time of the *Kap København* Formation and the most recent 100,000 years. Locally, ice expanded during colder times and ice retreated during warmer times, but data provide no comprehensive overviews of the ice sheet. This section (5.3.2) summarizes data especially from **Marine Isotope Stage (MIS)** 11 (about 440 ka; see Chapter 2, Paleoclimate Concepts, section 2.5, Chronology) to MIS 5 (about 130 ka), although dating uncertainties allow the possibility that some of the samples are older than MIS 11, and detailed consideration of MIS 5 is deferred to subsequent sections.

Glacial-interglacial cycles have been studied by examining the oxygen isotope composition of foraminifers in deep-sea cores, and we now have a fairly detailed picture of how glacial ice has expanded and retreated during the past 2 m.y. or so (the Quaternary period). Figure 5.4 shows the four most recent glacial-interglacial cycles: peaks represent interglacial periods (relatively high sea levels) and troughs represent glacial periods (relatively low sea levels). Glacial periods in the oxygen isotope record are called "**stages**" and are numbered back in time with even numbers; interglacial stages are numbered back in time with odd numbers. Thus, the present interglacial is MIS 1 and the preceding glacial period is MIS 2.

5.3.2A Far-Field Sea-Level Indications

Other far-field evidence supports the concept that during Marine Isotope Stage 11 sea level was higher than at present.

In the absence of clear and well-dated records proximal to the *Greenland Ice Sheet*, records of global sea level that may be related to changes on *Greenland* are of interest. If we consider only the past few glacial cycles, it is most likely that sea level was as high as or higher than present during previous interglacial times (MIS 5, 7, 9, and 11; Figure 5.4). Under the assumption that any glacial-isostatic-adjustment contributions to these relative highstands of sea level were small, and thus that highstands of sea level were primarily related to changes in ice volume, the amplitudes of the various highstands of sea level provide a measure of the long-term mass balance of the *Greenland Ice Sheet* and other contemporaneous ice masses.

Far from the *Greenland Ice Sheet*, some fragmentary and poorly dated deposits suggest a higher-than-present sea-level stand during MIS 11, about 400 ka. Sea-level history of MIS 11 (about 362–420 ka) is of particular interest to paleoclimatologists because the Earth-Sun orbital geometry during that interglacial epoch is similar to the configuration during the current interglacial (Berger and Loutre, 1991). (As noted in Chapter 2, Chronology, age assignments to Marine Isotope Stages may differ in different usages; both age ranges and Marine Isotope Stage names are given here for information, not as definitions.)

Hearty et al. (1999) proposed that marine deposits found in a cave on the tectonically stable island of Bermuda date to the MIS 11 interglacial epoch. These marine deposits are about 21 m above modern sea level, and they contain coral pebbles that have been dated by U-series techniques. Hearty et al. (1999) interpreted the deposits to date to about 400 ka, although the coral pebbles were dated older than 500 ka. The authors' interpretation is based primarily on an overlying deposit that dates to about 400 ka. Although the underlying marine deposit appears to record an old sea stand markedly higher than present, the chronology is still uncertain.

An Alaskan marine deposit is also found at altitudes of up to 22 m (Kaufman et al., 1991), similar to altitudes of the cave deposit on Bermuda. The deposit, representing what has been called the "Anvilian marine transgression," extends along the *Seward Peninsula* and *Arctic Ocean* coast of *Alaska*. This part of *Alaska* is tectonically stable. It is landward of Pelukian (MIS 5 (about 74–130 ka)) marine deposits. Amino-acid ratios in mollusks (Kaufman and Brigham-Grette, 1993) show that the Anvilian deposit is easily distinguishable from last-interglacial (locally called Pelukian) deposits, but it is younger than deposits thought to be of **Pliocene** age (about 1.8–5.3 Ma). Kaufman et al. (1991) reported that basaltic lava overlies deposits of the Nome River glaciation, which in turn overlie Anvilian marine deposits. An average of several analyses on the lava yields an age of 470±190 ka. Within the broad limits permitted by this age, and using reasonable rates of changes in the amino-acid ratios of marine mollusks, Kaufman et al. (1991) proposed that the Anvilian marine transgression dates to about 400 ka and correlates with MIS 11.

Other far-field evidence supports the concept that during MIS 11 sea level was higher than at present. Oxygen-isotope and faunal data from the Cariaco Basin off Venezuela provide independent evidence of a higher-than-present sea level during MIS 11 (Poore and Dowsett, 2001). If the Bermudan cave deposits and the Anvilian marine deposits of *Alaska* prove to be genuine manifestations of a high sea stand about 400 ka, the implication for climate history is that all of the GREENLAND ICE SHEET (Willerslev et al., 2007; see section 5.3.2B, below), all of the West Antarctic ice sheet, and part of the East Antarctic ice sheet would have disappeared at this time (these ice masses are generally accepted as the most vulnerable); preservation of the GREENLAND ICE SHEET would require much more loss from the East Antarctic ice sheet, which is widely considered to be relatively stable (e.g., Huybrechts and de Wolde, 1999).

Until recently, no reliably dated emergent marine deposits from MIS 9 (about 303–331 ka) had been found on tectonically stable coasts, although coral reefs of this age have been recognized for some time on the tectonically rising island of Barbados (Bender et al., 1979). Stirling et al. (2001) reported that well-preserved fringing reefs are found on Henderson Island in the southeastern Pacific Ocean. Reef elevations on this tectonically stable island are as high as about 29 m above sea level, and U-series dates between about 334±4 and 293±5 ka correlate with MIS 9. Despite the good preservation of the corals and the reefs they are found in, and the reliable U-series ages, it is uncertain how high sea level was at this time. Although Henderson Island is geologically stable, it is experiencing slow uplift (less than 0.1 m/1,000 yr) due to volcanic loading by the emplacement of nearby Pitcairn Island. A correction for maximum uplift rate, therefore, could put the MIS 9 ancient level estimate below present sea level. Multer et al. (2002) reported U-series ages of about 370 ka for a coral (*Montastrea annularis*) from a fossil reef drilled at a locality called Pleasant Point in Florida Bay. This coral showed clear evidence of open-system conditions (i.e., it was not completely chemically isolated from its surroundings since formation, a requirement for the measured age to be accurate), and the age is probably closer to 300–340 ka, if we use the correction scheme of Gallup et al. (1994). If so, the age suggests that during MIS 9, sea level was close to but not much above the present level.

As with MIS 9, several MIS 7 (about 190–241 ka) reef or terrace records have been found on tectonically rising coasts (Bender et al., 1979; Gallup et al., 1994; Edwards et al., 1997), but far fewer have been found on tectonically relatively stable coasts. However, two recent reports show evidence of MIS 7 sea-level high stands on tectonically stable islands. One is a pair of U-series ages of about 200 ka from coral-bearing marine deposits about 2 m above sea level on Bermuda (Muhs et al., 2002). The other is a single coral age from the Florida Keys (Muhs et al., 2004). They collected samples of near-surface *Montastrea annularis* corals in quarry spoil piles on Long Key. Analysis of a single sample shows an apparent age of 235±4 ka. The higher-than-modern initial $^{234}U/^{238}U$ value indicates a probable bias to an older age by about 7 ka; thus, the true age may be closer to about 220–230 ka, if we again use the Gallup et al. (1994) correction scheme. If valid, these data suggest that sea level may have stood close to its present level during the interglacial period MIS 7. Much more study is needed to confirm these preliminary ages, however.

If the Bermudan cave deposits and the Anvilian marine deposits of Alaska prove to be genuine manifestations of a high sea stand about 400 ka, the implication for climate history is that all of the Greenland Ice Sheet, all of the West Antarctic ice sheet, and part of the East Antarctic ice sheet would have disappeared at this time.

Taken together, these data point to MIS 11 as a time in which sea level likely was notably higher than at present, although the data are sufficiently sparse that stronger conclusions are not warranted. If so, melting of GREENLAND ice seems likely, mostly on the basis of elimination: GREENLAND meltwater is thought to be able to supply much of the sea-level rise needed to explain the observations, and the alternative—extracting an additional 7 m of sea-level rise through melting in East Antarctica—is not considered as likely. Marine Isotope Stages 9 and 7 seem to have had sea levels similar to modern ones.

Taken together, these data point to Marine Isotope Stage 11 as a time in which sea level likely was notably higher than at present.

5.3.2b Ice-Sheet Indications

The cold MIS 6 ice age (about 130–188 ka) may have produced the most extensive ice in *Greenland* (Wilken and Meinert, 2006). Recently described glacial deposits in East *Greenland* support this view (Adrielsson and Alexanderson, 2005), although more-extensive, older deposits are known locally (Funder et al., 2004). Funder et al. (1998) reconstructed thick ice (greater than 1,000 m) during MIS 6 in areas of *Jameson Land* (East *Greenland*) that now are ice-free. However, no confident ice-sheet-wide reconstructions based on paleoclimatic data are available for MIS 6 ice.

Both northwest and East *Greenland* preserve widespread marine deposits from early in the MIS 5 interglacial (the interglacial previous to the present one) (about 74–130 ka), and particularly from the warmest subdivision of MIS 5, called MIS 5e (about 123 ka). Depression of the land from the weight of MIS 6 ice allowed incursion of seawater as ice melted during the transition to MIS 5e. The resulting deposits were not reworked by the subsequent incursion of seawater during the transition from the most recent glaciation (MIS 2, which peaked about 24 ka or slightly more recently) to the modern interglacial (MIS 1, less than 11 ka). Thus, seawater moved farther inland during the transition from MIS 6 (glacial) to MIS 5 (interglacial) than during the transition from MIS 2 (most recent glacial) to MIS 1 (current interglacial).

Several hypotheses can explain this difference. Perhaps most simply, there may have been more ice on *Greenland* causing greater isostatic depression during MIS 6 than during MIS 2. However, if some or all of the older deposits survived being overridden by cold-based ice of MIS 2, additional possibilities exist. Because isostatic uplift occurs while ice is thinning but before the ice margin melts enough to allow incursion of seawater, perhaps the MIS 6 ice melted faster and allowed incursion of seawater over more-depressed land than was true for MIS 2 ice. Additionally, at the time during MIS 6 that ice in *Greenland* receded and thus allowed incursion of sea-water, global sea level might have been higher than during the corresponding part of MIS 2 (perhaps because of relatively earlier melting of MIS 6 ice on North America or elsewhere beyond *Greenland*). More-detailed modeling of glacial isostatic adjustment will be required to test these hypotheses. Nonetheless, the leading hypothesis seems to be that ice was more extensive in MIS 6 than in MIS 2.

A particularly interesting new result comes from analysis of materials found in ice cores from the deepest part of the ice sheet. Willerslev et al. (2007) attempted to amplify deoxyribonucleic acid (DNA) in three samples:

1. silty ice at the base of the *Greenland Ice Sheet* from the *Dye-3* drill site (on the southern dome of the ice sheet) and the *GRIP* drill site (at the crest of the main dome of the ice sheet),
2. "clean" ice just above the silty ice of these sites, and
3. the *Kap København* formation.

The *Kap København*, clean-ice, and *GRIP* silty samples did not yield identifiable quantities of DNA, but it was possible to prepare extensive materials from the *Dye-3* silty ice. The lack of DNA at *Kap København* probably indicates post-depositional changes, perhaps during room-temperature storage following collection. The lack of DNA at *GRIP* demonstrates that there is no important wind-blown source. In turn, this indicates that the material at *Dye-3* has a relatively local source. The *Dye-3* material indicates a northern boreal forest, compared with the tundra environment that exists in coastal sites at the same latitude and lower elevation today. The taxa indicate mean July temperatures then above 10°C and minimum winter temperatures above −17°C at an elevation of about 1 km above sea level (allowing for isostatic rebound following ice melting). Dating of this warm, reduced-ice time is uncertain, but a tentative age of 450–800 ka is probably consistent with the indications of high sea level in MIS 11.

Nishiizumi et al. (1996) reported on radioactive cosmogenic isotopes in rock core collected from beneath the ice at the ***GISP2*** site (central *Greenland*, 28 km west of the *GRIP* site at the *Greenland* summit). Joint analysis of beryllium-10 and aluminum-26 indicated a few-millennia-long interval of exposure to cosmic rays (hence ice cover of thickness less than 1 m or so) about 500±200 ka. This information is consistent with, and thus provides further support for, the DNA results of Willerslev et al. (2007). This work was presented at a scientific meeting and in an abstract but not in a refereed scientific journal, and thus it is subject to lower confidence than is other evidence discussed in this report.

No long, continuous climate records from GREENLAND itself are available for the time interval occupied by the boreal forest at DYE-3 reported by Willerslev et al. (2007). Marine-sediment records from around the NORTH ATLANTIC point toward MIS 11, at about 440 ka, as the most likely time of anomalous warmth. Owing to orbital forcing factors (reviewed in Droxler et al., 2003), this interglacial seems to have been anomalously long compared with those before and after. As discussed above, indications of sea level above modern level exist for this interval (Kindler and Hearty, 2000), but much uncertainty remains (see Rohling et al., 1998; Droxler et al., 2003). Records of sea-surface-temperature in the NORTH ATLANTIC indicate that MIS 11 temperatures were similar to those from the current interglacial (**Holocene**) within 1°–2°C; slightly cooler, similar, or slightly warmer conditions have all been reported (e.g., McManus et al., 1999; Bauch et al., 2000; Helmke et al., 2003; Kandiano and Bauch, 2003; de Abreu et al., 2005). The longer of these records show no other anomalously warm times within the age interval most consistent with the Willerslev et al. (2007) dates. (Notice, however, that during MIS 5e locally higher temperatures are indicated in GREENLAND than are indicated in the far-field sea-surface temperatures. Thus, the absence of warm temperatures far from the ice sheet does not guarantee the absence of warm temperatures close to the ice sheet; see section 5.3.3, below.) The independent indications of high global sea level during MIS 11, as discussed above in section 5.3.2A, and of major GREENLAND ICE SHEET shrinkage or loss at that time, are mutually consistent.

The GREENLAND ICE SHEET is thought to complete most of its response to a step forcing in climate within a few millennia (e.g., Alley and Whillans, 1984; Cuffey and Clow, 1997). Thus, any of the interglacials during the last 420,000 years was long enough for the ice sheet to have completed most of its response to the end-of-ice-age forcings (although smaller forcings during the interglacials may have precluded a completely steady state). Thus, it is not obvious how a longer-yet-not-warmer interglacial, as suggested by MIS 11 indicators in the NORTH ATLANTIC away from GREENLAND, would have caused notable or even complete loss of the GREENLAND ICE SHEET, although this result cannot be ruled out completely. Many possible interpretations remain: greater GREENLAND warming in MIS 11 than indicated by marine records from well beyond the ice sheet, large age error in the Willerslev et al. (2007) estimates, great warmth at DYE-3 yet a reduced but persistent GREENLAND ICE SHEET nearby, and others. One possible interpretation is that the threshold for notable shrinkage or loss of GREENLAND ice is just 1°–2°C above the temperature reached during MIS 5e, thus falling within the error bounds of the data.

The data strongly indicate that GREENLAND's ice was notably reduced, or lost, sometime after ice coverage became extensive and large ice ages began, while temperatures surrounding GREENLAND were not grossly higher than they have been recently. The rate of mass loss within the warm period is unconstrained; the long interglacial at MIS 11 allows the possibility of very slow loss or much faster loss. If the cosmogenic isotopes in the *GISP2* rock core are interpreted at face value, then the time over which ice was absent was only a few millennia.

The data strongly indicate that Greenland's ice was notably reduced, or lost, sometime after ice coverage became extensive and large ice ages began, while temperatures surrounding Greenland were not grossly higher than they have been recently.

5.3.3 Marine Isotope Stage 5e

5.3.3A Far-Field Sea-Level Indications

Investigators studying sea-level history have paid most attention to sea level during the last interglacial, MIS 5 (about 71–122 ka), and specifically to MIS 5e (about 123 ka). The evidence of past sea level during MIS 5e along tectonically stable coasts is summarized here (Muhs, 2002). Sea-level high stand during MIS 5e is best estimated from coral reef and marine deposits now above sea level at sites in Australia, the Bahamas, Bermuda, and the Florida Keys.

On the coast and islands of tectonically stable Western Australia, emergent coral reefs and marine deposits now 2–4 m above sea level are widespread and well-preserved. U-series ages of the fossil corals at mainland localities and Rottnest Island range from 128±1 to 116±1 ka (Stirling et al., 1995, 1998). The main period of last-interglacial coral growth was a restricted interval from about 128–121 ka (Stirling et al., 1995, 1998). Because the highest corals are about 4 m above sea level at present but grew at some unknown depth below sea level, 4 m is a minimum for the amount of last-interglacial sea-level rise.

The islands of the Bahamas are tectonically stable, although they may be slowly subsiding owing to carbonate loading on the Bahamian platform. Fossil reefs in the Bahamas are well preserved (Chen et al., 1991), reefs have elevations up to 5 m above sea level, and many corals are in growth position. On San Salvador Island, reef ages range from 130.3±1.3 to 119.9±1.4 ka. The sea level record of the Bahamas is particularly valuable because many reefs contain the coral *Acropora palmata*, a species that almost always lives within the upper 5 m of the water column (Goreau, 1959). Thus, fossil reefs containing this species place a fairly precise constraint on the former water depth.

As discussed above (section 5.3.2A), Bermuda is tectonically stable. Bermuda does not host MIS 5e fossil reefs, but numerous coral-bearing marine deposits fringe the island. A number of U-series ages of corals from Bermuda range from about 119 ka to about 113 ka (Muhs et al., 2002). The deposits are found 2–3 m above present sea level, although overlying wind-blown sand prevents precise estimates of where the former shoreline lay.

A summary of additional sites indicate a sea-level rise of 4 m to more than 6 m during MIS 5e.

The Florida Keys, not far from the Bahamas, are also tectonically stable. Fruijtier et al. (2000) reported ages for corals from Windley Key, Upper Matecumbe Key, and Key Largo that, when corrected for high initial $^{234}U/^{238}U$ values (Gallup et al., 1994), are in the range of 130–121 ka. The last-interglacial MIS 5 reef on Windley Key is 3–5 m above present sea level, on Grassy Key it is 1–2 m above sea level, and on Key Largo it is 3–4 m above modern sea level.

The collective evidence from Australia, the Bahamas, Bermuda, and the Florida Keys shows that sea level was above its present stand during MIS 5e. On the basis of measurements of the reefs themselves, sea level then was at least 4–5 m higher than sea level now. An additional correction should be applied for the water depth at which the various coral species grew. Most coral species found in the Bahamas, Bermuda, and the Florida Keys require water depths of at least a few meters for optimal growth, and many live tens of meters below the ocean surface. For example, *Montastrea annularis*, the most common coral found in MIS 5e reefs of the Florida Keys, has an optimum growth depth of 3–45 m and can live as deep as 80 m (Goreau, 1959). A minimum rise in sea level is calculated thusly: fossil reefs are 3 m above present sea level, and the most conservative estimate of the depth at which they grew is 3 m. Thus, the MIS 5e sea level was at least 6 m higher than modern-day sea level (Figures 5.5, 5.6). A summary of additional sites led Overpeck et al. (2006) to indicate a sea-level rise of 4 m to more than 6 m during MIS 5e.

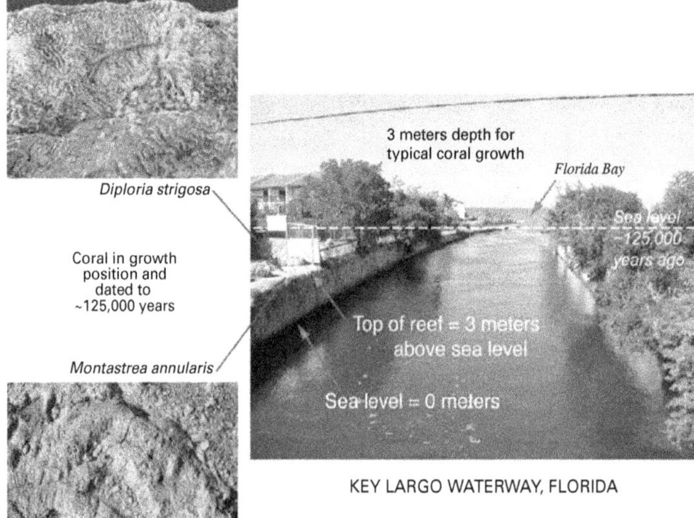

Figure 5.5. Last-interglacial (MIS 5e) reef and corals on Key Largo, Florida, their elevations, probable water depths, and estimated paleo-sea level. Photographs by D.R. Muhs.

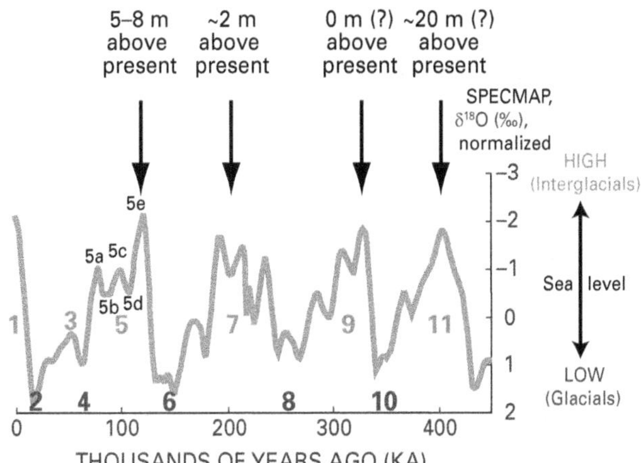

Figure 5.6. Oxygen isotope data from the SPECMAP record (Imbrie et al., 1984), with indications of sea-level stands for different interglacials, assuming minimal glacial isostatic adjustments to the observed reef elevations. Numbers identify Marine Isotope Stages (MIS) 1 through 11.

Existing estimates generally presume that glacial isostatic adjustment have not notably affected the sites at the key times. The data set, and the accuracy of the dates (also see Thompson and Goldstein, 2005) are becoming sufficient to support, in future work, improved corrections for glacial isostatic adjustment.

The implications of a 4-m- to more than 6-m-sea-level highstand during the last interglacial are as follows:

1. all or most of the *Greenland Ice Sheet* would have melted; or
2. all or most of the West Antarctic ice sheet would have melted; or
3. parts of both would have melted.

Both ice sheets may indeed have melted in part, but greater melting is likely from *Greenland* (Overpeck et al., 2006), as described in section 5.3.3c, below.

5.3.3b Conditions in Greenland

Paleoclimate data provide strong evidence for notable warmth on and around *Greenland* during MIS 5e, with peak temperatures occurring ~130 ka. As summarized by CAPE (2006), terrestrial data indicate peak summertime temperatures ~4°C above recent in northwest *Greenland* and ~5°C above recent in East *Greenland* (and thus 2°–4°C above the mid-Holocene warmth (~6 ka); Funder et al., 1998, and see below), with near-shore marine conditions 2°–3°C above recent in East *Greenland*. Climate-model simulations by Otto-Bliesner et al. (2006) show that the strong summertime increase of sunshine (**insolation**) in MIS 5e as compared with now caused strong warming, which was amplified by ice-albedo and other **feedbacks**. Simulated summertime warming around *Greenland* exhibited local maxima of 4°–5°C in those northwestern and eastern coastal regions for which terrestrial and shallow-marine summertime data are available and show matching warmings; elsewhere over *Greenland* and surroundings, typical warmings of ~3°C were simulated.

The sea-level record in East Greenland (*Scoresby Sund*) indicates a two-step inundation at the start of MIS 5e. Of the possible interpretations, Funder et al. (1998) favored one in which early deglaciation of the coastal region of *Greenland* preceded much of the melting of non-*Greenland* land ice, so that early coastal flooding after deglaciation of isostatically depressed land was followed by uplift and then by flooding attributable to sea-level rise as that far-field land ice melted. Additional testing of this idea would be very interesting, because it suggests that the *Greenland Ice Sheet* has responded rapidly to climate forcing in the past.

Much of the evidence of climate change in *Greenland* comes from ice-core records. As discussed next, these changes cannot be estimated independent of a discussion of the ice sheet, because of the possibility of thickness change. Hence, the changes in the ice sheet are discussed before additional evidence bearing on forcing and response.

The implications of a 4 m to more than 6 m sea-level highstand during the last interglacial are as follows:

1. all or most of the Greenland Ice Sheet would have melted; or
2. all or most of the West Antarctic ice sheet would have melted; or
3. parts of both would have melted.

5.3.3C Ice-Sheet Changes

The *Greenland Ice Sheet* during MIS 5e covered a smaller area than it does now. How much smaller is not known with certainty. The most compelling evidence is the absence of pre-MIS 5e ice in the ice cores from south, northwest, and East *Greenland* (the locations *Dye-3*, *Camp Century*, and *Renland* drilling sites, respectively). In all of these cores, the climate record extends through the entire last glacial epoch and then terminates at the bed in a layer of ice deposited in a much warmer climate (Koerner, 1989; Koerner and Fisher, 2002). This basal ice is most likely MIS 5e ice. Moreover, the composition of this ice is not an average of glacial and interglacial values, as would be expected if it were a mixture of ices from earlier cold and warm climates. Instead, the ice composition exclusively indicates a climate considerably warmer than that of the Holocene. (One cannot entirely eliminate the possibility that each core independently bottomed on a rock that had been transported up from the bed, and that older ice lies beneath each rock, but this seems highly improbable.)

Unfortunately the picture is cloudy—not unlike the basal ice itself, which has a small amount of silt and sand dispersed through it.

At *Dye-3*, the oxygen isotope composition of this basal ice layer is reported as $\delta^{18}O = -23‰$, which means that it is 23‰ (or 2.3%) lighter than standard mean ocean water. Moreover, a value of $\delta^{18}O = -30‰$ is reported for modern snowfall in the source region (up-flow from the site of *Dye-3*). At *Camp Century*, a value of $\delta^{18}O = -25‰$ is reported for basal ice; a value of $\delta^{18}O = -31.5‰$ is reported in the source region (see Table 2 of Koerner, 1989). These changes of about 7‰ are much larger than the MIS 5e-to-MIS 1 climatic signal (about 3.3‰, according to the central *Greenland* cores; see below in this section). Thus, the MIS 5e ice at *Dye-3* and *Camp Century* not only indicates a warmer climate but also a much lower source elevation: the ice sheet was re-growing when these MIS 5e ices were deposited.

The central Greenland cores do reveal two important facts: MIS 5e was warmer than MIS 1, and the elevation in the center of the ice sheet was similar to that of the modern ice sheet.

In combination, these two observations (absence of pre-MIS 5e ice, and anomalously low-elevation sources of the basal ice) indicate that the *Greenland* margin had retreated considerably during MIS 5e. Of greatest importance is that retreat of the margin northward past *Dye-3* implies that the southern dome of the ice sheet was nearly or completely gone.

In this context it is useful to understand the genesis of the basal ice layer, and the layer at *Dye-3* in particular. Unfortunately the picture is cloudy—not unlike the basal ice itself, which has a small amount of silt and sand dispersed through it, making it opaque. This silty basal layer is about 25 m thick (Souchez et al., 1998). Overlying it is "clean" (not notably silty) ice that appears to be typical of polar ice sheets. Its total gas content and gas composition indicate that the ice formed by normal densification of firn in a cold, dry environment. The oxygen isotope composition of this clean ice is –30.5‰. The bottom 4 m of the silty ice is radically different; its oxygen isotope value is –23‰, and its gas composition indicates substantial alteration by water. The total gas content of this basal silty ice is about half that of normal cold ice formed from solid-state transformation of firn, the carbon dioxide content is 100 times normal, and the oxygen/nitrogen ratio is less than 20% that of normal cold ice. This basal silty layer may be superimposed ice (ice formed by refreezing of meltwater in snow on a glacier or ice sheet, as Koerner (1989) suggested for the entire silty layer), or it may be non-glacial snowpack, or it may be a remnant of segregation ice in **permafrost** (permafrost commonly contains relatively "clean" although still impure lenses of ice, called segregation ice).

In any case, the upper 21 m of the silty ice may be explained as a mixture of these two end members (Souchez et al. 1998). As they deform, ice sheets do mix ice layers by small-scale structural folding (e.g., Alley et al., 1995b), by interactions between rock particles, by grain-boundary diffusion, and possibly by other processes. Unfortunately, there is no way to distinguish rigorously how much this ice really is a mixture of these end-member components and how much of it is warm-climate (presumably MIS 5e) normal ice-sheet ice. The difficulty is that the bottom layer is not itself well mixed (its gas composition is highly variable), so a mixing model for the middle layer uses an essentially arbitrary composition for one end member. Souchez et al., (1998) used the composition at the top of the bottom layer for their mixing calculations, but it could just as well be argued that the composition here is determined by exchange with the overlying layer and is not a fixed quantity.

As discussed in section 5.3.2B, above, in a recent study, Willerslev et al. (2007) examined biological molecules in the silty ice from

Dye-3, including DNA and amino acids. They concluded that organic material contained in that *Dye-3* ice originated in a boreal forest (remnants of diagnostic plants and insects were identified). This environment implies a very much warmer climate than at the present margin in *Greenland* (e.g., July temperatures at 1 km elevation above 10°C), and hence it also suggests a great antiquity for this material; no evidence suggests that MIS 5e in *Greenland* was nearly this warm. Indeed, Willerslev et al. (2007) also inferred the age of the organic material and the age of exposure of the rock particles, using several methods. They concluded that a 450–800 ka age is most likely, although uncertainties in all four of their dating techniques prevented a definitive statement. This conclusion suggests that the bottom ice layer (the source of rock material in the overlying mixed layer) is much older than MIS 5e.

This evidence admits of two principal interpretations. One is that this material survived the MIS 5e deglaciation by being contained in permafrost. The second is that the MIS 5e deglaciation did not extend as far north as the *Dye-3* site, and that local topography allowed ice to persist, isolated from the large-scale flow. This latter hypothesis (apparently favored by Willerslev et al., 2007) does not explain the several-hundred-thousand-year hiatus within the ice, however, or the purely interglacial composition of the entire basal ice, both of which favor the permafrost interpretation. (Both hypotheses can be modified slightly to allow short-distance ice-flow transport to the *Dye-3* site; e.g., Clarke et al., 2005.)

Ice-sheets can also slide at their margins. Sliding near the modern margin of the *Greenland Ice Sheet* (e.g., Joughin et al., 2008a) provides a way to rapidly re-establish the ice sheet in deglaciated regions and to preserve soil or permafrost materials as the ice re-grows, as described next. Marginal regions of the *Greenland Ice Sheet* are thawed at the bottom and slide over the materials beneath (e.g., Joughin et al., 2008a)—on a thin film of water or possibly thicker water or soft sediments. During a time of cooling, sliding advances the ice margin more rapidly than would be possible if the ice were frozen to the bed. Furthermore, the sliding will bring to a given point ice that was deposited elsewhere and at higher elevation; subsequently, that ice may freeze to the bed. As discussed below (section 5.3.5B), widespread evidence shows a notable advance of the ice-sheet margin during the last few millennia. Regions near the ice-sheet margin, and icebergs calving from that margin, now contain ice that was deposited somewhere in the accumulation zone at higher elevation and that slid into position (e.g., Petrenko et al., 2006). Were sliding not present, one might expect that re-glaciation of a site such as *Dye-3* would have required cooling until the site became an accumulation zone, followed by slow buildup of the ice sheet.

In contrast to all the preceding information from south-, northwest-, and east-*Greenland* ice cores, the ice cores from central *Greenland* (the *GISP2* and *GRIP* cores; Suwa et al., 2006) and north-central *Greenland* (the ***NGRIP*** core) do contain MIS 5e ice that is normal, cold-environment, ice-sheet ice. Unfortunately, none of these cores contains a complete or continuous MIS 5e chronology. Layering of the *GISP2* and *GRIP* cores is disrupted by ice flow (Alley et al., 1995b) and, in the *NGRIP* core, basal melting has removed the early part of MIS 5e and any older ice (Dahl-Jensen et al., 2003). The central *Greenland* cores do reveal two important facts: MIS 5e was warmer than MIS 1 (oxygen isotope ratios were 3.3‰ higher than modern ones), and the elevation in the center of the ice sheet was similar to that of the modern ice sheet, although the ice sheet was probably slightly thinner in MIS 5e (within a few hundred meters of elevation, based on the total gas content). Thus, if we consider also evidence from the other cores, the ice sheet shrank substantially under a warm climate, but it persisted in a narrower, steeper form.

Thus, if we consider also evidence from the other cores, the ice sheet shrank substantially under a warm climate, but it persisted in a narrower, steeper form.

What climate conditions were responsible for driving the ice sheet into this configuration? The answer is not clear. None of the paleoclimate proxy information is continuous over time, both precipitation and temperature changes are important, and some factors related to ice flow are poorly constrained. Cuffey and Marshall (2000; also see Marshall and Cuffey, 2000) were the first to address this question using the information from the central *Greenland* cores as constraints. In particular, Cuffey and Marshall (2000) noted that oxygen isotope ratios were at least 3.3‰ higher during MIS 5e, and they used this value to constrain the climate forcing on an ice sheet model. Because the isotopic composition depends on the elevation of the ice-sheet surface as well as on temperature change at a constant elevation, these analyses generated both climate histories

and ice-sheet histories. Results depended critically on the isotopic sensitivity parameter relating isotopic composition to temperature and on the way past accumulation rates are estimated, which have large uncertainties. Furthermore, there was no attempt to model increased flow in response to changes of calving margins, or increased flow in response to production of surface meltwater (see Lemke et al., 2007). Thus, the ice sheet model was conservative; a given climatic temperature change produced a smaller response in the modeled ice sheet than is expected in nature.

In the reconstruction favored by Cuffey and Marshall (isotopic sensitivity α = 0.4‰ per °C), the southern dome of *Greenland* completely melted after a sustained (for at least 2,000 years) climate warming (mean annual, but with summer most important) of approximately 7°C higher than present. In a different scenario (sensitivity α = 0.67‰ per °C), the southern ice sheet margin did not retreat past *Dye-3* after a sustained warming of 3.5°C. Thus an intermediate scenario (sustained warming of 5°–6°C) is required, in this view, to cause the margin to retreat just to *Dye-3*. Given the conservative representation of ice dynamics in the model, a smaller sustained warming would in fact be sufficient to accomplish such a retreat. How much smaller is not known, but it could be quite small. Outflow of ice can increase by a factor of two in response to modest changes in air and ocean temperatures at the calving margins (see Lemke et al., 2007).

Mass balance depends on numerous variables that are not modeled, introducing much uncertainty. Examples of these variables are storm-scale weather controls on the warmest periods within summers, similar controls on annual snowfall, and increased warming due to exposure of dark ground as the ice sheet retreats. In contrast to the under-representation of ice dynamics, however, no major observations show that the models are fundamentally in error with respect to surface mass-balance forcings.

A hint of a serious error is, however, provided by the record of accumulation rate from central *Greenland*. During the past about 11,000 years (MIS 1) variations in snow accumulation and in temperature show no consistent correlation (Cuffey and Clow, 1997; Kapsner et al., 1995), whereas most models assume that snowfall (and hence accumulation) will increase with temperature. This lack of correlation suggests that models are over-predicting the extent to which increased snowfall will partly balance increased melting in a warmer climate. If this MIS 1 situation in central *Greenland* applied to much of the ice sheet in MIS 5e, then models would require less warming to match the reconstructed ice-sheet footprint. Again, the real ice sheet appears to be more vulnerable than the model ones. We refer to this observation as only a "hint" of a problem, however, because snowfall on the center of *Greenland* may not represent snowfall over the whole ice sheet, for which other climatological influences come into play.

The climate forcing for the Cuffey and Marshall (2000) ice dynamics model, like that of most recent models that explore *Greenland*'s glacial history, is driven by a single **paleoclimate record**, the isotope-based surface temperature at the Summit ice core sites. From this information, temperature and precipitation fields are derived and then combined to obtain a mass balance forcing over space and time, which is then applied to the entire ice sheet. This approach can be criticized for eliminating all local-scale climate variability, but few observations would allow such variability to be adequately specified.

Recent efforts to estimate the minimum MIS 5e ice volume for *Greenland* have much in common with the Cuffey and Marshall (2000) approach, but they focus on adding observational constraints that optimize the model parameters. For example, the new ability to model the movement of materials passively entrained in ice sheets (Clarke and Marshall, 2002) now allows the predicted and observed isotope profiles at ice core sites to be compared. By using these capabilities, Tarasov and Peltier (2003) produced new estimates of MIS 5e ice volume that were constrained by the measured ice-temperature profiles at *GRIP* and *GISP2* and by the $\delta^{18}O$ profiles at *GRIP*, *GISP2*, and *North GRIP*. Their conservative estimate is that the *Greenland Ice Sheet* contributed enough meltwater to cause a 2.0–5.2 m rise in MIS 5e sea level; the more likely range is 2.7–4.5 m—lower than the 4.0–5.5 m estimate of Cuffey and Marshall (2000).

Ice-core sites closer to the ice sheet margins, such as *Camp Century* and *Dye-3*, better constrain ice extent than do the central *Greenland* sites (Lhomme et al., 2005). These authors added a tracer transport capability to the model used by Marshall and Cuffey (2000) and attempted to optimize the model fit to the isotope profiles at *GRIP*, *GISP2*, *Dye-3* and *Camp Century*. For now, their estimate of a 3.5–4.5 m maximum MIS 5e sea-level rise attributable to meltwater from the *Greenland Ice Sheet* is the most comprehensive estimate based on this technique (Lhomme et al., 2005).

The discussion just previous rested on interpretation of paleoclimatic data from the central *Greenland* ice cores to drive a model to match the inferred ice-sheet "footprint" (and sometimes other indicators) and thus learn volume changes in relation to temperature changes. An alternative approach is to use what we know about climate forcings to drive a coupled ocean-atmosphere climate model and then test the output of that model against paleoclimatic data from around the ice sheet. If the model is successful, then the modeled conditions can be used over the ice sheet to drive an ice-sheet model to match the reconstructed ice-sheet footprint. From response to forcing changes we then learn volume changes. This latter approach avoids the difficulty of inferring the **"α" parameter** relating isotopic composition of ice to temperature, and of assuming a relation between temperature and snow accumulation, although this latter approach obviously raises other issues. The latter approach was used by Otto-Bliesner et al. (2006; also see Overpeck et al., 2006).

The primary forcings of Arctic warmth during MIS 5e are the seasonal and latitudinal changes in solar insolation at the top of the atmosphere associated with periodic, cyclical changes in Earth's orbit (Berger, 1978). Earth's orbit varies in its **obliquity** (the inclination of Earth's spin axis to the orbital plane, which peaked at about 130 ka), **eccentricity** (the out-of-roundness of Earth's elliptical orbit around the Sun), and **precession** (the timing of closest approach to the Sun on the elliptical orbit relative to hemispheric seasons). The net effect of these factors was anomalously high summer insolation in the Northern Hemisphere during the first half of this interglacial (about 130–123 ka) (Otto-Bliesner et al., 2006; Overpeck et al., 2006). Atmosphere-Ocean General Circulation Models of the climate (AOGCMs) have used the MIS 5e seasonal and latitudinal insolation changes to calculate both the seasonal temperatures and precipitation of the atmosphere, as well as changes to sea ice and ocean temperatures. These models simulate approximately correct sensitivity to the MIS 5e orbital forcing. They reproduce the proxy-derived summer warmth for the Arctic of up to 5°C, and they place the largest warming over northern *Greenland*, northeast *Canada*, and *Siberia* (CAPE, 2006; Jansen et al., 2007).

In one of the models that has been extensively analyzed, the NCAR CCSM (National Center for Atmospheric Research Community Climate System Model), the orbitally induced warmth of MIS 5e caused loss of snow and sea ice, which in turn caused positive albedo feedbacks that reduced reflection of sunlight (Otto-Bliesner et al., 2006). The insolation anomalies increased sea-ice melting early in the northern spring and summer seasons, and those anomalies reduced the extent of Arctic sea ice from April into November. The simulated reduced summer sea ice allowed the *North Atlantic* to warm, particularly along coastal regions of the Arctic and the surrounding waters of *Greenland*. Feedbacks associated with the reduced sea ice around *Greenland* and decreased snow depths on *Greenland* further warmed *Greenland* during the summer months. In combination with simulated precipitation rates, which overall were not substantially different from present rates, the simulated mass balance of the *Greenland Ice Sheet* resulting from the model was negative. Then, as now, the surface of the ice sheet melted primarily in the summer.

The NCAR CCSM model has a mid-range climate sensitivity among comprehensive atmosphere-ocean models; that is, this model generates mid-range warming in response to doubling of CO_2 or other specified forcing (Kiehl and Gent, 2004). Temperatures and precipitation produced by the NCAR CCSM model for 130 ka were then used to drive an ice-flow model. (The model used an updated version of that used by Cuffey and Marshall (2000), and thus it also lacked representations of some physical processes that would accelerate ice-sheet response and increase sensitivity to climate change.) The ice-flow model simulated the likely configuration of the MIS 5e Greenland Ice Sheet, for comparison with paleoclimatic data on ice-sheet configuration. In this model, the Greenland Ice Sheet proved sensitive to the warmer summer temperatures when melting was taking place. Increased melting outweighed the increase in snowfall. For all but the summit of Greenland and isolated coastal sites, increased rates of melting and the extended ablation season led to a negative mass balance in response to the orbitally induced changes in temperature and snowfall. As the simulated ice sheet retreated for several millennia, the loss of ice mass lowered the surface of the Greenland Ice Sheet, which amplified the negative mass-balance and accelerated retreat. The Greenland Ice Sheet responded to the seasonal orbital forcings because it is particularly sensitive to warming in summer and autumn, rather than in winter when temperatures are too cold for melting. The modeled Greenland Ice Sheet melted in response to both direct effects (warmer atmospheric temperatures) and indirect effects (reduction of its altitude and size).

The model suggests that the Greenland Ice Sheet contributed enough meltwater to account for 1.9–3.0 m of sea-level rise for several millennia during the last interglacial.

The simulated MIS 5e Greenland Ice Sheet was a steep-sided ice sheet in central and northern Greenland (Otto-Bliesner et al., 2006) (Figure 5.7). The model did not incorporate feedbacks associated with the exposure of bedrock as the ice sheet retreated, potential meltwater-driven or ice-shelf-driven ice-dynamical processes, or time-evolving orbital forcing, so the model was probably less sensitive and more slowly responsive to warming than the real ice sheet, as noted just above. The lateral extent of the modeled minimal Greenland Ice Sheet was constrained by ice core data (see above). If the Greenland Ice Sheet's southern dome did not survive the peak interglacial warmth, as suggested by those data (Koerner and Fisher, 2002; Lhomme et al., 2005), then the model suggests that the Greenland Ice Sheet contributed

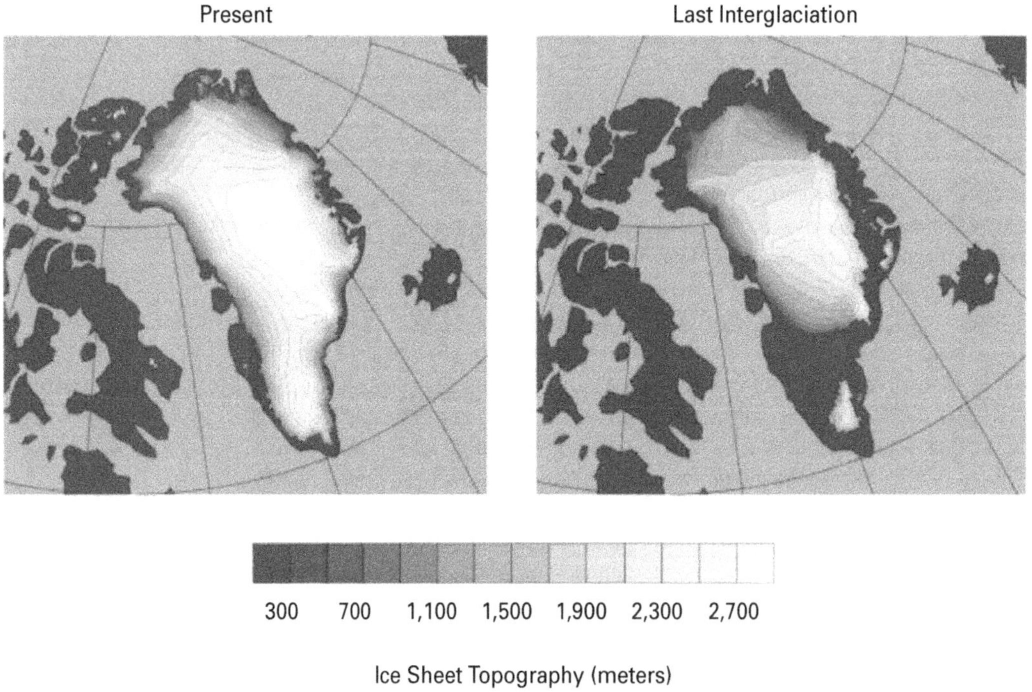

Figure 5.7. Modeled configuration of the Greenland Ice Sheet today (left) and in MIS 5e (right), from Otto-Bliesner et al. (2006). [Reprinted by permission of American Association for the Advancement of Science.]

enough meltwater to account for 1.9–3.0 m of sea-level rise (another 0.3–0.4 m rise was produced by meltwater from ice on *Arctic Canada* and *Iceland*) for several millennia during the last interglacial. The evolution through time of the *Greenland Ice Sheet*'s retreat and the linked rate at which sea level rose cannot be constrained by paleoclimatic observational data or current ice-sheet models. Furthermore, because the ice-sheet model was forced by conditions appropriate for 130 ka rather than being forced by more realistic, slowly time-varying conditions, the details of the modeled time-evolution of the *Greenland Ice Sheet* are not expected to exactly match reality. Sensitivity studies that set melting of the *Greenland Ice Sheet* at a more rapid rate than suggested by the ice-sheet model indicate that the meltwater added to the *North Atlantic* was not sufficient to induce oceanic and other climate changes that would have inhibited melting of the *Greenland Ice Sheet* (Otto-Bliesner et al., 2006).

The atmosphere-ocean modeling driven by known forcings produces reconstructions that match many data from around *Greenland* and the Arctic. The earlier work of Cuffey and Marshall (2000) had found that a very warm and snowy MIS 5e, or a more modest warming with less increase in snowfall, could be consistent with the data, and the atmosphere-ocean model favors the more modest temperature change. (The results of the different approaches, although broadly compatible, do not agree in detail, however.) The Otto-Bliesner et al. (2006) modeling leads to a somewhat smaller sea-level rise from melting of the *Greenland Ice Sheet* than does the earlier work of Cuffey and Marshall (2000). A temperature rise of 3°–4°C and a sea-level rise of 3–4 m may be consistent with the data, with notable uncertainties.

As much as 4–5 m of ice may have been removed from the Greenland Ice Sheet during MIS 5e, in response to climatic temperature changes of perhaps 2°–7°C.

Considering all of the efforts summarized above, as little as 1–2 m or as much as 4–5 m of ice may have been removed from the *Greenland Ice Sheet* during MIS 5e, in response to climatic temperature changes of perhaps 2°–7°C. At least the higher numbers for the warming are based on estimates that include the feedbacks from melting of the ice sheet. Central values in the 3–4 m and 3°–4°C range may be appropriate.

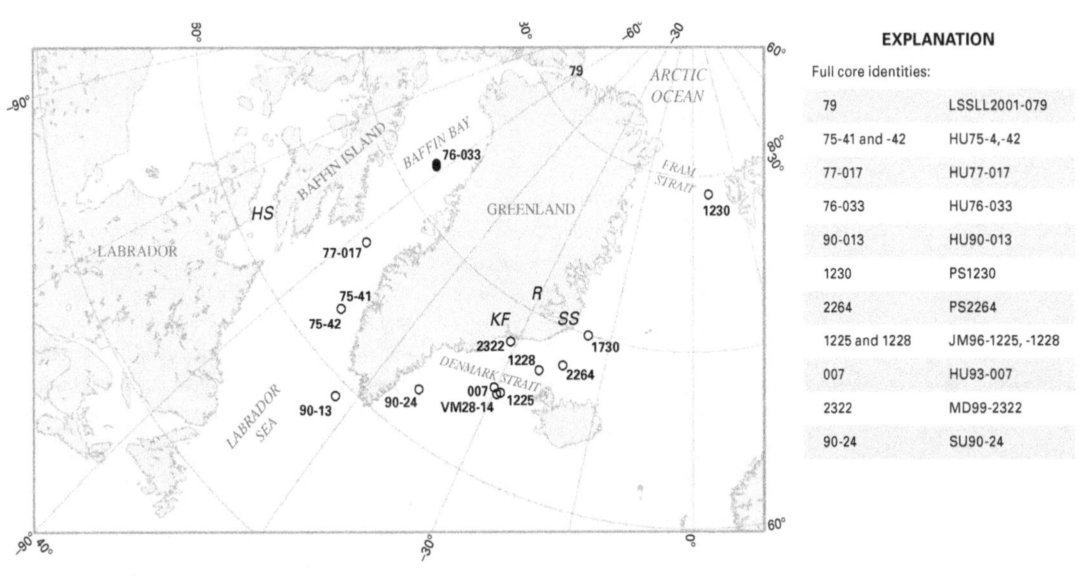

EXPLANATION

Full core identities:

79	LSSLL2001-079
75-41 and -42	HU75-4,-42
77-017	HU77-017
76-033	HU76-033
90-013	HU90-013
1230	PS1230
2264	PS2264
1225 and 1228	JM96-1225, -1228
007	HU93-007
2322	MD99-2322
90-24	SU90-24

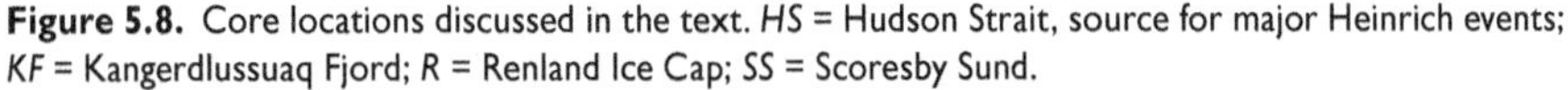

Figure 5.8. Core locations discussed in the text. *HS* = Hudson Strait, source for major Heinrich events; *KF* = Kangerdlussuaq Fjord; *R* = Renland Ice Cap; *SS* = Scoresby Sund.

5.3.4 Post-MIS 5e Cooling to the Last Glacial Maximum (LGM, or MIS 2)

5.3.4A Climate Forcing

Both climate and ice-sheet reconstructions become more confident for times younger than MIS 5e. The climatic records derived from ice cores are especially good. The *Greenland* ice cores, primarily from the *GRIP*, *NGRIP*, and *GISP2* cores but also from *Camp Century*, *Dye-3*, and *Renland* cores, provide what are probably the most reliable paleoclimatic records of any sites on Earth (e.g., Cuffey et al., 1995; Dahl-Jensen et al., 1998; Johnsen et al., 2001; Jouzel et al., 1997; Severinghaus et al., 1998).

The paleoclimate information derived from near-field marine records is less robust. Because sediment accumulated rapidly in depositional centers adjacent to glaciated margins, relatively few cores span all of the last 130,000 years. In core HU90-013 (Figure 5.8) from the Eirik Drift (Stoner et al., 1995), rapid sedimentation buried the sediments from MIS 5e to about 13 m depth. At that site, the $\delta^{18}O$ of **planktonic foraminiferal** shells changes markedly from MIS 5e to 5d. The change, of close to 1.5‰, is consistent with cooling as well as ice growth on land, and it is associated with a rapid increase in magnetic susceptibility that indicates delivery of glacially derived sediments.

The broad picture, which is based on ice-core, far-field and near-field marine records, and more, indicates the following for climatic conditions most relevant to the *Greenland Ice Sheet*:

- a general cooling from MIS 5e (about 123 ka) to MIS 2 (coldest temperatures were at about 24 ka in *Greenland*; Alley et al., 2002),
- warming to the mid-Holocene/MIS 1 a few millennia ago,
- cooling into the **Little Ice Age** of one to a few centuries ago,
- and then a bumpy warming (see section 5.3.5B, below).

The cooling trend from MIS 5e involved temperature minima in MIS 5d, 5b, and 4 before reaching the coldest of these minima in MIS 2, with maxima in MIS 5c, 5a, and 3.

Throughout the cooling from MIS 5e to MIS 2, and the subsequent warming into MIS 1 (the Holocene), shorter-lived "millennial" events occurred. During these events, central *Greenland* warmed abruptly—roughly 10°C in a few years to decades, cooled gradually, then cooled more abruptly, gradually warmed slightly, and then repeated the sequence (Figure 5.9) (also see Alley, 1998). The abrupt coolings were usually spaced about 1,500 years apart, although longer intervals are often observed (e.g., Alley et al., 2001; Braun et al., 2005).

Marine sediment cores from around the *North Atlantic* and beyond show temperature histories closely tied to those recorded in *Greenland* (Bond et al., 1993). Indeed, the *Greenland* ice cores appear to have recorded quite clearly the template for millennial climate oscillations around much of the planet (although that template requires a modified seesaw in far-southern regions (Figure 5.9) (Stocker and Johnsen, 2003)).

Closer to the ice sheet, marine cores display strong oscillations that correlate in time with that template, but with more complexity in the response (Andrews, 2008). Figure 5.10*A* shows data from a transect of cores (Andrews, 2008) and compares the marine near-surface isotopic variations with $\delta^{18}O$ data from the *Renland* ice core, just inland from *Scoresby Sund* (Johnsen et al., 1992a; 2001) (Figure 5.8). The complexity observed in this comparison likely arises because of the rich nature of the marine indicators. As noted in section 5.2.1C, above, the oxygen isotope composition of surface-dwelling foraminiferal shells becomes lighter when the temperature increases and also when meltwater supply is increased to the system (or meltwater removal is reduced). If cooling is caused by freshwater-induced reduction in the formation of deep water, then one may observe either heavier or lighter isotopic ratios, depending on whether the core primarily reflects the temperature change or the freshwater change. Some of the signals in Figure 5.10*A* likely involve delivery of additional meltwater (which could have had various sources, such as melting of icebergs) to the vicinity of the core during colder times.

The slower tens-of-millennia cycling of the climate records is well explained by features of Earth's orbit and by associated influences of Earth-system response to the orbital features (especially changes in atmospheric $\mathbf{CO_2}$ and other **greenhouse gases**, ice-albedo feedbacks,

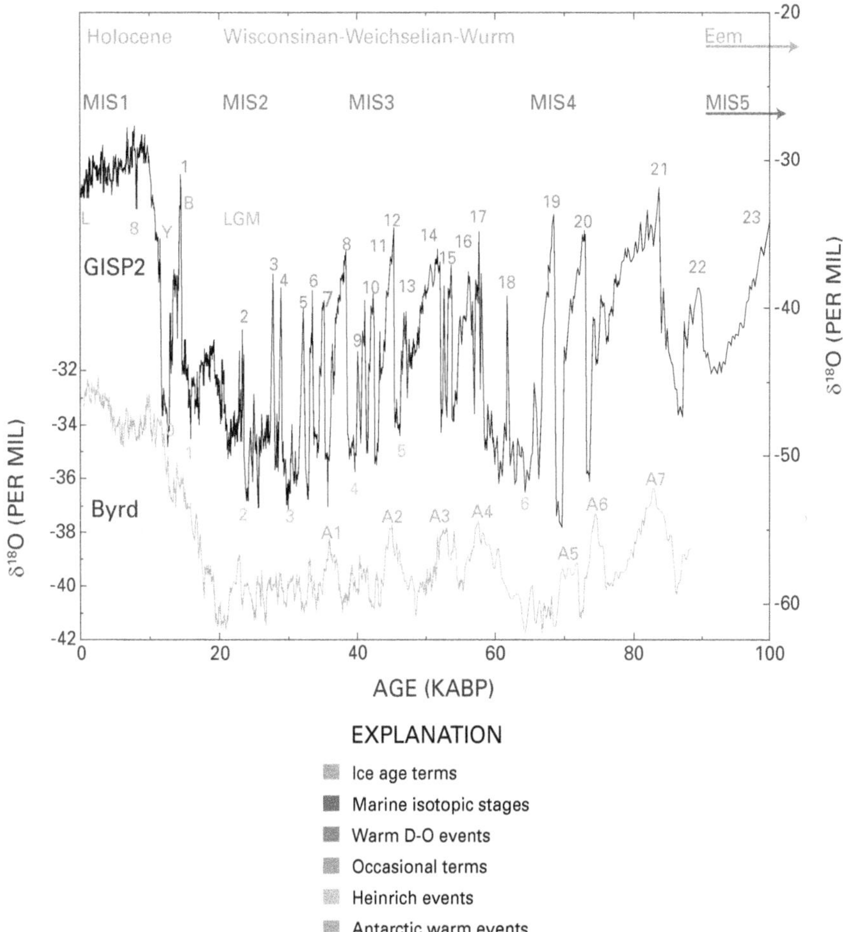

Figure 5.9. Ice-isotopic records ($\delta^{18}O$, a proxy for temperature; less-negative values indicate warmer conditions) from GISP2, Greenland (Grootes and Stuiver, 1997) (scale on right) and Byrd Station, Antarctica (scale on left), as synchronized by Blunier and Brook (2001), with various climate-event terminology indicated. Ice age terms are shown in blue (top); the classical Eemian/Sangamonian is slightly older than shown here, as is the peak of marine isotope stage (MIS, shown in purple) 5, known as 5e. Referring specifically to the GISP2 curve, the warm Dansgaard-Oeschger events or stadial events, as numbered by Dansgaard et al. (1993), are indicated in red; Dansgaard-Oeschger event 24 is older than shown here. Occasional terms (L = Little Ice Age, 8 = 8k event, P = Preboreal Oscillation (PBO), Y = Younger Dryas, B = Bølling-Allerød, and LGM = Last Glacial Maximum) are shown in pink. Heinrich events are numbered in green just below the GISP2 isotopic curve, as placed by Bond et al. (1993). The Antarctic warm events A1–A7, as identified by Blunier and Brook (2001), are indicated for the Byrd record. Modified from Alley (2007). [Copyright 2007, reprinted with permission of the Proceedings of the National Academy of Sciences.]

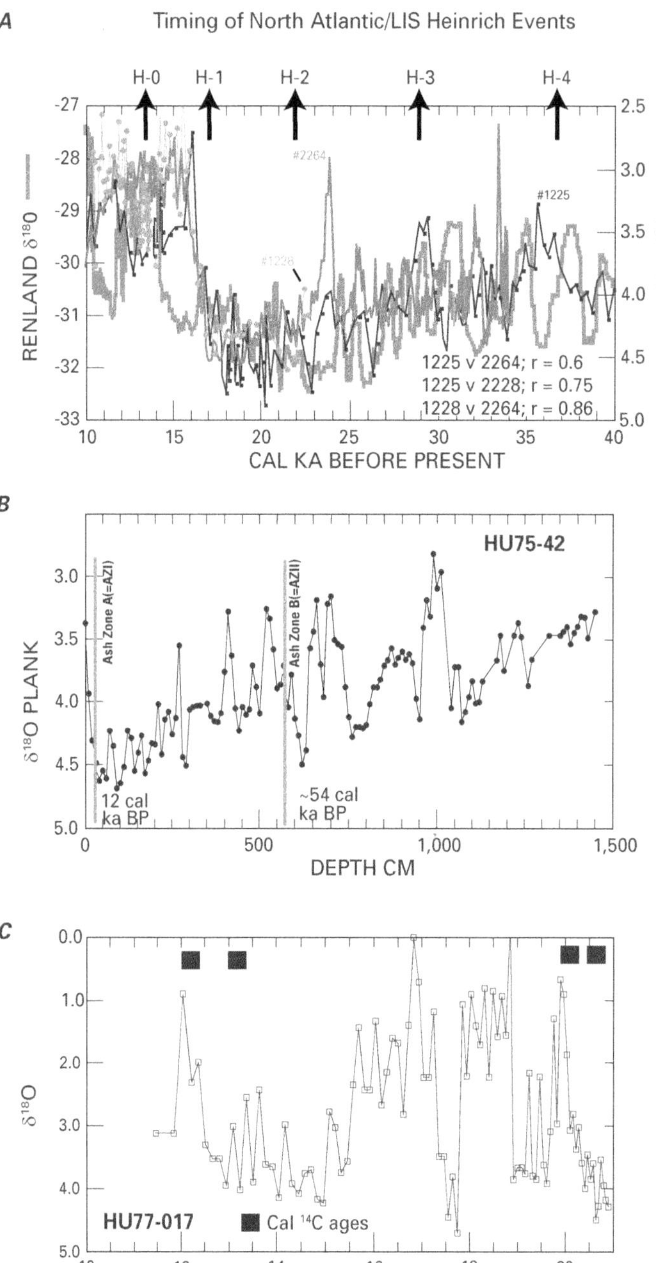

Figure 5.10. *A*) Variations in near-surface plankton $\delta^{18}O$ values from a series of sediment cores north to south of Denmark Strait (see Fig. 6.8), namely: PS2264, JM96-1225 and 1228 plotted with the $\delta^{18}O$ from the Renland Ice Cap. (Correlation coefficients between sediment cores shown in lower right); *B*) $\delta^{18}O$ variations in cores HU75-42 (NW Labrador Sea); *C*) $\delta^{18}O$ variations in cores HU77-017 from north of the Davis Strait.

and effects of changing dust loading), and strongly modulated by the response of the large ice sheets (e.g., Broecker, 1995). The faster changes are rather clearly linked to switches in the behavior of the *North Atlantic* (e.g., Alley, 2007): colder intervals mark times of more-extensive wintertime sea ice, and warmer intervals mark times of lesser sea ice (Denton et al., 2005). These links are in turn coupled to changes in deep-water formation in the *North Atlantic* and thus to **"conveyor-belt" circulation** (e.g., Broecker, 1995; Alley, 2007). (Note that a fully quantitative mechanistic understanding of forcing and response of these faster changes is still being developed; e.g., Stastna and Peltier, 2007.)

Of particular interest relative to the ice sheets is the observation that iceberg-rafted debris is much more abundant throughout the *North Atlantic* during some cold intervals, called **Heinrich events** (Figure 5.9). The material in this debris is largely tied to sources in Hudson Bay and Hudson Strait at the mouth of Hudson Bay, and thus to the North American *Laurentide Ice Sheet*, but it also contains other materials from almost everywhere around the *North Atlantic* (Hemming, 2004).

5.3.4b Ice-Sheet Changes

With certain qualifications, the behavior of the *Greenland Ice Sheet* during this interval was closely tied to the climate: the ice sheet expanded with cooling and retreated with warming. Records are generally inadequate to assess response to millennial changes, and dating is typically sufficiently uncertain that lead-or-lag relations cannot be determined with high confidence, but colder temperatures were accompanied by more-extensive ice.

Furthermore, with some uncertainty, the larger footprint of the *Greenland Ice Sheet* during colder times corresponded with a larger ice volume. This conclusion emerges both from limited data on total gas content of ice cores (Raynaud et al., 1997) indicating small changes in thickness, and from physical understanding of the ice-flow response to changing temperature, accumulation rate, ice-sheet extent, and other changes in the ice. As described in section 5.1.2, above, the retreat of ice-sheet margins tends to thin central regions, whereas the advance of margins tends to thicken central regions. Moreover, because ice thickness in central regions is relatively insensitive to changes in accumulation rate (or other factors), marginal changes largely dominate the ice-volume changes.

The best records of ice-sheet response during the cooling into MIS 2 are probably those from the *Scoresby Sund* region of East *Greenland* (Funder et al., 1998). These records indicate

- ice advances during the coolings of MIS 5d and 5b that did not fully fill the *Scoresby Sund* fjord,
- retreats during the relatively warmer MIS 5c and 5a (although 5c and 5a were colder than MIS 5e or MIS 1; e.g., Bennike and Bocher, 1994),
- advance to the mouth of *Scoresby Sund*, probably during MIS 4,
- and remaining there into MIS 2, building the extensive moraine at the mouth of the *Sund*.

Whether ice advanced beyond the mouth of the *Sund* during this interval remains unclear. Most reconstructions place the ice edge very close to the mouth (e.g., Dowdeswell et al., 1994a; Mangerud and Funder, 1994). However, the recent work of Hakansson et al. (2007) indicates wet-based ice on the south side of the mouth of the *Sund* at a site that is 250 m above modern sea level at the Last Glacial Maximum (MIS 2). Such a position almost certainly requires ice advance past the mouth. Seismic studies and cores on the *Scoresby Sund* trough-mouth fan offshore indicate that, on the southern portion of the fan, debris flows have been deposited fairly recently, whereas on the northern portion this activity pre-dates MIS 5 (O'Cofaigh et al., 2003). It is not clear how such debris flow activity occurred unless the ice had advanced well onto the shelf (O'Cofaigh et al., 2003).

To the south of *Scoresby Sund*, at *Kangerdlugssuaq*, ice extended to the edge of the continental shelf during about 31–19 ka (Andrews et al., 1997, 1998a; Jennings et al., 2002a). These data, combined with widespread geomorphic evidence that ice reached the shelf break around south *Greenland*, are then the primary evidence for extensive ice cover of this age in southern *Greenland* (Funder et al., 2004; Weidick et al., 2004).

In the *Thule* region of northwestern *Greenland*, the data are consistent both with the broad climate picture (the MIS 5e to MIS 2 sequence)

Colder intervals mark times of more-extensive wintertime sea ice, and warmer intervals mark times of lesser sea ice. These links are in turn coupled to changes in deep-water formation in the North Atlantic and thus to "conveyor-belt" circulation.

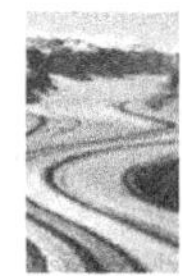

and with ice-sheet response as in *Scoresby Sund* (advances in colder MIS 5d, 5b, 4 (about 59–73 ka) and especially MIS 2, retreats in warmer 5c and 5a, possibly in MIS 3 (about 24–59 ka), and surely in MIS 1; see Figure 5.6 for general chronology) (Kelly et al., 1999). However, the dating is not secure enough to insist on much beyond the warmth of MIS 5e (marked by retreated ice), the cold of MIS 2 (marked by notably expanded ice), and the ice's subsequent retreat.

Ice from Greenland did merge with that from Ellesmere Island, thus joining the great Greenland Ice Sheet with the Innuitian sector of the North American Laurentide Ice Sheet.

The extent of ice at the **Last Glacial Maximum** also remains in doubt in the northwestern part of the *Greenland Ice Sheet*. The submarine moraines at the edge of the continental shelf are poorly dated. Ice from *Greenland* did merge with that from *Ellesmere Island*, thus joining the great *Greenland Ice Sheet* with the **Innuitian sector** of the North American *Laurentide Ice Sheet* (England, 1999; Dyke et al., 2002). However, whether ice advanced to the edge of the continental shelf in widespread regions to the north and south of the merger zone is poorly understood (Blake et al., 1996; Kelly et al., 1999). A recent reconstruction (Funder et al., 2004) favored the view that **grounded ice** advanced to the shelf edge in the northwest and merged there with North American ice, and that the merged ice spread to the northeast and southwest along what is now *Nares Strait* to feed ice shelves extending toward the *Arctic Ocean* and *Baffin Bay*. If the grounded ice had been more extensive than this, we would expect to find marine deposits from just after the Last Glacial Maximum at higher elevation than they are observed. The more extensive ice would have depressed the land more and allowed the ocean to advance farther inland following deglaciation, and then subsequent uplift would have raised the resulting marine deposits higher. Such high marine deposits are not found where they would be expected (e.g., just south of *Smith Sund*) in northwest *Greenland*. But, a trade-off does exist between slow retreat and small retreat in controlling the upper limit of the moraine deposits. This trade-off has been explored by some workers (e.g., Huybrechts, 2002; Tarasov and Peltier, 2002), but the relative sea-level data are not as sensitive to the earlier part (about 24 ka) as to the later, and so strong conclusions are not available.

The broad picture of ice advance in cooling conditions and ice retreat in warming conditions is quite clear.

Thus, the broad picture of ice advance in cooling conditions and ice retreat in warming conditions is quite clear. Remaining issues include the extent of advance onto the continental shelf (and if it was limited, why), and the rates and times of response.

We will look first at ice extent. The generally accepted picture has been one of expansion to the edge of the continental shelf in the south, much more limited expansion in the north, and a transition somewhere between *Kangerdlugssuaq* and *Scoresby Sund* on the east coast (Dowdeswell et al., 1996). On the west coast, the moraines that typically lie 30–50 km beyond the modern coastline (and even farther along troughs) are usually identified with MIS 2. The shelf-edge moraines (usually called Hellefisk moraines and usually roughly twice as far from the modern coastline as the presumably MIS 2 moraines) are usually identified with MIS 6, although few solid dates are available (Funder and Larsen, 1989). On the east coast, the evidence from the mouth of *Scoresby Sund* and the trough-mouth fan, noted above in this section, opens the possibility of more-extensive ice there than is indicated by the generally accepted picture; ice may have extended to the mid-shelf or the shelf edge. Similarly, the work of Blake et al. (1996) in *Greenland*'s far northwest may indicate that ice reached the shelf edge. The indications of Blake et al. (1996) are geomorphically consistent with wet-based ice. The increasing realization that cold-based ice is sometimes extensive yet geomorphically inactive (e.g., England, 1999) further complicates interpretations. No evidence overturns the conventional view of expansion to the shelf-edge in the south, expansion to merge with North American ice in the northwest, and expansion onto the continental shelf but not to the shelf-edge elsewhere. Thus, this interpretation is probably favored, but additional data would clearly be of interest.

Glaciological understanding indicates that ice sheets almost always respond to climatic or other environmental forcings (such as sufficiently large sea-level change). The most prominent exception may be advance to the edge of the continental shelf under conditions that would allow further advance if a huge topographic step in the sea floor were not present. (Similarly, ice may not respond to relatively small climate changes, such as during the advance stage of the **tidewater-glacier cycle** (Meier and Post, 1987)). If this assessment is accurate, and if the *Greenland Ice Sheet* at the time of the Last Glacial Maximum terminated somewhere on the continental shelf rather than at the shelf edge around part of the coastline, then glaciological understanding indicates that the ice sheet should have responded to short-lived climate changes.

The near-field marine record is consistent with such fluctuations, as discussed next. However, owing to the complexity of the controls on the paleoclimatic indicators, unambiguous interpretations are not possible.

Several marine sediment cores extend back through MIS 3 and even into MIS 4 (the cores were obtained from *Baffin Bay*, the *Eirik Drift* off southwestern *Greenland*, the *Irminger* and *Blosseville Basins* (e.g., cores SU90-24 and PS2264, Figure 5.8), and from the *Denmark Strait*) (Figure 5.8). In many of those cores, the $\delta^{18}O$ of near-surface planktic foraminifers varies widely during MIS 3. These variations were initially documented by Fillon and Duplessy (1980) in cores HU75-041 and -042 from south of *Davis Strait* (Figures 5.8 and 5.10*B*), and this documentation preceded the recognition of large millennial oscillations (**Dansgaard-Oeschger** or **D-O events**; Johnsen at al., 1992b, Dansgaard et al., 1993) in the *Greenland* ice core records. In addition, Fillon and Duplessy (1980) also contributed information on the down-core numbers of volcanic-ash (tephra) shards in these two cores. These authors identified "Ash Zone B" in core HU75-042, which is correlated with the North Atlantic Ash Zone II, for which the current best-estimate age is about 54 ka (Figure 5.10*B*; it is associated with the end of **interstadial** 15 as identified by Dansgaard et al., 1993). Subsequent work, especially north and south of *Denmark Strait*, has also shown large oscillations in planktonic foraminiferal $\delta^{18}O$ (Elliott et al., 1998; Hagen, 1999; van Kreveld et al., 2000; Hagen and Hald, 2002). As noted in section 5.3.4A, above, and shown in Figure 5.10*A*, the transect of cores appears to show both climate forcing and ice-sheet response in the millennial oscillations, although strong conclusions are not possible.

Cores from the *Scoresby Sund* and *Kangerdlugssuaq* trough mouth fans, two of the major outlets of the eastern *Greenland Ice Sheet*, also have distinct layers that are rich in ice-rafted debris (Stein et al., 1996; Andrews et al., 1998a; Nam and Stein, 1999). Cores HU93030-007 and MD99-2260 from the *Kangerdlugssuaq* trough-mouth fan (Dunhill, 2005) (Figure 5.8) consist of alternating layers with more and less ice-rafted debris that overlie a massive debris flow. Material above the debris flow is dated about 35 ka. The debris-rich layers have radiocarbon dates that are approximately coeval with Heinrich events 3 and 2 (Figure 5.9). On the *Scoresby Sund* trough-mouth fan, Stein et al (1996) also recorded intervals rich in ice-rafted debris that they quantified by counting the number of clasts greater than 2 mm as observed on X-rays. Although these cores are not as well dated as many from sites south of the Scotland-Greenland Ridge, they do indicate that such debris was delivered to the fan in pulses that may be approximately coeval with the *North Atlantic* Heinrich events.

Although several reports have invoked the Iceland Ice Sheet as a major contributor to North Atlantic sediment (Bond and Lotti, 1995; Elliot et al., 1998; Grousset et al., 2001), Farmer et al. (2003) and Andrews (2008) have questioned this assertion. They argue that the eastern *Greenland Ice Sheet* has been an ignored source of ice-rafted debris in the eastern *North Atlantic* south of the Scotland-Greenland Ridge. In particular, Andrews (2008) argued that the data from *Iceland* and *Denmark Strait* precluded any Icelandic contribution for Heinrich event 3. As noted by Huddard et al (2006), the area of the Iceland Ice Sheet during the Last Glacial Maximum was only 200,000 km^2 with an annual loss of ~600 km³, and only ~150 km³ of this loss was associated with calving. This is less than one-half the estimated calving rate of the present day *Greenland Ice Sheet* (Reeh, 1985).

The marine evidence from the western margin of the *Greenland Ice Sheet* for fluctuations of the ice sheet during MIS 3 is confounded by two facts: there are no published chronologies from the trough-mouth fan off *Disko Island*, and the stratigraphic record from *Baffin Bay* consists of glacially derived sediments from the *Greenland Ice Sheet* and from the *Laurentide Ice Sheet* including its Innuitian section (Dyke et al., 2002). Evidence for major ice-sheet events during MIS 3 is abundant, as is seen throughout *Baffin Bay* in layers rich in carbonate clasts transported from adjacent continental rocks (Aksu, 1985; Andrews et al., 1998b; Parnell et al., 2007) (Figure 5.11).

Core PS1230 from *Fram Strait*, which records the export of sediments from ice sheets around the *Arctic Ocean* (Darby et al., 2002), shows ice-rafted debris intervals associated with major contributions from north *Greenland* about 32, 23, and 17 ka. These debris intervals correspond closely in timing with ice-rafted debris events from the Arctic margins of the *Laurentide Ice Sheet.*

The fact that ice-rafted debris does not directly indicate ice-sheet behavior presents a continuing difficulty. Iceberg rafting of debris at an offshore site may increase owing to several possible factors:

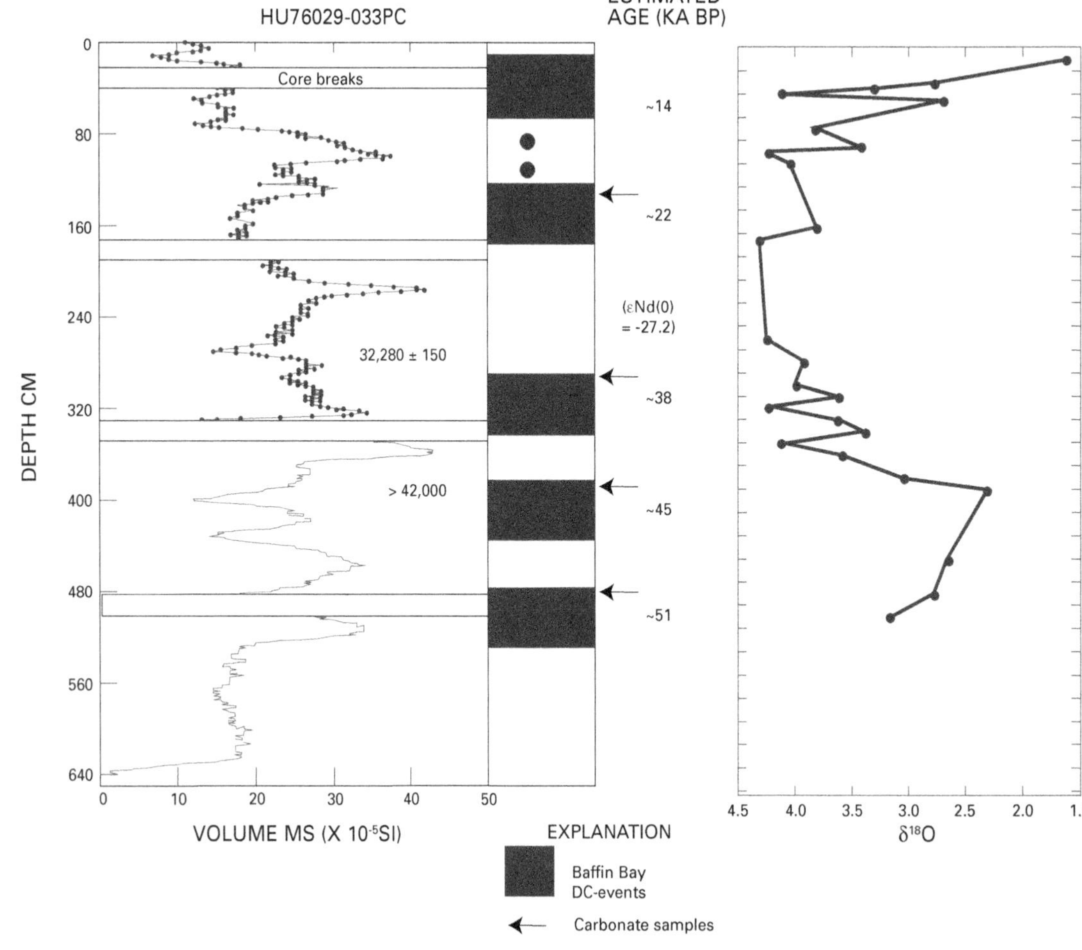

Figure 5.11. Variations in detrital carbonate (pieces of old rock) in core HU76-033 from Baffin Bay (Figure 5.8) showing down-core variations in magnetic susceptibility and $\delta^{18}O$.

- faster flow of ice from an adjacent ice sheet;
- flow of ice containing more clasts;
- loss of an ice shelf (most ice shelves experience basal melting, tending to remove debris in the ice, so ice-shelf loss would allow calving of bergs bearing more debris);
- cooling of ocean waters that allows icebergs—and their debris—to reach a site;
- loss of extensive coastal sea ice that allows icebergs to reach sites more rapidly (Reeh, 2004);
- alterations in currents or winds that control iceberg drift tracks;
- or other changes.

The very large changes in volume of incoming sediment from the North American *Laurentide Ice Sheet* during Heinrich events (Hemming, 2004) are generally interpreted to be true indicators of ice-dynamical changes (e.g., Alley and MacAyeal, 1994), but even that is debated (e.g., Hulbe et al., 2004). Thus, the marine-sediment record is consistent with *Greenland* fluctuations in concert with millennial variability during the cooling into MIS 2. Moreover, trained observers have interpreted the records as indicating millennial oscillations of the *Greenland Ice Sheet* in concert with climate, but those fluctuations cannot be demonstrated uniquely.

5.3.5 Ice-Sheet Retreat from the Last Glacial Maximum (MIS 2)

5.3.5a Climatic History and Forcing

The coldest conditions recorded in *Greenland* ice cores since MIS 6 were reached about 24 ka, which corresponds closely in time with the minimum in local midsummer sunshine and with Heinrich Event H2. The suite of sediment cores from *Denmark Strait* (Figures 5.8 and 5.10A) plus data from other sediment cores (VM28-14 and HU93030-007) indicate that the most extreme values indicating Last Glacial Maximum in $\delta^{18}O$ of marine foraminifera occurred ~18–20 ka (slightly younger than the Last Glacial Maximum values in the ice cores) with values of 4.6‰ indicating cold, salty waters.

The "orbital" warming signal in ice-core records and other climate records is fairly weak until perhaps 19 ka or so (Alley et al., 2002). The very rapid onset of warmth about 14.7 ka (the **Bølling** interstadial) is quite prominent. However, more than a third of the total deglacial warming was achieved before that abrupt step, and that pre-14.7 ka orbital warming was interrupted by Heinrich event H1. Bølling warmth was followed by general cooling (punctuated by two prominent but short-lived cold events, usually called the Older Dryas and the Inter-Allerød cold period), before faster cooling led into the **Younger Dryas** about 12.8 ka. Gradual warming then occurred through the Younger Dryas, followed by a step warming at the end of the Younger Dryas about 11.5 ka. This abrupt warming was followed by ramp warming to above recent values by 9 ka or so, punctuated by the short-lived cold event of the **Preboreal** Oscillation about 11.2–11.4 ka (Björck et al., 1997; Geirsdottir et al., 1997; Hald and Hagen, 1998; Fisher et al., 2002; Andrews and Dunhill, 2004; van der Plicht et al., 2004; Kobashi et al., 2008), and followed by the short-lived cold event about 8.3–8.2 ka (the "8k event"; e.g., Alley and Ágústsdóttir, 2005).

The cold times of Heinrich events H2, H1, the Younger Dryas, the 8k event, and probably other short-lived cold events including the Preboreal Oscillation are linked to greatly expanded wintertime sea ice in response to decreases in near-surface salinity and to the strength of the overturning circulation in the *North Atlantic* (see review by Alley, 2007). The cooling associated with these oceanic changes probably affected summers in and around *Greenland* (but see Bjorck et al., 2002 and Jennings et al., 2002a), but the changes were largest in wintertime (Denton et al., 2005).

Peak Holocene warmth was followed by cooling (with oscillations) into the Little Ice Age.

Peak MIS 1/Holocene summertime warmth before and after the 8.2-ka event was, for roughly millennial averages, ~1.3°C above late Holocene values in central *Greenland*, based on frequency of occurrence of melt layers in the *GISP2* ice core (Alley and Anandakrishnan, 1995), with mean-annual changes slightly larger although still smaller than ~2°C (and with correspondingly larger wintertime changes); other indicators are consistent with this interpretation (Alley et al., 1999). Indicators from around *Greenland* similarly show mid-Holocene warmth, although with different sites often showing peak warmth at slightly different times (Funder and Fredskild, 1989). Peak Holocene warmth was followed by cooling (with oscillations) into the Little Ice Age. The ice-core data indicate that the century- to few-century-long anomalous cold of the Little Ice Age was ~1°C or slightly more (Johnsen, 1977; Alley and Koci, 1990; Cuffey et al., 1994).

The Greenland Ice Sheet lost about 40% of its area and a notable fraction of its volume after the peak of the last glaciation about 24–19 ka.

5.3.5b Ice-Sheet Changes

The *Greenland Ice Sheet* lost about 40% of its area (Funder et al., 2004) and a notable fraction of its volume (see below; also Elverhoi et al., 1998) after the peak of the last glaciation about 24–19 ka. These losses are much less than those of the warmer *Laurentide* and *Fennoscandian Ice Sheets* (essentially complete loss) and much more than those in the colder Antarctic.

The time of onset of retreat from the Last Glacial Maximum is poorly defined because most of the evidence is now below sea level. Funder et al. (1998) suggested that the ice was most extended in the *Scoresby Sund* area from about 24,000 to about 19,000 ka, on the basis of a comparison of marine and terrestrial data. This interval started at the coldest time in *Greenland* ice cores (which corresponds with the millennial Heinrich event H2) and extends to roughly the time when sea-level rise became notable because many ice masses around the world retreated (e.g., Peltier and Fairbanks, 2006).

Extensive deglaciation that left clear records is typically more recent. For example, a core from *Hall Basin* (core 79, Figure 5.8), the northernmost of a series of basins that lie between northwest *Greenland* and *Ellesmere Island*, has a date on hand-picked foraminifers of about 16.2 ka. This date implies that the land ice flowing to the *Arctic Ocean* had retreated by this time (Mudie et al., 2006). At *Sermilik Fjord* in southwest *Greenland*, retreat from the shelf preceded about 16 ka (Funder, 1989c). The ice was at the modern coastline or back into the fjords along much of the coast by approximately Younger Dryas time (13–11.5 ka, but with no implication that this position is directly linked to the climatic anomaly of the Younger Dryas) (Funder, 1989c; Marienfeld, 1992b; Andrews et al., 1996; Jennings et al., 2002b; Lloyd et al., 2005; Jennings et al., 2006). In the Holocene, the marine evidence of ice-rafted debris from the east-central *Greenland* margin (Marienfeld, 1992a; Andrews et al., 1997; Jennings et al., 2002a; Jennings et al., 2006) shows a tripartite record with early debris inputs, a middle-Holocene interval with very little such debris, and a late Holocene (neoglacial) period that spans the last 5–6 ka of steady delivery of such debris (Figure 5.12).

Along most of the *Greenland* coast, radiocarbon dates much older than the end of Younger Dryas time are rare, likely because of persistent cover by the *Greenland Ice Sheet*. Radiocarbon dates become common near the end of the Younger Dryas and especially during the Preboreal interval, and they remain common for all younger ages, indicating deglaciation (Funder, 1989a,b,c). The term "Preboreal" typically refers to the millennium-long interval following the Younger Dryas; the Preboreal Oscillation is a shorter-lived cold event within this interval, but the terminology has sometimes been used loosely in the literature. Owing to uncertainty about the radiocarbon "reservoir" age of the waters in which mollusks lived and other issues, it typically is not possible to assess whether a given date traces to the Preboreal Oscillation or the longer Preboreal. These uncertainties typically preclude linking a particular date with Preboreal or with Younger Dryas.

Given the prominence of the end of the Younger Dryas cold event in ice-core records (it was marked by a temperature increase of about 10°C in about 10 years; Severinghaus et al., 1998), it may seem surprising at first that widespread moraines abandoned in response to that warming have not been identified with confidence. Part of the difficulty is solved by the hypothesis of Denton et al. (2005), who argued that most of the warming occurred in winter. Björck et al. (2002) and Jennings et al. (2002a) argued for notable summertime warmth in *Greenland* during the Younger Dryas, but the work of Denton et al.

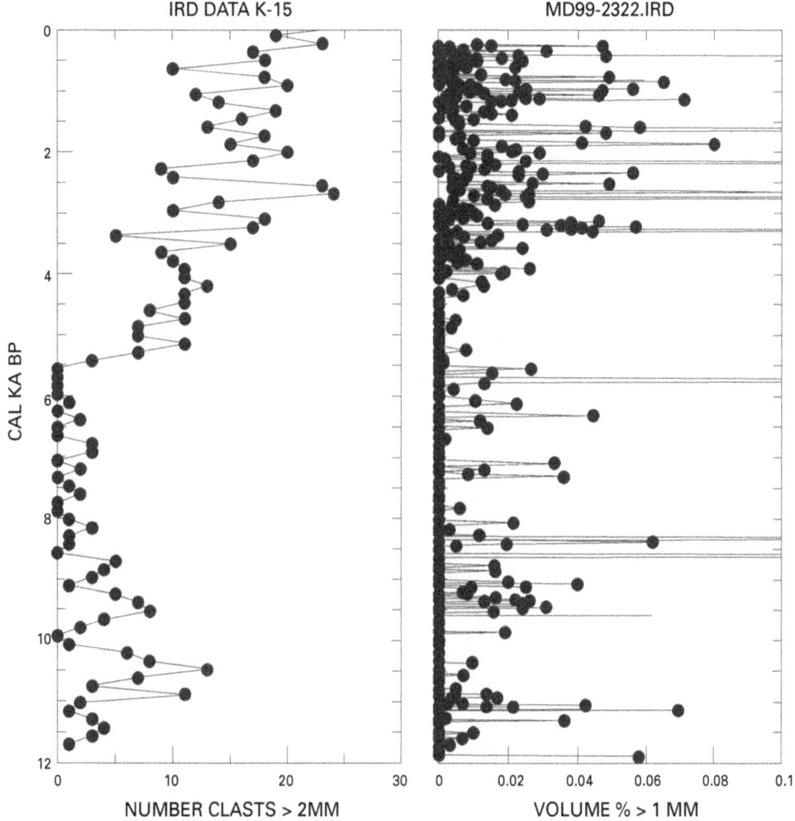

Figure 5.12. Inputs of ice-rafted debris in two cores off East Greenland (see Figure 5.8). The data from K-15 are based on counts of clasts >2 mm counted in 2-cm-thick depth increments from X-radiographs, whereas the data from MD9999-2322 (Jennings et al., 2006) are the log of weights of the >1-mm sediment fraction.

(2005) and Lie and Paasche (2006) indicates that at least some warming or lengthening of the melt season probably occurred at the end of the Younger Dryas. The terminal Younger Dryas warming then would be expected to have affected glacier and ice-sheet behavior.

All ice-core records from *Greenland* show clearly that the temperature drop into the Younger Dryas was followed by a millennium of slow warming before the rapid warming at the end (Johnsen et al., 2001; North Greenland Ice Core Project Members, 2004). The slow warming perhaps reflected rising mid-summer insolation (a function of Earth's orbit) during that time. The Younger Dryas was certainly long enough for coastal mountain glaciers to reflect both the cooling into the event and the warming during the event before the terminal step. The ice-sheet margin probably would have been influenced by these changes as well (as discussed in section 5.3.4B, above, and in this section below). If the ice margin did advance with the cooling into the Younger Dryas, and did retreat during the Younger Dryas and its termination, then moraine sets would be expected from near the start of the Younger Dryas and from the cooling of the Preboreal Oscillation after the Younger Dryas (perhaps with minor moraines marking small events during the latter Younger Dryas retreat). Because so much of the ice-sheet margin was marine at the start of the Younger Dryas, events of that age would not be recorded well.

Much study has focused on the spectacular late-glacial moraines of the *Scoresby Sund* region of East *Greenland* (Funder et al., 1998; Denton et al., 2005). Funder et al. (1998) suggested that the last resurgence of glaciers in the region, known as the Milne Land Stade, was correlated with the Preboreal Oscillation, although a Younger Dryas age for at least some of the moraines, perhaps with both Preboreal Oscillation and Younger Dryas present, cannot be excluded (Funder et al., 1998; Denton et al., 2005). Data and modeling remain sufficiently sketchy that strong conclusions do not seem warranted, but the available results are consistent with rapid response of the ice to forcing, with warming causing retreat.

Retreat of the ice sheet from the coastline passed the position of the modern ice margin about 8 ka and continued well inland.

Retreat of the ice sheet from the coastline passed the position of the modern ice margin about 8 ka and continued well inland, perhaps more than 10 km in West GREENLAND (Funder, 1989c), up to 20 km in north GREENLAND (Funder, 1989b), and perhaps as much as 60 km in parts of south GREENLAND (Tarasov and Peltier, 2002). Reworked marine shells and other organic matter of ages 7–3 ka found on the ice surface and in younger moraines document this retreat (Weidick et al., 1990; Weidick, 1993). In West GREENLAND, the general retreat from the coast was interrupted by intervals during which moraines formed, especially about 9.5–9 ka and 8.3 ka (Funder, 1989c). These moraines are not all of the same age and are not, in general, directly traceable to the short-lived 8k cold event about 8.3–8.2 ka (Long et al., 2006). Timing of the onset of late Holocene readvance is not tightly constrained. Funder (1989c) suggested about 3 ka for west GREENLAND, the approximate time when relative sea-level fall (from isostatic rebound of the land) switched to begin a relative sea-level rise of about 5 m (perhaps in part a response to depression of the land by the advancing ice load). Similar considerations place the onset of readvance somewhat earlier in the south, where relative sea-level fall switched to a relative rise of about 10 m beginning about 8–6 ka (Sparrenbom et al., 2006a,b).

The late Holocene advance culminated in different areas at different times, especially in the mid-1700s, 1850–1890, and near 1920 (Weidick et al., 2004). Since then, ice has retreated from this maximum.

Evidence of relative sea-level changes is consistent with this history (Funder, 1989d; Tarasov and Peltier, 2002, 2003; Fleming and Lambeck, 2004). Flights of raised beaches or other marine indicators are observed on many coasts of GREENLAND, and they lie as much as 160 m above modern sea level in West GREENLAND.

Fleming and Lambeck (2004) used an iterative technique to reconstruct the ice-sheet volume over time to match relative sea-level curves. They obtained an ice-sheet volume at the time of the Last Glacial Maximum about 42% larger than modern (3.1 m of additional sea-level equivalent in the ice sheet, compared with the modern value of 7.3 m of sea-level equivalent; interestingly, Huybrechts (2002) obtained a model-based estimate of 3.1 m of excess ice at the Last Glacial Maximum). Fleming and Lambeck (2004) estimated that 1.9 m of the 3.1 m of excess ice during the Last Glacial Maximum persisted at the end of the Younger Dryas. In their reconstruction, ice of the Last Glacial Maximum terminated on the continental shelf in most places, but it extended to or near the shelf edge in parts of southern GREENLAND, northeast GREENLAND, and in the far northwest where the GREENLAND ICE SHEET coalesced with the Innuitian ice from North America. Ice along much of the modern coastline was more than 500 m thick, and it was more than 1500 m thick in some places. Mid-Holocene retreat of about 40 km behind the present margin before late Holocene advance was also indicated. Rigorous error limits are not available, and modeling of the Last Glacial Maximum did not include the effects of the Holocene retreat behind the modern margin, so additional uncertainty is introduced.

In the ICE5G model, Peltier (2004) (with a GREENLAND ICE SHEET history based on Tarasov and Peltier, 2002) found that the relative sea-level data were inadequate to constrain GREENLAND ice-sheet volume accurately. In particular, these constraints provide only a partial history of the ice-sheet footprint and no information on the small—but nonzero—changes inland. Thus, Tarasov and Peltier (2002, 2003) and Peltier (2004) chose to combine ice-sheet and glacial isostatic adjustment modeling with relative-sea-level observations to derive a model of the ice-sheet geometry extending back to the Eemian (MIS 5e, about 125–130 ka). The previous ICE4G reconstruction had been characterized by an excess ice volume during the Last Glacial Maximum, relative to the present, of 6 m; this volume is reduced to 2.8 m in ICE5G. Later shrinkage of the GREENLAND ICE SHEET largely occurred in the last 10 ka in the ICE5G reconstruction, and proceeded to a mid-Holocene (7-6 ka) volume about 0.5 m less than at present, before regrowth to the modern volume.

The 20th century warmed from the Little Ice Age to about 1930, sustained warmth into the 1960s, cooled, and then warmed again since about 1990 (e.g., Box et al., 2006). The earlier warming caused marked ice retreat in many places (e.g., Funder, 1989a, b, c), and retreat and mass loss are now widespread (e.g., Alley et al., 2005). Study of declassified satellite images shows that at least for *Helheim Glacier* in the southeast of *Greenland*, the ice was in a retreated position in 1965, advanced after that during a short-lived cooling, and has again switched to retreat (Joughin et al., 2008b). This latest phase of retreat is consistent with global positioning system-based inferences of rapid melting in the southeastern sector of the *Greenland Ice Sheet* (Khan et al., 2007). It is also consistent with GRACE satellite gravity observations, which indicate a mean mass loss in the period April 2002–April 2006 equivalent to 0.5 mm/yr of globally uniform sea-level rise (Velicogna and Wahr, 2006).

As discussed in section 5.2.2E, above, geodetic measurements of **perturbation**s in Earth's rotational state can also help constrain the recent ice-mass balance. Munk (2002) suggested that length-of-day and true-polar-wander data were well fit by a model of ongoing glacial isostatic adjustment, and that this fit precluded a contribution from the *Greenland Ice Sheet* to recent sea-level rise. Mitrovica et al. (2006) reanalyzed the rotation data and applied a new theory of true polar wander induced by glacial isostatic adjustment. They found that an anomalous 20th-century contribution of as much as about 1 mm/yr of sea-level rise is consistent with the data; the partitioning of this value into signals from melting of mountain glaciers, Antarctic ice, and the *Greenland Ice Sheet* is non-unique. Interestingly, Mitrovica et al. (2001) analyzed a set of robust tide-gauge records and found that the geographic trends in the glacial isostatic adjustment–corrected rates suggested a mean 20th century melting of the *Greenland Ice Sheet* equivalent to about 0.4 mm/yr of sea-level rise.

5.4 DISCUSSION

Glaciers and ice sheets are highly complex, and they are controlled by numerous climatic factors and by internal dynamics. Textbooks have been written on the controls, and no complete list is possible. The attribution of a given ice-sheet change to a particular cause is generally difficult, and it requires appropriate modeling and related studies.

It remains, however, that in the suite of observations as a whole, the behavior of the *Greenland Ice Sheet* has been more closely tied to temperature than to anything else. The *Greenland Ice Sheet* shrank with warming and grew with cooling. Because of the generally positive relation between temperature and precipitation (e.g., Alley et al., 1993), the ice sheet has tended to grow with reduced precipitation (snowfall) and to shrink when the atmospheric mass supply increased, so precipitation changes cannot have controlled ice-sheet behavior. However, local or regional events may at times have been controlled by precipitation.

The ice sheet has tended to grow with reduced precipitation (snowfall) and to shrink when the atmospheric mass supply increased, so precipitation changes cannot have controlled ice-sheet behavior.

The hothouse world of the dinosaurs and into the Eocene occurred with no evidence of ice reaching sea level in *Greenland*. The long-term cooling that followed is correlated in time with appearance of ice in *Greenland*.

Once ice appeared, paleoclimatic archives record fluctuations that closely match not only local but also widespread records of temperature, because local temperatures correlate closely with more-widespread temperatures. Because any ice-albedo feedback or other feedbacks from the *Greenland Ice Sheet* itself are too weak to have controlled temperatures far beyond *Greenland*, the arrow of causation cannot have run primarily from the ice sheet to the widespread climate.

One must consider whether something controlled both the temperature and the ice sheet, but this possibility appears unlikely. The only physically reasonable control would be sea level, in which warming caused melting of ice beyond *Greenland*, and the resultant sea-level rise forced retreat of the *Greenland Ice Sheet* by floating marginal regions and speeding iceberg calving and ice-flow spreading. However, data point to times when this explanation is not sufficient. There at least is a suggestion at MIS 6 that *Greenland* deglaciation led strong global sea-level rise, as described in section 5.3.2B, above. Ice expanded from MIS 5e to MIS 5d from a reduced ice sheet, which would have had little contact with the sea. Much of the retreat from the MIS 2 maximum took place on land, although fjord glaciers did contact the sea. Ice re-expanded after the mid-Holocene warmth against a baseline of very little change in sea level but in general with slight sea-level rise—opposite to expectations if sea-level controls

At many other times the ice-sheet size changed in the direction expected from sea-level control as well as from temperature control, because trends in temperature and sea level were broadly correlated. Strictly on the basis of the paleoclimatic record, it is not possible to disentangle the relative effects of sea-level rise and temperature on the ice sheet. However, it is notable that terminal positions of the ice are marked by sedimentary deposits; although erosion in *GREENLAND* is not nearly as fast as in some mountain belts such as coastal *ALASKA*, notable sediment supply to **grounding lines** continues. And, as shown by Alley et al. (2007a), such sedimentation tends to stabilize an ice sheet against the effects of relative rise in sea level. Although a sea-level rise of tens of meters could overcome this stabilizing effect, the ice would need to be nearly unaffected for many millennia by other environmental forcings, such as changing temperature, to allow that much sea-level rise to occur and control the response (Alley et al., 2007a). Strong temperature control on the ice sheet is observed for recent events (e.g., Zwally et al., 2002; Thomas et al., 2003; Hanna et al., 2005; Box et al., 2006) and has been modeled (e.g., Huybrechts and de Wolde, 1999; Huybrechts, 2002; Toniazzo et al., 2004; Ridley et al., 2005; Gregory and Huybrechts, 2006).

Thus, it is clear that many of the changes in the ice sheet were forced by temperature. In general, the ice sheet responded oppositely to that expected from changes in precipitation, retreating with increasing precipitation. Events explainable by sea-level forcing but not by temperature change have not been identified. Sea-level forcing might yet prove to have been important during cold times of extensively advanced ice; however, the warm-time evidence of Holocene and MIS 5e changes that cannot be explained by sea-level forcing indicates that temperature control was dominant.

Temperature change may affect ice sheets in many ways, as discussed in section 5.1.2. Warming of summertime conditions increases meltwater production and runoff from the ice-sheet surface, and that warming may increase basal lubrication to speed mass loss by iceberg calving into adjacent seas. Warmer ocean waters (or more-vigorous circulation of those waters) can melt the undersides of ice shelves, which reduces friction at the ice-water interface and so increases flow speed and mass loss by iceberg calving. In general, the paleoclimatic record is not yet able to separate these influences, which leads to the broad use of "temperature" in discussing ice-sheet forcing. In detail, ocean temperature will not exactly correlate with atmospheric temperature, so the possibility may exist that additional studies could quantify the relative importance of changes in ocean and in air temperatures.

Most of the forcings of past ice-sheet behavior considered here have been applied slowly. Orbital changes in sunshine, greenhouse-gas forcing, and sea level have all varied on 10,000-year timescales. Purely on the basis of paleoclimatic evidence, it is generally not possible to separate the ice-volume response to incremental forcing from the continuing response to earlier forcing. In a few cases, sufficiently high time resolution and sufficiently accurate dating are available to attempt this separation for ice-sheet area. At least for the most recent events during the last decades of the 20th century and into the 21st century, ice-marginal changes have tracked forcing, with very little lag. The data on ice-sheet response to earlier rapid forcing, including the Younger Dryas and Preboreal Oscillation, remain sketchy and preclude strong conclusions, but results are consistent with rapid temperature-driven response.

A summary of many of the observations is given in Figure 5.13, which shows changes in ice-sheet volume in response to temperature forcing from an assumed "modern" equilibrium (before the warming of the last decade or two). Error bars cannot be placed with confidence. A discussion of the plotted values and error bars is given in the caption to Figure 5.13. Some of the ice-sheet change may have been caused directly by temperature and some by sea-level effects correlated with temperature; the techniques used cannot separate them (nor do modern models allow complete separation; Alley et al., 2007a). However, as discussed above in this section, temperature likely dominated, especially during warmer times when contact with the sea was reduced because of ice-sheet retreat. Again, no rates of change are implied. The large error bars on Figure 5.13 remain disturbing, but general covariation of temperature forcing and sea-level change from *GREENLAND* is indicated. The decrease in sensitivity to temperature with decreasing temperature also is physically reasonable; if the ice sheet were everywhere cooled to well below the freezing point, then a small

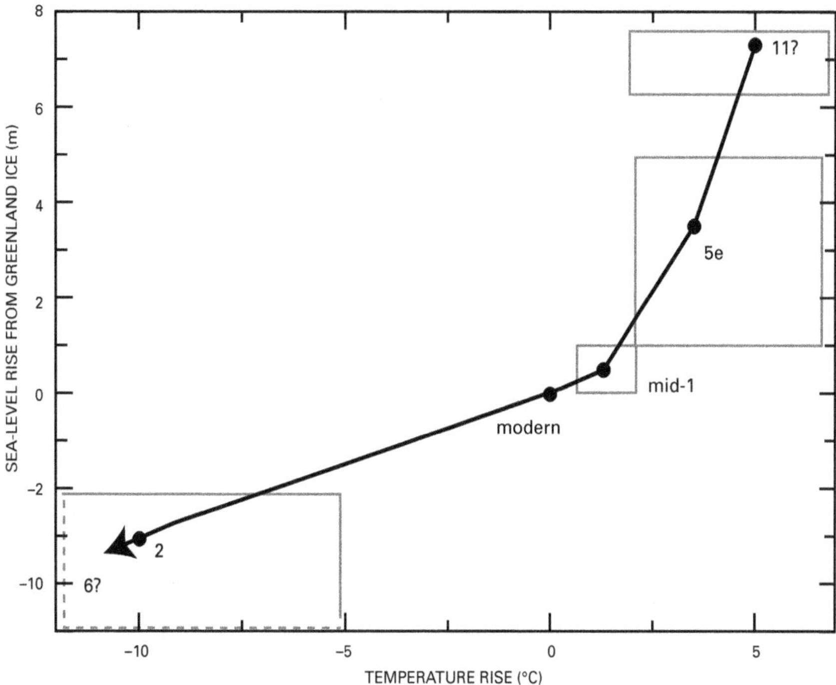

Figure 5.13. A best-guess representation of the dependence of the volume of the Greenland Ice Sheet on temperature. Large uncertainties should be understood, and any ice-volume changes in response to sea-level changes correlated with temperature changes are included (although, as discussed in the text, temperature changes probably dominated forcing, especially at warmer temperatures when the reduced ice sheet had less contact with the sea). Recent values of temperature and ice volume (perhaps appropriate for 1960 or so) are assigned 0,0. The Last Glacial Maximum was probably ~6°C colder than modern for global average (e.g., Cuffey and Brook, 2000; data and results summarized in Jansen et al., 2007). Cooling in central Greenland was ~15°C (with peak cooling somewhat more; Cuffey et al., 1995). Some of the central Greenland cooling was probably linked to strengthening of the temperature inversion that lowers near-surface temperatures relative to the free troposphere (Cuffey et al., 1995). A cooling of ~10°C is thus plotted. The ice-volume-change estimates of Peltier (2004; ICE5G) and Fleming and Lambeck (2004) are used, with the upper end of the uncertainty taken to be the ICE4G estimate (see Peltier, 2004), and somewhat arbitrarily set as 1 m on the lower side. The arrow indicates that the ice sheet in MIS 6 was more likely than not slightly larger than in MIS 2, and that some (although inconsistent) evidence of slightly colder temperatures is available (e.g., Bauch et al., 2000). The mid-Holocene result from ICE5G (Peltier, 2004) of an ice sheet smaller than modern by ~0.5 m of sea-level equivalent is plotted; the error bars reflect the high confidence that the mid-Holocene ice sheet was smaller than modern, with similar uncertainty assumed for the other side. Mid-Holocene temperature is taken from the Alley and Anandakrishnan (1995) summertime melt-layer history of central Greenland, with their 0.5°C uncertainty on the lower side, and a wider uncertainty on the upper side to include larger changes from other indicators (which are probably weighted by wintertime changes that have less effect on ice-sheet mass balance, and so are not used for the best estimate; Alley et al., 1999). As discussed in sections 5.3.3B and C, MIS 5e (the Eemian) is plotted with a warming of 3.5°C and a sea-level rise of 3.5 m. The uncertainties on sea-level change come from the range of data-constrained models discussed in 6.3.3C. The temperature uncertainties reflect the results of Cuffey and Marshall (2000) on the high side, and the lower values simulated over Greenland by Otto-Bliesner et al. (2006). Loss of the full ice sheet is also plotted, to reflect the warmer conditions that may date to MIS 11 if not earlier, and perhaps also to the Pliocene times of the Kap København Formation. Very large warming is indicated by the paleoclimatic data from Greenland, but much of that warming probably was a feedback from loss of the ice sheet itself (Otto-Bliesner et al., 2006). Data from around the North Atlantic for MIS 11 and other interglacials do not show significantly higher temperatures than during MIS 5e, allowing the possibility that sustaining MIS 5e levels for a longer time led to loss of the ice sheet. Slight additional warming is indicated here, within the error bounds of the other records, based on assessment that MIS 5e was sufficiently long for much of the ice-sheet response to have been completed, so that additional warmth was required to cause additional retreat. The volume of ice possibly persisting in highlands even after loss of central regions of the ice sheet is poorly quantified; 1 m is indicated.

the ice sheet. Similarly, the advance of Helheim Glacier after the 1960s occurred with a slightly rising global sea level and probably a slightly rising local sea level.

Many of the changes in the ice sheet were forced by temperature.

At many other times the ice-sheet size changed in the direction expected from sea-level control as well as from temperature control, because trends in temperature and sea level were broadly correlated. Strictly on the basis of the paleoclimatic record, it is not possible to disentangle the relative effects of sea-level rise and temperature on the ice sheet. However, it is notable that terminal positions of the ice are marked by sedimentary deposits; although erosion in *Greenland* is not nearly as fast as in some mountain belts such as coastal *Alaska*, notable sediment supply to **grounding lines** continues. And, as shown by Alley et al. (2007a), such sedimentation tends to stabilize an ice sheet against the effects of relative rise in sea level. Although a sea-level rise of tens of meters could overcome this stabilizing effect, the ice would need to be nearly unaffected for many millennia by other environmental forcings, such as changing temperature, to allow that much sea-level rise to occur and control the response (Alley et al., 2007a). Strong temperature control on the ice sheet is observed for recent events (e.g., Zwally et al., 2002; Thomas et al., 2003; Hanna et al., 2005; Box et al., 2006) and has been modeled (e.g., Huybrechts and de Wolde, 1999; Huybrechts, 2002; Toniazzo et al., 2004; Ridley et al., 2005; Gregory and Huybrechts, 2006).

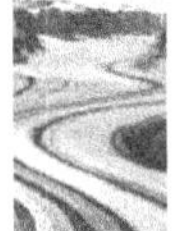

Thus, it is clear that many of the changes in the ice sheet were forced by temperature. In general, the ice sheet responded oppositely to that expected from changes in precipitation, retreating with increasing precipitation. Events explainable by sea-level forcing but not by temperature change have not been identified. Sea-level forcing might yet prove to have been important during cold times of extensively advanced ice; however, the warm-time evidence of Holocene and MIS 5e changes that cannot be explained by sea-level forcing indicates that temperature control was dominant.

At least for the most recent events during the last decades of the 20th century and into the 21st century, ice-marginal changes have tracked forcing, with very little lag.

Temperature change may affect ice sheets in many ways, as discussed in section 5.1.2. Warming of summertime conditions increases meltwater production and runoff from the ice-sheet surface, and that warming may increase basal lubrication to speed mass loss by iceberg calving into adjacent seas. Warmer ocean waters (or more-vigorous circulation of those waters) can melt the undersides of ice shelves, which reduces friction at the ice-water interface and so increases flow speed and mass loss by iceberg calving. In general, the paleoclimatic record is not yet able to separate these influences, which leads to the broad use of "temperature" in discussing ice-sheet forcing. In detail, ocean temperature will not exactly correlate with atmospheric temperature, so the possibility may exist that additional studies could quantify the relative importance of changes in ocean and in air temperatures.

Most of the forcings of past ice-sheet behavior considered here have been applied slowly. Orbital changes in sunshine, greenhouse-gas forcing, and sea level have all varied on 10,000-year timescales. Purely on the basis of paleoclimatic evidence, it is generally not possible to separate the ice-volume response to incremental forcing from the continuing response to earlier forcing. In a few cases, sufficiently high time resolution and sufficiently accurate dating are available to attempt this separation for ice-sheet area. At least for the most recent events during the last decades of the 20th century and into the 21st century, ice-marginal changes have tracked forcing, with very little lag. The data on ice-sheet response to earlier rapid forcing, including the Younger Dryas and Preboreal Oscillation, remain sketchy and preclude strong conclusions, but results are consistent with rapid temperature-driven response.

A summary of many of the observations is given in Figure 5.13, which shows changes in ice-sheet volume in response to temperature forcing from an assumed "modern" equilibrium (before the warming of the last decade or two). Error bars cannot be placed with confidence. A discussion of the plotted values and error bars is given in the caption to Figure 5.13. Some of the ice-sheet change may have been caused directly by temperature and some by sea-level effects correlated with temperature; the techniques used cannot separate them (nor do modern models allow complete separation; Alley et al., 2007a). However, as discussed above in this section, temperature likely dominated, especially during warmer times when contact with the sea was reduced because of ice-sheet retreat. Again, no rates of change are implied. The large error bars on Figure 5.13 remain disturbing, but general covariation of temperature forcing and sea-level change from *Greenland* is indicated. The decrease in sensitivity to temperature with decreasing temperature also is physically reasonable; if the ice sheet were everywhere cooled to well below the freezing point, then a small warming would not cause melting and the ice sheet would not shrink.

5.5 SUMMARY

Paleoclimatic data show that the *GREENLAND ICE SHEET* has changed greatly with time. Physical understanding indicates that many environmental factors can force changes in the size of an ice sheet. Comparison of the histories of important forcings and of ice-sheet size implicates cooling as causing ice-sheet growth, warming as causing shrinkage, and sufficiently large warming as causing loss. The evidence for temperature control is clearest for temperatures similar to or warmer than recent temperatures (the last few millennia). Snow accumulation rate is inversely related to ice-sheet volume (less ice when snowfall is higher), and thus the snow-accumulation rate in general is not the leading control on ice-sheet change. Rising sea level tends to float marginal regions of ice sheets and force retreat, so the generally positive relation between sea level and temperature means that typically both reduce the volume of the ice sheet. However, for some small changes during the most recent millennia, marginal fluctuations in the ice sheet have been opposed to those expected from local relative sea-level forcing but in the direction expected from temperature forcing. These fluctuations, plus the tendency of ice-sheet margins to retreat from the ocean during intervals of shrinkage, indicate that sea-level change is not the dominant forcing at least for temperatures similar to or above those of the last few millennia. High-time-resolution histories of ice-sheet volume are not available, but the limited paleoclimatic data consistently show that short-term and long-term responses to temperature change are in the same direction. The best estimate from paleoclimatic data is thus that warming will shrink the *GREENLAND ICE SHEET*, and that warming of a few degrees is sufficient to cause ice-sheet loss. Tightly constrained numerical estimates of the threshold warming required for ice-sheet loss are not available, nor are rigorous error bounds, and rate of loss is very poorly constrained. Numerous opportunities exist for additional data collection and analyses that would reduce these uncertainties.

The best estimate from paleoclimatic data is thus that warming will shrink the Greenland Ice Sheet, and that warming of a few degrees is sufficient to cause ice-sheet loss.

History of Arctic Sea Ice

Lead Author: Leonid Polyak*, Ohio State University, Columbus, OH

Contributing Authors: John Andrews, University of Colorado, Boulder, CO; Julie Brigham-Grette*, University of Massachusetts, Amherst, MA; Dennis Darby, Old Dominion University, Norfolk, VA; Arthur Dyke, Geological Survey of Canada, Ottowa, Ontario, Canada; Svend Funder, University of Copenhagen, Denmark; Marika Holland, National Center for Atmospheric Research, Boulder, CO; Anne Jennings, University of Colorado, Boulder, CO James Savelle, McGill University, Montréal, Quebec, Canada; Mark Serreze, University of Colorado, Boulder, CO; Eric Wolff, British Antarctic Survey, Cambridge, United Kingdom

*SAP 1.2 Federal Advisory Committee Member

ABSTRACT

The volume and areal extent of Arctic sea ice is rapidly declining, and to put that decline into perspective we need to know the history of Arctic sea ice in the geologic past. Sedimentary proxy records from the Arctic Ocean floor and from the surrounding coasts can provide clues. Although incomplete, existing data outline the development of Arctic sea ice during the last several million years. Some data indicate that sea ice consistently covered at least part of the Arctic Ocean for no less than 13–14 million years, and that ice was most widespread during the last approximately 2 million years in relation with Earth's overall cooler climate. Nevertheless, episodes of considerably reduced ice cover or even a seasonally ice-free Arctic Ocean probably punctuated even this latter period. Ice diminished episodically during warmer climate events associated with changes in Earth's orbit on the time scale of tens of thousands of years. Ice cover in the Arctic began to diminish in the late 19th century, and this shrinkage has accelerated during the last several decades. Shrinkages that were both similarly large and rapid have not been documented during at least the last few thousand years, although the paleoclimatic record is sufficiently sparse that similar events might have been missed. Orbital changes have made ice melting less likely than during the previous millennia since the end of the last ice age, making the recent changes especially anomalous. Improved reconstructions of sea-ice history would help clarify just how anomalous these recent changes are.

6.1 INTRODUCTION

The most defining feature of the surface of the *Arctic Ocean** and adjacent seas is its **sea ice** cover, which waxes and wanes with the seasons, and which also changes in extent and thickness on interannual and longer time scales. These changes in ice cover are related to **climate**, notably temperature changes (e.g., Smith et al., 2003), and themselves affect atmospheric and hydrographic conditions in high latitudes (Kinnard et al., 2008; Steele et al., 2008). Observations during the past several decades document substantial retreat and thinning of the Arctic sea ice cover: retreat is accelerating, and it is expected to continue. The *Arctic Ocean* may become seasonally ice free as early as 2040 (Holland et al., 2006a; Comiso et al., 2008; Stroeve et al., 2008). A reduction in sea ice will promote Arctic warming through a **feedback** mechanism between ice and its reflectivity (the ice-albedo feedback mechanism), and this reduction will thus influence weather systems in the northern high and perhaps middle latitudes. Changes in ice cover and freshwater flux out of the *Arctic Ocean* will also affect oceanic circulation of the *North Atlantic*, which has profound influence on climate in Europe and North America (Seager et al., 2002; Holland et al., 2006b). Furthermore, continued retreat of sea ice will accelerate coastal erosion owing to increased wave action. Ice loss will modify the *Arctic Ocean* food web and its large predators, such as polar bears and seals, that depend on the ice cover. These changes, in turn, will affect indigenous human populations that harvest such species. All of these possibilities make it important to know how fast Arctic ice will diminish and the consequences of that reduction, a task that requires thorough understanding of the natural variability of ice cover in the recent and longer term past.

Observations during the past several decades document substantial retreat and thinning of the Arctic sea ice cover: retreat is accelerating, and it is expected to continue.

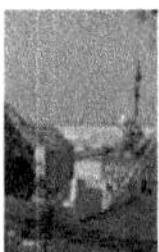

*For bold terms, refer to Glossary; for italic terms, refer to Plate 1; for geologic ages, refer to Plate 2.

6.2 BACKGROUND OF ARCTIC SEA-ICE COVER

6.2.1 Ice Extent, Thickness, Drift, and Duration

Arctic sea ice cover grows to its maximum extent by the end of winter and shrinks to a minimum in September. For the period of reliable satellite observations (1979–2007), extremes in Northern Hemisphere ice extent are 16.44×106 square kilometers (km^2) for March 1979 and 4.28×106 km^2 for September 2007 (*http://nsidc.org/data/seaice_index/*; Stroeve et al., 2008). Ice extent is defined as the region of the ocean of which at least 15% is covered by ice. The ice cover can be broadly divided into a perennial ice zone where ice is present throughout the year and a seasonal ice zone where ice is present only seasonally. A considerable fraction of Arctic sea ice is perennial, which differs strongly from Antarctic sea ice which is nearly all seasonal. Ice concentrations in the perennial ice zone typically exceed 97% in winter but fall to 85–95% in summer. Sea ice concentrations in the seasonal ice zone are highly variable, and in general (but not always) they decrease toward the southern sea ice margin.

The thickness of sea ice, which varies markedly in both space and time, can be described by a probability distribution. For the *Arctic Ocean* as a whole, the peak of this distribution (as thick as the ice ever gets) is typically cited at about 3 meters (m) (Serreze et al., 2007b), but growing evidence (discussed below) suggests that during recent decades not only is the area of sea ice shrinking, but that it is also thinning substantially. Although many different types of sea ice can be defined, the two basic categories are first-year ice, which represents a single year's growth, and multiyear ice, which has survived one or more melt seasons. Undeformed first-year ice typically is as much as 2 m thick. Although in general multiyear ice is thicker (greater than 2 m), first-year ice that is locally pushed into ridges can be as thick as 20–30 m.

Under the influence of winds and ocean currents, the Arctic sea ice cover is in nearly constant motion. The large-scale circulation principally consists of the Beaufort Gyre, a mean annual clockwise motion in the western *Arctic Ocean* with a drift speed of 1–3 centimeters per second, and the Transpolar Drift, the movement of ice from the coast of *Siberia* east across the pole and into the *North Atlantic* by way of *Fram Strait*, which lies between northern *Greenland* and *Svalbard*. Ice velocities in the Transpolar Drift increase toward *Fram Strait*, where the mean drift speed is 5–20 centimeters per second (Figure 6.1) (Thorndike, 1986; Gow and Tucker, 1987). About 20% of the total ice area of the *Arctic Ocean* is discharged each year through *Fram Strait*, the majority of which is multiyear ice. This ice subsequently melts in the northern *North Atlantic*, and since the ice is relatively fresh compared with sea water, this melting adds freshwater to the ocean in those regions.

6.2.2 Influences on the Climate System

Seasonal changes in the amount of heat at the surface (net surface heat flux) associated with sea ice modulate the exchange and transport of energy in the atmosphere. Ice, as sheets or as sea ice, reflects a certain percentage of incoming solar radiation back into the atmosphere. The albedo (reflectivity) of ice cover ranges from 80% when it is freshly snow covered to around 50% during the summer melt season (but lower in areas of ponded ice). This high reflectivity contrasts with the dark ocean sur-

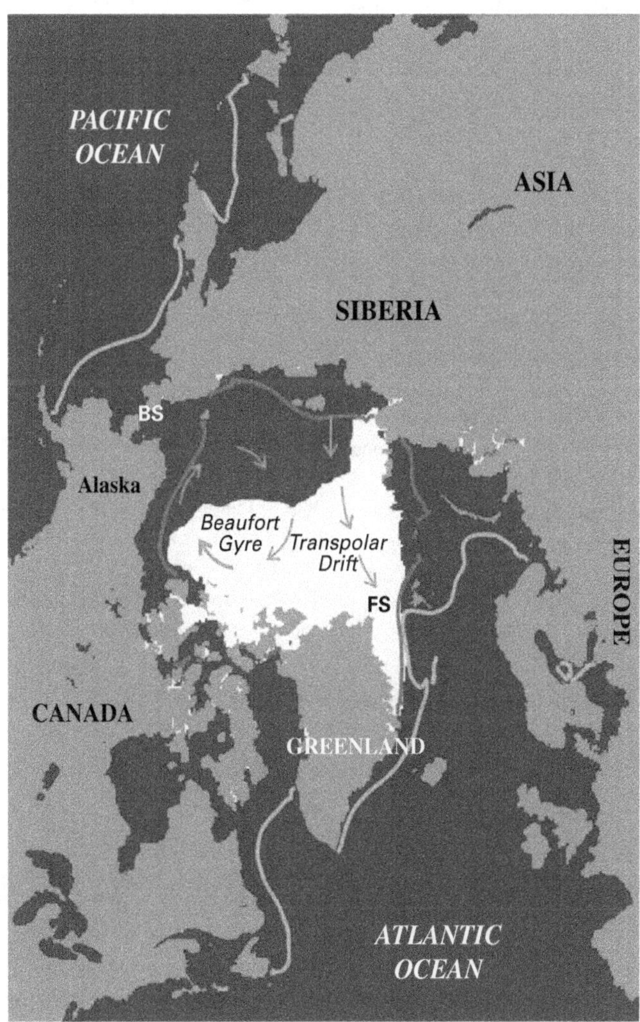

Figure 6.1. Overview map of the Arctic and adjacent regions showing the extent of sea ice: magenta and blue lines = September and March medians for 1979–2000, respectively; white field = September 2007 extent. Major circulation systems are schematically shown by green arrows. BS = Bering Strait, FS = Fram Strait. (Courtesy National Snow and Ice Data Center, Boulder, CO.)

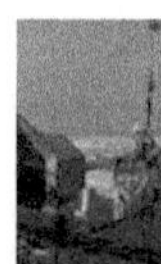

The extent of sea ice has diminished in every month and most obviously in September, for which the trend for the period 1979–2007 is 10% per decade.

face, which has an albedo of less than 10%. Ice's high albedo and its large surface area, coupled with the solar energy used to melt ice and to increase the sensible heat content of the ocean, keep the Arctic atmosphere cool during summer. This cooler polar atmosphere helps to maintain a steady poleward transport of atmospheric energy (heat) from lower latitudes into the Arctic. During autumn and winter, energy derived from incoming solar radiation is small or nonexistent in polar areas. However, heat loss from the surface adds heat to the atmosphere, and it reduces the requirements for atmospheric heat to be transported poleward into the Arctic (Serreze et al., 2007a).

Model experiments have addressed potential changes in the regional and large-scale aspects of atmospheric circulation that are associated with loss of sea ice. The models commonly use ice conditions that have been projected through the 21st century (see following section). Magnusdottir et al. (2004) found that a reduced area of winter sea ice in the *North Atlantic* modified the modeled circulation in the same way as the **North Atlantic Oscillation**; declining ice promotes a negative North Atlantic Oscillation response: storm tracks are weaker and shifted to the south. Many observations show that sea ice in this region affects the development of mid- and high-latitude cyclones because of the strong horizontal temperature gradients along the ice margin (e.g., Tsukernik et al., 2007). Singarayer et al. (2006) forced a model by combining the area of sea ice in 1980–2000 and projected reductions in sea ice until 2100. In one simulation, mid-latitude storm tracks were intensified and they increased winter precipitation throughout western and southern Europe. Sewall and Sloan (2004) found that reduced ice cover led to less rainfall in the American west. In summary, although these and other simulations point to the importance of sea ice on climate outside of the Arctic, different models may produce very different results. Coordinated experiments that use a suite of models are needed to help to reduce uncertainty.

Climate models also indicate that changes in the melting of and export of sea ice to the *North Atlantic* can modify large-scale ocean circulation (e.g., Delworth et al., 1997; Mauritzen and Hakkinen, 1997; Holland et al., 2001). In particular, exporting more freshwater from the Arctic increases the stability of the upper ocean in the northern *North Atlantic*. This stability may suppress convection, leading to reduced formation of *North Atlantic* deep water and weakening of the Atlantic meridional overturning circulation. This suppression may have far-reaching climate consequences. The considerable freshening of the *North Atlantic* since the 1960s has an Arctic source (Peterson et al., 2006). Total Arctic freshwater output to the *North Atlantic* is projected to increase through the 21st century, and decreases in the export of sea ice will be more than balanced by the export of liquid freshwater (derived from the melting of Arctic ice and increased net precipitation). However, less ice may melt in the *Greenland-Iceland-Norwegian* (GIN) seas because less ice is moved through *Fram Strait* into those seas. These changes may increase vertical instability in the ocean regions where deep water forms and counteract the tendency of a warmer climate to increase ocean stability (Holland et al., 2006b). However, this possible instability may be mitigated somewhat if less sea ice accumulates in the *Greenland-Iceland-Norwegian* seas. Additionally, as discussed by Levermann et al. (2007), the reduction in sea ice may help to stabilize the Atlantic meridional overturning circulation by removing the insulating ice cover which, perhaps counterintuitively, limits the amount of heat lost by the ocean to the atmosphere. Thus, sea ice may help to maintain the formation of deep water in the *Greenland-Iceland-Norwegian* seas. Overall, a smaller area of sea ice influences the Atlantic meridional overturning circulation in sometimes competing ways. How they will ultimately affect future climate is not yet certain.

6.2.3 Recent Changes and Projections for the Future

On the basis of satellite records, the extent of sea ice has diminished in every month and most obviously in September, for which the trend for the period 1979–2007 is 10% per decade (Figure 6.2). [Satellite records originated in the National Snow and Ice Data Center (*http:/nsidc.org/data/seaice_index/*) and combine information from the Nimbus-7 Scanning Multichannel Microwave Radiometer (October 1978–1987) and the Defense Meteorological Satellite Program Special Sensor

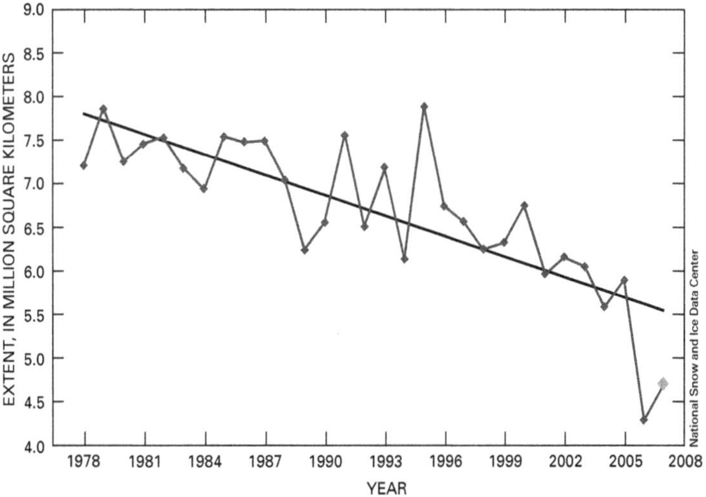

Figure 6.2. Extent of Arctic sea ice in September, 1979–2007. The linear trend (trend line shown in blue) including 2007 shows a decline of 10% per decade (data from National Snow and Ice Data Center, Boulder, Colorado).

Distribution of ice of various thickness suggests that this loss of older ice translates to a decrease in mean thickness for the Arctic from 2.6 m in March 1987 to 2.0 m in 2007.

Microwave/Imager (1987–present.)] Conditions in 2007 serve as an exclamation point on this ice loss (Comiso et al., 2008; Stroeve et al., 2008). The average September ice extent in 2007 of 4.28 million km^2 was not only the least ever recorded but also 23% lower than the previous September record low of 5.56 million km^2 set in 2005. The difference in areas corresponds with an area roughly the size of Texas and California combined. On the basis of an extended sea ice record, it appears that the area of ice in September 2007 is only half of its area in 1950–70 [estimated by use of the Hadley Centre sea ice and sea surface temperature data set (HadISST) (Rayner et al., 2003)].

Many factors may have contributed to this ice loss (as reviewed by Serreze et al., 2007b), such as general Arctic warming (Rothrock and Zhang, 2005), extended summer melt (Stroeve et al., 2006), effects of the changing phase of the **Northern Annular Mode** and the North Atlantic Oscillation. These and other atmospheric patterns have flushed some older, thicker ice out of the Arctic and left thinner ice that is more easily melted out in summer (e.g., Rigor and Wallace, 2004; Rothrock and Zhang, 2005; Maslanik et al., 2007a), changed ocean heat transport (Polyakov et al., 2005; Shimada et al., 2006), and increased recent spring cloud cover that augments the longwave radiation flux to the surface (Francis and Hunter, 2006). Strong evidence for a thinning ice cover comes from an ice-tracking algorithm applied to satellite and buoy data, which suggests that the area of the *Arctic Ocean* covered by predominantly older (and hence generally thicker) ice (ice 5 years old or older) decreased by 56% between 1982 and 2007. Within the central *Arctic Ocean*, the coverage of old ice has declined by 88%, and ice that is at least 9 years old (ice that tends to be sequestered in the Beaufort Gyre) has essentially disappeared. Examination of the distribution of ice of various thickness suggests that this loss of older ice translates to a decrease in mean thickness for the Arctic from 2.6 m in March 1987 to 2.0 m in 2007 (Maslanik et al., 2007b).

The role of **greenhouse gas forcing** on the observed ice loss finds strong support from the study of Zhang and Walsh (2006). These authors show that for the period 1979–1999, the multi-model mean trend projected by models discussed in the **Intergovernmental Panel on Climate Change** Fourth Assessment Report (**IPCC-AR4,** 2007) is downward, as are trends from most individual models. However, Stroeve et al. (2007) find that few or none (depending on the time period of analysis) of the September trends from the IPCC-AR4, 2007 runs are as large as observed. If the multi-model mean trend is assumed to be a reasonable representation of change forced by increased concentrations of greenhouse gases, then 33–38% of the observed September trend from 1953 to 2006 is externally forced and that percentage increases to 47–57% from 1979 to 2006, when both the model mean and observed trend are larger.

Although this trend argues that natural variability has strongly contributed to the observed trend, Stroeve et al. (2006) concluded that, as a group, the models underestimate the sensitivity of sea ice cover to **forcing** by greenhouse gases. Overly thick ice assumed by many of the models appears to provide at least a partial explanation.

The Intergovernmental Panel on Climate Change Fourth Assessment Report (IPCC-AR4) models driven with the SRES A1B emissions scenario (in which CO_2 reaches 720 parts per million (ppm), in comparison to the current value of 380 ppm, by the year 2100), point to complete or nearly complete loss (less than 1×10^6 km^2) of September sea ice anywhere from year 2040 to well beyond the year 2100, depending on the model and particular run (ensemble member) for that model. Even by the late 21st century, most models project a thin ice cover in March (Serreze et al., 2007b). However, given the findings just discussed, the models as a group may be too conservative—predict a later rather than earlier date—when the *Arctic Ocean* will be ice-free in summer.

These recent reductions in the extent and thickness of ice cover and the projections for its further shrinkage necessitate a comprehensive investigation of the longer term history of Arctic sea ice. To interpret present changes we need to understand the Arctic's natural variability.

Abrupt change in future Arctic ice conditions is difficult to model. For instance, the extent of end-of-summer ice is sensitive to ice thickness in spring (simulations based on the Community Climate System Model, version 3 (Holland et al., 2006a)). If the ice is already thin in the spring, then a "kick" associated with natural climate variability might make it melt rapidly in the summer owing to ice-albedo feedback. In the Community Climate System Model, version 3 events, anomalous ocean heat transport acts as this trigger. In one ensemble member, the area of September ice decreases from about 6×106 km^2 to 2×106 km^2 in 10 years, resulting in an essentially ice-free September by 2040. This result is not just an artifact of Community Climate System Model, version 3: a number of other climate models show similar rapid ice loss.

The past distribution of sea ice is recorded in sediments preserved on the sea floor and in deposits along many Arctic coasts.

These recent reductions in the extent and thickness of ice cover and the projections for its further shrinkage necessitate a comprehensive investigation of the longer term history of Arctic sea ice. To interpret present changes we need to understand the Arctic's natural variability. A special emphasis should be placed on the times of change such as the initiation of seasonal and then perennial ice and the periods of its later reductions.

6.3 TYPES OF PALEOCLIMATE ARCHIVES AND PROXIES FOR THE SEA-ICE RECORD

The past distribution of sea ice is recorded in sediments preserved on the sea floor and in deposits along many Arctic coasts. Indirect information on sea-ice extent can be derived from cores drilled in **glaciers** and **ice sheets** such as the *Greenland Ice Sheet*. Ice cores record atmospheric precipitation, which is linked with air-sea exchanges in surrounding oceanic areas. Such paleoclimate information provides a context within which the patterns and effects of the current and future ice-reduced state of the Arctic can be evaluated.

6.3.1 Marine Sedimentary Records

The most complete and spatially extensive records of past sea ice are provided by sea-floor sediments from areas that are or have been covered by floating ice. Sea ice affects deposition of such sediments directly or indirectly through physical, chemical, and biological processes. These processes and, thus, ice characteristics can be reconstructed from a number of sediment proxies outlined below.

Sediment cores that represent the long-term history of sea ice embracing several million years are most likely to be found in the deep, central part of the *Arctic Ocean* where the sea floor was not eroded during periods of lower sea-level (and larger ice sheets). On the other hand, rates of sediment deposition in the central *Arctic Ocean* are generally low, on the order of centimeters or even millimeters per thousand years (Backman et al., 2004; Darby et al., 2006), so that sedimentary records from these areas may not capture short-term variations in paleoenvironments. In contrast, cores from Arctic continental margins usually represent a much shorter time interval, less than 20 thousand years (**k.y.**) since the **Last Glacial Maximum**, but they sometimes provide high-resolution records that capture events on century or even decadal time scales. Therefore, investigators need sediment cores from both the central basin and continental margins of the *Arctic Ocean* to fully characterize sea-ice history and its relation to **climate change**.

Until recently, and for logistical reasons, most cores relevant to the history of sea ice cover were collected from low-Arctic marginal seas, such as the *Barents Sea* and the *Norwegian-Greenland Sea*. There, modern ice conditions allow for easier ship operation, whereas sampling in the central *Arctic Ocean* requires the use of heavy icebreakers. Recent advances in drilling the floor of the *Arctic Ocean*—notably the first deep-sea drilling in the central *Arctic Ocean* (**ACEX**: Backman et al., 2006) and the 2005 Trans-Arctic Expedition (**HOTRAX**: Darby et al., 2005)— provide new, high-quality material from the *Arctic Ocean* proper with which to characterize variations in ice cover during the late **Cenozoic** (the last few million years).

A number of sediment proxies have been used to predict the presence or absence of sea ice in down-core studies. The most direct proxies are derived from sediment that melts out or drops from sea ice owing to the following sequence of processes:

1. sediment is entrained in sea ice,
2. this ice is transported by wind and surface currents to the sites of interest, and
3. sediment is released and deposited.

The size of sediment grains is commonly analyzed to identify ice-rafted debris. The entrainment of sediments in sea ice mostly occurs along the shallow continental margins during periods of ice freeze-up and is largely restricted to silt and clay-size sediments and rarely contains grains larger than 0.1 millimeters (mm) (Lisitzin, 2002; Darby, 2003). Coarser ice-rafted debris is mostly transported by floating icebergs rather than by regular sea ice (Dowdeswell et al., 1994; Andrews, 2000). A small volume of coarse grains are shed from steep coastal cliffs onto land-fast ice. To link sediment with sea ice may require investigations other than measurement of grain size: for example, examination of shapes and surface textures of quartz grains will help distinguish sea-ice-rafted and iceberg-rafted material (Helland and Holmes, 1997; Dunhill et al., 1998). Detailed grain-size distributions say something about ice conditions. For example, massive accumulation of silt-size grains (mostly larger than 0.01 mm) may indicate the position of an ice margin where melting ice is the source of most sediment (Hebbeln, 2000).

Some indicators (sediment **provenance** indicators) help to establish the source of sediment and thus help to track ice drift. Especially telling is sediment carrying some diagnostic peculiarity that is foreign to the site of deposition and that can be explained only by ice transport—such as the particular composition of iron-oxide sand grains, which can be matched with an extensive data base of source areas around the *Arctic Ocean* (Darby, 2003). Bulk sediment analyzed by quantitative methods such as X-ray diffraction can also be used in those instances where minerals that are "exotic" relative to the composition of the nearest terrestrial sources are deposited. Quartz in *Iceland* marine cores (Moros et al., 2006; Andrews and Eberl, 2007) and dolomite (limestone rich in magnesium), in sediments deposited along eastern *Baffin Island* and Labrador are two examples (Andrews et al., 2006).

Sediment cores commonly contain skeletons of microscopic organisms (for example **foraminifers**, **diatoms**, and **dinocysts**). These findings are widely used for deciphering the past environments in which these organisms lived. Some marine planktonic organisms live in or on sea ice or are otherwise associated with ice. Their skeletons in bottom sediments indicate the condition of ice cover above the study site. Other organisms that live in open water can be used to identify intervals of diminished ice. Remnants of ice-related algae such as diatoms and dinocysts have been used to infer changes in the length of the ice-cover season (Koç and Jansen, 1994; de Vernal and Hillaire-Marcel, 2000; Mudie et al., 2006; Solignac et al., 2006). To quantify the relationship between these organisms and paleoenvironment, three major research steps are required. The first is to develop a database of the percent compositions in a certain group of organisms from water-column or surficial sea-floor samples that span a wide environmental range. Second, various statistical methods must be used to express the relationship (usually called "transfer functions") between these compositions and key environmental parameters, such as sea-ice duration and summer surface temperatures. Finally, after sediment cores are analyzed and transfer functions are developed on the modern data sets, they are then applied to the temporal (i.e., down-core) data. The usefulness of the transfer functions, however, depends upon the accuracy of the environmental data, which is commonly quite limited in Arctic areas.

Sediment cores commonly contain skeletons of microscopic organisms. These findings are widely used for deciphering the past environments in which these organisms lived.

Bottom dwelling (benthic) organisms in polar seas are also affected by ice cover because it controls what food can reach the sea floor. The particular suite of benthic organisms preserved in sediments can help to identify ice-covered sites. For instance, environments within the pack ice produce very little organic matter, whereas environments on the margin of the ice produce a great deal. Accordingly, species of bottom-dwelling organisms that prefer relatively high fluxes of fresh organic matter can indicate, for the Arctic shelves, the location of the ice margin (Polyak et al., 2002; Jennings et al., 2004). In the central *Arctic Ocean*, benthic **foraminifers** and **ostracodes** also offer a good potential for identifying ice conditions (Cronin et al., 1995; Wollenburg and Kuhnt, 2000; Polyak et al., 2004).

The composition of organic matter in sediment, including specific organic compounds (**biomarkers**), can also be used to reconstruct the environment in which it formed. For instance, a specific biomarker, IP25, can be associated with diatoms living in sea ice (Belt et al., 2007). The method has been tested by the analysis of sea-floor samples from the *Canadian Arctic* and is being further applied to down-core samples for characterization of past ice conditions.

It is important to understand that although all of the above **proxies** have a potential for identifying the former presence of or the seasonal duration of sea-ice cover, each of them has limitations that complicate interpretations based on a single proxy. For instance, by use of a dinocyst transfer function from East *Greenland*, it was estimated that the sea-ice duration is about 2–3 months (Solignac et al., 2006) when in reality it is closer to 9 months (Hastings, 1960). Agreement among many proxies is required for a confident inference about variations in sea-ice conditions. A thorough understanding of sea-ice history depends on the refining of sea-ice proxies in sediment taken from strategically selected sites in the *Arctic Ocean* and along its continental margins.

6.3.2 Coastal Records

In many places along the Arctic and subarctic coasts, evidence of the extent of past sea ice is recorded in coastal-plain sediments, marine terraces, ancient barrier island sequences, and beaches. Deposits in all of these formerly marine environments are now above water owing to relative changes in sea level caused by eustatic, glacioisostatic, or **tectonic** factors. Although these coastal deposits represent a limited time span and geographic distribution, they provide critical information that can be compared with marine sediment records. The primary difference between coastal and sea-floor records is in the type of fossils recovered. Notably, the spacious coastal exposures (as compared with sediment cores) enable large paleontological material such as plant remains, driftwood, whalebone, and relatively large mollusks to be recovered. These items contribute valuable information about past sea-surface and air temperatures, the northward expansions of subarctic and more temperate species, and the seasonality of past sea-ice cover. For example, fossils preserved in these sequences document the dispersals of coastal marine biota between the Pacific, Arctic, and *North Atlantic* regions, and they commonly carry telling evidence of ice conditions. Plant remains in their turn provide a much-needed link to documented information about past vegetation on land throughout Arctic and subarctic regions. The location of the northern tree line that is presently controlled by the July 7°C mean isotherm is a critical paleobotanic indicator for understanding ice conditions in the Arctic. Nowhere in the Arctic do trees exist near shores lined with perennial sea ice; they thrive only in southerly reaches of regions of seasonal ice. The combination of spatial relationships between marine and terrestrial data allows a comprehensive reconstruction of past climate.

6.3.3 Coastal Plains and Raised Marine Sequences

A number of coastal plains around the Arctic are blanketed by marine sediment sequences laid down during high sea levels. Although these sequences lie inland of coastlines that today are bordered by perennial or by seasonal sea ice, they commonly contain packages of fossil-rich sediments that provide an exceptional record of earlier warm periods. The most well-documented sections are those preserved along the eastern and northern coasts of *Greenland* (Funder et al., 1985, 2001), the eastern *Canadian Arctic* (Miller et al., 1985), *Ellesmere Island* (Fyles et al., 1998), *Meighen Island* (Matthews, 1987; Matthews and Overden, 1990; Fyles et al., 1991), *Banks Island* (Vincent,

1990; Fyles et al., 1994), the *North Slope of Alaska* (Carter et al., 1986; Brigham-Grette and Carter, 1992); the *Bering Strait* (Kaufman and Brigham-Grette, 1993; Brigham-Grette and Hopkins, 1995), and in the western *Eurasian Arctic* (Funder et al., 2002) (Figure 6.3). In nearly all cases the primary evidence used to estimate the extent of past sea ice is in situ molluscan and microfossil assemblages. These assemblages, from many sites, coupled with evidence for the northward expansion of tree line during **interglacial** intervals (e.g., Funder et al., 1985; Repenning et al., 1987; Bennike and Böcher, 1990; CAPE, 2006), provide an essential view of past sea-ice conditions with direct implications for sea surface temperatures, sea ice extent, and seasonality.

6.3.4 Driftwood

The presence or absence of sea ice may be inferred from the distribution of tree logs, mostly spruce and larch found in raised beaches along the coasts of *Arctic Canada* (Dyke et al., 1997), *Greenland* (Bennike, 2004), *Svalbard* (Haggblom, 1982), and *Iceland* (Eggertsson, 1993). Coasts with the highest numbers of driftwood probably were once near a sea-ice margin, whereas coasts hosting more modest amounts were near either too much ice or too much open water—neither of which deliver much driftwood. Most of the logs found are attributed to a northern Russian source, although some can be traced to northwest *Canada* and *Alaska*. Logs can drift only about 1 year before they become waterlogged and sink (Haggblom, 1982). The logs are probably derived from rivers flooded by spring snowmelt, which bring sediment and trees onto **landfast ice** around the margin of the Arctic Basin. In areas other than *Iceland*, the glacial isostatic uplift of the land has led to a staircase of raised beaches hosting various numbers of logs with time. An extensive database catalogs these variations in the beaching of logs during the present interglacial (**Holocene**). These variations have been associated with the growth and disappearance of landfast sea ice (which restricts the beaching of driftwood) and changes in atmospheric circulation with resulting changes in ocean surface circulation (Dyke et al., 1997).

6.3.5 Whalebone

Reconstructions of sea-ice conditions in the *Canadian Arctic Archipelago* have to date been derived mainly from the distribution in space and time of marine mammal bones in raised marine deposits (Dyke et al., 1996, 1999; Fisher et al., 2006). Several large marine mammals have strong affinities for sea ice: polar bear, several species of seal, walrus, narwhal, beluga (white)

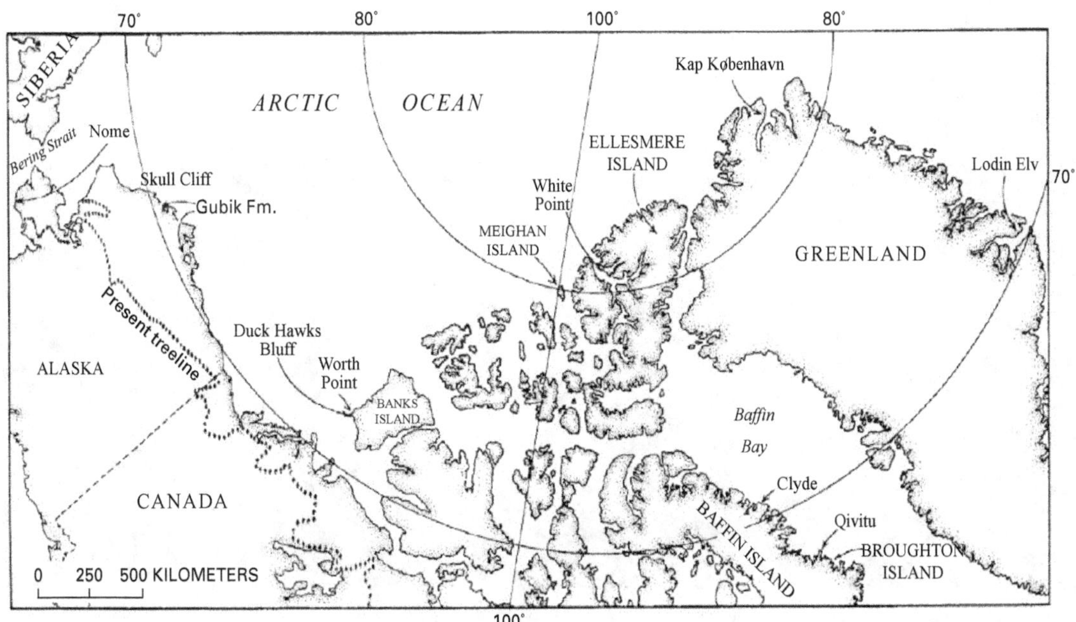

Figure 6.3. Key marine sedimentary sequences exposed at the coasts of Arctic North America and Greenland.

whale, and bowhead (Greenland right) whale. Of these, the bowhead has left the most abundant, hence most useful, fossil record, followed by the walrus and the narwhal. **Radiocarbon** dating of these remains has yielded a large set of results, largely available through Harington (2003) and Kaufman et al. (2004).

Former sea-ice conditions can be reconstructed from bowhead whale remains because seasonal migrations of the whale are dictated by the **oscillations** of the sea-ice pack. The species is thought to have had a strong preference for ice-edge environments since the **Pliocene** (2.6–5.3 million years ago (Ma)), perhaps because that environment allows it to escape from its only natural predator, the killer whale. The Pacific population of bowheads spends winter and early spring along the ice edge in the *Bering Sea* and advances northward in the summer ice into the Canadian Beaufort Sea region along the western edge of the *Canadian Arctic Archipelago*. The Atlantic population spends winter and early spring in the northern *Labrador Sea* between southwest *Greenland* and northern Labrador and advances northward in summer into the eastern channels of the *Canadian Arctic Archipelago*. In normal summers, the Pacific and Atlantic bowheads are prevented from meeting by a large, persistent, plug of sea ice that occupies the central region of the *Canadian Arctic Archipelago*; i.e., the central part of the Northwest Passage (Figure 6.4). Both populations retreat southward upon autumn freeze-up.

However, the ice-edge environment is hazardous, especially during freeze-up, and individuals or pods may become entrapped (as has been observed today). Detailed measurements of fossil bowhead skulls (a proxy of age) now found in raised marine deposits allow a reconstruction of their lengths (Dyke et al., 1996; Savelle et al., 2000). The distribution of lengths compares very closely with the length distribution of the modern Beaufort Sea bowhead population (Figure 6.5), indicating that the cause of death of many bowheads in the past was a catastrophic process that affected all ages indiscriminately. This process can be best interpreted as ice entrapment.

6.3.6 Ice Cores

Among paleoenvironmental **archives**, ice cores from glaciers and ice sheets have a particular strength as a direct recorder of atmospheric composition, especially in the polar regions, at a fine time resolution. The main issue is whether ice cores contain any information about the past extent of sea ice. Such information may be

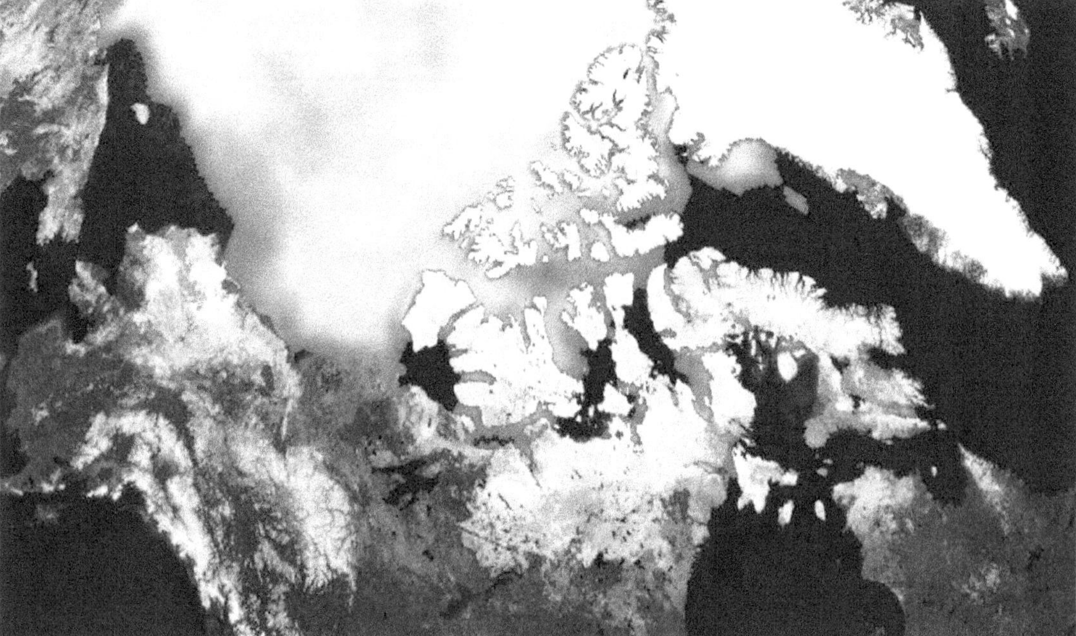

Figure 6.4. Arctic sea ice conditions in September 1996. These conditions were typical for late 20th century summer sea ice in the *Canadian Arctic Archipelago*. [Data from the National Snow and Ice Data Center using NASA Goddard Space Flight Center Scientific Visualization Studio Blue Marble.]

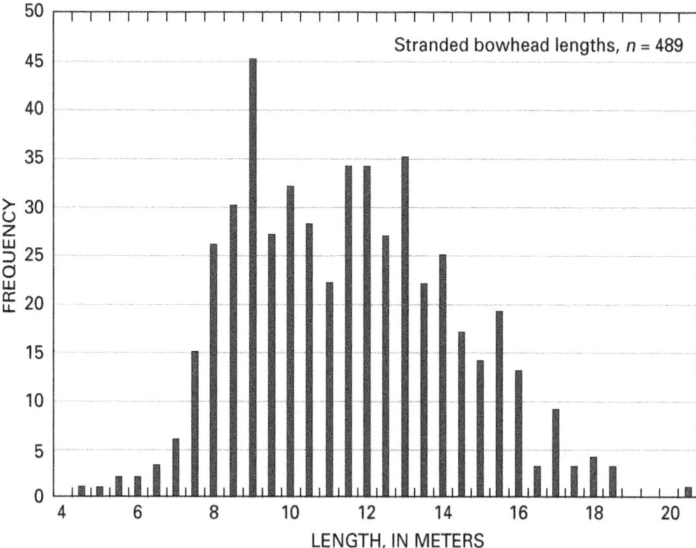

Figure 6.5. The reconstructed lengths of Holocene bowhead whales based on skull measurements (485 animals) and mandible measurements (an additional 4 animals) (Savelle, et al., 2000). This distribution is very similar to the lengths of living Pacific bowheads, indicating that past strandings affected all age classes. [Reproduced by permission of Arctic Institute of North America.]

inferred indirectly: for example, one can imagine that higher temperatures recorded in an ice core are associated with reduced sea ice. However, the real goal is to find a chemical indicator whose concentration is mainly controlled by past sea-ice extent (or by a combination of ice extent and other climate characteristics that can be deduced independently). Any such indicator must be transported for relatively long distances, as by wind, from the sea ice or the ocean beyond. Such an indicator frozen into ice cores would then allow ice cores to give an integrated view throughout a region for some time average, but the disadvantage is that atmospheric transport can then determine what is delivered to the ice.

The ice-core proxy that has most commonly been considered as a possible sea ice indicator is sea salt, usually estimated by measuring a major ion in sea salt, sodium (Na). In most of the world's oceans, salt in sea water becomes an aerosol in the atmosphere by means of a bubble bursting at the ocean surface, and formation of the aerosol is related to wind speed at the ocean surface (Guelle et al., 2001). Expanding sea ice moves the source region (open ocean) further from ice core sites, so that a first assumption is that a more-extensive sea-ice cover should lead to less sea salt in an ice core.

A statistically significant inverse relationship between annual average sea salt in the Penny Ice Cap ice core (*Baffin Island*) and the spring sea ice coverage in *Baffin Bay* (Grumet et al., 2001) was found for the 20th century, and it has been suggested that the extended record could be used to assess the extent of past sea ice in this region. However, the correlation coefficient in this study was low, indicating that only about 7% of the variability in the abundance of sea salt was directly linked to variability in position of sea ice. The inverse relationship between sea salt and sea-ice cover in *Baffin Bay* was also reported for a short core from *Devon Island* (Kinnard et al., 2006). However, more geographically extensive work is needed to show whether these records can reliably reconstruct past sea ice extent.

For *Greenland*, the use of sea salt in this way seems even more problematic. Sea salt in aerosol and snow throughout the *Greenland* plateau tends to peak in concentration in the winter months (Whitlow et al., 1992; Mosher et al., 1993), when sea ice extent is largest, which already suggests that other factors are more important than the proximity of open ocean. Most authors carrying out statistical analyses on sea salt in *Greenland* ice cores in recent years have found relationships with aspects of atmospheric

circulation patterns rather than with sea ice extent (Fischer, 2001; Fischer and Mieding, 2005; Hutterli et al., 2007). Sea-salt records from *Greenland* ice cores have therefore been used as general indicators of storminess (inducing production of sea salt aerosol) and transport strength (Mayewski et al., 1994; O'Brien et al., 1995), rather than as sea-ice proxies.

An alternative interpretation has arisen from study of Antarctic aerosol and ice cores, where the sea ice surface itself can be a source of large amounts of sea-salt aerosol in coastal Antarctica (Rankin et al., 2002). It has then been argued that, although sea salt concentrations and fluxes may be dominated by transport effects on a year-to-year basis, they could be used as an indicator of regional sea ice extent for Antarctica over longer time periods (Fischer et al., 2007a; Wolff et al., 2003). An Antarctic sea ice record covering 740 k.y. has been presented on this basis, showing extended sea ice at times of low temperature (Wolff et al., 2006). The obvious question arises as to whether this inverted model of the relationship between sea salt and sea ice might also be applicable in the Arctic (Rankin et al., 2005). Current ideas about the source of sea-ice relate it to the production of new, thin ice. In the regions around *Greenland* and the nearby islands, much of the sea ice is old ice that has been advected, rather than new ice. It therefore seems unlikely that the method can easily be applied under present conditions (Fischer et al., 2007). The complicated geometry of the oceans around *Greenland* compared with the radial symmetry of Antarctica also poses problems in any interpretation. It is possible that under the colder conditions of the last glacial period, new ice produced around *Greenland* may have led to a more dominant sea-ice source, opening up the possibility that there may be a sea ice record available within this period. However, there is no published basis on which to rely at the moment (2008), and the balance of importance between salt production and salt transport in the Arctic needs further investigation.

One other chemical (methanesulfonic acid, MSA) has been used as a sea-ice proxy in the Antarctic (e.g., Curran et al., 2003). However, studies of MSA in the Arctic do not yet support any simple statistical relationship with sea ice there (Isaksson et al., 2005).

In summary, sea salt in ice cores has the potential to add a well-resolved and regionally integrated picture of the past extent of sea ice extent. At one site weak statistical evidence supports a relationship between sea ice extent and sea salt. However, the complexities of aerosol production and transport mean that no firm basis yet exists for using sea salt in ice cores to estimate past sea-ice extent in the Arctic. Further investigation is warranted to establish whether such proxies might be usable: investigators need a better understanding of the sources of proxies in the Arctic region, further statistical study of the modern controls on their distribution, and modeling studies to assess proxies' sensitivity to major changes in sea-ice extent.

6.3.7 Historical Records

Historical records may describe recent paleoclimatic processes such as weather and ice conditions. The longest historical records of ice cover exist from ice-marginal areas that are more accessible for shipping, as exemplified by a compilation for the Barents Sea covering four centuries in variable detail (Vinje, 1999, 2001). Systematic records of the position of sea-ice margin around the *Arctic Ocean* have been compiled for the period since 1870 (Walsh, 1978; Walsh and Chapman, 2001). These sources vary in quality and availability with time. More reliable observational data on ice concentrations for the entire Arctic are available since 1953, and the most accurate data from satellite imagery is available since 1972 (Cavalieri et al., 2003).

Seas around *ICELAND* provide a rare opportunity to investigate the ice record in a more distant past because *ICELAND* has for 1,200 years recorded observations of drift ice (i.e., sea ice and icebergs) following the settlement of the island in approximately 870 AD (Koch, 1945; Bergthorsson, 1969; Ogilvie, 1984; Ogilvie et al., 2000). This long record has facilitated efforts to quantify the changes in the extent and duration of drift ice around the *ICELAND* coasts during the last 1,200 years (Koch, 1945; Bergthorsson, 1969). During times of extreme drift-ice incursions, ice wraps around *ICELAND* in a clockwise motion. Ice commonly develops off the northwest and north coasts and only occasionally extends into southwest *ICELAND* waters (Ogilvie, 1996). Historical sources have been used to construct a sea-ice index that compares well with springtime temperatures at a climate station in northwest *ICELAND* (Figure 6.6).

6.4 HISTORY OF ARCTIC SEA-ICE EXTENT AND CIRCULATION PATTERNS

6.4.1 Pre-Quaternary History (Prior to ~2.6 Ma)

The shrinkage of the perennial ice cover in the Arctic and predictions that it may completely disappear within the next 50 years or even sooner (Holland et al., 2006a; Stroeve et al., 2008) are especially disturbing in light of recent discoveries that sea ice in the Arctic has persisted for the past 2 million years and may have originated several million years earlier (Darby, 2008; Krylov et al., 2008). Until recently, evidence of long-term (million-year scale) climatic history of the north polar areas was limited to fragmentary records from the Arctic periphery.

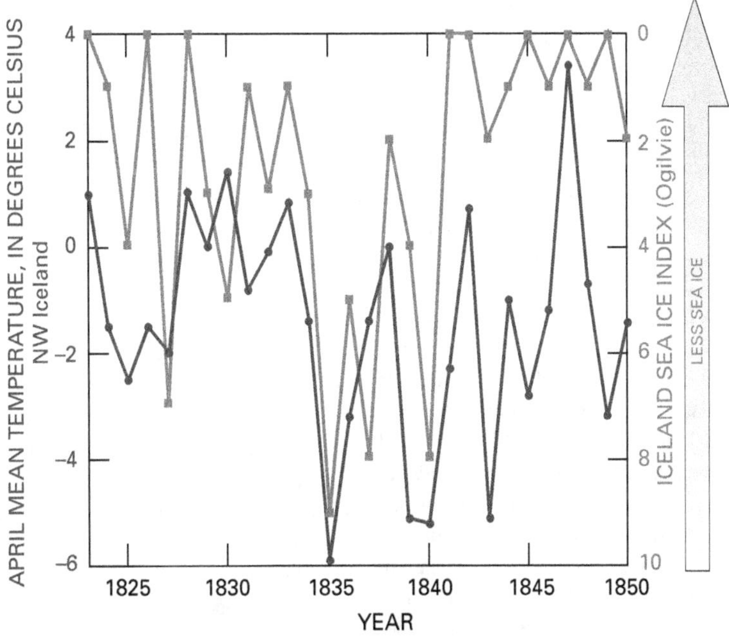

Figure 6.6. The sea-ice index on the Iceland shelf plotted against springtime air temperatures in northwest Iceland that are affected by the distribution of ice in this region (from Ogilvie, 1996). The two correlate well.

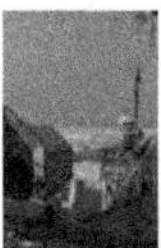

About 50 Ma, during the Eocene Optimum, the Arctic Ocean was considerably warmer than it is today, as much as 24°C.

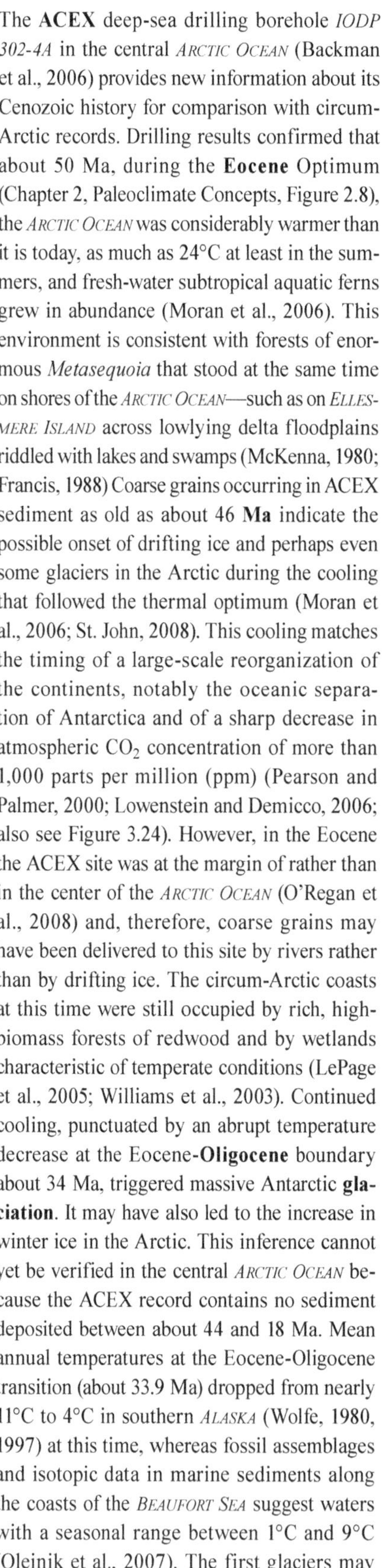

The **ACEX** deep-sea drilling borehole *IODP 302-4A* in the central *Arctic Ocean* (Backman et al., 2006) provides new information about its Cenozoic history for comparison with circum-Arctic records. Drilling results confirmed that about 50 Ma, during the **Eocene** Optimum (Chapter 2, Paleoclimate Concepts, Figure 2.8), the *Arctic Ocean* was considerably warmer than it is today, as much as 24°C at least in the summers, and fresh-water subtropical aquatic ferns grew in abundance (Moran et al., 2006). This environment is consistent with forests of enormous *Metasequoia* that stood at the same time on shores of the *Arctic Ocean*—such as on *Ellesmere Island* across lowlying delta floodplains riddled with lakes and swamps (McKenna, 1980; Francis, 1988) Coarse grains occurring in ACEX sediment as old as about 46 **Ma** indicate the possible onset of drifting ice and perhaps even some glaciers in the Arctic during the cooling that followed the thermal optimum (Moran et al., 2006; St. John, 2008). This cooling matches the timing of a large-scale reorganization of the continents, notably the oceanic separation of Antarctica and of a sharp decrease in atmospheric CO_2 concentration of more than 1,000 parts per million (ppm) (Pearson and Palmer, 2000; Lowenstein and Demicco, 2006; also see Figure 3.24). However, in the Eocene the ACEX site was at the margin of rather than in the center of the *Arctic Ocean* (O'Regan et al., 2008) and, therefore, coarse grains may have been delivered to this site by rivers rather than by drifting ice. The circum-Arctic coasts at this time were still occupied by rich, high-biomass forests of redwood and by wetlands characteristic of temperate conditions (LePage et al., 2005; Williams et al., 2003). Continued cooling, punctuated by an abrupt temperature decrease at the Eocene-**Oligocene** boundary about 34 Ma, triggered massive Antarctic **glaciation**. It may have also led to the increase in winter ice in the Arctic. This inference cannot yet be verified in the central *Arctic Ocean* because the ACEX record contains no sediment deposited between about 44 and 18 Ma. Mean annual temperatures at the Eocene-Oligocene transition (about 33.9 Ma) dropped from nearly 11°C to 4°C in southern *Alaska* (Wolfe, 1980, 1997) at this time, whereas fossil assemblages and isotopic data in marine sediments along the coasts of the *Beaufort Sea* suggest waters with a seasonal range between 1°C and 9°C (Oleinik et al., 2007). The first glaciers may have developed in *Greenland* about the same time, on the basis of coarse grains interpreted as iceberg-rafted debris in the *North Atlantic* (Eldrett et al., 2007). Sustained, relatively warm conditions lingered during the early **Miocene** (about 23–16 Ma) when cool-temperate *Metasequoia* dominated the forests of northeast *Alaska* and the Yukon (White and Ager, 1994; White et al., 1997), and the central *Canadian Arctic* Islands were covered in mixed conifer-hardwood forests similar to those of southern Maritime *Canada* and New England today. Such forests and associated wildlife would have easily tolerated seasonal sea ice, but they would not have survived the harshness of perennial ice cover on the adjacent ocean (Whitlock and Dawson, 1990).

Marked changes in Arctic climate in the middle Miocene were concurrent with global cooling and the onset of Antarctic reglaciation.

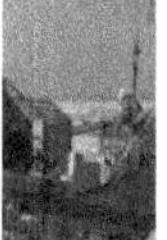

A large unconformity (a surface in a sequence of sediments that represents missing deposits, and thus missing time) in the ACEX record prevents us from characterizing sea-ice conditions between about 44–18 Ma (Backman et al., 2008). Sediments overlying the unconformity contain little ice-rafted debris, and they indicate a smaller volume of sea ice in the *Arctic Ocean* at that time (St. John, 2008). Marked changes in Arctic climate in the middle Miocene were concurrent with global cooling and the onset of Antarctic reglaciation (Chapter 2, Figure 2.8). These changes may have been promoted by the opening of the *Fram Strait* between the Eurasian and *Greenland* margins about 17 Ma, which allowed the modern circulation system in the *Arctic Ocean* to develop (Jakobsson et al., 2007). Resultant cooling led to a change from pine-redwood-dominated to larch-spruce-dominated floodplains and swamps at the Arctic periphery at about 16 Ma as recorded, for example, on *Banks Island* by extensive peats with stumps in growth position (Fyles et al., 1994; Williams, 2006). A combination of cooling and increased moisture from the *North Atlantic* caused ice masses on and around *Svalbard* to grow and icebergs to discharge into the eastern *Arctic Ocean* and the *Greenland* Sea at about 15 Ma (Knies and Gaina, 2008). The source of sediment in the central *Arctic Ocean* changed between 13–14 Ma and indicates the likelihood that sea ice was now perennial (Krylov et al., 2008), although the ice's geographic distribution and persistence is not yet understood. Evidence of perennial ice can be found in even older sediments, starting from at least 14 Ma (Darby, 2008). Several pulses of more-abundant-than-normal ice-rafted debris in the late Miocene ACEX record indicate further growth of sea ice (St. John, 2008). This interpretation is consistent with a cooling climate indicated by the spread of pine-dominated forests in northern *Alaska* (White et al., 1997).

On the other hand, paleobotanical evidence also suggests that throughout the late Miocene and most of the Pliocene in at least some intervals perennial ice was severely restricted or absent. Thus, extensive braided-river deposits of the *Beaufort* Formation (early to middle Pliocene, about 5.3–3 Ma) that blanket much of the western *Canadian Arctic Archipelago* enclose abundant logs and other woody detritus. These deposits contain more than 100 vascular plants, such as pine (2 and 5 needles) and birch, and are dominated at some locations by spruce and larch (Fyles, 1990; Devaney, 1991). Although these floral remains indicate overall **boreal** conditions cooler than in the Miocene, extensive perennial sea ice is not likely to have existed in the adjacent Beaufort Sea during this time. This inference is consistent with the presence of the bivalve Icelandic Cyprine (*Arctica islandica*) in marine sediments capping the Beaufort Formation on *Meighen Island* at 80°N and dated to the peak of Pliocene warming, about 3.2 Ma (Fyles et al., 1991). Foraminifers in Pliocene deposits in the Beaufort-Mackenzie area are also characteristic of boreal but not yet high-Arctic waters (McNeil, 1990), whereas the only known pre-**Quaternary** foraminiferal evidence from the central *Arctic Ocean* indicates seasonally ice-free conditions in the early Pliocene about 700 km north of the Alaskan coast (Mullen and McNeil, 1995).

Cooling in the late Pliocene profoundly reorganized the Arctic system: tree line retreated from the Arctic coasts (White et al., 1997; Matthews and Telka, 1997), **permafrost** formed (Sher et al., 1979; Brigham-Grette and Carter, 1992), and continental ice masses grew around the *Arctic Ocean*—for example, the *Svalbard* ice sheet advanced onto the outer shelf (Knies et al., 2002) and between 2.9–2.6 Ma ice sheets began to grow in North America (Duk-Rodkin et al., 2004). The ACEX cores record especially large volumes of ice-rafted debris in the *Arctic Ocean* around 2 Ma (St. John, 2008). Despite the overall cooling, extensive warm intervals during the late Pliocene and the initial **stages** of the Quaternary (about 2.4–3 Ma) are repeatedly documented at the Arctic periphery from northwest *Alaska* to northeastern *Greenland* (Feyling-Hanssen et al., 1983; Funder et al., 1985, 2001; Carter et al., 1986; Bennike and Böcher, 1990; Kaufman, 1991; Brigham-Grette and Carter, 1992). For example, beetle and plant macrofossils in the nearshore high-energy sediments of the upper *Kap København* Formation on northeast *Greenland*, dated about 2.4 Ma, mimic paleoenvironmental conditions similar to those of southern Labrador today (Funder et al., 1985, 2001; Bennike and Böcher, 1990). At the same time, marine conditions were distinctly Arctic but, analogous with present-day faunas along the Russian coast, open water must have existed for 2 or 3 months in the summer. These results imply that summer sea ice in the entire *Arctic Ocean* was probably much reduced.

A more complete history of perennial versus seasonal sea ice and ice-free intervals during the past several million years requires additional sedimentary records distributed throughout the *Arctic Ocean* and a synthesis of sediment and paleobiological evidence from both land and sea. This history will provide new clues about the stability of the Arctic sea ice and about the sensitivity of the *Arctic Ocean* to changing temperatures and other climatic features such as snow and vegetation cover.

6.4.2 Quaternary Variations (the past 2.6 m.y.)

Quaternary climate history is composed of cold intervals (glacials) when very large ice sheets formed in northern Eurasia and North America and of interspersed warm intervals (interglacials), such as the present one, referred to as the Holocene (which began about 11.5 thousand years ago (ka)).

The Quaternary period of Earth's history during the past 2.6 million years (**m.y.**) or so is characterized by overall low temperatures and especially large swings in climate regime (Chapter 2, Paeloclimate Concepts, Figure 2.9). These swings are related to changes in **insolation** (incoming solar radiation) modulated by Earth's orbital parameters with periodicities of tens to hundreds of thousand years (see Chapter 2 for more detail). During cold periods when large ice masses are formed, such as during the Quaternary, these variations are amplified by powerful feedbacks due to changes in the albedo (reflectivity) of Earth's surface and concentration of greenhouse gases in the atmosphere. Quaternary climate history is composed of cold intervals (glacials) when very large ice sheets formed in northern Eurasia and North America and of interspersed warm intervals (interglacials), such as the present one, referred to as the Holocene (which began about 11.5 thousand years ago (**ka**). Temperatures at Earth's surface during some interglacials were similar to or even somewhat warmer than those of today; therefore, climatic conditions during those times can be used as approximate analogs for the conditions predicted by climate models for the 21st century (Otto-Bliesner et al., 2006; Goosse et al., 2007). One of the biggest questions in this respect is to what degree sea-ice cover was reduced in the Arctic during those warm intervals. This issue is insufficiently understood because interglacial deposits at the Arctic margins are exposed only in fragments (CAPE, 2006) and because sedimentary records from the *Arctic Ocean* generally have only low resolution. Even the age assigned to sediments that appear to be interglacial is commonly problematic because of the poor preservation of fossils and various stratigraphic complications (e.g., Backman et al., 2004). A better understanding has begun to emerge from recent collections of sediment cores from strategic sites drilled in the *Arctic Ocean*, such as **ACEX** (Backman et al., 2006) and **HOTRAX** (Darby et al., 2005). The severity of ice conditions (widespread, thick, perennial ice) during glacial stages is indicated by of the extreme rarity of biological remains in cool-climate sediment layers and possible non-deposition intervals due to especially solid ice (Polyak et al., 2004; Darby et al., 2006; Cronin et al., 2008). In contrast, interglacials are characterized by higher marine productivity that indicates reduced ice cover. In particular, planktonic foraminifers typical of subpolar, seasonally open water lived in the area north of *Greenland* during the last interglacial (**Marine Isotope Stage (MIS)** 5e), 120–130 **ka** (Figure 6.7; Nørgaard-Pedersen et al., 2007a,b). Given that this area is presently characterized by especially thick and widespread ice, most of the *Arctic Ocean* may have been free of summer ice cover in the interval between 120–130 ka. Investigators need to carefully examine correlative sediments throughout the *Arctic Ocean* to determine how widespread were these low-ice or possibly ice-free conditions. Some intervals in sediment cores from various sites in the central Arctic have been reported to contain subpolar microfauna (e.g., Herman, 1974; Clark et al., 1990), but their age was not well constrained. New sediment core studies are needed to place these intervals in the coherent stratigraphic context and to reconstruct corresponding ancient ice conditions.

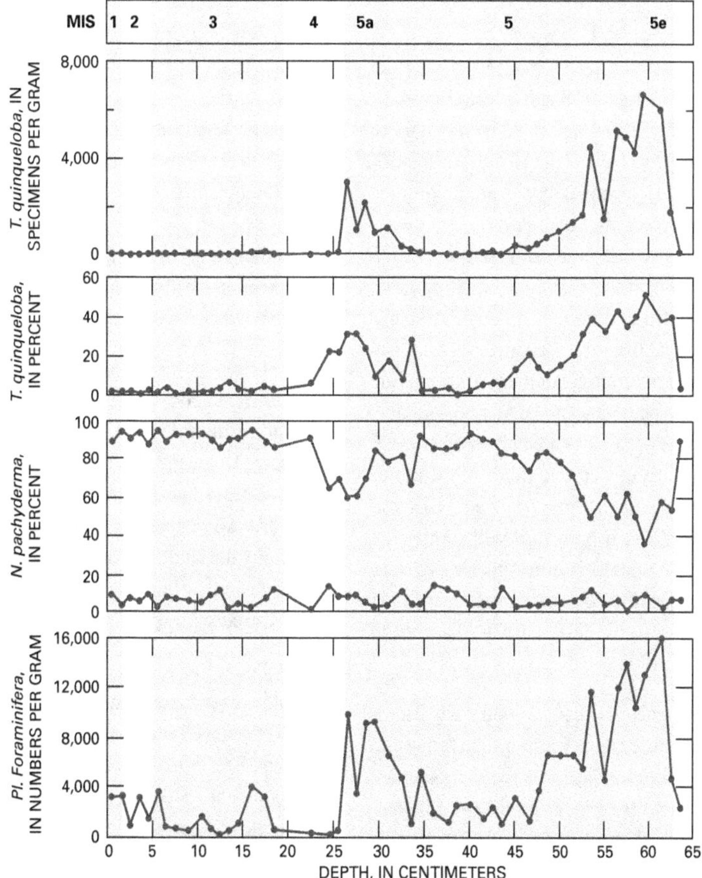

Figure 6.7. Planktonic foraminiferal record, core GreenICE-11, north of Greenland (from Nørgaard-Pedersen et al., 2007b). Note high numbers of a subpolar planktonic foraminifer *T. quinqueloba* during the last interglacial, Marine Isotope Stage (MIS) 5e; they indicate warm temperatures or reduced-ice conditions (or both) north of GREENLAND at that time. [Copyright 2007 Amercian Geophysical Union, reproduced by permission American Geophysical Union.]

This task is especially important because only those records from the central *Arctic Ocean* can provide direct evidence for ocean-wide ice-free water.

Some coastal exposures of interglacial deposits such as MIS 11 (about 400 ka) and 5e (about 120–130 ka) also indicate water temperatures warmer than present and, thus, reduced ice. For example, deposits of the last interglacial on the Alaskan coast of the *Chukchi Sea* (the so-called Pelukian transgression) contain some fossils of species that are limited today to the northwest Pacific, whereas inter-tidal snails found near *Nome*, just slightly south of the *Bering Strait*, suggest that the coast here may have been annually ice free (Brigham-Grette and Hopkins, 1995; Brigham-Grette et al., 2001). On the Russian side of the *Bering Strait*, formaninifer assemblages suggest that coastal waters were fairly warm, like those in the Sea of Okhotsk and Sea of Japan (Brigham-Grette et al., 2001). Deposits of the same age along the northern Arctic coastal plain show that at least eight mollusk species extended their distribution ranges well into the *Beaufort Sea* (Brigham-Grette and Hopkins, 1995). Deposits near *Barrow* include at least one mollusk and several ostracode species known now only from the *North Atlantic*. Taken together, these findings suggest that during the peak of the last interglacial, about 120–130 ka, the winter limit of sea ice did not extend south of the *Bering Strait* and was probably located at least 800 km north of historical limits (such as on Figure 6.1), whereas summer sea-surface temperatures were warmer than present through the *Bering Strait* and into the *Beaufort Sea*.

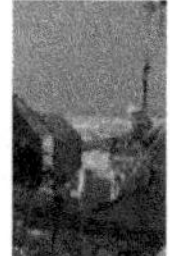

6.4.3 The Holocene (the most recent 11.5 k.y.)

The present interglacial that has lasted approximately 11.5 k.y. is characterized by much more paleoceanographic data than earlier warm periods, because Holocene deposits are ubiquitous on continental shelves and along many coastlines. Owing to relatively high sedimentation rates at continental margins, ice drift patterns can be constructed on sub-millennial scales from some sedimentary records. Thus, the periodic influx of large numbers of iron oxide grains from specific sources, as into the *Siberian* margin-to-sea-floor area north of *Alaska*, has been linked to a certain mode of the atmospheric circulation pattern (Darby and Bischof, 2004). If this link is proven, it will signify the existence of longer term atmospheric cycles in the Arctic than the decadal Arctic Oscillation observed during the last century (Thompson and Wallace, 1998).

Early Holocene temperatures were warmer than today and the Arctic contained less ice. This climate is consistent with a higher intensity of insolation that peaked about 11 ka owing to Earth's orbital variations.

Many proxy records indicate that early Holocene temperatures were warmer than today and that the Arctic contained less ice. This climate is consistent with a higher intensity of insolation that peaked about 11 ka owing to Earth's orbital variations. Evidence of warmer temperatures appears in many paleoclimatic records from the high Arctic—*Svalbard* and northern *Greenland*, northwestern North America, and eastern *Siberia* (Kaufman et al., 2004; Blake, 2006; Fisher et al., 2006; Funder and Kjær, 2007). Decreased sea-ice cover in the western Arctic during the early Holocene has also been inferred from high sodium concentrations in the *Penny Ice Cap* of *Baffin Island* (Fisher et al., 1998) and the *Greenland Ice Sheet* (Mayewski et al., 1994), although the implications of salt concentration is yet to be defined. Areas that were affected by the extended melting of the **Laurentide Ice Sheet**, especially the northeastern sites in North America and the adjacent *North Atlantic*, show more complex patterns of temperature and ice distribution (Kaufman et al., 2004).

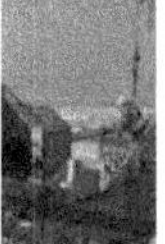

An extensive record has been compiled from bowhead whale findings along the coasts of the *Canadian Arctic Archipelago* straits (Dyke et al., 1996, 1999; Fisher et al., 2006). Understanding the dynamics of ice conditions in this region is especially important for modern-day considerations because ice-free, navigable straits through the *Canadian Arctic Archipelago* will provide new opportunities for shipping lanes. The current set of radiocarbon dates on bowheads from the *Canadian Arctic Archipelago* coasts is grouped into three regions: western, central, and eastern (Figure 6.8). The central region today is the area of normally persistent summer sea ice; the western region is within the summer range of the Pacific bowhead; the eastern region is within the summer range of the Atlantic bowhead. These three graphs allow us to draw the following conclusions:

1. Bowhead bones have been most commonly found in all three regions in early Holocene (10–8 ka) deposits. At that time Pacific and Atlantic bowheads were able to intermingle freely along the length of the Northwest Passage indicating at least periodically ice-free summers.

2. Following an interval (8–5 ka) containing fewer bones, abundant bowhead bones have been found in deposits in the eastern channels during the middle Holocene (5–3 ka). At times, the Atlantic bowheads penetrated the central region, particularly 4.5–4.2 ka. The Pacific bowhead apparently did not extend its range at this time.

3. A final peak of bowhead bones dated about 1.5–0.75 ka has been found in all three regions, suggesting an open Northwest Passage during at least some summers. During this interval the bowhead-hunting Thule Inuit (Eskimo) expanded eastward out of the *Bering Sea* region and ultimately spread to *Greenland* and Labrador.

4. The decline of bowhead abundances during the last few centuries is evident in all three graphs. Thule bowhead hunters abandoned the high *Arctic* of *Canada* and *Greenland* during the **Little Ice Age** cooling (around 13th to 19th centuries) and Thule living in more-southern Arctic regions increasingly focused on alternate resources.

On the basis of the summer ice melt record of the *Agassiz Ice Cap* (Fisher et al., 2006), summer temperatures that accompanied the early Holocene bowhead maximum are estimated at

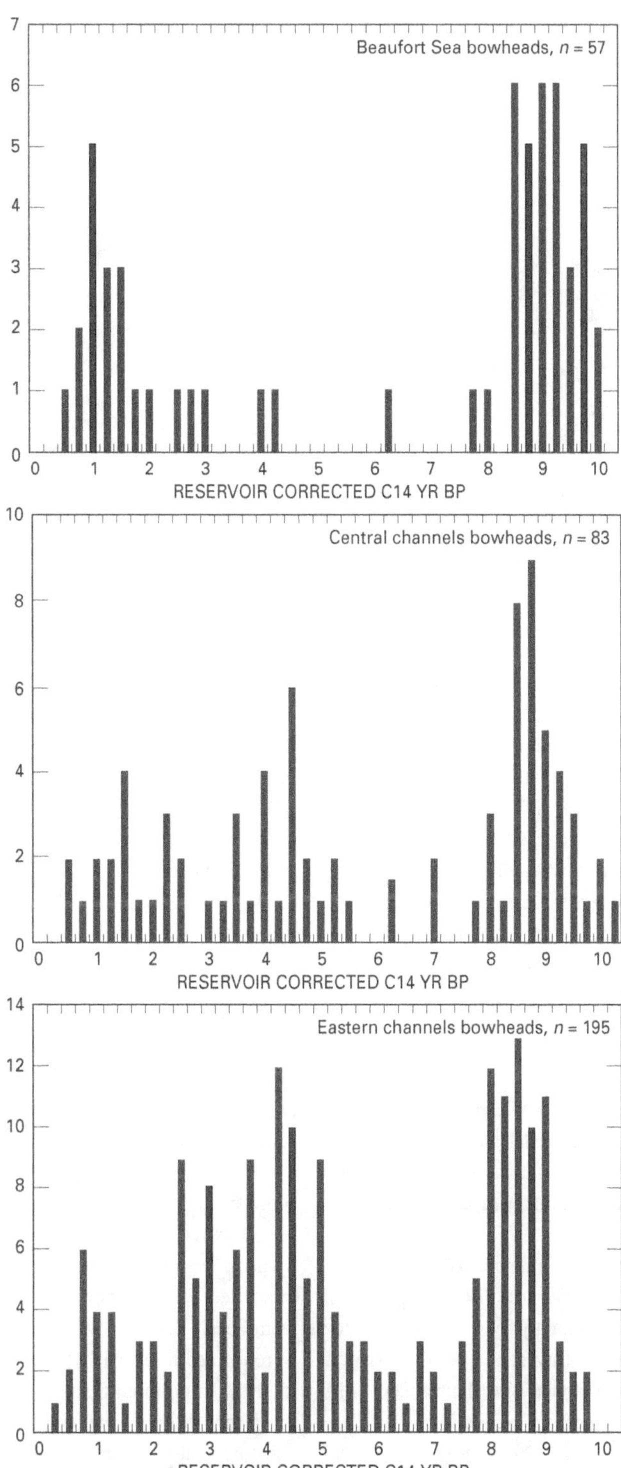

Figure 6.8. Distribution of radiocarbon ages (in thousands of years) of bowhead whales in three regions of the Canadian Arctic Archipelago (data from Dyke et al., 1996; Savelle et al., 2000). [Reproduced by permission of Arctic Institute of North America.]

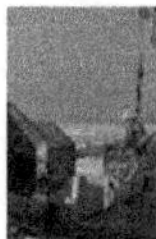

about 3°C above mid-20th century conditions, when July mean daily temperatures along the central Northwest Passage (Figure 6.4) were about 5°C. Unless other processes, such as a different ocean circulation pattern, were also forcing greater summer sea-ice clearance in the early Holocene, the value of 3°C is an upper bound on the amount of warming necessary to clear the Northwest Passage region of summer sea ice. At times during the middle and late Holocene (especially 4.5–4.2 ka) the threshold condition was approached and, at least briefly, met, as indicated by Atlantic bowhead bones in the central channels. The threshold condition for clearance of ice from the Northwest Passage was crossed in summer 2007. Whether this open passage will become a regular feature and what the consequences might be for Pacific-Atlantic exchanges of biota remains to be seen.

The bowhead record can be compared with the distribution of driftwood. Dated driftwood from raised marine beaches along the Arctic coasts of North America, notably around the margins of *Baffin Bay* (Blake, 1975), has been used to infer changes in the transport of sea ice from the *Arctic Basin* (Dyke et al., 1997) (Figure 6.9). The ratio of larch (mainly from *Russia*) to spruce (mainly from northwest *Canada*) driftwood declines sharply about 7 ka. This abrupt shift might have been caused by the intensity of ice drift from the *Arctic Ocean* or changes in its trajectories (Tremblay et al., 1997), or it might reflect changes in the composition or extent of forests. The delivery of driftwood, which probably was borne on the East Greenland Current, peaked during the middle Holocene, possibly in conjunction with less ice cover in the *Arctic Ocean*. Levac et al. (2001) estimated the duration of sea-ice cover during the Holocene in northern *Baffin Bay* (southern reach of *Nares Strait* between *Ellesmere Island* and northwest *Greenland*) based on transfer functions of dinocyst assemblages. The present-day duration of the ice cover in this area is about 8 months, whereas the predicted duration for the Holocene ranges between 7 and 10–12 months. An interval of minimal sea-ice cover existed until about 4.5 ka, whereas afterwards the sea-ice cover was considerably more extensive (Figure 6.10).

Along the North *Greenland* coasts, isostatically raised staircases of wave-generated beach ridges (Figure 6.11) document seasonally open water (Funder and Kjær, 2007). Large numbers of striated boulders in and on the marine sediments also indicate that the ocean was open enough for icebergs to drift along the shore and drop their loads. Presently the North *Greenland* coastline is permanently surrounded by pack ice, and rare icebergs are locked up in sea ice. Radiocarbon-dated mollusk shells from beach ridges show that the beach ridges were formed in the early Holocene, within the interval from about 8.5–6 ka, which is progressively shorter from south to north. These wave-generated shores and abundant iceberg-deposited boulders indicate the possibility that the adjacent Arctic Ocean was free of sea ice in summer at this time.

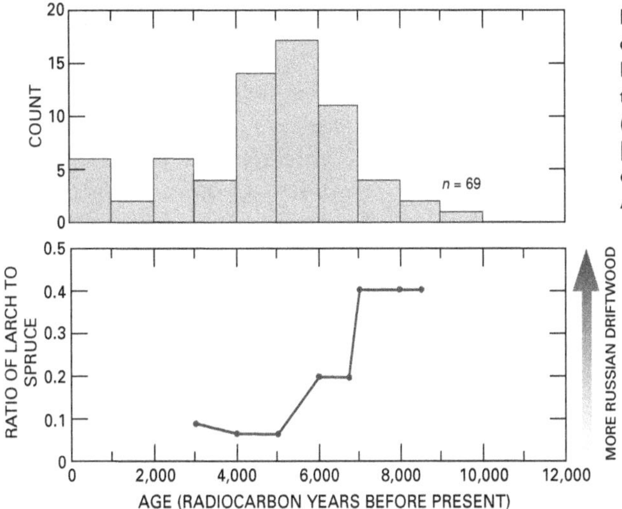

Figure 6.9. Distribution of radiocarbon ages of Holocene driftwood on the shores of Baffin Bay (from Dyke et al., 1997). [Reproduced by permission of Arctic Institute of North America.]

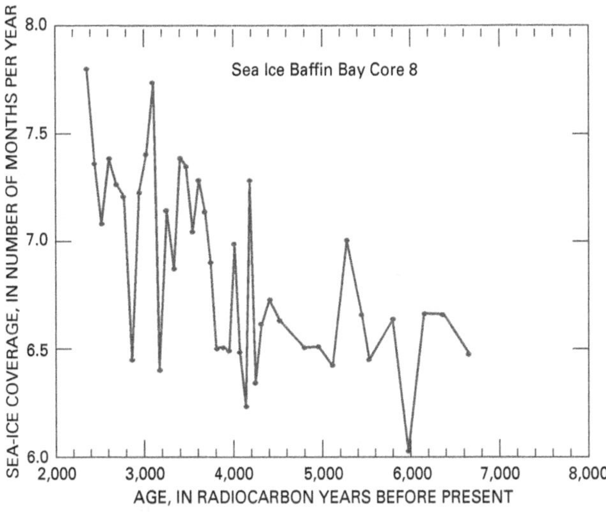

Figure 6.10. Reconstruction of the duration of ice cover (months per year) in northern Baffin Bay during the Holocene based on dinocyst assemblages (modified from Levac et al., 2001).

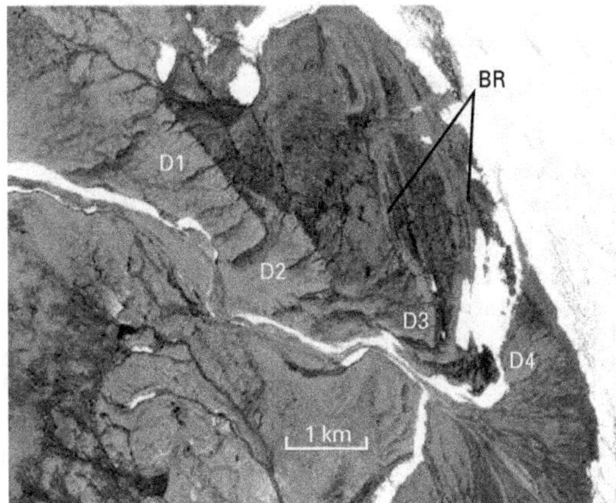

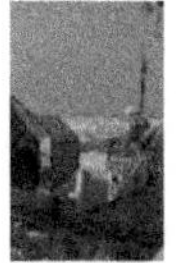

Figure 6.11. Aerial photo (left) of wave-generated beach ridges (BR) at Kap Ole Chiewitz, 83°25'N, northeast Greenland. D1–D4 are raised deltas. The oldest, D1, is dated to ~10 ka while D4 is the modern delta. Only D3 is associated with wave activity. The period of beach ridge formation is dated to ca. 8.5–6 ka. The photo on the right shows the upper beach ridge. (Funder, S. and K. Kjær, 2007) [Copyright 2007 American Geophysical Union, reproduced by permission of American Geophysical Union.]

A somewhat different history of ice extent in the Holocene emerges from the northern *North Atlantic* and *Nordic seas*, exemplified by the *Iceland* margin. A 12,000 year record of quartz content in shelf sediment, which is used in this area as a proxy for the presence of drift ice (Eiriksson et al., 2000), has been produced for a core (MD99-2269) from the northern *Iceland* shelf. The record has a resolution of 30 years per sample (Moros et al., 2006); these results are consistent with data obtained from 16 cores across the northwestern *Iceland* shelf (Andrews, 2007). These data show a minimum in quartz and, thus, ice cover at the end of deglaciation, whereas the early Holocene area of ice increased and then reached another minimum around 6 ka, after which the content of quartz steadily rose (Figure 6.12). The lagged Holocene optimum in the *North Atlantic* in comparison with high Arctic records can be explained by the nature of oceanic controls on ice distribution. In particular, the discharge of glacial meltwater from the remains of the *Laurentide Ice Sheet* slowed the warming in the *North Atlantic* region in the early Holocene (Kaufman et al., 2004). Additionally, oceanic circulation seesawed between the eastern and western regions of the *Nordic seas* throughout much of the Holocene. For example, in the *Norwegian Sea* the Holocene ice-rafting peaked in the mid-Holocene, 6.5–3.7 ka (Risebrobakken et al., 2003), and changes in Earth's orbit forced decreasing summer temperatures and decreased seasonality (Moros et al., 2004). By contrast, the middle Holocene is a relatively warm period off East *Greenland*, and it received a strong subsurface current of Atlantic Water around 6.5–4 ka, while ice-rafted debris was low (Jennings et al., 2002). These patterns are consistent with modern marine and atmospheric temperatures that commonly change in opposite directions on the eastern and western side of the *North Atlantic* ("seesaw effect" of van Loon and Rogers, 1978).

The Neoglacial cooling of the last few thousand years is considered overall to be related to decreasing summer insolation (Koç and Jansen, 1994). However, high-resolution climate records reveal greater complexity in the system—changes in seasonality and links with conditions in low latitudes and southern high latitudes (e.g., Moros et al., 2004). Variations in the volumes of ice-rafted debris indicate several cooling and warming intervals during Neoglacial time, similar to the so-called "**Little Ice Age**" and "**Medieval Climate Anomaly**" cycles of greater and lesser areas of sea ice (Jennings and Weiner, 1996; Bond et al., 1997; Jennings et al., 2002; Moros et al., 2006). Polar Water excursions have been reconstructed as multi-century to decadal-scale variations superimposed on the Neoglacial cooling at several sites in the subarctic *North Atlantic* (Jennings et al., 2002; Andersen et al., 2004; Giraudeau et al., 2004). In contrast, a decrease in drift ice during the Neoglacial is documented for areas influenced by the *North Atlantic* Current, possibly indicating a warming in the eastern *Nordic Seas* (Moros et al., 2006). A seesaw climate pattern has been evident between seas adjacent to West *Greenland* and Europe. For instance, warm periods in Europe around 800–100 BC and 800–1300 AD (Roman and the Medieval Climate Anomalies) were cold periods on West *Greenland* because little warm Atlantic Water fed into the West *Greenland* Current. Moreover, a cooling interval in western Europe (during the Dark Ages) correlated with increased meltwater—and thus warming—on West *Greenland* (Seidenkrantz et al., 2007).

Bond et al. (1997, 2001) suggested that cool periods manifested as past expansions of drift ice and ice-rafted debris (most notably, hematite-stained quartz grains) in the *North Atlantic* punctuated deglacial and Holocene records at intervals of about 1,500 years and that these drift ice events were a result of climates that cycled independently of glacial influence. Bond et al. (2001) concluded that peak volumes of Holocene drift ice resulted from southward expansions of polar waters that correlated with times of reduced solar output. This conclusion suggests that variations in the Sun's output is linked to centennial- to millennial-scale variations in Holocene climate through effects on production of *North Atlantic* deep water. However, continued investigation of the drift ice signal indicates that although the variations reported by Bond et al. (2001) may record a solar influence on climate, they likely do not pertain to a simple index of drift ice (Andrews et al., 2006). In addition, those cooling events prior to the Neoglacial interval may stem from deglacial meltwater forcing rather than from southward drift of Arctic ice (Jennings et al., 2002; Giraudeau et al., 2004). In an effort to test the idea of solar forcing of 1,500-year cycles in

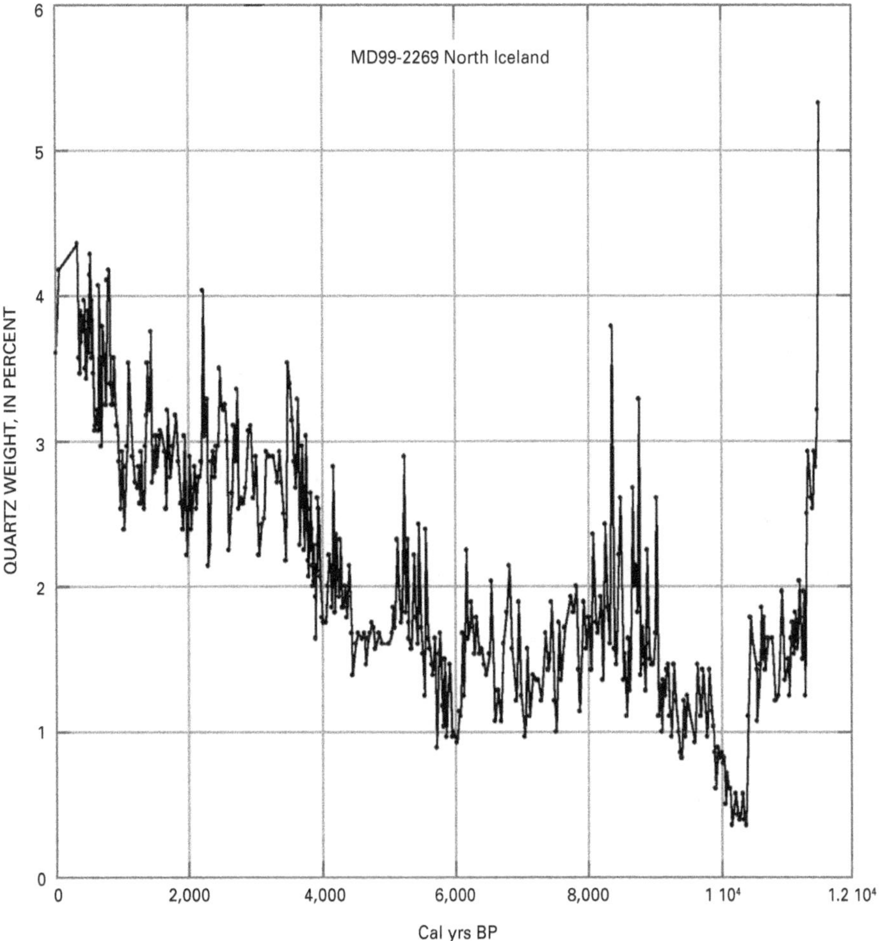

Figure 6.12. Variations in the percentage of quartz (a proxy for drift ice) in Holocene sediments from the northern Iceland shelf (from Moros et al., 2006). BP, before present. [Copyright 2006 American Geophysical Union, reproduced by permission of American Geophysical Union.]

Holocene climate change, Turney et al. (2005) compared Irish tree-ring-derived chronologies and radiocarbon activity, a proxy for solar activity, with the Holocene drift-ice sequence of Bond et al. (2001). They found a dominant 800-year cycle in moisture, reflecting atmospheric circulation changes during the Holocene but no link with solar activity.

Despite many records from the Arctic margins indicating considerably reduced ice covering the early Holocene, no evidence of the decline of perennial ice cover has been found in sediment cores from the central *Arctic Ocean*. *Arctic Ocean* sediments contain some ice-rafted debris interpreted to arrive from distant shelves requiring more than 1 year of ice drift (Darby and Bischof, 2004). One explanation is that the true record of low-ice conditions has not yet been found because of low sedimentation rates and stratigraphic uncertainties. Additional investigation of cores by use of many proxies with highest possible resolution is needed to verify the distribution of ice in the Arctic during the warmest phase of the current interglacial.

6.4.4 Historical Period

Arctic **paleoclimate records** that contain proxies such as lake and marine sediments, trees, and ice cores indicate that from the mid-19th to late 20th century the Arctic warmed to the highest temperatures in at least four centuries (Overpeck et al., 1997). Subglacial material exposed by retreating glaciers in the *Canadian Arctic* indicates that modern temperatures are warmer than any time in at least the past 1,600 years (Anderson et al., 2008). Paleoclimatic proxy records of the last two centuries agree well with hemispheric and global data (including instrumental measurements) (Mann et al., 1999; Jones et al., 2001). The composite record of ice conditions for Arctic ice margins since 1870 shows a steady retreat of seasonal ice since 1900; in addition, the retreat of both seasonal and annual ice has accelerated during the last 50 years (Figure 6.13) (Kinnard et al., 2008). The latter observations are the most reliable for the entire data set and are based on satellite imagery since 1972. The rate of ice-margin retreat during the most recent decades is spatially variable, but the overall trend in ice is down. The current decline of the Arctic sea-ice cover is much larger than expected from decadal-scale climatic and hydrographic variations (e.g., Polyakov et al., 2005; Steele et al., 2008). The recent warming and associated ice shrinkage are especially anomalous because orbitally driven insolation has been decreasing steadily since its maximum at 11 ka, and it is now near its minimum in the 21 k.y. **precession** cycle (e.g., Berger and Loutre, 2004), which should lead to cool summers and extensive sea ice.

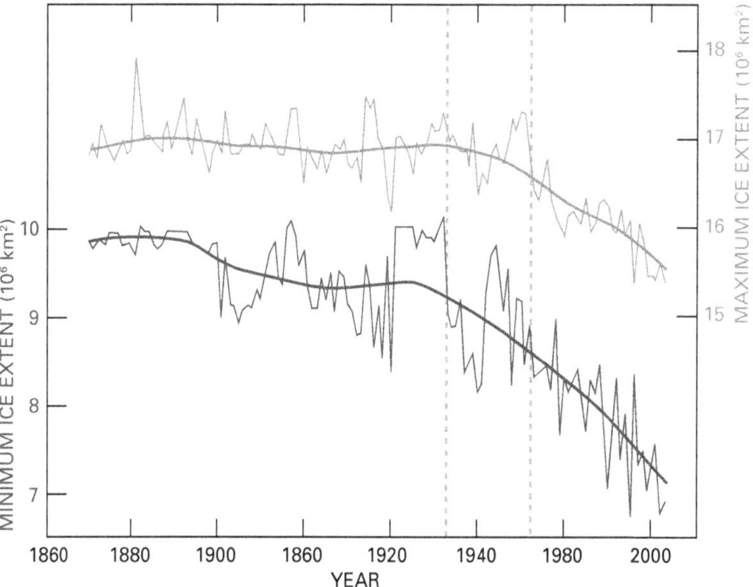

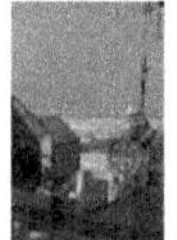

Figure 6.13. Total sea-ice extent time series, 1870–2003 (from Kinnard et al., 2008). Green lines = maximal extent. Red lines = minimal extent. Thick lines are robust spline functions that highlight low-frequency changes. Vertical dotted lines separate the three periods for which data sources changed fundamentally: earliest, 1870–1952, observations of differing accuracy and availability; intermediate, 1953–1971, generally accurate hemispheric observations; most recent, 1972–2003, satellite period, best accuracy and coverage. [Copyright 2008 American Geophysical Union, reproduced by permission of American Geophysical Union.]

6.5 SUMMARY

Geological data indicate that the history of Arctic sea ice is closely linked with temperature changes. Sea ice in the *ARCTIC OCEAN* may have appeared as early as 46 Ma, after the onset of a long-term climatic cooling related to a reorganization of the continents and subsequent formation of large ice sheets in polar areas. Year-round ice in the Arctic possibly developed as early as 13–14 Ma, in relation to a further overall cooling in climate and the establishment of the modern hydrographic circulation in the *ARCTIC OCEAN*. Nevertheless, extended seasonally ice-free periods were likely until the onset of large-scale Quaternary glaciations in the Northern Hemisphere approximately 2.5 Ma. These glaciations were likely to have been accompanied by a fundamental increase in the extent and duration of sea ice. Ice may have been less prevalent during Quaternary interglacials, and the *ARCTIC OCEAN* even may have been seasonally ice free during the warmest interglacials (owing to changes in insolation modulated by variations in Earth's orbit that operate on time scales of tens of thousands to a hundred thousand years). Reduced-ice conditions are inferred, for example, for the previous interglacial and for the onset of the current interglacial, about 130 and 10 ka. These low-ice periods can be used as ancient analogs for future conditions expected from the marked ongoing loss of Arctic ice cover. On time scales of hundreds and thousands of years, patterns of ice circulation vary somewhat; this feature is not yet well understood, but large periodic reductions in ice cover at these time scales are unlikely. Recent historical observations suggest that ice cover has consistently shrunk since the late 19th century, and that shrinkage has accelerated during the last several decades. Shrinkage that was both similarly large and rapid has not been documented for at least the last few thousand years, although the paleoclimatic record is sufficiently sparse that similar events might have been missed. The recent ice loss does not seem to be explainable by natural climatic and hydrographic variability on decadal time scales, and this loss is remarkable for occurring when reduction in summer sunshine from orbital changes has caused sea-ice melting to be less likely than in the previous millennia since the end of the last ice age. The recent changes thus appear notably anomalous; improved reconstructions of sea-ice history would help clarify just how anomalous these changes are.

The composite record of ice conditions for Arctic ice margins since 1870 shows a steady retreat of seasonal ice since 1900; in addition, the retreat of both seasonal and annual ice has accelerated during the last 50 years.

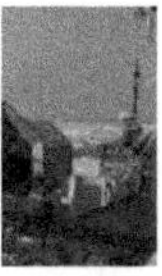

Key Findings and Recommendations

Lead Authors: Richard B. Alley*, Pennsylvania State University, University Park, PA; Julie Brigham-Grette*, University of Massachusetts, Amherst , MA; Gifford H. Miller*, University of Colorado, Boulder, CO; Leonid Polyak*, Ohio State University, Columbus, OH; James W.C. White, University of Colorado, Boulder, CO

*SAP 1.2 Federal Advisory Committee Member

INTRODUCTION

Paleoclimatic data provide a highly informative if incomplete history of Arctic climate. Temperature history is especially well recorded, and it commonly allows researchers to accurately reconstruct changes and rates of changes for particular seasons. Precipitation (rain or snow) and the extent of ice on land and sea are some of the many other climate variables that have also been reconstructed. The data also provide insight into the histories of many possible causes of the climate changes and feedback processes that amplify or reduce the resulting changes. Comparing climate with possible causes allows scientists to generate and test hypotheses, and those hypotheses then become the basis for projections of future changes.

Arctic data show changes on numerous time scales and indicate many causes and important feedback processes. Changes in greenhouse gases appear to have been especially important in causing climate changes (Chapter 2, section 2.4; Chapter 3, sections 3.4.1 and 3.4.4; Chapter 4, sections 4.4.1 and 4.4.2). Global climate changes have been notably amplified in the Arctic (Chapter 3, section 3.5), and warmer times have melted ice on land and sea (Chapter 6).

Statistically valid confidence levels often can be attached to scientific findings, but those confidence levels commonly require many independent samples from a large population. Such a standard can be applied to paleoclimatic data in only some cases, whereas in other cases the necessary archives or interpretative tools are not available. However, expert judgment can also be used to assess confidence. The key findings here cannot all be evaluated rigorously using parametric statistics, but on the basis of assessment by the authors, all of the key findings are at least "likely" as used by the Intergovernmental Panel on Climate Change (more than 66% chance of being correct); the authors believe that the most of the findings are "very likely" (more than a 90% chance of being correct).

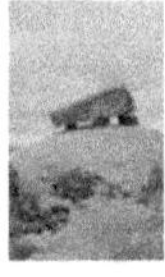

7.1 SUMMARY OF KEY FINDINGS

Chapter 3 Temperature and Precipitation History of the Arctic

The Arctic of 65 million years ago (Ma) was much warmer than in recent decades; forests grew in all land regions and neither perennial sea ice nor the Greenland Ice Sheet were present. Gradual but bumpy cooling has dominated since, with the falling atmospheric CO_2 concentration apparently the most important contributor to the cooling, although with possible additional contributions from changing continental positions and their effects on atmospheric or oceanic circulation. Warm "bumps" during the general cooling trend include the relatively abrupt Paleocene-Eocene Thermal Maximum about 55 Ma, apparently caused by an increase in greenhouse gas concentrations, and a more gradual warming in the middle Pliocene (about 3 Ma) of uncertain cause.

Arctic amplification of temperature changes thus appears to be a consistent feature of the Earth system.

Around 2.7 Ma, cooling reached the threshold for extensive development of continental ice sheets throughout the North American and Eurasian Arctic. Periodic growth and shrinkage of the ice during hundreds of thousands of years indicate strong control by periodic changes in Northern Hemisphere sunshine caused by cyclic variations in Earth's orbit. Recent work suggests that, in the absence of human influence, the current interglacial would continue for a few tens of thousands of years before the start of a new ice age. The large temperature differences between glacial and interglacial periods, although driven by Earth's orbital cycles and the globally synchronous response, reflect the effects of strong positive feedbacks, such as changes in atmospheric concentrations of CO_2 and other greenhouse gases and in the extent of reflective snow and ice.

The cooling into and warming out of the most recent glacial, which peaked about 21 ka, were punctuated by numerous abrupt climate changes, with millennial persistence of conditions between jumps requiring years to decades. These events were very large around the North Atlantic but much smaller elsewhere in the Arctic and beyond. Large changes in the extent of sea ice in the North Atlantic were probably responsible, linked to changes in regional and global patterns of ocean circulation. Freshening of the North Atlantic also favored formation of sea ice.

Such abrupt changes also occurred in the current interglacial (the Holocene), but they ended as the Laurentide Ice Sheet on Canada melted away. Arctic temperatures in the Holocene broadly responded to orbital changes with warmer temperatures during the early to middle Holocene, when there was more summer sunshine. Warming generally led to northward migration of vegetation and to shrinkage of ice on land and sea. Small oscillations in climate during the Holocene, such as the Medieval Climate Anomaly and the Little Ice Age, were linked to variations in the sun-blocking effect of particles from explosive volcanoes and perhaps to small variations in solar output or in ocean circulation or other factors. The warming from the Little Ice Age appears to have begun for largely natural reasons, but there is now high scientific confidence that human contributions, and especially increasing concentrations of CO_2, have come to dominate the warming (Jansen et al., 2007).

Comparison of summertime temperature anomalies for the Arctic and for lower latitudes, averaged over at least millennia for key climatic intervals of the past, shows that Arctic changes were larger than those in lower latitudes. This more pronounced response applies to intervals that were both warmer and colder than in recent decades. Arctic amplification of temperature changes thus appears to be a consistent feature of the Earth system.

Chapter 4 Past Rates of Climate Change in the Arctic

Changes in climate have many causes, occur at different rates, and are sustained for different intervals. Changes in atmospheric composition, along with changes in atmospheric and oceanic circulations linked to tectonic processes over tens of millions of years, have led to large climate changes, including conditions so warm that the Arctic was ice-free in winter and so cold that large Arctic regions remained ice-covered year-round. Features of Earth's orbit acting for tens of thousands of years have rearranged sunshine on the planet and paced the growth and shrinkage of great ice-age ice sheets. Anomalously cold single years have resulted from the influence of large, explosive volcanoes, with slightly anomalous decades in response to the random variations in the frequency of occurrence of such explosive volcanoes.

As observed in Greenland or more generally around the Arctic, the more-persistent of these causes of climate change have produced larger climate changes, but at lower average rates. When compared with this general trend, the regional effects around the North Atlantic of abrupt climate changes linked to shifts in ocean circulation have been anomalously rapid; however, the globally averaged temperature effects of those abrupt climate changes were not anomalously large. And, relative to this general trend of larger climate changes occurring more slowly, human-linked Arctic perturbations of the most recent decades do not appear anomalously rapid or large, but model-projected changes summarized by the IPCC may become anomalously large and rapid.

Interpretation of these observations is complicated by lack of a generally accepted way of formally assessing the effects or importance of size versus rate versus persistence of climate change. The report here relied much more heavily on ice-core data from Greenland than would be ideal in assessing Arctic-wide changes. Existing techniques described in this report offer substantial opportunities for generation and synthesis of additional data that could extend the available results. If widely applied, such research could remove the over-reliance on Greenland data.

Chapter 5 History of the Greenland Ice Sheet

Paleoclimate data show that the volume of the Greenland Ice Sheet has changed greatly in the past, affecting global sea level. Physical understanding indicates that many environmental factors can force changes in ice-sheet size. Comparing histories of important forcings with ice-sheet size implicates cooling as causing ice-sheet growth, warming as causing shrinkage, and sufficiently large warming as causing compete or almost complete loss. The evidence for temperature control is clearest for temperatures similar to or warmer than those occurring in the last few millennia. The available evidence shows that Greenland had less ice when snowfall was higher, indicating that snowfall rate is not the leading control on ice-sheet size. Rising sea level tends to float marginal regions of ice sheets and force their retreat, so the generally positive relation between sea level and temperature means that, typically, both have pushed the ice sheet in the same direction. However, for some small changes during the most recent millennia, marginal fluctuations in the ice sheet have been opposed to those expected from local relative sea-level forcing but in the direction expected from temperature forcing. This, plus the tendency for shrinkage to pull ice-sheet margins out of the ocean, indicates that sea-level change has not been the dominant forcing at least for temperatures similar to or greater than those of the last few millennia.

Histories of ice-sheet volume in fine time detail are not available, but the limited paleoclimatic data at least agree that short-term and long-term responses to temperature change have been in the same direction. The best estimate from paleoclimatic data is thus that warming shrinks the Greenland Ice Sheet, and warming of a few degrees is sufficient to cause ice-sheet loss. Figure 6.13 shows a threshold for ice-sheet removal from sustained summertime warming of 5°C, with a range of uncertainties from 2° to 7°C, but tightly constrained numerical estimates are not available, nor are rigorous error bounds, and the available data poorly constrain the rate of loss. Numerous opportunities exist for additional data collection and analyses that would reduce the uncertainties.

Chapter 6 History of Arctic Sea Ice

Geological data indicate that the history of Arctic sea ice is closely linked with temperature changes. Sea ice in the Arctic Ocean may have appeared in response to long-term cooling as early as 46 Ma. Year-round sea ice in the Arctic possibly developed as early as 13–14 Ma, before the opening of the Bering Strait at 5.5 Ma. Nevertheless, extended seasonally ice-free periods probably occurred until about 2.5 Ma. They ended with a large increase in the extent and duration of sea-ice cover that more or less coincided with the onset of extensive glaciation on land (within the considerable dating uncertainties). Some data suggest that ice reductions marked subsequent interglacials and that the Arctic Ocean may have been seasonally ice-free during the warmest events. For example, reduced-ice conditions are inferred for the last interglacial and the onset of the current interglacial, about 130 and 10 ka.

Limited data suggest poorly understood variability in ice circulation for centuries to millennia, but without strong periodic behavior on these time scales. Historical observations indicate that ice cover in the Arctic began to diminish in the late 19th century, and that this shrinkage has accelerated during the last several decades. Shrinkages that were both similarly large and rapid have not been documented for at least the last few thousand years, although the paleoclimatic record is sufficiently sparse that similar events might have been missed. Orbital changes have made ice melting less likely than during the previous millennia since the end of the last ice age, making the recent changes especially anomalous. Improved reconstructions of sea-ice history would help clarify just how anomalous these recent changes are.

HIGHLIGHTS OF KEY FINDINGS

- In the past, Arctic temperature changes have been larger than correlative globally averaged changes. Past climate forcings discussed in this report and/or the background states are quite different from those associated with the increasing of well-mixed greenhouse gases in the current climate, so that the magnitude of past summer responses cannot be literally translated into what we should expect this coming century. However, there is every indication that the Arctic climate has been quite sensitive to climate forcing in the past, and that it will respond strongly in the future.
- Arctic temperatures have changed greatly but slowly in response to long-lasting causes and by lesser amounts but more rapidly in response to other causes. Human-forced changes of the most recent decades do not appear notably anomalous in rate or size for their duration when they are compared with the fastest of these natural changes, but projections for future human-caused changes include the possibility of anomalously large and rapid changes.
- The Greenland Ice Sheet has consistently grown with cooling and shrunk with warming, and an average annual warming of a few degrees (about 5°C, with uncertainties between about 2° and 7°C) or more has been sufficient to completely or almost completely remove the ice sheet if maintained long enough; the rate of that removal is poorly known. Reduction in the size of the Greenland Ice Sheet in the past has resulted in a corresponding rise in sea level.

- Warming has decreased sea ice, which in turn strongly magnifies warming, and seasonally ice-free conditions and even year-round ice-free conditions have occurred in response to sufficiently large but poorly quantified forcing.
- Although major climate changes have typically affected the whole Arctic, important regional differences have been common; a full understanding of Arctic climatology and paleoclimatology requires regionally resolved studies.

7.2 RECOMMENDATIONS

Paleoclimatic data on the Arctic are generated by numerous international investigators who study a great range of archives throughout the vast reaches of the Arctic. The value of this diversity is evident in this report. Many of the key results of this report rest especially on the outcomes of community-based syntheses, such as the CAPE Project, and on multiply replicated and heavily sampled archives, such as the central Greenland deep ice cores. Results from the ACEX deep coring in Arctic Ocean sediments were appearing as this report was being written; these results were quite valuable and will become more so with synthesis and replication, including comparison with land-based as well as marine records. The number of questions answered, and raised, by this one new data set shows how sparse the data are on many aspects of Arctic paleoclimate change.

Future research should maintain and expand the diversity of investigators, techniques, archives, and geographic locations, while promoting development of community-based syntheses and multiply replicated, heavily sampled archives; only through breadth and depth can the remaining uncertainties be reduced while confidence in the results is improved.

The questions asked of this study by the CCSP are relevant to public policy and require answers. The answers provided here are, we hope, useful and informative. However, we recognize that despite the contributions of numerous community members to this report, in many cases a basis was not available in the refereed scientific literature to provide answers with the accuracy and precision desired by policymakers.

Future research activities in Arctic paleoclimate should address in greater detail the policy-relevant questions that motivated this report.

Paleoclimatic data provide very clear evidence of past changes in important aspects of the Arctic climate system. The ice of the Greenland Ice Sheet, smaller glaciers and ice caps, the Arctic Ocean, and soils are shown to be vulnerable to warming, and Arctic ecosystems are strongly affected by changing ice and climate. National and international studies generally project rapid warming in the future. If this warming occurs, the paleoclimatic data indicate that melting of ice and associated effects will follow, with implications for ecosystems and economies.

The results presented here should be utilized in the design of monitoring, process, and model-projection studies of Arctic change and linked global responses.

GLOSSARY

Italicized terms within a definition refer to other entries in this glossary. Terms appearing in this glossary appear in bold type in the body of this *SAP* at the first occurrence in each chapter. All definitions supplied in this glossary refer to the use of these terms within the context of paleoclimate science.

8 ka cold event

A prominent, abrupt cold event that took place approximately 8,200 *ka*; most clearly seen in North Atlantic *climate* records. The event persisted approximately 200 to 400 years before the *climate* returned to the warm conditions of the early *Holocene*.

^{137}Cs

A radioactive isotope of the element cesium utilized in dating modern sediments. It has a half-life of approximately 30 years. ^{137}Cs is a by-product of nuclear weapons testing (in conjunction with *$^{239,240}Pu$* and *^{241}Am*). Its concentration in the environment peaked in 1963 AD, and profiles of ^{137}Cs (and some times in conjunction with *^{241}Am* and *$^{239,240}Pu$*) through lake and marine sediments are used to constrain the ages of young sedimentary *archives*.

^{210}Pb

A radioactive isotope of the element lead used in dating young sediments. It is one of the isotopes in the decay chain of uranium 238 and it has a half-life of approximately 22 years. ^{210}Pb that forms in the atmosphere as a by-product of radon decay is washed into lakes and estuaries. The level of ^{210}Pb in sediments that is above the background concentration of ^{210}Pb from in situ uranium-series decay is used to determine sedimentation rates the past 150 years.

$^{239,240}Pu$

Radioactive isotopes of the element plutonium utilized in dating modern sediments. ^{239}Pu has a half-life of 24,110 years and prior to the production of nuclear weapons was virtually nonexistent in nature. It is one of the two fissile materials used in nuclear weapons and some nuclear reactors. ^{240}Pu has a half-life of approximately 6,600 years; it is a by-product of the manufacture of ^{239}Pu and is produced in nuclear reactors as part of the fuel cycle. About 10,000 kilograms of plutonium were released into the atmosphere during atmospheric nuclear weapons testing during the post-World War II years through the 1970s, peaking in 1963, and it became part of the stratigraphic record as fallout from these tests. Detection of peak plutonium concentrations in a sample therefore indicates that the sample being analyzed dates from 1963 AD.

^{241}Am

A radioactive isotope of the synthetic element americium utilized in dating recent sediments. It has a half-life of approximately 432 years and is a byproduct of plutonium production as well as a component in fallout from nuclear weapons. It is also currently used in tiny quantities in smoke detectors. Its concentration in the environment peaked in 1963 AD; therefore its detection in peak amounts (especially in conjunction with *^{137}Cs* and *$^{239,240}Pu$*) indicates that the sample being analyzed dates from that time period.

α parameter

The relation between a change in stable isotope composition of oxygen or hydrogen in precipitation or in accumulated snow and the associated change in temperature, usually expressed as per-mil (‰) per degree. The isotopic composition in the comparison is the difference between the heavy:light ratio of the specified species and the corresponding ratio in a specified standard, divided by the ratio in the standard.

$\delta^{18}O$

A measure of the ratio of the stable isotopes of oxygen, ^{18}O:^{16}O, in water or a biomineral. The definition is $\delta^{18}O(‰)=10^3[(R_{sample}/R_{standard})-1]$, where $R_x=(^{18}O)/(^{16}O)$ is the ratio of isotopic composition of a sample compared with that of an established standard, such as ocean water. It is commonly used as a measure of the temperature of precipitation, the temperature of ocean surface waters, or the volume of freshwater sequestered as ice on the continents, and as an indicator of processes that show isotopic fractionation.

δD

A measure of the ratio of the stable isotopes of hydrogen, ^2H:^1H, in water. The definition is $\delta D(‰)=10^3[(R_{sample}/R_{standard})-1]$, where $R_x=(^2H)/(^1H)$ is the ratio of isotopic composition of a sample compared with that of an established standard, such as ocean water. It is commonly used as a measure of the temperature of precipitation, and when compared to the *$\delta^{18}O$* in the same water sample provides information on sources of water vapor or the extent of evaporation during transport or after precipitation. "D" is the chemical abbreviation for deuterium, the name given to hydrogen that contains one extra neutron.

Accelerator mass spectrometer (AMS)
An analytical tool that permits the detection of isotopes of the elements to very low concentrations by accelerating the ions of the substance being analyzed to very high kinetic energies (energy of motion) prior to mass analysis.

ACEX (Arctic Coring Expedition)
A multi-national scientific research effort to better understand both the *climate* history of the Arctic region and the role that the Arctic has played and continues to play in the Earth's ongoing climatic variations; work is based on recovery and analysis of sediment cores from the *ARCTIC OCEAN*.

Alkenone
Long-chain organic compound produced by certain *phytoplankton*, which biosynthetically control the number of carbon-carbon double bonds in response to the water temperature. The survival of this temperature signal in marine sediment sequences provides a time-resolved record of sea surface temperatures that reflect past *climates*.

Amplification (with respect to *climate*)
Phenomenon by which an observed change in a *climate* parameter in a particular area of the Earth is larger in magnitude than the global average. *Climate* amplification is typically connected to a *climate feedback mechanism*.

Analogue (climatology)
Generally used to describe a *climate* state that is reasonably well known and that is similar to or has the same characteristics as the *climate* of a particular ancient time period under study.

Anthropogenic
Effects, processes, objects, or materials that are derived primarily from human activities, as opposed to those occurring in natural environments without human influence.

Archives
Sources of information about the past.

Arctic amplification
The result of interactive positive *feedback* mechanisms in the Arctic. Owing to interactive *feedback* primarily from *sea ice* and snow cover, *greenhouse-gas*-induced warming is expected to be accelerated in the Arctic region in comparison with that for the Northern Hemisphere or entire globe. This effect is referred to as Arctic amplification.

Bed
The materials on which a *glacier* or *ice sheet* rests. These materials may be solid rock or unconsolidated sediment. The term is sometimes applied to water between the ice and rock materials, but usually it is reserved for the rock materials.

Benthic foraminifera
see *foraminifer*

Biomarkers
Residual organic molecules indicating the existence, past or present, of living organisms with specific *climate* or environmental constraints.

Biome
An ecological community of organisms adapted to a particular *climate* or environment; that community dominates the large geographic area in which it occurs.

Bølling
A term used primarily in Europe for a warm interval (*interstadial*) of late-*glacial* time centered at about 12,500 years ago when *climate* warmed sufficiently to permit northward extension of vegetation on land, and sea level rose approximately 20 meters relative to the colder period immediately preceding it.

Boreal
Pertaining to the northern regions of the Northern Hemisphere (from Boreas, god of the North Wind in Greek mythology).

Boundary condition
In *climate* science this term refers to a prescribed state of Earth's surface at a particular point in time, often at the start of a *climate model* experiment. Examples include the topography of Earth or the extent of *sea ice*.

Boundary current
Ocean currents whose dynamics are determined by a coastline. For example, the Gulf Stream is a warm, fast moving, and strong western boundary current along the east coast of North America.

Calving
The breaking off of ice from the front of a *glacier* that, typically, extends into a lake or sea; in the sea, calved ice forms icebergs.

Calving flux
The rate at which ice breaks off the front of a *glacier*. Most typically, calving flux will be expressed as either the rate of mass loss per unit width of the *glacier* per unit time (e.g., kilogram per meter per second (kg/m/s)) or the rate of volume loss per unit width per unit time (e.g., cubic meter per meter per second ($m^3/m/s$), which is also square meter per second (m^2/s)).

CAPE Project
Circum-Arctic PaleoEnvironments Project. A research program within the International Geosphere-Biosphere Program (IGBP)–Past Global Changes (PAGES), the focus of which is integration of paleo-environmental research on terrestrial environments and adjacent margins covering the last 250,000 years of Earth history.

Carbon dioxide (CO_2)
An atmospheric *greenhouse gas* with many natural and *anthropogenic* sources, it is the second most abundant *greenhouse gas* in the atmosphere after water vapor. Natural sources of carbon dioxide include animal and plant respiration, release at the sea surface, and volcanic eruptions. *Anthropogenic* sources include the combustion of fossil fuels, biomass burning, and specialized industrial production processes. It is the principal *anthropogenic greenhouse gas* that affects Earth's radiative balance.

Carbon ketones
Functional chemical groups characterized by a carbonyl group (C=O) linked to two other carbon atoms.

CCSP
United States Climate Change Science Program; a consortium of federal agencies carrying out scientific research in the field of *climate change*. The primary objective of the CCSP is to provide the best science-based knowledge possible to support public discussion and government- and private-sector decisions about the risks and opportunities associated with changes in *climate* and in related environmental systems. See also *U.S. Climate Change Science Program*.

Cenozoic
The period of Earth's history encompassing the past 65 million years. The Cenozoic is subdivided into seven series or epochs: (oldest to most recent) *Paleocene*, *Eocene*, *Oligocene*, *Miocene*, *Pliocene*, *Pleistocene*, and *Holocene* (the current epoch).

Chlorinated fluorocarbon compounds (CFCs)
A family of man-made chemical compounds composed of carbon, hydrogen, chlorine, and fluorine. With respect to *climate change*, this term usually refers to manufactured CFCs used as refrigerants, aerosol propellants, and solvents and in insulation. When released into the lower atmosphere, these compounds act as *greenhouse gases*. However, because they are not destroyed in the lower atmosphere, CFCs drift into the upper atmosphere where, given suitable conditions, they break down ozone. Prior to industrialization these gases did not exist in the atmosphere; they now exist in concentrations of several hundred parts per trillion.

CH_4
see *methane*

Chironomids
The informal taxonomic name for non-biting members of the Diptera (true flies) family of insects commonly known as midges.

Climate
The average weather over a particular region of the Earth. Climate originates in recurring meteorological phenomena that result from specific modes of atmospheric circulation. The averaging period is conventionally a 30-year interval as promulgated by the World Meteorological Organization (WMO). Typical characteristics include mean seasonal temperature and precipitation, storm frequency, and wind velocity.

Climate change
A statistically significant variation in either the mean state of the *climate* or the mean variability of the *climate* that persists for an extended period (typically 10 years or more). Climate change may result from such factors as changes in solar activity, long-period changes in the Earth's orbital elements (*eccentricity*, *obliquity*, *precession of equinoxes*), natural internal processes of the *climate* system, or *anthropogenic forcing* (for example, increasing atmospheric concentrations of *carbon dioxide* and other *greenhouse gases*).

Climate feedback mechanisms
Processes that amplify the effects of a change in the controls on global temperature. *Feedbacks* are said to be positive when they increase the size of the original response or negative when they cause it to decrease.

CO_2
see *carbon dioxide*

Coccolith
Microscopic skeletal plates of calcium carbonate formed by certain species of marine *phytoplankton* (see *coccolithophorid algae*).

Coccolithophorid algae
Tiny, photosynthesizing, single-celled marine algae (protists) and *phytoplankton*, that are distinguished by special calcium carbonate plates called *coccoliths*. *Coccoliths* serve as important marine paleoclimate proxies relevant to past characteristics of the ocean's surface layer.

Continental drift
The slow motion of the continents on the surface of the Earth. Continents ride on underlying segments of the Earth's crust

that fit together like pieces of a jigsaw puzzle and are in constant motion, sliding over, under, past, or away from each other at their boundaries. The underlying physics of plate motions is referred to as *plate tectonics* and encompasses an understanding of the deep internal structure and motions of the Earth.

Continentality
Characteristic of regions near the centers of large continents, where daily and seasonal variations of temperature and precipitation are relatively large compared with lands closer to the oceans (maritime lands) where such variations are moderated by the adjacent oceans. Continentality increases inland, away from ocean coastlines.

Conveyor belt circulation
Colloquial term for that part of the modern ocean currents (circulation) in which near-surface waters of the Atlantic flow northward, sink into the deep ocean, then flow southward, circulate around Antarctica, flow northward again but now in the deep parts of the Pacific and Indian Oceans, mix up to near the surface, and return to the surface flow of the Atlantic. The term is especially applied to that part of this globe-girdling circulation in the Atlantic.

Crenarcheota (taxonomy)
Microscopic water-living organisms belonging to the kingdom of Archaea originally thought to thrive only under extreme conditions of heat, acidity, and high sulfur concentrations. However, recent studies indicate a much broader environmental distribution and pelagic (surface dwelling) Crenarcheota are now understood to be probably the most abundant group of Archea on Earth.

Cretaceous
The period of Earth history in the Mesozoic prior to the *Cenozoic* encompassing approximately 145 million to 65 million years ago. It was a time characterized by a relatively warm *climate*, abundant diversification of terrestrial plants and animals, and the appearance of the first flowering plants and some insect groups and mammals. The termination of the Cretaceous was marked by a mass extinction and the loss of about 60% of all terrestrial and marine species, theorized to have been caused by Earth's collision with one or more large extraterrestrial objects such as an asteroid (Chicxulub Crater, Yucatan, Mexico).

Dansgaard-Oeschger event (D-O event)
Widespread *climate* event seen as anomalously warm times in the northern hemisphere and especially around the *NORTH ATLANTIC OCEAN*, during most recent ice age (from about 110,000 to 11,500 years ago), with large and rapid terminations and very large and rapid onsets, often persisting for a few centuries.

Deep-water formation
The sinking of water from near the surface into the depths of the ocean, followed by lateral movement of that water. In the modern world, this process occurs only in restricted regions in the *NORTH ATLANTIC OCEAN* and around Antarctica.

Dendroclimatology
The science of determining past *climates* from trees (primarily tree rings).

Diachronous
"Cutting across time"; said of a single geologic unit whose age differs depending on the location in which it is found. Such deposits are formed when the location of active deposition migrates, such as during the gradual melting of an ice sheet or the inland advance of seawater. Synonymous with *time-transgressive*.

Diatom
Photosynthesizing microscopic aquatic algae (protists), typically single celled, that produce silica skeletons which are commonly preserved in both marine and freshwater sediments.

Diffusion
General name for the motion of mass or energy from regions of higher concentration to regions of lower concentration through a large number of small events that do not depend directly on each other. For instance, in a room with absolutely no wind, a new type of gas released in one corner will eventually spread throughout the room by the random motions of the individual molecules, and this spreading is called diffusion.

Dinocyst
The dormant reproductive cells of *dinoflagellates*, microscopic planktonic algae (protists).

Dinoflagellates
Microscopic primarily marine planktonic algae (protists).

D-O event
see *Dansgaard-Oeschger event*

Driving stress
As used in glaciology, the gravitational impetus for the flow of ice as it spreads under its own weight. The driving stress is calculated as the product of the ice density, ice thickness, ice surface slope, and the acceleration of gravity. *Glaciers* that are thicker or have a steeper surface thus have a greater tendency to spread or flow.

Eccentricity

Out of roundness (ellipticality) of the Earth's orbit around the sun. The magnitude of Earth's orbital eccentricity completes a full cycle about every 100,000 years and varies between a minimum departure from circularity of 0.0034 to a maximum departure of 0.058.

Elastic

Characterized by experiencing changes in shape or size in response to applied stress, but returning to the original shape or size when the stress is removed.

Elastic deformation

Changes in shape or size experienced by a material or body in response to applied stress that will be reversed when the stress is removed. See *elastic*.

Eocene

The geological epoch spanning 55.8 *Ma* to 22.9 *Ma.*

Equilibrium line

An imaginary line on the upper surface of a *glacier*, separating the accumulation zone (the region in which mass supply to that surface exceeds mass loss) from the ablation zone (the region in which mass supply is less than mass loss). (Mass supply is typically dominated by snowfall and mass loss by runoff of meltwater, although drifting snow, *sublimation*, and other processes may contribute.) Almost always, the accumulation zone is higher in elevation than the ablation zone.

Equilibrium line altitude

The elevation above sea level of the *equilibrium line.*

Far field

The region at a sufficiently great distance from the source of a disturbance that some physical processes known to be important near the disturbance are no longer important because their influence has dropped greatly with increasing distance. For example, the initial growth of the ice sheet on *GREENLAND* lowered sea level globally (because water that evaporated from the ocean was stored in the ice sheet), but the weight of the ice pushed *GREENLAND* down farther than the globally averaged lowering of the sea surface; thus, sea level rose in the *near field* just beyond the growing ice sheet where sinking under the ice weight was important, whereas sea level fell in the far field where the influence of the weight of the growing ice sheet was small.

Feedback

A process or phenomenon that serves to intensify or weaken an initial change or impulse.

Firn

Old snow during transformation to *glacier* ice. The name firn is often applied to any snow on a *glacier* that is more than one year old. Firn becomes *glacier* ice when the interconnected pore spaces of the firn become isolated from the atmosphere above to form bubbles.

Foraminifer (benthic, deep-sea)

Microscopic single-celled marine organisms (protists) that live either on the sea floor or in near-surface waters (planktic) and that secrete calcium-carbonate shells in equilibrium with the sea water. The analysis of the stable isotopes contained in foraminifer shells found in sea floor sediment cores is the most commonly used method for determining ocean paleotemperatures and past changes in ice volume.

Forcing

With respect to *climate*, processes and factors external to the *climate* system which, when changed, generate a compensatory change in the *climate* system. Examples of *climate* forcings include variability in solar output, in the amount of sunshine received by a region of the Earth due to orbital changes, volcanic eruptions that inject particles and gases into the atmosphere, and changes in the positions of continents.

Gigaton

In the International System of Measurement (Système International d'unités, or SI), a gigaton is 1,000,000,000 tons (10^9 tons, or 1 billion tons in U.S. usage); a ton is 1,000 kilograms, and 1 kilogram is the mass equivalent of 2.2 pounds.

GISP2

Acronym for the Greenland Ice Sheet Project 2 location and ice core in central *GREENLAND* (see Plate 1 for location). Deep drilling at this site began in 1989 and was completed to bedrock at a depth of 3,053 meters in 1993.

Glacial (interval) or **glaciation**

An interval of time during the past 2.6 million years in the Earth's history when the average global temperature was colder than it is currently and during which ice sheets expanded substantially in the northern hemisphere.

Glacial isostatic adjustment

Changes in the shape and elevation of Earth's surface in response to growth and shrinkage of *glaciers* and ice sheets. For example, just as the surface of a water bed sinks beneath someone who sits on it but bulges up around that person, adding the load of an ice sheet causes sinking of the Earth's surface beneath and near the ice sheet but bulging up beyond (*peripheral bulge*). Changes in global sea level associated with loss of that water stored in an ice sheet or gain of water as an ice sheet melts also cause rising or sinking of the seabed beneath. Taken together, these changes are glacial isostatic adjustment.

Glacier
A mass of ice that persists for many years and notably deforms and flows under the influence of gravity. The term is especially applied to relatively small ice masses that flow down the sides of mountains, but it may also be applied to a fast-moving region of a larger ice mass or even to the larger ice mass itself.

Greenhouse gas
Gaseous constituents of the atmosphere that absorb and emit radiation at specific wavelengths within the spectrum of infrared radiation emitted by Earth's surface, the atmosphere, and clouds. The primary greenhouse gases in the atmosphere are water vapor (H_2O), *carbon dioxide* (CO_2), *nitrous oxide* (N_2O), *methane* (CH_4), and ozone (O_3), all of which have many natural and *anthropogenic* sources.

Grounded ice
Ice that remains on land and is not floating. The term is especially applied to nonfloating portions of *glaciers*, *ice caps*, or *ice sheets* that flow into lakes or seas and could have floating portions.

Grounding line
The geographic area where a body of land-based ice begins to float as it moves into a body of water.

Heinrich event
Interval of anomalously rapid deposition of sand-sized and coarser materials in the open *NORTH ATLANTIC OCEAN*, formed by an anomalously rapid supply of icebergs carrying debris. Six or seven events are identified during the most recent ice-age cycle (from about 110,000 to 11,500 years ago), and older events have occurred as well. Characteristics of the debris in most of the events indicate that ice in Hudson Bay was a dominant source. Large and widespread *climate* anomalies were associated with the Heinrich events, such as cool conditions in the north and especially around the North Atlantic, and warmth in the far south.

Holocene
The current geologic epoch that began about 11,500 years ago when the *climate* warmed at the end of the most recent *glacial* period. The most recent epoch (subdivision) of the *Quaternary* period.

HOTRAX
Healy-Oden Trans-Arctic Expedition, 2005. The first full scientific crossing of the *ARCTIC OCEAN* by surface vessels from the Pacific to the Atlantic side by way of the North Pole. The crossing was a joint effort of the United States icebreaker Healy and the Swedish icebreaker Oden. HOTRAX geological and geophysical teams recovered a series of sediment cores and seismic-reflection records across the *ARCTIC OCEAN* to investigate its late *Cenozoic paleoclimatology* and the deep stratigraphy and structure of the Arctic basin. Other studies included the variability and provenance of sea-ice as well as physical and chemical oceanographic investigation of the deep-water masses.

Hot-spot volcanic chain
Linear array of volcanoes produced by a single source, especially seen as lines of islands in the ocean. A "hot spot" is a rising column of hot rock that rises from relatively deep in the Earth. The upper, cold layer of the Earth involved in *continental drift* typically moves horizontally much faster than a hot-spot does. The hot spot will poke through the overlying layer and form a volcano, then that volcano ceases to erupt as it is carried away by the drifting layer, while the hot spot pokes through to make a new volcano. The Hawaiian Islands are the younger part of such a hot-spot volcanic chain, which also includes the generally undersea Emperor Seamounts to the northwest of Hawaii.

Ice cap
A flowing mass of ice (*glacier*), moving away from a central dome or ridge, and notably smaller than an otherwise-similar *ice sheet*, which normally is of continental or subcontinental scale.

Ice dynamical model
As used here, a representation of the physical behavior of a *glacier*, *ice cap* or *ice sheet*, developed with the use of a computer to solve mathematical equations approximating the important physical processes.

Ice sheet
A flowing mass of ice (*glacier*), moving away from a central dome or ridge, normally of continental or subcontinental scale and notably larger than an otherwise similar *ice cap*.

Ice shelf
A floating extension of a *glacier*, *ice cap*, or *ice sheet*, nourished in part by flow from nonfloating (*grounded*) ice. An ice shelf may gain or lose mass on its upper surface (usually by snowfall or melting) or lower surface (usually by freezing or melting). Normally, an ice shelf loses mass into the adjacent water body by iceberg *calving*. The term "ice shelf" is sometimes applied to relatively small ice masses that largely or completely lack flow from adjacent *grounded ice* and that thus are nourished by snowfall above or freezing beneath; these "ice shelves" typically are thicker and more persistent than features called *sea ice*, but they could be classified as *sea ice*.

Ice stream
A faster moving "jet" of ice flanked by slower flowing parts of an *ice sheet* or *ice cap*.

Innuitian sector
The *ice sheet* that covered the Queen Elizabeth Islands of northern and northeastern *CANADA*. The term was originally proposed as the Innuitian Ice Sheet (Blake Jr., W., 1970. Studies of *glacial* history in Arctic Canada. Canadian Journal of Earth Sciences, **7**, 634-664), and was applied to the ice mass that formed during the most recent *glaciation*. The Innuitian Ice Sheet was joined to the *Laurentide Ice Sheet* to the south and to the *GREENLAND ICE SHEET* to the east when the *ice sheets* were largest; the term "Innuitian Sector of the *Laurentide Ice Sheet*" is often used. The term is also often applied to relict ice in the indicated region from earlier *glaciations*.

Insolation
The amount of sunshine, measured in watts per square meter (W/m^2), on one unit of horizontal surface. With respect to *climate* studies, insolation is typically evaluated at the Earth's surface. The intrinsic latitudinal differences in the amount of sunshine that reaches the Earth's surface (e.g., at the equator and at the poles) depend on the seasons, but the total global value does not.

Interannual variability
Changes in a measured value from year to year. As an example, during the last 30 years, globally averaged surface temperatures have increased, with high statistical confidence. However, events such as an El Niño cause the average temperature for a year to plot off of the line that best represents the whole 30-year history. The difference between the annual average temperature and the best-fit line changes from year to year in response to this interannual variability.

Interglacial (interval) or **interglaciation**
An interval of time during the past 2.6 million years in Earth's history when the average global temperature was as warm or warmer than it is currently and during which *ice sheets* contracted substantially in the northern hemisphere.

Intergovernmental Panel on Climate Change (IPCC)
A multinational group of experts in the field of *climate change* established (by the World Meteorological Organization and the United Nations Environmental Program) to provide decision-makers and other interested persons with an objective source of information about *climate change*. The IPCC does not conduct any research nor does it monitor *climate*-related data or parameters. Its role is to assess on a comprehensive, objective, open, and transparent basis the latest scientific, technical, and socio-economic literature produced worldwide relevant to the understanding of the risk of human-induced *climate change*, its observed and projected effects, and options for adaptation and mitigation.

IPCC
see *Intergovernmental Panel on Climate Change*

Interstadial
A warmer period of time within an ice age marked by a temporary retreat of ice.

Irradiance (solar)
The amount of intrinsic radiant energy emitted by the sun over all wavelengths that falls each second on 1 square meter ($W/m^2/s$) outside the Earth's atmosphere. The current average value of solar irradiance is approximately 1,367 watts per square meter. Small variations in irradiance attributable to a variety of internal solar process have been observed and have had small but detectable effects on global temperature during the past 65 million years.

Isochron
A line on a map or a chart connecting all points at which an event or phenomenon occurred simultaneously or which represent the same time value or time difference. In sediment or sediment core analysis, a point of known age that can be identified in multiple locations that ties the datasets derived from the analyses to a common point in time.

ka
Kiloannum; thousands of years ago (a point in time).

k.y.
Thousands of years (a time interval).

Landfast ice
Ice that is anchored to the shore or ocean bottom, typically over shallow ocean shelves at continental margins; landfast ice is defined by the fact that it does not move with the winds or currents.

Last Glacial Maximum (LGM)
The period of time from approximately 30,000 to 15,000 years ago, characterized by significantly lower global temperatures and maximum extent of *ice sheets* during the last *glaciation* prior to the current warm period. The LGM occurred within *Marine isotope stage 2*.

Laurentide Ice Sheet
Name proposed by Flint (Flint, R.F., 1943, Growth of the North American *ice sheet* during the Wisconsin age. Geological Society of America Bulletin, **54**, 325-362) for the great *ice sheet* that covered much of northern North America east of the Rocky Mountains during the most recent ice age (from about 110,000 to 11,500 years ago). Use of the term is widely extended to include older *ice sheets* that occupied the same general area.

LGM
see *Last Glacial Maximum*

Little Ice Age
A period of time during the last millennium (approximately 1500 to 1850 AD) during which summers globally, but particularly in the higher latitudes of the Northern Hemisphere, were colder than during the preceding millennium or the 20th century. The Little Ice Age is widely manifested by the advance of mountain *glaciers* and *ice caps*, as well as by periodic crop failures, especially in northwest Europe.

Ma
Mega-annum; millions of years ago (a point in time).

Marine Isotope Stage (MIS)
A subdivision of recent geologic time, identified by number (e.g., Marine Isotope Stage 1, or Marine Isotope Stage 8); Marine Isotope Stage 1 includes today, and numbers increase with increasing age. The marine isotope *stages* were defined from the oxygen-isotope ratios of shells that accumulated on the ocean floor and were collected in sediment cores; shells that grew in cooler water, or at times when more water was stored on land in ice sheets, are isotopically heavier. Intervals of warmer water or smaller ice are labeled with odd numbers (Marine Isotope *Stage* 1, or 5), and times of colder water or larger ice have even numbers. Marine Isotope *Stages* average a few tens of thousands of years long, but different *stages* have different durations.

Mass flux
Rate at which material passes an observational site. The mass flux added to the surface of a *glacier* by snowfall may be reported as the ice added to an area during a time interval and thus measured in kilograms per square meter per second ($kg/m^2/s$) or equivalent units; the mass flux per unit width for flow of a *glacier* may be reported as kilograms per meter per second (kg/m/s).

Medieval Climate Anomaly
An interval between AD 1000 and 1300 in which some Northern Hemisphere regions were warmer than during the *Little Ice Age* that followed.

Meristematic
The tissue in all plants consisting of undifferentiated cells (meristematic cells) and found in zones of the plant where growth can take place.

Methane (CH_4)
An atmospheric greenhouse gas with many natural and anthropogenic sources, chief of which are decomposition of organic matter in the absence of oxygen (e.g., in wetlands and landfills), animal digestion and animal waste, and the production and distribution of natural gas, oil, and coal. It is the third most abundant atmospheric greenhouse gas after water vapor and carbon dioxide. The current lower-atmospheric concentration of methane at middle latitudes in the northern hemisphere is approximately 1,847 parts per billion and is stable. This concentration is substantially above the pre-industrial level of about 730 parts per billion.

Milankovitch cycles (time scales)
The Milankovitch or astronomical theory of *climate change* is an explanation for cyclical changes in the seasons which result from cyclical changes in the earth's orbit around the sun. The theory is named for Serbian astronomer Milutin Milankovitch, who calculated the slow changes in the earth's orbit by careful measurements of the position of the stars and through equations using the gravitational pull of other planets and stars. He determined that the earth "wobbles" in its orbit. The earth's "tilt" causes seasons, and changes in the tilt of the earth change the strength of the seasons. The seasons can also be accentuated or modified by the *eccentricity* (degree of roundness) of the orbital path around the sun and by the *precession* effect, the position of the solstices in the annual orbit. Together, the periods of these orbital motions (40,000 years for tilt, 90,000–100,000 years for *eccentricity*, and approximately 26,000 years for *precession*) have become known as Milankovitch cycles and their associated periodicities as Milankovitch time scales.

Milankovitch forcing
The changes in seasonal and latitudinal distribution of solar energy striking the Earth, which are caused by cyclical features of Earth's orbit, which cause *climate changes*, and which were described by the mathematician Milutin Milankovitch.

Millennial
Occurring or repeating every thousand years.

Miocene
The geological epoch spanning 23 *Ma* to 5.3 *Ma*.

MIS
see *marine isotope stage*

Model
With respect to *climate* studies, a computer program designed to mimic a natural process or system of processes with the aim of aiding in understanding how the process or system behaves. The representation of the *climate* system is based on mathematical equations governing the behavior of the various components of the *climate* system and includes treatment of key physical processes and interactions.

Moraine
Landforms (typically ridges) composed of sediment deposited at or near the edge of a *glacier*; a moraine provides an outline of all or part of a *glacier* at some time. (Note that the term "ground moraine" is sometimes used for a blanket of sediment deposited beneath a *glacier*, and the term "medial moraine" can be used for a band of debris on the surface of a *glacier* marking the junction of confluent flows; however, moraine normally is used as given in the main definition here.)

m.y.
Millions of years (a time interval).

N_2O
see *nitrous oxide*

NAO
see *North Atlantic Oscillation*

Near field
The region sufficiently close to the source of a disturbance that some physical processes must be considered that are unimportant at greater distance from the disturbance in the *far field*. For example, the initial growth of the ice sheet on Greenland lowered the sea level globally (because water evaporated from the ocean was stored in the ice sheet), but the weight of the ice depressed Greenland more than the globally averaged lowering of the sea surface; thus, sea level rose in the near field just beyond the growing ice sheet where sinking under the ice weight was important, whereas sea level fell in the *far field* where the influence of the weight of the growing ice sheet was small.

Negative feedback
In *climate* studies, a process that acts to decrease the magnitude of the *climate*'s response to an initial *forcing*.

NGRIP
North Greenland Ice Sheet Project location and ice core (see Plate 1 for location). Deep drilling at the *NGRIP* site began in 1999 and was completed to bedrock at 3,094 meters in 2003.

Nitrous oxide (N_2O)
An atmospheric *greenhouse gas*. It is the fourth most abundant *greenhouse gas* after water vapor, *carbon dioxide*, and *methane*. Natural sources include many biological sources in soil and water, primarily through bacterial breakdown of nitrogen in soils and in Earth's oceans. Primary human-related sources of N_2O are agricultural soil management, animal manure management, sewage treatment, mobile and stationary combustion of fossil fuel, and the manufacture of adipic and nitric acid. The pre-industrial value of N_2O in the atmosphere was approximately 265 parts per billion; it has increased monotonically since that time. The current atmospheric concentration is approximately 319 parts per billion.

North Atlantic Oscillation (NAO)
A large-scale see-saw in barometric pressure between the vicinity of Iceland and the Azores. It corresponds with fluctuations in the strength of the main westerly winds across the North Atlantic Ocean and is the primary wintertime weather-maker for the North Atlantic region of the eastern United States and Canada, Greenland, and Europe. When this pressure difference is large the NAO is said to be in positive phase; when it is small the NAO is said to be in negative phase.

Northern Annular Mode (NAM)
One of the two hemispheric-scale patterns of climate variability (the other is the Southern Annular Mode). The NAM is characterized by a seesaw in the strength of the west-east atmospheric flow along approximately 55°N and 35°N. It is accompanied by displacements of atmospheric mass between the Arctic basin and the mid-latitudes centered about 45°N.

Obliquity
The angle between the rotational axis of the Earth and a line perpendicular to the plane containing Earth's orbit about the sun. Earth's obliquity varies predictably from 22.1° to 24.5° in a 41,000 year cycle. The fact that Earth's axis of rotation is not perpendicular to the plane of its orbit around the sun (i.e., Earth's obliquity is not zero) is the origin of the seasons.

Oligocene
The geological epoch spanning 33.9 *Ma* to 23 *Ma.*

Orbitally paced
Phenomena that are synchronous with cyclical features of the Earth's orbit are described as orbitally paced.

Oscillation (*climate*)
A cyclical change in value between two different states. The *North Atlantic Oscillation* is a particularly important cyclic variation in atmospheric pressure over the North Atlantic region that is the primary wintertime weather-maker in the North Atlantic region.

Ostracode
A microfossil group of bivalved crustaceans that secrete a calcareous shell commonly preserved in sediments in the Arctic region. Because many ostracode species have ecological limits controlled by temperature, salinity, oxygen, food, and other factors, they can provide an important tool for paleoceanographic reconstruction in the Arctic.

Outlet glacier
A jet of ice flowing from an *ice sheet* or *ice cap*. Usage may be imprecise, but in general outlet glacier is the preferred term when the sides of fast-flowing ice are controlled prominently by bedrock (which usually is visible above the ice surface but also includes cases in which the fast-flowing ice occupies a deep bedrock trough but is flanked by a thin layer of slower-flowing ice); *ice stream* is usually applied when bedrock control is weak and the faster flowing ice is flanked by a considerable thickness of slower flowing ice.

Paleoceanographic archives
Sources of information about the past *climate* originating in records from the deep ocean, typically derived from an analysis of the stable isotopes of oxygen contained in the shells of marine microorganisms.

Paleocene
The earliest geological epoch of the *Cenozoic* spanning 65.5 *Ma* to 55.8 *Ma*.

Paleocene-Eocene Thermal Maximum (PETM)
An abrupt, intense, and short-lived warm event that began ~55 *Ma* and lasted approximately 170 *k.y.* The initial warming took place in <10,000 years. The PETM probably represents the warmest global temperatures of the past 65 million years (the *Cenozoic*). It was characterized by a massive release to the atmosphere of isotopically light carbon from a biogenic source, possibly sea-floor gas hydrates on the continental shelf. Low- and mid-latitude surface and deep ocean waters warmed as much as 4°–8°C at this time, and sea-surface temperatures near the North Pole warmed from approximately 18°C to more than 23°C.

Paleoclimate reconstruction
The determination of past states of Earth's *climate* (prior to historical or instrumental records) created by interpreting the *climate* signals contained in natural recorders such as tree rings, ice cores, deep sea and lake sediments, and cave deposits. Also, a reconstruction of past *climates* based on a *model* that uses paleoclimate data.

Paleoclimatology
The science of reconstructing the past *climate* of Earth.

Paleorecord, paleoclimate record
A data set constructed from a direct or indirect (*proxy*) recorder of *climate*. At their most useful, these records contain *climate* information that is unambiguous, continuous, and capable of being dated at a level of resolution sufficient to reveal *climate changes* at the scale of interest of the study.

Paleothermometers
A *climate proxy* (physical, biological, or chemical) preserved in geological *archives* that provides either qualitative or quantitative estimates of past temperatures.

Perfectly plastic behavior
A *model* for material behavior in which no permanent deformation occurs in response to small applied stress but, when stress is raised to the strength of the material, arbitrarily large and rapid deformation results, such that the stress cannot be raised above that strength. Perfect plasticity provides a useful approximation of real material behavior in some cases.

Peripheral bulge
A raised region encircling the region pushed down by the weight of an ice sheet or other large mass placed on the surface of the Earth. See *glacial isostatic adjustment* above.

Permafrost
Ground that is permanently frozen below its uppermost layer, which thaws in summer.

Perturbation
A change or deviation from the predicted, average, or otherwise anticipated stable state; typically caused by a force or process outside the perturbed system.

Phytoplankton
Microscopic algae which inhabit the illuminated surface waters of both marine and freshwater bodies.

Planktonic foraminifera
see *foraminifer*

Plate tectonics
The theory of the Earth that describes the outermost layer of Earth as comprising a series of rigid pieces or "plates" on which the continents ride that are in constant motion relative to each other and that interact with each other at their boundaries. Plate boundaries are typically the site of substantial seismic and volcanic activity.

Pleistocene
The geological epoch spanning 2.6 *Ma* to 11,477 years ago. The Pleistocene was characterized by multiple cyclical episodes of cold and warm times during which *ice sheets* and *glaciers* grew and shrank in response to global temperature changes initiated by *climate forcings* originating from cyclical changes in Earth's orbit around the sun.

Pliocene
The geological epoch spanning 5.3 *Ma* to 2.6 *Ma*.

Positive feedback
In *climate* studies, a process that acts to increase the magnitude of the *climate*'s response to an initial *forcing*.

Preboreal
Originally, the term applied to the approximately millennium-long interval occurring just after the end of the *Younger Dryas*, which is now known to have ended about 11,500 years before present. During the Preboreal interval, a short-lived cold event occurred between about 11,400 and 11,200 years before present. This event is often referred to as the Preboreal *Oscillation*.

Precession (of the equinoxes)
The wobble of the Earth's rotational axis expressed in degrees of arc. The Earth executes a complete precessional cycle once every 19,000 to 23,000 years.

Provenance
The geological term for the site of origin of rock material that has since been transported elsewhere. The provenance of much of the material deposited in the North Atlantic during *Heinrich events* is the Hudson Bay region of Canada.

Proxy
In paleoclimate studies, an indirect indicator of *climate* from which a record of change can be reconstructed once the relationship between the proxy and the desired parameter (e.g., temperature, precipitation) is understood. Many *paleoclimate reconstructions* are based on proxy records.

Prymnesiophyte algae
Typically unicellular, photosynthetic algae primarily found in marine environments. This term includes the *cocolithophorid algae*.

Quaternary
The geologic subdivision of the *Cenozoic* encompassing the past approximately 2.6 million years.

Radiocarbon reservoir age
The number of years old (the age) of carbon-14 (radiocarbon) incorporated by a sample when it formed. In radiocarbon dating, the simplest approach is to use the carefully reconstructed history of radiocarbon abundance in the atmosphere, together with the known half-life of radio-carbon and the measured abundance of radiocarbon in a sample today, to estimate how long it has been since the sample formed. However, the radiocarbon in some environments contains less radiocarbon than would be expected based on equilibrium with the atmosphere, causing the simplest possible approach to overestimate the time since a sample formed, and motivating the use of a correction for the radiocarbon reservoir age. For example, water near the surface of the oceans exchanges radiocarbon with the atmosphere and then sinks into the deep ocean, remaining there for roughly one millennium before returning to the surface and exchanging radiocarbon again. While the water is deep in the ocean and out of contact with the atmosphere, some of the radiocarbon in the water decays. A creature living in the deep ocean will thus incorporate less radiocarbon than an equivalent creature living at the same time near the ocean surface. This difference in initial radiocarbon abundance in the samples would lead to an error in estimating the age of the deep dweller if not corrected for; the correction is the radiocarbon reservoir age.

Radiogenic isotopes
Atomic species produced by radioactive decay.

Rifting
As used here, the geological process associated with *plate tectonics* (which is the science of drifting continents; see *tectonic forces*) by which continents are split apart to make ocean basins.

SAP
Synthesis and Assessment Product; one of the 21 technical reports sponsored by the *U.S. Climate Change Science Program* that discuss aspects of *climate change*.

Sea ice
Any form of ice found at sea that has originated from the freezing of sea water (in contrast to floating ice at sea that has originated from *glaciers* on land).

Sea level equivalent (SLE)
As used here, a measure of a mass of ice, calculated as the rise in global sea level that would result if the ice were melted and the resulting water spread uniformly over the world's oceans

Shelf break
The continental shelf, the undersea extension of a continent, ends at the shelf break, where the continental slope begins its steep drop into the deep ocean.

Sill
As used here, a narrow, shallow sea-floor region connecting continents or islands and separating two deeper basins.

SLE
see *sea level equivalent*

Speleothem
Mineral deposits (most commonly calcium carbonate) found in caves where water seeping through cracks in a cave's surrounding bedrock carries dissolved compounds that precipitate when the solution reaches an air-filled cave. Speleothems accumulate slowly, often spanning decades to millennia.

Stage
In paleoclimate studies, a time term for a major subdivision of a *glacial* epoch. See *marine isotope stage* above.

Step forcing
A rapid jump from one sustained level to another in some environmental feature that controls a system. For an ice sheet after a long interval at one temperature, a rapid warming to a new and sustained level would constitute a step forcing on the ice sheet system.

Stochastic
Randomly determined; involving or containing a random variable. A stochastic process is one in which the current state does not fully determine the next.

Striated
Scratched. *Glaciers* typically entrain loose rocks at their base; the moving ice then drags those rocks over bedrock, which may be scratched (striated) by the entrained rocks; striations may occur in sets of parallel marks.

Striated boulders
Boulders scratched by being dragged across other rocks by the passage of overlying, moving *glacier* ice.

Sublimation
Evaporation of water molecules directly from ice without first melting the ice to make water and then evaporating the water. Water molecules also can condense on ice directly (forming frost or hoarfrost, for example), and this process is often referred to as a negative rate of sublimation

Tectonic
Related to the large features and movements of geology. The outer, colder layer of Earth is broken into a few large plates, which drift around carrying the continents (hence the term *continental drift*). The interactions of these plates give rise to most of Earth's mountain ranges, volcanoes, and earthquakes, and these motions and interactions are called tectonic.

Tectonic forces
Forces internal to the Earth that cause segments of its crust to move in ways that build mountains and open or close oceans.

Tephra
Anything thrown by eruption of a volcano.

Tetraether lipids
Biomarkers produced by *Crenarcheota* that preserve well in marine and lacustrine sediments and that have been used to reconstruct past temperature changes in surface waters.

TEX_{86} index
A *proxy* related to surface water temperatures for marine and lacustrine (lake) systems, the index is based on membrane *tetraether lipids* of *Crenarcheota*, with 86 carbon atoms.

Thermohaline; thermohaline circulation
Deep circulation of the global ocean that is driven by density gradients established by differences in the temperature ("thermo") and salinity ("haline") of the water masses. Salty surface waters that lose heat in the polar regions become denser than underlying waters and sink, establishing a global network of deep ocean currents.

Tidewater glacier
A mountain *glacier* that terminates in the ocean.

Tidewater glacier cycle
The typically centuries-long behavior of *tidewater glaciers* that consists of recurring periods of advance alternating with rapid retreat and punctuated by periods of stability

Till
A mixed deposit of unconsolidated clay, silt, sand, gravel, and boulders deposited directly by and underneath a *glacier*. Till deposits that remain behind after the *glacier* has melted or retreated are characterized by a lack of stratification or layering.

Time-transgressive
Said of a single geologic unit whose age differs depending on location in which it is found. This nature is characteristic of geologic units created by processes that require a substantial time during which the location of active deposition migrates, such as the melting of an ice sheet or the recession of a shoreline. Synonymous with *diachronous*.

Troposphere
The lowest layer of the atmosphere closest to Earth's surface. It extends from the surface up to approximately 7 kilometers at the poles and about 17 kilometers in the equatorial regions. The troposphere is characterized by decreasing temperature with increasing height, significant vertical air movement, and appreciable water vapor content.

Trough-mouth fans
Undersea deposits of sediment on a slope, narrow at the top and wider at the bottom (hence fan-shaped) that develop near the downslope ends (or mouths) of submarine canyons (or troughs) that cross the *continental shelf* and descend the continental slope. Rapid rates of sedimentation in trough-mouth fans makes them good sources of sediment cores for paleoclimatic analyses.

Tundra
A treeless landscape on *permafrost* (ground that is permanently frozen below the uppermost layer, which thaws in summer), today restricted to high-latitude and high-altitude areas. The dominant vegetation are low-growing lichens, mosses, and stunted shrubs.

$U^{k'}_{37}$ index
The relative abundances of long-chain C_{37} *alkenones* in marine sediment that serve as a *proxy* for past sea-surface temperatures.

U.S. Climate Change Science Program (CCSP)
A consortium of Federal agencies that investigate *climate*. The primary objective of the CCSP is to provide the best science-based knowledge possible to support public discussion and government- and private-sector decisions about the risks and opportunities associated with changes in *climate* and in related environmental systems.

Viscoelastic deformation
The general term for change in shape or volume (deformation) of materials in response to applied stress. It involves changes that will be reversed (returning the material to its original configuration) if the stress is removed (*elastic deformation*), and also as changes that are permanent and thus will not be reversed if the stress is removed (viscous deformation, broadly defined).

Viscoelastic structure
Distribution of the material properties of the planet that controls how it deforms in response to applied stress (*viscoelastic deformation*), especially referring to how these material properties vary with depth.

Yedoma
A frozen, organic-rich, wind-blown accumulation, dominantly of silt-sized particles (loess), with ice content of 50–90% by volume. Yedoma is a frozen reservoir of carbon that will, if melted, release a substantial volume of carbon to the atmosphere and contribute substantively to Earth's greenhouse effect. Yedoma covers more than one million square kilometers of Russia.

Yield strength
The stress required to cause permanent deformation of a material. In many materials, if the applied stress falls below some level (the yield strength) then *elastic deformation* occurs but no permanent or viscous deformation, whereas for higher stresses permanent deformation occurs.

Younger Dryas
A *climate* event that occurred just after the *Last Glacial Maximum*, between about 11,500 and 12,800 years before present (with uncertainties of a couple of centuries). The Younger Dryas was characterized by a return to cool conditions in the Northern Hemisphere following the initial post-*glacial* warmth and warm conditions in the far south, a southward shift of the tropical circulation, reduction in monsoonal rainfall in Africa and Asia, extended *sea ice* and reduced sinking of surface ocean waters in the North Atlantic, and a fast start (decades) and a very fast end (perhaps less than a decade) to the anomalous conditions.

ABBREVIATIONS AND ACRONYMS

°C	degree Centigrade (a measure of temperature)
%	percent; parts per hundred
‰	per mil; parts per thousand
$\delta^{18}O$	the ratio of the concentrations of ^{18}O to ^{16}O relative to a standard
δD	the ratio of the concentrations of deuterium to hydrogen relative to a standard
ΔT	delta-T, temperature difference from a predefined state
^{16}O	oxygen-16, the lighter, abundant isotope of oxygen
^{18}O	oxygen-18, the heavier, less abundant isotope of oxygen
AAR	amino acid racemization dating
ACEX	Arctic Coring Expedition
AIRS	Atmospheric Infrared Sounder
AMAP	Arctic Monitoring and Assessment Programme
AMS	accelerator mass spectrometer
AOGCM	atmosphere-ocean coupled global climate model
AR4	Assessment Report 4 (the fourth climate assessment report of the IPCC)
ASTER	Advanced Spaceborne Thermal Emission and Reflection Radiometer
AVHRR	Advanced Very High Resolution Radiometery
BP	before present
CAPE	CircumArctic PaleoEnvironments
CCSM	Community Climate System Model
CCSP	U.S. Climate Change Science Program
CFC	chlorofluorocarbon
CIMSS	Cooperative Institute for Meteorological Satellite Studies
CLIMAP	Climate: Long range Investigation, Mapping, and Prediction
cm	centimeter
cm/yr	centimeter per year
CO_2	carbon dioxide
CH_4	methane
CRN	cosmogenic radionuclide dating
D	deuterium
DNA	deoxyribonucleic acid
D-O	Dansgaard-Oeschger
ECS	equilibrium climate sensitivity
FAC	Federal Advisory Committee
FACA	Federal Advisory Committee Act
GISP2	U.S. Greenland Ice Sheet Project 2
GISS	NASA Goddard Institute for Space Studies
GRACE	Gravity Recovery and Climate Experiment
GPS	global positioning system
GRIP	joint European Greenland Ice Core Project
Gt/yr	gigaton per year
H_n	Heinrich event number
HOTRAX	Healy-Oden Trans-Arctic Expedition
HTM	Holocene Thermal Maximum
IPCC	Intergovernmental Panel on Climate Change
IRD	ice-rafted debris
ka	thousands of years ago
km^{-1}	per kilometer
km^2	square kilometers (a measure of area)
kg	kilogram
kg/m/s	kilogram per meter per second
$kg/m^2/s$	kilogram per square meter per second
k.y.	thousands of years
LGM	Last Glacial Maximum
LIA	Little Ice Age
LIG	the last interglacial
m	meter
m.y.	million years
m/yr	meters per year (a measure of accumulation rate)
mm	millimeter
mm/yr	millimeter per year
Ma	millions of years ago

MCA Medieval Climate Anomaly

MSA methanesulfonic acid

MWP Medieval Warm Period

MIS marine isotope stage

MOC meridional overturning circulation

N_2O nitrous oxide

NAO North Atlantic Oscillation

NASA National Aeronautics and Space Administration

NCAR National Center for Atmospheric Research

NGRIP North Greenland Ice Core Project

NOAA National Oceanographic and Atmospheric Administration

NRC National Research Council

OSL optically stimulated luminescence

PDF climate sensitivity probability density function

PETM Paleocene-Eocene Thermal Maximum

ppm parts per million

ppmv parts per million by volume

SAP Synthesis and Assessment Product

SLE sea level equivalent

SMOW standard mean ocean water

SPECMAP Spectral Variability in Global Climate Project

SRES Special Report on Emission Scenarios (IPCC)

TIMS thermal ionization mass spectrometry

TMF trough-mouth fan

TPW true polar wander

U-series uranium-series disequilibrium dating

W/m^2 watts per square meter

yr year

REFERENCES CITED

Chapter 1. Preface: Why and How to Use This Sythesis and Assessment Report

Arctic Climate Impact Assessment (**ACIA**), 2005: *Arctic Climate Impact Assessment.* Cambridge University Press, Cambridge and New York, 1042 pp.

Bradley, R.S., 1999: *Paleoclimatology: Reconstructing Climates of the Quaternary.* Academic Press, San Diego, CA, 610 pp.

Brigham-Grette, J. and D.M. Hopkins, 1995: Emergent-marine record and paleoclimate of the Last Interglaciation along the northwest Alaskan coast. *Quaternary Research*, **43**, 154-173.

CAPE-Last Interglacial Project Members, 2006: Last Interglacial Arctic warmth confirms polar amplification of climate change. *Quaternary Science Reviews.* **25**, 1383-1400.

Cicerone, R.J., E.J. Barron, R.E. Dickinson, I.F. Fung, J.E. Hansen, T.R. Karl, R.S. Lindzen, J.C. McWilliams, F.S. Rowland, and J.M. Wallace, 2001: *Climate Change Science—An Analysis of Some Key Questions.* National Academies Press, Washington, D.C., 42 pp.

Darby, D., M. Jakobsson, and L. Polyak, 2005: Icebreaker expedition collects key Arctic seafloor and ice data. *EOS, Transactions of the American Geophysical Union*, **86(52)**, 549-556.

Denton, G.H., R.B. Alley, G.C. Comer, and W.S. Broecker, 2005: The role of seasonality in abrupt climate change. *Quaternary Science Reviews*, **24(10-11)**, 1159-1182.

Dyke, A.S., J. Hooper, and J.M. Savelle, 1996: History of sea ice in the Canadian Arctic archipelago based on postglacial remains of the bowhead whale. *Arctic*, **49(3)**, 235-255.

Hemming, S.R., 2004: Heinrich events: Massive late Pleistocene detritus layers of the North Atlantic and their global climate imprint. *Reviews of Geophysics* **42(2)**, RG1005.

Holland, M.M., C.M. Bitz, and B. Tremblay, 2006: Future abrupt reductions in the summer Arctic sea ice. *Geophysical Research Letters*, **33**, L23503.

Intergovernmental Panel on Climate Change (IPCC), 2000: *Special Report on Emissions Scenarios* [Nebojsa Nakicenovic and Robert Swart, (eds.)]. Cambridge University Press, Cambridge and New York, 612 pp.

Intergovernmental Panel on Climate Change (IPCC), 2007: Summary for policymakers. In: *Climate Change 2007—The Physical Science Basis. Contribution of Working Group I to the Fourth Assessment Report of the Intergovernmental Panel on Climate Change* [Solomon, S., D. Qin, M. Manning, Z. Chen, M. Marquis, K.B. Averyt, M. Tignor and H.L. Miller (eds.)]. Cambridge University Press, Cambridge and New York, 18 pp.

Lemke, P., J. Ren, R.B. Alley, I. Allison, J. Carrasco, G. Flato, Y. Fujii, G. Kaser, P. Mote, R.H. Thomas, and T. Zhang, 2007: Observations—Changes in snow, ice and frozen ground. In: *Climate Change 2007—The Physical Science Basis. Contribution of Working Group I to the Fourth Assessment Report of the Intergovernmental Panel on Climate Change* [Solomon, S., D. Qin, M. Manning, Z. Chen, M. Marquis, K.B. Averyt, M. Tignor and H.L. Miller (eds.)]. Cambridge University Press, Cambridge and New York, pp. 339-378.

Meehl, G.A., T.F. Stocker, W.D. Collins, P. Friedlingstein, A.T. Gaye, J.M. Gregory, A. Kitoh, R. Knutti, J.M. Murphy, A. Noda, S.C.B. Raper, I.G. Watterson, A.J. Weaver and Z.-C. Zhao, 2007: Global climate projections. In: *Climate Change 2007—The Physical Science Basis. Contribution of Working Group I to the Fourth Assessment Report of the Intergovernmental Panel on Climate Change* [Solomon, S., D. Qin, M. Manning, Z. Chen, M. Marquis, K.B. Averyt, M. Tignor and H.L. Miller (eds.)]. Cambridge University Press, Cambridge and New York, pp. 747-845.

National Research Council (NRC), 2002: *Abrupt Climate Change, Inevitable Surprises.* National Academies Press, Washington, D.C., 230 pp.

National Research Council (NRC) Committee on the Assessment of U.S. Coast Guard Polar Icebreaker Roles and Future Needs, 2007: *Polar Icebreakers in a Changing World—An Assessment of U.S. Needs.* National Academies Press, Washington D.C., 122 pp.

Raynaud, D., J. Chappellaz, C. Ritx, and P. Martinerie, 1997: Air content along the Greenland Ice Core Project core: A record of surface climatic parameters and elevation in central Greenland. *Journal of Geophysical Research*, **102(C12)**, 26607-26613.

Richter-Menge, J., J. Overland, A. Proshutinsky, V. Roman-ovsky, L. Bengtsson, L. Brigham, M. Dyurgerov, J.C. Gascard, S. Gerland, R. Graversen, C. Haas, M. Karcher, P. Kuhry, J. Maslanik, H. Melling, W. Maslowski, J. Morison, D. Perovich, R. Przybylak, V. Rachold, I. Rigor, A. Shiklomanov, J. Stroeve, D. Walker, and J. Walsh, 2006: State of the Arctic report. NOAA OAR Special Report, NOAA/OAR/PMEL, Seattle, WA, 36 pp.

Seager, R., D.S. Battisti, J. Yin, N. Gordon, N. Naik, A.C. Clement, and M.A. Cane, 2002: Is the Gulf Stream responsible for Europe's mild winters? *Quarterly Journal of the Royal Meteorological Society*, **128B**, 2563-2586.

Serreze, M.C. and J.A. Francis, 2006: The Arctic amplification debate: *Climatic Change*, **76(3-4)**, 241-264.

Shipboard Scientific Party, 2005: *Arctic Coring Expedition (ACEX): paleoceanographic and tectonic evolution of the central Arctic Ocean.* IODP Preliminary Report, 302, *http://publications.iodp.org/preliminary_report/302/.*

U.S. Climate Change Science Program, 2007: *Our Changing Planet, The U.S. Climate Change Science Program for Fiscal Year 2007.* U.S. CCSP, Washington, D.C., 264 pp.

Chapter 2. Paleoclimate Concepts

Alley, R.B., C.A. Shuman, D.A. Meese, A.J. Gow, K.C. Taylor, K.M. Cuffey, J.J. Fitzpatrick, P.M. Grootes, G.A. Zielinski, M. Ram, G. Spinelli, and B. Elder, 1997: Visual-stratigraphic dating of the GISP2 ice core—Basis, reproducibility, and application. *Journal of Geophysical Research*, **102(C12)**, 26367-26381.

Ammann, C.M., F. Joos, D.S. Schimel, B.L. Otto-Bliesner, and R.A. Tomas, 2007: Solar influence on climate during the past millennium—Results from transient simulations with the NCAR Climate System Model. *Proceedings of the National Academy of Sciences of the United States of Amerrica*, **104**, 3713-3718.

Anderson, R.K., G.H. Miller, J.P. Briner, N.A. Lifton, and S.B. DeVogel, 2008: A millennial perspective on Arctic warming from ^{14}C in quartz and plants emerging from beneath ice caps. *Geophysical Research Letters*, **35**, L01502, doi:10.1029/2007GL032057.

Andrews, J.T. and A.S. Dyke, 2007: Late Quaternary in North America. In: *The Encyclopedia of Quaternary Sciences* [Elias, S. (ed.)]. Elsevier, Amsterdam, pp. 1095-1101.

Arrhenius, S., 1896: On the influence of carbonic acid in the air upon the temperature on the ground. *Philosophical Magazine*, **41**, 237-276.

Baliunas, S. and R. Jastrow, 1990: Evidence for long-term brightness changes of solar-type stars. *Nature*, **348**, 520-522.

Bard, E., G. Raisbeck, F. Yiou, and J. Jouzel, 2000: Solar irradiance during the last 1200 years based on cosmogenic nuclides. *Tellus B*, **52**, 985-992.

Barron, E.J. and W.M. Washington, 1982: Atmospheric circulation during warm geologic periods—Is the equator-to-pole surface-temperature gradient the controlling factor? *Geology*, **10**, 633-636.

Bassinot, F.C., 2007: Oxygen isotope stratigraphy of the oceans. In: *The Encyclopedia of Quaternary Sciences* [Elias, S. (ed.)]. Elsevier, Amsterdam, pp. 1740-1748.

Beer, J., M. Vonmoos, and R. Muscheler, 2006: Solar variability over the past several millennia. *Space Science Reviews*, **125(1-4)**, 67-79.

Beerling, D.J. and R.A. Berner, 2005: Feedbacks and the coevolution of plants and atmospheric CO_2. *Proceedings of the National Academy of Sciences of the United States of America*, **102**, 1302-1305.

Berger, A. and M.F. Loutre, 1992: Astronomical solutions for paleoclimate studies over the last 3 million years. *Earth and Planetary Science Letters*, **111**, 369-382.

Berner, R.A., 1991: A model for atmospheric CO_2 over Phanerozoic time. *American Journal of Science*, **291**, 339-376.

Billups, K., J.E.T. Channell, and J. Zachos, 2002: Late Oligocene to early Miocene geochronology and paleoceanography from the subantarctic South Atlantic. *Paleoceangraphy*, **17(1)**, 1004, doi:10.1029/2000PA000568 11 pp.

Bohaty, S.M. and J. Zachos, 2003: Significant southern ocean warming even in the late middle Eocene. *Geology*, **31(11)**, 1017-1020.

Booth, D.B., K.G. Troost, J.J. Clague, and R.B. Waitt, 2004: The Cordilleran Ice Sheet. In: *The Quaternary Period in the United States* [Gillespie, A.R., S.C. Porter, and B.F. Atwater (eds.)]. Elsevier, Amsterdam, pp. 17-43.

Bradley, R.S., 1999: *Paleoclimatology—Reconstructing Climate of the Quaternary*. Academic Press, San Diego, 613 pp.

Bralower, T.J., D.J. Thomas, J.C. Zachos, M.M. Hirschmann, U. Rohl, H. Sigurdsson, E. Thomas, and D.L. Whitney, 1995: High-resolution records of the late Paleocene thermal maximum and circum-Caribbean volcanism—Is there a causal link? *Geology*, **25**, 963-966.

Briffa, K.R., P.D. Jones, F.H. Schweingruber, and T.J. Osborn, 1998: Influence of volcanic eruptions on Northern Hemisphere summer temperature over the past 600 years. *Nature*, **393**, 450-455.

Callendar, G.S., 1938: The artificial production of carbon dioxide and its influence on temperature. *Quarterly Journal of the Royal Meteorological Society*, **64**, 223-237.

Camp, C.D. and K.K. Tung, 2007: Surface warming by the solar cycle as revealed by the composite mean difference projection. *Geophysical Research Letters*, **34**, L14703, doi:10.1029/2007GL030207.

Cook, E.R. and L.A. Kairiukstis (eds.), 1990: *Methods of Dendrochronology—Applications in the Environmental Sciences*. Kluwer Academic Press, Norwell, MA, 394 pp.

Cronin, T.M., 1999: *Principles of Paleoclimatology*. Columbia University Press, New York, 560 pp.

Crowley, T.J. and G.R. North, 199: *Paleoclimatology*. Oxford University Press, New York, 339 pp.

Cuffey, K.M. and G.D. Clow, 1997: Temperature, accumulation, and ice sheet elevation in central Greenland through the last deglacial transition. *Journal of Geophysical Research*, **102(C12)**, 26383-26396.

D'Arrigo, R.D. and G.C. Jacoby, 1999: Northern North American tree-ring evidence for regional temperature changes after major volcanic events. *Climatic Change*, **41**, 1-15.

D'Hondt, S., 2005: Consequences of the Cretaceous/Paleogene mass extinction for marine ecosystems. *Annual Review of Ecology, Evolution and Systematics*, **36**, 295-317.

De Silva, S.L. and G.A. Zielinski, 1998: Global influence of the AD 1600 eruption of Huaynaputina, Peru, *Nature*, **393**, 455-458.

Donnadieu, Y., R. Pierrehumbert, R. Jacob, and F. Fluteau, 2006: Modeling the primary control of paleogeography on Cretaceous climate. *Earth and Planetary Science Letters*, **248**, 426-437.

Dyke, A.S. and Prest, V.K., 1987: Late Wisconsinan and Holocene history of the Laurentide Ice Sheet. *Géographie Physique et Quaternaire*, **41**, 237-263.

Ehhalt, D., M. Prather, F. Dentener, R. Derwent, E. Dlugokencky, E. Holland, I. Isaksen, J. Katima, V. Kirchhoff, P. Matson, P. Midgley, M. Wang, 2001: Atmospheric chemistry and Greenhouse Gases. In: *Climate Change 2001—The Scientific Basis. Contribution of Working Group I to the Third Assessment Report of the Intergovernmental Panel on Climate Change* [Houghton, J.T., Y. Ding, D.J. Griggs, M. Noguer, P.J. van der Linden, X. Dai, K. Maskell, and C.A. Johnson (eds.)]. Cambridge University Press, Cambridge and New York, 881 pp.

Emiliani, C., 1955: Pleistocene temperatures. *Journal of Geology*, **63**, 538-578.

Fairbanks, R.G., R.A. Mortlock, T.C. Chiu, L. Cao, A. Kaplan, T.P. Guilderson, T.W. Fairbanks, and A.L. Bloom, 2005: Marine radiocarbon calibration curve spanning 0 to 50,000 years B.P. based on paired $^{230}Th/^{234}U/^{238}U$ and ^{14}C dates on pristine corals. *Quaternary Science Reviews*, **24**, 1781-1796.

Fischer, E.M., J. Luterbacher, E. Zorita, S.F.B. Tett, C. Casty, and H. Wanner, 2007: European climate response to tropical volcanic eruptions over the last half millennium. *Geophysical Research Letters*, **34**, L05707.

Fleitmann, D., S.J. Burns, and M. Mudelsee, 2003: Holocene forcing of the Indian monsoon recorded in a stalagmite from southern Oman. *Science*, **300**, 1737-1739.

Forster, P., V. Ramaswamy, P. Artaxo, T. Berntsen, R. Betts, D.W. Fahey, J. Haywood, J. Lean, D.C. Lowe, G. Myhre, J. Nganga, R. Prinn, G. Raga, M. Schulz, and R. Van Dorland, 2007: Changes in atmospheric constituents and in radiative forcing. In: *Climate Change 2007—The Physical Science Basis. Contribution of Working Group I to the Fourth Assessment Report of the Intergovernmental Panel on Climate Change* [Solomon, S., D. Qin, M. Manning, Z. Chen, M. Marquis, K.B. Averyt, M. Tignor, and H.L. Miller (eds.)]. Cambridge University Press, Cambridge and New York, pp. 131-234.

Foukal, P., C. Frohlich, H. Spruit, and T.M.L. Wigley, 2006: Variations in solar luminosity and their effect on the Earth's climate. *Nature*, **443**, 161-166.

Friedrich, M., S. Remmelel, B. Kromer, J. Hofmann, M. Spurk, K.F. Kaiser, C. Orcel, and M. Kuppers, 2004: The 12,460-year Hohenheim oak and pine tree-ring chronology from central Europe—A unique annual record for radiocarbon calibration and paleoenvironment reconstructions. *Radiocarbon*, **46**, 1111-1122.

Fritts, H., 1976: *Tree Rings and Climate*. Academic Press, London, 567 pp.

Fröhlich, C. and J. Lean, 2004: Solar radiative output and its variability: evidence and mechanisms. *Astronomy and Astrophysics Review*, **12**, 273-320.

Gallup, C., H. Cheng, F.W. Taylor, and R.L. Edwards, 2002: Direct determination of the time of sea level change during Termination II. *Science*, **295**, 310-313.

Gao, C., L. Oman, A. Robock, and G.L. Stenchikov, 2007: Atmospheric volcanic loading derived from bipolar ice cores—Accounting for the spatial distribution of volcanic deposition. *Journal of Geophysical Research*, **112**, doi:10.1029/2006JD007461.

Hansen, J., A. Lacis, D. Rind, G. Russell, P. Stone, I. Fung, R. Ruedy, and J. Lerner, 1984: Climate sensitivity—Analysis of feedback mechanisms. In: *Climate Processes and Climate Sensitivity* [Hansen, J.E. and T. Takahashi (eds.)]. American Geophysical Union Geophysical Monograph 29, Maurice Ewing Vol. 5, pp. 130-163.

Hansen, J., Mki. Sato, R. Ruedy, K. Lo, D.W. Lea, and M. Medina-Elizade, 2006: Global temperature change. *Proceedings of the National Academy of Sciences of the United States of America*, **103**, 14288-14293, doi:10.1073/pnas.0606291103.

Hegerl, G.C., F.W. Zwiers, P. Braconnot, N.P. Gillett, Y. Luo, J.A. Marengo Orsini, N. Nicholls, J.E. Penner, and P.A. Stott, 2007: Understanding and attributing climate change. In: *Climate Change 2007—The Physical Science Basis. Contribution of Working Group I to the Fourth Assessment Report of the Intergovernmental Panel on Climate Change* [Solomon, S., D. Qin, M. Manning, Z. Chen, M. Marquis, K.B. Averyt, M. Tignor, and H.L. Miller (eds.)]. Cambridge University Press, Cambridge and New York, pp. 663-745.

Heusser, L.E., M. Lyle, and A. Mix, 2000: Vegetation and climate of the northwest coast of North America during the last 500 k.y.—High-resolution pollen evidence from the northern California margin. In: Lyle, M., Richter, C., and Moore, T.C. Jr. (eds.)]. *Proceedings of the Ocean Drilling Program, Scientific Results*, **167**, 217-226.

Higgins, J.A. and D.P. Schrag, 2006: Beyond methane: Towards a theory for the Paleocene-Eocene Thermal Maximum. *Earth and Planetary Science Letters*, **245**, 523-537.

Hovan, S.A., D.K. Rea, and N.G. Pisias, 1991: Late Pleistocene continental climate and oceanic variability recorded in northwest Pacific sediments. *Paleoceanography*, **6**, 349-370.

Huybers, P., 2007: Glacial variability over the last two million years—An extended depth derived age model, continuous obliquity pacing, and the Pleistocene progression. *Quaternary Science Reviews*, **26**, 37-55.

Imbrie, J., A. Berger, E.A. Boyle, S.C. Clemens, A. Duffy, W.R. Howard, G. Kukla, J. Kutzbach, D.G. Martinson, A. McIntyre, A.C. Mix, B. Molfino, J.J. Morley, L.C. Peterson, N.G. Pisias, W.L. Prell, M.E. Raymo, N.J. Shackleton, and J.R. Toggweiler. 1993: On the structure and origin of major glaciation cycles. 2. The 100,000-year cycle. *Paleoceanography*, **8**, 699-735.

Imbrie, J., J.D. Hays, D.G. Martinson, A. McIntyre, A.C. Mix, J.J. Morley, N.G. Pisias, W.L. Prell, and N.J. Shackleton, 1984: The orbital theory of Pleistocene climate—Support from a revised chronology of the marine $\delta^{18}O$ record. In: *Milankovitch and climate—Understanding the Response to Astronomical Forcing* [Berger, A., J. Imbrie, J. Hays, G. Kukla, and B. Saltzman]. D.Reidel Publishing Company, Dordrecht, pp. 269-305.

Intergovernmental Panel on Climate Change (IPCC), 2007: Summary for policymakers. In: *Climate Change 2007—The Physical Science Basis. Contribution of Working Group I to the Fourth Assessment Report of the Intergovernmental Panel on Climate Change* [Solomon, S., D. Qin, M. Manning, Z. Chen, M. Marquis, K.B. Averyt, M. Tignor and H.L. Miller (eds.)]. Cambridge University Press, Cambridge and New York, pp. 1-18.

Jansen, E., J. Overpeck, K.R. Briffa, J.-C. Duplessy, F. Joos, V. Masson-Delmotte, D. Olago, B. Otto-Bliesner, W.R. Peltier, S. Rahmstorf, R. Ramesh, D. Raynaud, D. Rind, O. Solomina, R. Villalba, and D. Zhang, 2007: Palaeoclimate. In: *Climate Change 2007—The Physical Science Basis. Contribution of Working Group I to the Fourth Assessment Report of the Intergovernmental Panel on Climate Change* [Solomon, S., D. Qin, M. Manning, Z. Chen, M. Marquis, K.B. Averyt, M. Tignor and H.L. Miller (eds.)]. Cambridge University Press, Cambridge and New York, pp. 434-497.

Jiang, D., H. Wang, Z. Ding, X. Lang, and H. Drange, 2005: Modeling the middle Pliocene climate with a global atmospheric general circulation model. *Journal of Geophysical Research*, **110**, D14107, doi:10.1029/2004JD005639.

Johnson, W.H., A.K. Hansel, E.A. Bettis III, P.F. Karrow, G.J. Larson, T.V. Lowell, and Schneider, A.F., 1997: Late Quaternary temporal and event classifications, Great Lakes region, North America. *Quaternary Research*, **47**, 1-12.

Jouzel, J., V. Masson-Delmotte, O. Cattani, G. Dreyfus, S. Falourd, G. Hoffmann, B. Minster, J. Nouet, J. M. Barnola, J. Chappellaz, H. Fischer, J. C. Gallet, S. Johnsen, M. Leuenberger, L. Loulergue, D. Luethi, H. Oerter, F. Parrenin, G. Raisbeck, D. Raynaud, A. Schilt, J. Schwander, E. Selmo, R. Souchez, R. Spahni, B. Stauffer, J.P. Steffensen, B. Stenni, T.F. Stocker, J.L. Tison, M. Werner, and E.W. Wolff, 2007: Orbital and millennial Antarctic climate variability over the past 800,000 years. *Science*, **317**, 793-796, doi:10.1126/science.1141038.

Kiehl, J. T. and K.E. Trenberth, 1997: Earth's annual global mean energy budget. *Bulletin of the American Meteorological Society*, **78**, 197-208.

Kump, L.R. and R.B. Alley, 1994: Global chemical weathering on glacial timescales. In: *Material Fluxes on the Surface of the Earth* [Hay, W.W. (ed.)]. National Academies Press, Washington, D.C., pp. 46-60.

Kump, L.R., J.F. Kasting, and R.G. Crane, 2003: *The Earth System*. Prentice Hall, New York, 2nd ed., 432 pp.

Ladurie, E.L., 1971: *Times of Feast, times of Famine—A History of Climate Since the Year 1000*. Farrar, Straus and Giroux, New York, 426 pp.

Lamb, H.H., 1982: *Climate History and the Modern World*. Routledge, New York, 433 pp.

LaMarche, V.C. Jr. and K.K. Hirschboeck, 1984: Frost rings in trees as records of major volcanic eruptions. *Nature*, **307**, 121-126.

Lear, C.H., Y. Rosenthal, H.K. Coxall, and P.A. Wilson, 2004: Late Eocene to Miocene ice sheet dynamics and the global carbon cycle. *Paleoceanography*, **19(4)**, PA4015, doi:10.1029/2004PA001039.

Le Treut, H., R. Somerville, U. Cubasch, Y. Ding, C. Mauritzen, A. Mokssit, T. Peterson, and M. Prather, 2007: Historical overview of climate change. In: *Climate Change 2007—The Physical Science Basis. Contribution of Working Group I to the Fourth Assessment Report of the Intergovernmental Panel on Climate Change* [Solomon, S., D. Qin, M. Manning, Z. Chen, M. Marquis, K.B. Averyt, M. Tignor and H.L. Miller (eds.)]. Cambridge University Press, Cambridge and New York, pp. 93-127.

Lisiecki, L.E. and M.E. Raymo, 2005: A Pliocene-Pleistocene stack of 57 globally distributed benthic $\delta^{18}O$ records. *Paleoceanography*, **20**, PA1003, doi:10.1029/2004PA001071 (see also *http://www.lorraine-lisiecki.com/stack.html*).

Lisiecki, L.E. and M.E. Raymo, 2007: Plio-Pleistocene climate evolution—Trends and transitions in glacial cycle dynamics. *Quaternary Science Reviews*, **26**, 56-69.

Livermore, R., C.D. Hillenbrand, M. Meredith, and G. Eagles, 2007: Drake Passage and Cenozoic climate—An open and shut case? *Geochemistry, Geophysics, Geosystems*, **8**, Article Q01005.

Loutre, M.F., D. Paillard, F. Vimeux, and E. Cortijo, 2004: Does mean annual insolation have the potential to change the climate? *Earth and Planetary Science Letters*, **221**, 1-14.

Maclennan, J., M. Jull, D. McKenzie, L. Slater, and K. Gronvold, 2002: The link between volcanism and deglaciation in Iceland. *Geochemistry, Geophysics, Geosystems*, **3**, Article 1062.

Meese, D.A., A.J. Gow, R.B. Alley, G.A. Zielinski, P.M. Grootes, M. Ram, K.C. Taylor, P.A. Mayewski, and J.F. Bolzan, 1997: The Greenland Ice Sheet Project 2 depth-age scale—Methods and results. *Journal of Geophysical Research*, **102(C12)**, 26411-26423.

Milankovitch, M., 1920: Théorie Mathématique des Phénomènes thermiques Produits per la Radiation solaire. Gauthier-Villars, Paris, 340 pp.

Milankovitch, M., 1941: Kanon der Erdbestrahlung. Royal Serbian Academy Special Publication 132, section of Mathematical and Natural Sciences, v. 33 (published in English by the Israël Program for Scientific Translation for the U.S. Department of Commerce and the National Science Foundation, Washington, D.C., 1969).

Miller, K.G., M.A. Kominz, J.V. Browning, J.D. Wright, G.S. Mountain, M.E. Katz, P.J. Sugarman, B.S. Cramer, N. Christie-Blick, and S.F. Pekar, 2005: The Phanerozoic record of global sea-level change. *Science*, **312**, 1293-1298.

Muhs, D.R., K.R. Simmons, and B. Steinke, 2002: Timing and warmth of the last interglacial period—New U-series evidence from Hawaii and Bermuda and a new fossil compilation for North America. *Quaternary Science Reviews*, **21**, 1355-1383.

Muhs, D.R., T.W. Stafford, J.B. Swinehart, S.D. Cowherd, S.A. Mahan, C.A. Bush, R.F. Madole, and P.B. Maat, 1997: Late Holocene eolian activity in the mineralogically mature Nebraska sand hills. *Quaternary Research*, **48**, 162-176.

Muhs, D.R., J.F. Wehmiller, K.R. Simmons, and L.L. York, 2004: Quaternary sea level history of the United States. In: *The Quaternary Period in the United States* [Gillespie, A.R., S.C. Porter, and B.F. Atwater, (eds.)]. Elsevier, Amsterdam, pp. 147-183.

Muller, P.J., G. Kirst, G. Ruhland, I. von Storch, and A. Rosell-Melé, 1998: Calibration of the alkenone paleotemperature index UK-37' based on core-tops from the eastern South Atlantic and the global ocean (60°N–60°S). *Geochimica et Cosmochimica Acta*, **62**, 1757-1772.

Muscheler, R., R. Beer, P.W. Kubik, and H.A. Synal, 2005: Geomagnetic field intensity during the last 60,000 years based on Be-10 and Cl-36 from the Summit ice cores and C-14. *Quaternary Science Reviews*, **24**, 1849-1860.

Muscheler, R., F. Joos, J. Beer, et al., 2007: Solar activity during the last 1000 yr inferred from radionuclide records. *Quaternary Science Reviews*, **26**, 82-97.

Nakamura, N. and A.H. Oort, 1988: Atmospheric heat budgets of the polar regions. *Journal of Geophysical Research*, **93(D8)**, 9510-9524.

National Research Council, 2006: Surface temperature reconstructions for the last 2,000 years. National Academies Press, Washington, D.C., 160 pp.

North Greenland Ice Core Project Members (NGRIP), 2004: High-resolution record of Northern Hemisphere climate extending into the last interglacial period. *Nature*, **431**, 147-151.

Ogg, James (compiler), 2004: Overview of global boundary stratotype sections and points (GSSPs). International Commission on Stratigraphy, available online at *http://www.stratigraphy.org/gssp.htm*.

Oman, L., A. Robock, G. Stenchikov, G.A. Schmidt, and R. Ruedy, 2005: Climatic response to high latitude volcanic eruptions. *Journal of Geophysical Research*, **110**, D13103, doi:10.1029/2004JD005487.

Oman, L., A. Robock, G. Stenchikov, T. Thordarson, D. Koch, D.T. Shindell, and C. Gao, 2006: Modeling the Distribution of the Volcanic Aerosol Cloud from the 1783–1784 Laki Eruption. *Journal of Geophysical Research*, **111**, D12209, doi:10.1029/2005JD006899.

Oppenheimer, C., 2003: Climatic, environmental, and human consequences of the largest known historic eruption: Tambora volcano (Indonesia) 1815. *Progress in Physical Geography*, **27**, 230-259.

Pagani, M., M.A. Arthur, and K.H. Freeman, 1999: Miocene evolution of atmospheric carbon dioxide. *Paleoceanography*, **14**, 273-292.

Pearson, P.N., P.W. Ditchfield, J. Singano, K.G. Harcourt-Brown, C.J. Nicholas, R.K. Olsson, N.J. Shackleton, and M.A. Hall, 2001: Warm tropical sea surface temperatures in the Late Cretaceous and Eocene epochs. *Nature*, **413**, 481-487.

Peixoto, J.P. and A.H. Oort, 1992: *Physics of Climate*. American Institute of Physics, New York, 520 pp.

Pierrehumbert, R.T., H. Brogniez, and R. Roca, 2007: On the relative humidity of the atmosphere. In: *The Global Circulation of the Atmosphere* [Schneider, T. and A. Sobel (eds.)]. Princeton University Press, Princeton, New Jersey, pp. 143-185.

Prentice, I.C., G.D. Farquhar, M.J.R. Fasham, M.L. Goulden, M. Heimann, V.J. Jaramillo, H.S. Kheshgi, C. Le Quéré, R.J. Scholes, and D.W.R. Wallace, 2001: *The Scientific Basis. Contribution of Working Group I to the Third Assessment Report of the Intergovernmental Panel on Climate Change* [Houghton, J.T., Y. Ding, D.J. Griggs, M. Noguer, P.J. van der Linden, X. Dai, K. Maskell, and C.A. Johnson (eds.)]. Cambridge University Press, Cambridge and New York, 881 pp.

Rahmstorf, S. and H.J. Schellnhuber: 2006: *Der Klimawandel*. Beck Verlag, Munich, 144 pp.

Ramanathan, V., 1975: Greenhouse effect due to chlorofluorocarbons—Climatic implications. *Science*, **190**, 50-52.

Robock, Alan, 2000: Volcanic eruptions and climate. *Reviews of Geophysics*, **38**, 191-219.

Robock, Alan, 2007: Correction to "Volcanic eruptions and climate." *Reviews of Geophysics*, **45**, RG3005, doi:10.1029/2007RG000232.

Robock, Alan, Tyler Adams, Mary Moore, Luke Oman, and Georgiy Stenchikov, 2007: Southern Hemisphere atmospheric circulation effects of the 1991 Mount Pinatubo eruption. *Geophysical Research Letters*, **34**, L237, doi:10.1029/2007GL031403.

Royer, D.L., R.A. Berner, and J. Park, 2007: Climate sensitivity constrained by CO_2 concentrations over the past 420 million years. *Nature*, **446**, 530-532.

Ruddiman, W.F., 2006, Ice-driven CO_2 feedback on ice volume. *Climate of the Past*, **2**, 43-55.

Salzer, M.W. and M.K. Hughs, 2006: Bristlecone pine tree rings and volcanic eruptions over the last 5000 yr. *Quaternary Research*, **67**, 57-68.

Serreze, M.C., A.P. Barrett, A.G. Slater, M. Steele, J.L. Zhang, and K.E. Trenberth, 2007: The large-scale energy budget of the Arctic. *Journal of Geophysical Research*, **112**, D11122.

Shellito, C. J., Sloan, L.C., and M. Huber, 2003: Climate model sensitivity to atmospheric CO_2 levels in the early-middle Paleogene. *Palaeogeography, Palaeoclimatology, Palaeoecology*, **193**, 113-123.

Shindell, D.T., G.A. Schmidt, M.E. Mann, and G. Faluvegi, 2004. Dynamic winter climate response to large tropical volcanic eruptions since 1600. *Journal of Geophysical Research*, **109**, D05104.

Stenchikov, Georgiy, Kevin Hamilton, Alan Robock, V. Ramaswamy, and M. Daniel Schwarzkopf, 2004: Arctic Oscillation response to the 1991 Pinatubo eruption in the SKYHI GCM with a realistic Quasi-Biennial Oscillation, *Journal of Geophysical Research*, **109**, D03112, doi:10.1029/2003JD003699.

Stenchikov, Georgiy, Kevin Hamilton, Ronald J. Stouffer, Alan Robock, V. Ramaswamy, Ben Santer, and Hans-F. Graf, 2006: Arctic Oscillation response to volcanic eruptions in the IPCC AR4 climate models. *Journal of Geophysical Research*, **111**, D07107, doi:10.1029/ 2005JD006286.

Stevens, M.J. and G.R. North, 1996: Detection of the climate response to the solar cycle. *Journal of Atmospheric Science*, **53**, 2594-2608.

Stuiver, M., P.M. Grootes, and T.F. Braziunas, 1995: The GISP2 delta O-18 climate record of the past 16,500 years and the role of the sun, ocean, and volcanoes. *Quaternary Research*, **44**, 341-354.

Sun, J.M., Z.L. Ding, D. Rokosh, and N. Rutter, 1999: 580,000 year environmental reconstruction from eolian deposits at the Mu Us Desert margin, China. *Quaternary Science Reviews*, **18**, 1351-1364.

Szabo, B.J., K.R. Ludwig, D.R. Muhs, and K.R. Simmons, 1994: Thorium-230 ages of corals and duration of the last interglacial sea-level high stand on Oahu, Hawaii. *Science*, **266**, 93-96.

Thordarson, T., Miller, D.J., Larsen, G., Self, S., and Sigurdsson, H., 2001: New estimates of sulfur degassing and atmospheric mass-loading by the 934 AD Eldgjá eruption, Iceland. *Journal of Volcanology and Geothermal Research*, **108**, 33-54.

Walker, J.C.G., P.B. Hays, and J.F. Kasting, 1981: A negative feedback mechanism for the long-term stabilization of Earth's surface-temperature. *Journal of Geophysical Research*, **86(NC-10)**, 9776-9782.

Walter, F.M. and D.C. Barry, 1991: Pre- and main-sequence evolution of solar activity. In: *The Sun in Time* [Sonnett, C.P., M.S. Giampapa, and M.S. Matthews]. University of Arizona Press, Tuscon, pp. 633-657.

Winckler, G. and H. Fischer, 2006: 30,000 years of cosmic dust in Antarctic ice. *Science*, **313**, 491-491.

Winograd, I.J., T.B. Coplen, J.M. Landwehr, A.C. Riggs, K.R. Ludwig, B.J. Szabo, P.T. Kolesar, and K.M. Revesz, 1992: Continuous 500,000-year climate record from vein calcite in Devils Hole, Nevada. *Science*, **258**, 255-260.

Winograd, I.J., J.M. Landwehr, K.R. Ludwig, T.B. Coplen, and A.C. Riggs, 1997: Duration and structure of the past four interglaciations. *Quaternary Research*, **48**, 141-154.

Zachos, J., M. Pagani, L. Sloan, E. Thomas, K. Billups, 2001: Trends, rhythms, and aberrations in global climate 65 Ma to present. *Science*, **292(5517)**, 686-693.

Zielinski, G.A., P.A. Mayewski, L.D. Meeker, S. Whitlow, M.S. Twickler, M. Morrison, D.A. Meese, A.J. Gow, and R.B. Alley, 1994: Record of volcanism since 7000 B.C. from the GISP2 Greenland ice core and implications for the volcano-climate system. *Science*, **264**, 948-950.

Chapter 3. Temperature and Precipitation History of the Arctic

Abbott, M.B., B.P. Finney, M.E. Edwards, and K.R. Kelts, 2000: Lake-level reconstructions and paleohydrology of Birch Lake, central Alaska, based on seismic reflection profiles and core transects. *Quaternary Research*, **53**, 154-166.

Adkins, J.F., E.A. Boyle, L.D. Keigwin, E. Cortijio, 1997: Variability of the North Atlantic thermohaline circulation during the last interglacial period. *Nature*, **390**, 154-156.

Ager, T.A. and L.B. Brubaker, 1985: Quaternary palynology and vegetation history of Alaska. In: *Pollen Records of Late Quaternary North American Sediments* [Bryant, V.M., Jr. and R.G. Holloway (eds.)]. American Association of Stratigraphic Palynologists, Dallas, TX, pp. 353-384.

Aksu, A.E., 1985: Planktonic foraminiferal and oxygen isotope stratigraphy of CESAR cores 102 and 103—Preliminary results. In: *Initial Geological Report on CESAR—The Canadian Expedition to Study the Alpha Ridge, Arctic Ocean* [Jackson, H.R., P.J. Mudie, and S.M. Blasco (eds.)]. Geological Survey of Canada Paper 84-22, pp. 115-124.

Alfimov, A.V., D.I. Berman, and A.V. Sher, 2003: Tundra-steppe insect assemblages and the reconstruction of the Late Pleistocene climate in the lower reaches of the Kolyma River. *Zoologicheskiy Zhurnal*, **82**, 281-300 (in Russian).

Alley, R.B., 1991: Deforming-bed origin for southern Laurentide till sheets? *Journal of Glaciology*, **37(125)**, 67-76.

Alley, R.B., 2003: Paleoclimatic insights into future climate challenges. *Philosophical Transactions of the Royal Society of London, Series A*, **361(1810)**, 1831-1849.

Alley, R.B., 2007: Wally was right: Predictive ability of the North Atlantic "conveyor belt" hypothesis for abrupt climate change. *Annual Review of Earth and Planetary Sciences*, **35**, 241-272.

Alley, R.B. and A.M. Ágústsdóttir, 2005: The 8k event—Cause and consequences of a major Holocene abrupt climate change. *Quaternary Science Reviews*, **24**, 1123-1149.

Alley, R.B. and S. Anandakrishnan, 1995: Variations in melt-layer frequency in the GISP2 ice core—Implications for Holocene summer temperatures in central Greenland. *Annals of Glaciology*, **21**, 64-70.

Alley, R.B., E.J. Brook, and S. Anandakrishnan, 2002: A northern lead in the orbital band—North-south phasing of ice-age events. *Quaternary Science Reviews*, **21(1-3)**, 431-441.

Alley, R.B. and K.M. Cuffey, 2001: Oxygen- and hydrogen-isotopic ratios of water in precipitation—Beyond paleothermometry. In: *Stable Isotope Geochemistry* [Valley, J.W. and D. Cole (eds.)]. Mineralogical Society of America Reviews in Mineralogy and Geochemistry, **43**, 527-553

Alley, R.B., D.A. Meese, C.A. Shuman, A.J. Gow, K.C. Taylor, P.M. Grootes, J.W.C. White, M. Ram, E.D. Waddington, P.A. Mayewski, and G.A. Zielinski, 1993: Abrupt increase in snow accumulation at the end of the Younger Dryas event. *Nature*, **362**, 527-529.

Alley, R.B., C.A. Shuman, D.A. Meese, A.J. Gow, K.C. Taylor, K.M. Cuffey, J.J. Fitzpatrick, P.M. Grootes, G.A. Zielinski, M. Ram, G. Spinelli, and B. Elder, 1997: Visual-stratigraphic dating of the GISP2 ice core—Basis, reproducibility, and application. *Journal of Geophysical Research*, **102**, 26367-26381.

Ammann, C.M., F. Joos, D.S. Schimel, B.L. Otto-Bliesner, and R.A. Tomas, 2007: Solar influence on climate during the past millennium—Results from transient simulations with the NCAR Climate System Model. *Proceedings of the National Academy of Sciences of the United States of America.*, *www.pnas.org/cgi//doi/10.1073/pnas.0605064103.*

Andersen, K.K., A. Svensson, S. Johnsen, S.O. Rasmussen, M. Bigler, R. Röthlisberger, U. Ruth, M.L. Siggaard-Andersen, J.P. Steffensen, D. Dahl-Jensen, B.M. Vinther, and H.B. Clausen, 2006: The Greenland Ice Core Chronology 2005, 15–42 kyr. Part I—Constructing the time scale. *Quaternary Science Reviews*, **25**, 3246-3257.

Anderson, L., M.B. Abbott, and B.P. Finney, 2001: Holocene climate inferred from oxygen isotope ratios in lake sediments, central Brooks Range, Alaska. *Quaternary Research*, **55**, 313-321.

Anderson, L., M.B. Abbott, B.P. Finney, and S.J. Burns, 2005: Regional atmospheric circulation change in the North Pacific during the Holocene inferred from lacustrine carbonate oxygen isotopes, Yukon Territory, Canada. *Quaternary Research*, **64**, 1-35.

Anderson, L., M.B. Abbott, B.P. Finney, and S.J. Burns, 2007. Late Holocene moisture balance variability in the southwest Yukon Territory, Canada. *Quaternary Science Reviews*, **26(1-2)**, 130-141.

Anderson, N.J. and M.J. Leng, 2004: Increased aridity during the early Holocene in West Greenland inferred from stable isotopes in laminated-lake sediments. *Quaternary Science Reviews*, **23**, 841-849.

Anderson, P.M. and A.V. Lozhkin, 2001: The stage 3 interstadial complex (Karginskii/middle Wisconsinan interval) of Beringia—Variations in paleoenvironments and implications for paleoclimatic interpretations. *Quaternary Science Reviews*, **20**, 93-125.

Anderson, P.M., P.J. Bartlein, L.B. Brubaker, K. Gajewski, and J.C. Ritchie, 1989: Modern analogues of late Quaternary pollen spectra from the western interior of North America. *Journal of Biogeography*, **16**, 573-596.

Anderson, P.M., P.J. Bartlein, L.B. Brubaker, K. Gajewski, and J.C. Ritchie, 1991: Vegetation-pollen-climate relationships for the arcto-boreal region of North America and Greenland. *Journal of Biogeography*, **18**, 565-582.

Anderson, R.K., G.H. Miller, J.P. Briner, N.A. Lifton, and S.B. DeVogel, 2008: A millennial perspective on Arctic warming from ^{14}C in quartz and plants emerging from beneath ice caps. *Geophyical Research Letters*, **35**, L01502, doi:10.1029/2007GL032057.

Andreev, A.A., D.M. Peteet, P.E. Tarasov, F.A. Romanenko, L.V. Filimonova, and L.D. Sulerzhitsky, 2001: Late Pleistocene interstadial environment on Faddeyevskiy Island, East-Siberian Sea, Russia. *Arctic, Antarctic, and Alpine Research*, **33**, 28-35.

Andrews, J.T., L.M. Smith, R. Preston, T. Cooper, and A.E. Jennings, 1997: Spatial and temporal patterns of iceberg rafting (IRD) along the East Greenland margin, ca. 68°N, over the last 14 cal ka. *Journal of Quaternary Science*, **12**, 1-13.

Archer, D., 2007: Methane hydrate stability and anthropogenic climate change. *Biogeosciences*, **4**, 521-544.

Arctic Climate Impact Assessment (ACIA), 2004: *Impacts of a Warming Arctic—Arctic Climate Impact Assessment*. Cambridge University Press, Cambridge and New York, 140 pp.

Arctic Climate Impact Assessment (ACIA), 2005: *Arctic Climate Impact Assessment*. Cambridge University Press, Cambridge and New York, 1042 pp.

Arctic Monitoring and Assessment Programme (AMAP), 1998: AMAP Assessment Report Arctic Pollution Issues. AMAP, Oslo, Norway, 871 pp.

Astakhov, V.I., 1995: The mode of degradation of Pleistocene permafrost in West Siberia. *Quaternary International*, **28**, 119-121.

Backman, J., M. Jakobsson, R. Løvlie, L. Polyak, and L.A. Febo, 2004: Is the central Arctic Ocean a sediment starved basin? *Quaternary Science Reviews*, **23**, 1435-1454.

Backman, J., K. Moran, D.B. McInroy, L.A. Mayer, and the Expedition 302 scientists, 2006: *Proceedings of IODP, 302*. Edinburgh (Integrated Ocean Drilling Program Management International, Inc.). doi:10.2204/iodp.proc.302.2006.

Balco, G., C.W. Rovey, and O.H. Stone, 2005a: The First Glacial Maximum in North America. *Science*, **307**, 222.

Balco, G., O.H. Stone, and C. Jennings, 2005b: Dating Plio-Pleistocene glacial sediments using the cosmic-ray-produced radionuclides ^{10}Be and ^{26}Al. *American Journal of Science*, **305**, 1-41.

Ballantyne, A.P., N.L. Rybczynski, P.A. Baker, C.R. Harington, and D. White, 2006: Pliocene Arctic temperature constraints from the growth rings and isotopic composition of fossil larch. *Palaeogeography, Palaeoclimatology, Palaeoecology*, **242**, 188-200.

Barber, V.A. and B.P. Finney, 2000: Late Quaternary paleoclimatic reconstructions for interior Alaska based on paleolake-level data and hydrologic models. *Journal of Paleolimnology*, **24**, 29-41.

Bard, E., 2002: Climate shocks—Abrupt changes over millennial time scales. *Physics Today*, **December**, 32-38.

Barley, E.M., I.R. Walker, J. Kurek, L.C. Cwynar, R.W. Mathewes, K. Gajewski, and B.P. Finney, 2006: A northwest North American training set—Distribution of freshwater midges in relation to air temperature and lake depth. *Journal of Paleolimnology*, **36**, 295-314.

Barnekow, L. and P. Sandgren, 2001: Palaeoclimate and tree-line changes during the Holocene based on pollen and plant macrofossil records from six lakes at different altitudes in northern Sweden. *Review of Palaeobotany and Palynology*, **117**, 109-118.

Barron, E.J., P.J. Fawcett, W.H. Peterson, D. Pollard, and S.L. Thompson, 1995: A "simulation" of mid-Cretaceous climate. *Paleoceanography*, **8**, 785-798.

Barron, E.J., P.J. Fawcett, D. Pollard, and S. Thompson, 1993: Model simulations of Cretaceous climates—The role of geography and carbon-dioxide. *Philosophical Transactions of the Royal Society of London Series B-Biological Sciences*, **341(1297)**, 307-315.

Barry, R.G., M.C. Serreze, J.A. Maslanik, and R.H. Preller, 1993: The arctic sea-ice climate system—Observations and modeling. *Reviews of Geophysics*, **31(4)**, 397-422.

Bartlein, P.J., K.H. Anderson, P.M. Anderson, M.E. Edwards, C.J. Mock, R.S. Thompson, R.S. Webb, T. Webb, III, and C. Whitlock, 1998: Paleoclimate simulations for North America over the past 21,000 years: features of the simulated climate and comparisons with paleoenvironmental data. *Quaternary Science Reviews*, **17**, 549-585.

Bauch, D., J. Carstens, and G. Wefer, 1997: Oxygen isotope composition of living *Neogloboquadrina pachyderma* (sin.) in the Arctic Ocean. *Earth and Planetary Science Letters*, **146**, pp. 47-58.

Bauch, D., P. Schlosser, and R.G. Fairbanks, 1995: Freshwater balance and the sources of deep and bottom waters in the Arctic Ocean inferred from the distribution of $H_2^{18}O$. *Progress in Oceanography*, **35**, 53-80.

Bauch, H.A. and H. Erlenkeuser, 2003: Interpreting glacial-interglacial changes in ice volume and climate from subarctic deep water foraminiferal $\delta^{18}O$. In: *Earth's Climate and Orbital Eccentricity—The Marine Isotope Stage 11 Question*. [Droxler, A.W., R.Z. Poore, and L.H. Burckle (eds.)]. Geophysical Monograph Series 137, American Geophysical Union, Washington, D.C., pp. 87-102.

Bauch, H.A., H. Erlenkeuser, K. Fahl, R.F. Spielhagen, M.S. Weinelt, H. Andruleit, and R. Henrich, 1999: Evidence for a steeper Eemian than Holocene sea surface temperature gradient between Arctic and sub-Arctic regions. *Palaeogeography, Palaeoclimatology, Palaeoecology*, **145**, 95-117.

Bauch, H.A., H. Erlenkeuser, J.P. Helmke, and U. Struck, 2000: A paleoclimatic evaluation of marine oxygen isotope stage 11 in the high-northern Atlantic (Nordic seas). *Global and Planetary Change*, **24**, 27-39.

Bauch, H.A. and E.S. Kandiano, 2007: Evidence for early warming and cooling in North Atlantic surface waters during the last interglacial. *Paleoceanography*, **22**, PA1201, doi:10.1029/2005PA001252.

Beget, J.E., 2001: Continuous Late Quaternary proxy climate records from loess in Beringia. *Quaternary Science Reviews*, **20**, 499-507.

Behl, R.J. and J.P. Kennett, 1996: Brief interstadial events in the Santa Barbara Basin, NE Pacific, during the past 60 kyr. *Nature*, **379**, 243-246.

Bendle, J.A. and Rosell-Melé, A. 2004: Distributions of UK37 and UK37' in the surface waters and sediments of the Nordic Seas: implications for paleoceanography. *Geochemistry, Geophysics, Geosystems*, **5**, 3589-3600. Q11013, doi:10.1029/2004GC000741.

Bennike, O., K.P. Brodersen, E. Jeppesen, and I.R. Walker, 2004: Aquatic invertebrates and high latitude paleolimnology. In: *Long-Term Environmental Change in Arctic and Antarctic Lakes* [Pienitz, R., M.S.V. Douglas, and J.P. Smol (eds.)]. Springer, Dordrecht, pp. 159-186.

Berger, A. and M.F. Loutre, 1991: Insolation values for the climate of the last million years. *Quaternary Sciences Review*, **10(4)**, 297-317.

Berger, A. and M.F. Loutre, 2002: An Exceptionally long interglacial ahead? *Science*, **297**, 1287-1288.

Berger, A., M.F. Loutre, and J. Laskar, 1992: Stability of the astronomical frequencies over the Earth's history for paleoclimate studies. *Science*, **255(5044)**, 560-566.

Berner, R.A. and Z. Kothavala, 2001: GEOCARB III: A revised model of atmospheric CO_2 over Phanerozoic time. *American Journal of Science*, **301(2)**, 182-204.

Bice, K.L., D. Birgel, P.A. Meyers, K.A. Dahl, K.U. Hinrichs, and R.D. Norris, 2006: A multiple proxy and model study of Cretaceous upper ocean temperatures and atmospheric CO_2 concentrations. *Paleoceanography*, **21(2)**, PA2002.

Bigelow, N., L.B. Brubaker, M.E. Edwards, S.P. Harrison, I.C. Prentice, P.M. Anderson, A.A. Andreev, P.J. Bartlein, T.R. Christensen, W. Cramer, J.O. Kaplan, A.V. Lozhkin, N.V. Matveyeva, D.F. Murrary, A.D. McGuire, V.Y. Razzhivin, J.C. Ritchie, B. Smith, D.A. Walker, K. Gajewski, V. Wolf, B.H. Holmqvist, Y. Igarashi, K. Kremenetskii, A. Paus, M.F.J. Pisaric, and V.S. Volkova, 2003: Climate change and Arctic ecosystems—1. Vegetation changes north of 55°N between the last glacial maximum, mid-Holocene, and present. *Journal of Geophysical Research*, **108**, doi:10.1029/2002JD002558.

Bigler, C. and R.I. Hall, 2003: Diatoms as quantitative indicators of July temperature—A validation attempt at century-scale with meteorological data from northern Sweden. *Palaeogeography, Palaeoclimatology, Palaeoecology*, **189**, 147-160.

Bintanja, R. and R.S.W. van de Wal, 2008: North American ice-sheet dynamics and the onset of 100,000-year glacial cycles. *Nature*, **454**, 869-872. doi:10.1038/*Nature*07158.

Birks, H.H., 1991: Holocene vegetational history and climatic change in west Spitzbergen—Plant macrofossils from Skardtjorna, an arctic lake. *The Holocene*, **1**, 209-218.

Birks, H.J.B., 1998: Numerical tools in palaeolimnology—Progress, potentialities, and problems. *Journal of Paleolimnology*, **20**, 307-332.

Björk, G., J. Soöderkvist, P. Winsor, A. Nikolopoulos, and M. Steele, 2002: Return of the cold halocline layer to the Amundsen Basin of the Arctic Ocean—Implications for the sea ice mass balance. *Geophysical Research Letters*, **29(11)**, 1513, doi:10.1029/2001GL014157.

Boellstorff, J., 1987: North American Pleistocene stages reconsidered in light of probable Pliocene-Pleistocene continental glaciation. *Science*, **202(2004365)**, 305-307.

Bonan, G.B., D. Pollard, and S.L. Thompson, 1992: Effects of boreal forest vegetation on global climate. *Nature*, **359(6397)**, 716-718.

Box, J.E., D.H. Bromwich, B.A. Veenhuis, L.S. Bai, J.C. Stroeve, J.C. Rogers, K. Steffen, T. Haran, and S.H. Wang, 2006: Greenland ice sheet surface mass balance variability (1988–2004) from calibrated polar MM5 output. *Journal of Climate*, **19(12)**, 2783-2800.

Boyd, T.J., M. Steel, R.D. Muench, and J.T. Gunn, 2002: Partial recovery of the Arctic Ocean halocline. *Geophysical Research Letters*, **29(14)**, 1657, doi:10.1029/2001GL014047.

Braconnot, P., B. Otto-Bliesner, S. Harrison, F.S. Joussaume, J.-Y. Peterchmitt, A. Abe-Ouchi, M. Crucifix, E. Driesschaert, T. Fichefet, C.D. Hewitt, M. Kageyama, A. Kitoh, A. Laine, M.-F. Loutre, O. Marti, U. Merkel, G. Ramstein, P. Valdes, S.L. Weber, Y. Yu, and Y. Zhao, 2007: Results of PMIP2 coupled simulations of the Mid-Holocene and Last Glacial Maximum—Part 1: experiments and large-scale features. *Climates of the Past*, **3**, 261-277.

Bradley, R.S., 1990: Holocene paleoclimatology of the Queen Elizabeth Islands, Canadian high Arctic. *Quaternary Science Reviews*, **9**, 365-384.

Bradley, R.S., 1999: *Paleoclimatology—Reconstructing climates of the Quaternary*. Academic Press, New York, 2nd ed., 613 pp.

Bradley, R.S., 2000: Past global changes and their significance for the future. *Quaternary Science Reviews*, **19**, 391-402.

Bradley, R.S., K.R. Briffa, J. Cole, and T.J. Osborn, 2003a: The climate of the last millennium. In: *Paleoclimate, Global Change and the Future* [Alverson, K.D., R.S. Bradley, and T.F. Pedersen (eds.)]. Springer, Berlin, pp. 105-141.

Bradley, R.S., M.K. Hughes, and H.F. Diaz, 2003b: Climate in Medieval time. *Science*, **302**, 404-405, doi:10.1126/science.1090372.

Bradley, R.S. and P.D. Jones (eds.), 1992: *Climate Since AD 1500*. Routlege, London. 677 pp.

Brassell, S.C., G. Eglinton, I.T. Marlowe, U. Pflaumann, and M Sarnthein, 1986: Molecular stratigraphy—A new tool for climatic assessment. *Nature*, **320**, 129-133.

Bray, P.J., S.P.E. Blockey, G.R. Coope, L.F. Dadswell, S.A. Elias, J.J. Lowe, and A.M. Pollard, 2006: Refining mutual climatic range (MCR) quantitative estimates of paleotemperature using ubiquity analysis. *Quaternary Science Reviews*, **25(15-16)**, 1865-1876.

Brewer, S., J. Guiot, and F. Torre, 2007: Mid-Holocene climate change in Europe—A data-model comparison. *Climate of the Past*, **3**, 499-512.

Briffa, K. and E. Cook, 1990: Methods of Response Function Analysis. In: *Methods of Dendrochronology*. sect. 5.6 [Cook, E.R. and L.A. Kairiukstis (eds.)].

Briffa, K.R., T.J. Osborn, F.H. Schweingruber, I.C. Harris, P.D. Jones, S.G. Shiyatov, and E.A. Vaganov, 2001: Low-frequency temperature variations from a northern tree ring density network. *Journal of Geophysical Research*, **106**, 2929-2941.

Brigham, J.K., 1985: *Marine Stratigraphy and Amino Acid Geochronology of the Gubik Formation, Western Arctic Coastal Plain, Alaska*. Doctoral dissertation, University of Colorado, Boulder; U.S. Geological Survey Open-File Report 85–381, 21 pp.

Brigham-Grette, J. and L.D. Carter, 1992: Pliocene marine transgressions of northern Alaska—Circumarctic correlations and paleoclimate. *Arctic*, **43(4)**, 74-89.

Brigham-Grette, J. and D.M. Hopkins, 1995: Emergent-marine record and paleoclimate of the last interglaciation along the northwest Alaskan coast. *Quaternary Research*, **43**, 154-173.

Brigham-Grette, J., A.V. Lozhkin, P.M. Anderson, and O.Y. Glushkova, 2004: Paleoenvironmental conditions in western Beringia before and during the Last Glacial Maximum. In: *Entering America: Northeast Asia and Beringia Before the Last Glacial Maximum* [Madsen, D.B. (ed.)]. University of Utah Press, Salt lake City, Chapter 2, pp. 29-61.

Briner, J.P., N. Michelutti, D.R. Francis, G.H. Miller, Y. Axford, M.J. Wooller, and A.P. Wolfe, 2006: A multi-proxy lacustrine record of Holocene climate change on northeastern Baffin Island. *Quaternary Research*, **65**, 431-442.

Brinkhuis, H., S. Schouten, M.E. Collinson, A. Sluijs, J.S.S. Damsfte, G.R. Dickens, M. Huber, T.M. Cronin, J. Onodera, K. Takahashi, J.P. Bujak, R. Stein, J. van der Burgh, J.S. Eldrett, I.C. Harding, A.F. Lotter, F. Sangiorgi, H.V.V. Cittert, J.W. de Leeuw, J. Matthiessen, J. Backman, and K. Moran, (Expedition 302 Scientists), 2006: Episodic fresh surface waters in the Eocene Arctic Ocean. *Nature*, **441(7093)**, 606-609.

Broecker, W.S., 2001: Was the Medieval Warm Period global? *Science*, **291(5508)**, 1497-1499, doi:10.1126/science.291.5508.1497.

Broecker, W.S. and S. Hemming, 2001: Climate swings come into focus. *Nature*, **294:5550,** 2308-2309.

Broecker, W.S., D.M. Peteet, D. Rind, 1985: Does the ocean-atmosphere system have more than one stable mode of operation? *Nature*, **315(6014)**, 21-26.

Brouwers, E.M., 1987: On *Prerygocythereis vunnieuwenhuisei* Brouwers sp.nov. In: *A stereo-atlas of ostracode shells* [Bate, R.H., D.J. Home, J.W. Neale, and D.J. Siveter [eds.)]. British Micropalaeontological Society, London 14, Part **1**, 17-20.

Brown, J. and V. Romanovsky, 2008: Report from the International Permafrost Association—State of permafrost in the first decade of the 21st century. *Permifrost and Periglacial Processes*, **19(2)**, 255-260.

Buchardt, B. and L.A. Símonarson, 2003: Isotope palaeotemperatures from the Tjörnes beds in Iceland—Evidence of Pliocene cooling. *Palaeogeography, Palaeoclimatology, Palaeoecology*, **189**, 71-95.

Budyko, M.I., A.B. Ronov, and A.L. Yanshin, 1985: *The History of the Earth's Atmosphere*. Leningrad, Gidrometeoirdat, 209 pp. (in Russian; English translation: Springer, Berlin, 1987, 139 pp.)

Calkin, P.E., 1988: Holocene glaciation of Alaska (and adjoining Yukon Territory, Canada). *Quaternary Science Reviews*, **7**, 159-184.

CAPE Project Members, 2001: Holocene paleoclimate data from the Arctic—Testing models of global climate change. *Quaternary Science Reviews*, **20**, 1275-1287.

CAPE—Last Interglacial Project Members, 2006: Last Interglacial Arctic warmth confirms polar amplification of climate change. *Quaternary Science Reviews*, **25**, 1383-1400.

Carter, L.D., 1981: A Pleistocene sand sea on the Alaskan Arctic Coastal Plain. *Science*, **211(4480)**, 381-383.

Carter, L.D., J. Brigham-Grette, L. Marincovich, Jr., V.L. Pease, and U.S. Hillhouse, 1986: Late Cenozoic Arctic Ocean sea ice and terrestrial paleoclimate. *Geology*, **14**, 675-678.

Chandler, M., D. Rind, and R. Thompson, 1994a: A simulation of the Pliocene (3 Ma) climate using the GISS GCM and PRISM Northern Hemisphere boundary conditions. *Global and Planetary Change*, **9**, 197-219

Chandler, M.A., D. Rind, and R.S. Thompson, 1994b: Joint investigations of the middle Pliocene climate II: GISS GCM Northern Hemisphere results. *Global and Planetary Change*, **9**, 197-219, doi:10.1016/0921-8181(94)90016-7.

Chapin, F.S. III, M. Sturm, M.C. Serreze, J.P. Mcfadden, J.R. Key, A.H. Lloyd, T.S. Rupp, A.H. Lynch, J.P. Schimel, J. Beringer, W.L. Chapman, H.E. Epstein, E.S. Euskirchen, L.D. Hinzman, G. Jia, C.L. Ping, K.D. Tape, C.D.C. Thompson, D.A. Walker, and J.M. Welker, 2005: Role of land-surface changes in Arctic summer warming. *Science*, **310**, 657-660.

Chapman, M.R., N.J. Shackleton, and J.-C. Duplessy, 2000: Sea surface temperature variability during the last glacial-interglacial cycle—Assessing the magnitude and pattern of climate change in the North Atlantic. *Palaeogeography, Palaeoclimatology, Palaeoecology*, **157**, 1-25.

Chapman, W.L. and J.E. Walsh, 2007: Simulations of Arctic temperature and pressure by global coupled models. *Journal of Climate*, **20(4)**, 609-632.

Clark, P.U. and D. Pollard, 1998: Origin of the middle Pleistocene transition by ice sheet erosion of regolith. *Paleoceanography*, **13**, 1-9.

Clark, P.U., D. Archer, D. Pollard, J.D. Blum, J.A. Rial, V. Brovkin, A.C. Mix, N.G. Pisias, and M. Roy, 2006: The middle Pleistocene transition—Characteristics, mechanisms, and implications for long-term changes in atmospheric pCO_2. *Quaternary Science Reviews*, **25**, 3150-3184.

Clayden, S.L., L.C. Cwynar, G.M. MacDonald, and A.A. Velichko, 1997: Holocene pollen and stomates from a forest-tundra site on the Taimyr Peninsula, Siberia. *Arctic and Alpine Research*, **29**, 327-333.

Climate Long-Range Investigation Mapping and Prediction (CLIMAP) Project Members, 1981: Seasonal reconstructions of the Earth's surface at the last glacial maximum. *Geological Society of America Map and Chart Series* **MC-36**, 1-18.

Climate Long-Range Investigation Mapping and Prediction (CLIMAP) Project Members, 1984: The last interglacial ocean. *Quaternary Research*, **21**, 123-224.

Cockford, S.J. and S.G. Frederick, 2007: Sea ice expansion in the Bering Sea during the Neoglacial—Evidence from archeozoology. *The Holocene*, **17**, 699-706.

Cohen, A.S., 2003: *Paleolimnology—The History and Evolution of Lake Systems*. Oxford University Press, Oxford, U.K., 528 pp.

Colinvaux, P.A., 1964: The environment of the Bering Land Bridge. *Ecological Monographs*, **34**, 297-329.

Conte, M.H., M. Sicre, C. Rühlemann, J.C. Weber, S. Schulte, D. Schulz-Bull, and T. Blanz, 2006: Global temperature calibration of the alkenone unsaturation index ($U^{K'}_{37}$) in surface waters and comparison with surface sediments. *Geochemistry, Geophysics, Geosystems*, **7**, Q02005, doi:10.1029/2005GC001054.

Cronin, T.M, G.S. Dwyer, T. Kamiyac, S. Schwedea, and D.A. Willarda, 2003: Medieval Warm Period, Little Ice Age and 20th century temperature variability from Chesapeake Bay. *Global and Planetary Change*, **36**, 17-29.

Crowley, T.J., 1998: Significance of tectonic boundary conditions for paleoclimate simulations. In: *Tectonic Boundary Conditions for Climate Reconstructions* [Crowley, T.J. and K.C. Burke (eds.)]. Oxford University Press, New York, pp. 3-17.

Crowley, T.J., 1990: Are there any satisfactory geologic analogs for a future greenhouse warming? *Journal of Climatology*, **3**, 1282-1492.

Crowley, T.J., 2000: Causes of climate change over the past 1000 years. *Science*, **289**, 270-277.

Crowley, T.J., S.K. Baum, K.Y. Kim, G.C. Hegerl, and W.T. Hyde, 2003: Modeling ocean heat content changes during the last millennium. *Geophysical Research Letters*, **30**, 1932, doi:10.1029/2003GL017801.

Crowley, T.J. and T. Lowery, 2000: How warm was the Medieval warm period? *Ambio*, **29**, 51-54.

Cuffey, K.M. and G.D. Clow, 1997: Temperature, accumulation, and ice sheet elevation in central Greenland through the last deglacial transition. *Journal of Geophysical Research*, **102(C12)**, 26383-26396.

Cuffey, K.M., G.D. Clow, R.B. Alley, M. Stuiver, E.D. Waddington, and R.W. Saltus, 1995: Large Arctic temperature change at the Wisconsin-Holocene glacial transition. *Science*, **270**, 455-458.

D'Arrigo, R., R. Wilson, and G. Jacoby, 2006: On the long-term context for late twentieth century warming. *Journal of Geophysical Research*, **111**, D03103, doi:10.1029/2005JD006352.

Dahe, Q., J.R. Petit, J. Jouzel, and M. Stievenard, 1994: Distribution of stable isotopes in surface snow along the route of the 1990 International Trans-Antarctic Expedition. *Journal of Glaciology*, **40**, 107-118.

Dahl, S.O. and A. Nesje, 1996: A new approach to calculating Holocene winter precipitation by combining glacier equilibrium-line altitudes and pine-tree limits—A case study from Hardangerjokulen, central southern Norway. *The Holocene*, **6(4)**, 381-398.

Dahl-Jensen, D., K. Mosegaard, N. Gundestrup, G.D. Clow, S.J. Johnsen, A.W. Hansen, and N. Balling, 1998: Past temperature directly from the Greenland Ice Sheet. *Science*, **282**, 268-271.

Dansgaard, W., 1964: Stable isotopes in precipitation. *Tellus*, **16**, 436-468.

Dansgaard, W., J.W.C. White, and S.J. Johnsen, 1989: The abrupt termination of the Younger Dryas climate event. *Nature*, **339(6225)**, 532-534.

Delworth, T.L. and T.R. Knutson, 2000: Simulation of early 20th century global warming. *Science*, **287(5461)**, 2246-2250.

de Vernal, A., C. Hillaire-Marcel, and D.A. Darby, 2005: Variability of sea ice cover in the Chukchi Sea (western Arctic Ocean) during the Holocene. *Paleoceanography*, **20**, PA4018, doi:10.1029/2005PA001157.

Denman, K.L., G. Brasseur, A. Chidthaisong, P. Ciais, P.M. Cox, R.E. Dickinson, D. Hauglustaine, C. Heinze, E. Holland, D. Jacob, U. Lohmann, S. Ramachandran, P.L. da Silva Dias, S.C. Wofsy, and X. Zhang, 2007: Couplings between changes in the climate system and biogeochemistry. In: *Climate Change 2007—The Physical Science Basis. Contribution of Working Group I to the Fourth Assessment Report of the Intergovernmental Panel on Climate Change* [Solomon, S., D. Qin, M. Manning, Z. Chen, M. Marquis, K.B. Averyt, M. Tignor, and H.L. Miller (eds.)]. Cambridge University Press, Cambridge and New York, 996 pp.

Denton, G.H., R.B. Alley, G.C. Comer, and W.S. Broecker, 2005: The role of seasonality in abrupt climate change. *Quaternary Science Reviews*, **24(10-11)**, 1159-1182.

Digerfeldt, G., 1988: Reconstruction and regional correlation of Holocene lake-level fluctuations in Lake Bysjön, South Sweden. *Boreas*, **17**, 237-263.

Donnadieu, Y., R. Pierrehumbert, R. Jacob, and F. Fluteau, 2006: Modeling the primary control of paleogeography on Cretaceous climate. *Earth and Planetary Science Letters*, **248**, 426-437.

Douglas, M.S.V., 2007: Environmental change at high latitudes. In: *Geological and Environmental Applications of the Diatom—Pond Scum to Carbon Sink* [Starratt, S.W. (ed.)]. The Paleontological Society Papers, **13**, 169-179.

Douglas, M.S.V. and J.P. Smol, 1999: Freshwater diatoms as indicators of environmental change in the High Arctic. In: *The Diatoms—Applications for the Environment and Earth Sciences* [Stoermer, E. and J.P. Smol (eds.)]. Cambridge University Press, Cambridge and New York, 488 pp.

Douglas, M.S.V., J.P. Smol, and W. Blake, Jr., 1994: Marked post-18th century environmental change in high Arctic ecosystems. *Science*, **266**, 416-419.

Douglas, M.S.V., J.P. Smol, R. Pienitz, and P. Hamilton, 2004: Algal indicators of environmental change in arctic and antarctic lakes and ponds. In: *Long-term environmental change in Arctic and Antarctic lakes* [Pienitz, R., M.S.V. Douglas, and J.P. Smol (eds.)]. Springer, Dordrecht, 117-157.

Dowdeswell, J.A., J.O. Hagen, H. Björnsson, A.F. Glazovsky, W.D. Harrison, P. Holmlund, J. Jania, R.M. Koerner, B. Lefauconnier, C.S.L. Ommanney, and R.H. Thomas, 1997: The mass balance of circum-Arctic glaciers and recent climate change. *Quaternary Research*, **48**, 1-14.

Dowsett, H.J., 2007: The PRISM palaeoclimate reconstruction and Pliocene sea-surface temperature. In: *Deep-time Perspectives on Climate Change—Marrying the Signal from Computer Models and Biological Proxies* [Williams, M., A.M. Haywood, F.J. Gregory, and D.N. Schmidt (eds.)]. The Micropalaeontological Society, Special Publication, The Geological Society, London, pp. 459-480.

Dowsett, H.J., J.A. Barron, R.Z. Poore, R.S. Thompson, T.M. Cronin, S.E. Ishman, and D.A. Willard, 1999: *Middle Pliocene Paleoenvironmental Reconstruction—PRISM2*. U.S. Geological Survey Open-File Report 99–535, unpaged.

Dowsett, H.J., and eight others, 1994: Joint investigations of the middle Pliocene climate I—PRISM paleoenvironmental reconstructions. *Global and Planetary Change*, **9**, 169-195.

Droxler, A.W., R.B. Alley, W.R. Howard, R.Z. Poore, and L.H. Burckle, 2003: Unique and exceptionally long interglacial marine isotope stage 11—Window into Earth future climate. In: *Earth's Climate and Orbital Eccentricity: The Marine Isotope Stage 11 Question* [Droxler, A.W., R.Z. Poore and L.H. Burckle (eds.)]. Geophysical Monograph Series 137, American Geophysical Union, Washington, D.C., 1-14.

Droxler, A.W. and J.W. Farrell, 2000: Marine Isotope Stage 11 (MIS11)—New insights for a warm future. *Global and Planetary Change*, **24(1)**, 1-5.

Duk-Rodkin, A., R.W. Barendregt, D.G. Froese, F. Weber, R.J. Enkin, I.R. Smith, Grant D. Zazula, P. Waters, and R. Klassen, 2004: Timing and extent of Plio-Pleistocene glaciations in north-western Canada and east-central Alaska. In: *Quaternary Glaciations—Extent and Chronology, Part II, North America* [Ehlers, J. and P.L. Gibbard (eds.)]. Elsevier, Amsterdam, pp. 313-345.

Duplessy, J-C., E. Ivanova, I. Murdmaa, M. Paterne, and L. Labeyrie, 2001. Holocene paleoceanography of the northern Barents Sea and variations of the northward heat transport by the Atlantic Ocean. *Boreas*, **30**, 2-16.

Dyke, A.S., J. England, E. Reimnitz, and H. Jetté, 1997: Changes in driftwood delivery to the Canadian Arctic Archipelago—The hypothesis of postglacial oscillations of the Transpolar Drift. *Arctic*, **50**, 1-16.

Dyke, A.S., J. Hooper, and J.M. Savelle, 1996: A history of sea ice in the Canadian Arctic Archipelago based on the postglacial remains of the bowhead whale (*Balaena mysticetus*). *Arctic*, **49**, 235-255.

Dyke, A.S. and J.M. Savelle, 2001: Holocene history of the Bering Sea bowhead whale (*Balaena mysticetus*) in its Beaufort Sea summer grounds off southwestern Victoria Island, western Canadian Arctic. *Quaternary Research*, **55**, 371-379.

Dykoski, C.A., R.L. Edwards, H. Cheng, D. Yuan, Y. Cai, M. Zhang, Y. Lin, J. Qing, Z. An, and J. Revenaugh, 2005: A high-resolution, absolute-dated Holocene and deglacial Asian monsoon record from Dongge Cave, China. *Earth and Planetary Science Letters*, **233**, 71-86.

Edwards, M.E., N.H. Bigelow, B.P. Finney, and W.R. Eisner, 2000: Records of aquatic pollen and sediment properties as indicators of late-Quaternary Alaskan lake levels. *Journal of Paleolimnology*, **24**, 55-68.

Edwards, T.L., M. Crucifix, and S.P. Harrison, 2007: Using the past to constrain the future: how the palaeorecord can improve estimates of global warming. *Progress in Physical Geography*, **31**, 481-500.

Elias, S.A., 2007: Beetle records—Late Pleistocene North America. In: *Encyclopedia of Quaternary Science* [Elias, S.A. (ed.)]. Elsevier, Amsterdam, pp. 222-236.

Elias, S.A., K. Anderson, and J.T. Andrews, 1996: Late Wisconsin climate in the northeastern United States and southeastern Canada, reconstructed from fossil beetle assemblages. *Journal of Quaternary Science*, **11**, 417-421.

Elias, S.A., J.T. Andrews, and K.H. Anderson, 1999: New insights on the climatic constraints on the beetle fauna of coastal Alaska derived from the mutual climatic range method of paleoclimate reconstruction. *Arctic, Antarctic, and Alpine Research*, **31**, 94-98.

Elias, S.A. and J.V. Matthews, Jr., 2002: Arctic North American seasonal temperatures in the Pliocene and Early Pleistocene, based on mutual climatic range analysis of fossil beetle assemblages. *Canadian Journal of Earth Sciences*, **39**, 911-920.

Ellis, J.M. and P.E. Calkin, 1984: Chronology of Holocene glaciation, central Brooks Range, Alaska. *Geological Society of America Bulletin*, **95**, 897-912.

EPICA Community Members, 2004. Eight glacial cycles from an Antarctic ice core. *Nature*, **429**, 623-628.

Epstein, S., H. Buchsbaum, H. Lowenstam, and H.C. Urey, 1953: Revised carbonate-water isotopic temperature scale. *Geological Society of America Bulletin*, **64**, 1315-1325.

Erez, J. and B. Luz, 1982: Temperature control of oxygen-isotope fractionation of cultured planktonic foraminifera. *Nature*, **297**, 220-222.

Eronen, M., P. Zetterberg, K.R. Briffa, M. Lindholm, J. Meriläinen, and M. Timonen, 2002: The supra-long Scots pine tree-ring record for Finnish Lapland—Part 1, chronology construction and initial inferences. *The Holocene*, **12(6)**, 673-680.

Esper, J., E.R. Cook, and F.H. Schweingruber, 2002: Low-frequency signals in long tree-ring chronologies for reconstructing past temperature variability. *Science*, **295**, 2250-2253.

Esper, J. and F.H. Schweingruber, 2004: Large-scale treeline changes recorded in Siberia. *Geophysical Research Letters*, **31**, L06202, doi:10.1029/2003GL019178.

Fairbanks, R.G., 1989: A 17,000-year glacio-eustatic sea level record—Influence of glacial melting rates on the Younger Dryas event and deep-ocean circulation. *Nature*, **343**, 612-616.

Farrera, I., S.P. Harrison, I.C. Prentice, G. Ramstein, J. Guiot, P.J. Bartlein, R. Bonnelle, M. Bush, W. Cramer, U. von Grafenstein, K. Holmgren, H. Hooghiemstra, G. Hope, D. Jolly, S.-E. Lauritzen, Y. Ono, S. Pinot, M. Stute, and G. Yu, 1999: Tropical climates at the Last Glacial Maximum—A new synthesis of terrestrial palaeoclimate data. I. Vegetation, lake-levels and geochemistry. *Climate Dynamics*, **15**, 823-856.

Finney, B., K. Rühland, J.P. Smol, and M.-A. Fallu, 2004: Paleolimnology of the North American subarctic. In: *Long-Term Environmental Change in Arctic and Antarctic Lakes* [Pienitz, R., M.S.V. Douglas, and J.P. Smol (eds.)]. Springer, Dordrecht, pp. 269-318.

Fisher, D.A., 1979: Comparison of 100,000 years of oxygen isotope and insoluble impurity profiles from the Devon Island and Camp Century ice cores. *Quaternary Research*, **11**, 299-304.

Fisher, D.A. and R.M. Koerner, 2003: Holocene ice core climate history, a multi-variable approach. In: *Global Change in the Holocene* [Mackay, A., R. Battarbee, J. Birks, and F. Oldfield (eds.)]. Arnold, London, pp. 281-293.

Fisher, D., R.M. Koerner, N. Reeh, 1995. Holocene climatic records from Agassiz Ice Cap, Ellesmere Island, NWT, Canada. *The Holocene*, **5(1)**, 19-24,doi:10.1177/095968369600600401.

Fisher, D.A., R.M. Koerner, J.C. Bourgeois, G. Zielinski, C. Wake, C.U. Hammer, H.B. Clausen, N. Gundestrup, S. Johnsen, K. Goto-Azuma, T. Hondoh, E. Blake, and M. Gerasimoff, 1998: Penny Ice Cap cores, Baffin Island, Canada, and the Wisconsinan Foxe Dome connection—Two states of Hudson Bay ice cover. *Science*, **279**, 692-695.

Fisher, D.A., C. Wake, K. Kreutz, K. Yalcin, E. Steig, P. Mayewski, L. Anderson, J. Aheng, S. Rupper, C. Zdanowicz, M. Demuth, M. Waskiewicz, D. Dahl-Jensen, K. Goto-Azuma, J.B. Bourgeois, R.M. Koerner, J. Sekerka, E. Osterberg, M.B. Abbott, B.P. Finney, and S.J. Burns, 2004: Stable isotope records from Mount Logan, Eclipse ice cores and nearby Jellybean Lake. Water cycle of the North Pacific over 2000 years and over five vertical kilometres—Sudden shift and tropical connections. *Géographie Physique et Quaternaire*, **58**, 9033-9048.

Flowers, G.E., H. Björnsson, Á. Geirsdóttir, G.H. Miller, J.L. Black, and G.K.C. Clarke, 2008: Holocene climate conditions and glacier variation in central Iceland from physical modelling and empirical evidence. *Quaternary Science Reviews*, **27**, 797-813.

Flowers, G.E., H. Björnsson, Á. Geirsdóttir, G.H. Miller, and G.K.C. Clarke, 2007: Glacier fluctuation and inferred climatology of Langjökull ice cap through the Little Ice Age, *Quaternary Science Reviews*, **22**, 2337-2353.

Francis, J.E., 1988: A 50-million-year-old fossil forest from Strathcona Fiord, Ellesmere Island, Arctic Canada—Evidence for a warm polar climate. *Arctic*, **41(4)**, 314-318.

Fricke, H.C. and J.R. O'Neil, 1999: The correlation between $^{18}O/^{16}O$ ratios of meteoric water and surface temperature—Its use in investigating terrestrial climate change over geologic time. *Earth and Planetary Science Letters*, **170**, 181-196.

Fritts, H.C., 1976: *Tree Rings and Climate*. London Academic Press, 576 pp.

Fronval, T. and E. Jansen, 1997: Eemian and early Weichselian (140–60 ka) paleoceanography and paleoclimate in the Nordic seas with comparisons to Holocene conditions. *Paleoceanography*, **12**, 443-462.

Funder, S., 1989: Quaternary geology of East Greenland. In: *Quaternary Geology of Canada and Greenland* [Fulton, R.J. (ed.)]. Geological Society of America Decade of North American *Geology*, **K1**, 756-763.

Funder, S., O. Bennike, J. Böcher, C. Israelson, K.S. Petersen, and L.A. Simonarson, 2001: Late Pliocene Greenland—The Kap København Formation in North Greenland. *Bulletin of the Geological Society of Denmark*, **48**, 117-134.

Funder, S., I. Demidov, and Y. Yelovicheva, 2002: Hydrography and mollusc faunas of the Baltic and the White Sea–North Sea seaway in the Eemian. *Palaeogeography, Palaeoclimatology, Palaeoecology*, **184**, 275-304.

Fyles, J.G., L. Marincovich, Jr., J.V. Mathews, Jr., and R. Barendregt, 1991: Unique mollusc find in the Beaufort Formation (Pliocene) Meighen Island, Arctic Canada. In: *Current Research, Part B, Geological Survey of Canada, Paper 91-1B*, pp. 461-468.

Gajewski, K. and G.M. MacDonald, 2004: Palynology of North American Arctic Lakes. In: *Long-Term Environmental Change in Arctic and Antarctic Lakes* [R. Pienitz, M.S.V. Douglas, and J.P. Smol (eds.)]. Springer, Netherlands, pp. 89-116.

Gard, G., 1986: Calcareous nannofossil biostratigraphy north of 80° latitude in the eastern Arctic Ocean. *Boreas*, **15**, 217-229.

Gard, G., 1987: Late Quaternary calcareous nannofossil biostratigraphy and sedimentation patterns—Fram Strait, Arctica. *Paleoceanography*, **2**, 219-229.

Gard, G., 1993: Late Quaternary coccoliths at the North Pole—Evidence of ice-free conditions and rapid sedimentation in the central Arctic Ocean. *Geology*, **21**, 227-230.

Geirsdóttir, Á., G.H. Miller, Y. Axford, and S. Olafsdottir, in press, Holocene and latest Pleistocene climate and glacier fluctuations in Iceland. *Quaternary Science Reviews*.

Gervais, B.R., G.M. MacDonald, J.A. Snyder, and C.V. Kremenetski, 2002: *Pinus sylvestris* treeline development and movement on the Kola Peninsula of Russia—Pollen and stomate evidence. *Journal of Ecology*, **90**, 627-638.

Gildor, H., E. Tziperman, and J.R. Toggweiler, 2002: Sea ice switch mechanism and glacial-interglacial CO_2 variations. *Global Biogeochemical Cycles*, **16**, 10.1029/2001GB001446.

Goetcheus, V.G. and H.H. Birks, 2001: Full-glacial upland tundra vegetaion preserved under tephra in the Beringia National Park, Seward Peninsula, Alaska. *Quaternary Science Reviews*, **20(1-3)**, 135-147.

Goetz, S.J., M.C. Mack, K.R. Gurney, J.T. Randerson, and R.A. Houghton, 2007: Ecosystem responses to recent climate change and fire disturbance at northern high latitudes—Observations and model results contrasting northern Eurasia and North America. *Environmental Research Letters*, **2**, 045031, doi:10.1088/1748-9326/2/4/045031.

Goodfriend, G.A., J. Brigham-Grette, G.H. Miller, 1996: Enhanced age resolution of the marine quaternary record in the Arctic using aspartic acid racemization dating of bivalve shells. *Quaternary Research*, **45**, 176-187.

Goosse, H., H. Renssen, A. Timmermann, and R.A. Bradley, 2005: Internal and forced climate variability during the last millennium—A model-data comparison using ensemble simulations. *Quaternary Science Reviews*, **24**, 1345-1360.

Gradstein, F.M., J.G. Ogg, and A.G. Smith (eds.), 2004: *A Geologic Time Scale*. Cambridge University Press, Cambridge and New York, 589 pp.

Grice, K., W.C.M. Klein Breteler, S. Schoten, V. Grossi, J.W. de Leeuw, and J.S. Sinninge Damsté, 1998: Effects of zooplankton herbivory on biomarker proxy records. *Paleoceanography*, **13**, 686-693.

Grootes, P.M., M. Stuiver, J.W.C. White, S. Johnsen, and J. Jouzel, 1993: Comparison of oxygen isotope records from the GISP2 and GRIP Greenland Ice cores. *Nature*, **366**, 522-555.

Grove, J.M., 1988: *The Little Ice Age*. Methuen, London, 498 pp.

Grudd, H., K.R. Briffa, W. Karlén, T.S. Bartholin, P.D. Jones, and B. Kromer, 2002: A 7400-year tree-ring chronology in northern Swedish Lapland—Natural climatic variability expressed on annual to millennial timescales. *The Holocene*, **12**, 657-666.

Gudina, V.D. Kryukov, L.K. Levchuk and L.A. Sudkov, 1983: Upper-Pleistocene sediments in north-eastern Taimyr. *Bulletin of Commission on Quaternary Researches*, **52**, 90-97 (in Russian).

Haeberli, W., G.D. Cheng, A.P. Gorbunov, and S.A. Harris, 1993: Mountain permafrost and climatic change. *Permafrost and Periglacial Processes*, **4(2)**, 165-174.

Hammarlund, D., L. Barnekow, H.J.B. Birks, B. Buchardt, and T.W.D. Edwards, 2002: Holocene changes in atmospheric circulation recorded in the oxygen-isotope stratigraphy of lacustrine carbonates from northern Sweden. *The Holocene*, **12**, 339-351.

Hammarlund, Dan, Svante Björcka Bjørn Buchardt, Carsten Israelsonç, and Charlotte T. Thomsen, 2003: Rapid hydrological changes during the Holocene revealed by stable isotope records of lacustrine carbonates from Lake Igelsjön, southern Sweden. *Quaternary Science Reviews*, **22(2-4)**, 353-370.

Hannon, G.E. and M.J. Gaillard, 1997: The plant-macrofossil record of past lake-level changes. *Journal of Paleolimnology*, **18**, 15-28.

Hantemirov, R.M. and S.G. Shiyatov, 2002: A continuous multimillennial ring-width chronology in Yamal, northwestern Siberia. *The Holocene*, **12(6)**, 717-726.

Harrison, S., P. Braconnot, C. Hewitt, and R.J. Stouffer, 2002: Fourth International Workshop of the Palaeoclimate Modelling Inter- comparison Project (PMIP): Launching PMIP Phase II. *Eos, Transactions of the American Geophysical Union*, **83**, 447-447.

Harrison, S.P., D. Jolly, F. Laarif, A. Abe-Ouchi, B. Dong, K. Herterich, C. Hewitt, S. Joussaume, J.E. Kutzbach, J. Mitchell, N. de Noblet, and P. Valdes, 1998: Intercomparison of simulated global vegetation distributions in response to 6 kyr BP orbital forcing. *Journal of Climat*e, **11**, 2721-2742.

Haywood, A.M., P. Dekens, A.C. Ravelo, and M. Williams, 2005: Warmer tropics during the mid-Pliocene? Evidence from alkenone paleothermometry and a fully coupled ocean-atmosphere GCM. *Geochemistry, Geophysics, Geosystems*, **6**, Q03010, doi:10.1029/2004GC000799.

Haywood, A.M. and P.J. Valdes, 2004: Modeling Pliocene warmth—Contribution of atmosphere, oceans and cryosphere. *Earth and Planetary Science Letters*, **218**, 7363-7377.

Haywood, A.M. and P.J. Valdes, 2006: Vegetation cover in a warmer world simulated using a dynamic global vegetation model for the Mid-Pliocene. *Palaeogeography, Palaeoclimatology, Palaeoecology*, **237**, 412-427.

Helmke, J.P. and H.A. Bauch, 2003: Comparison of glacial and interglacial conditions between the polar and subpolar North Atlantic region over the last five climatic cycles. *Paleoceanograph*, **18(2)**, 1036, doi:10.1029/2002PA000794.

Henrich, R. and K.-H. Baumann, 1994: Evolution of the Norwegian current and the Scandinavian ice sheets during the past 2.6 m.y.—Evidence from ODP Leg 104 biogenic carbonate and terrigenous records. *Palaeogeography, Palaeoclimatology, Palaeoecology*, **108**, 75-94.

Herbert, T.D., 2003: Alkenone paleotemperature determinations. In: *Treatise in Marine Geochemistry* [Elderfield, H. and K.K. Turekian (eds.)]. Elsevier, Amsterdam, pp. 391-432.

Heusser, L. and J. Morley, 1996: Pliocene climate of Japan and environs between 4.8 and 2.8 Ma—A joint pollen and marine faunal study. *Marine Micropaleontology*, **27**, 85-106.

Hewitt, C.D. and J.F.B. Mitchell, 1998: A fully coupled GCM simulation of the climate of the mid-Holocene. *Geophysical Research Letters*, **25**, 361-364.

Hoffmann, G., M. Werner, and M. Heimann, 1998: Water isotope module of the ECHAM atmospheric general circulation model—A study on timescales from days to several years. *Journal of Geophysical Research*, **103**, 16871-16896.

Holland, M.M. and C.M. Bitz, 2003: Polar amplification of climate change in coupled models. *Climate Dynamics*, **21**, 221-232.

Holland, M.M., C.M. Bitz, and B. Tremblay, 2006: Future abrupt reductions in the summer Arctic sea ice. *Geophysical Research Letters*, **33**, L23503, doi:10.1029/2006GL028024.

Hu, F.S. and Shemesh, A. 2003. A biogenic-silica $\delta^{18}O$ record of climatic change during the last glacial-interglacial transition in southwestern Alaska. *Quaternary Research*, **59**, 379-385.

Hu, F.S., Ito, E., Brown, T.A., Curry, B.B. and Engstrom, D.R. 2001. Pronounced climatic variations in Alaska during the last two millennia. *Proceedings of the National Academy of Sciences of the United States of America*, **98(19)**, 10552-10556.

Huber, C., M. Leuenberger, R. Spahni, J. Flückiger, J. Schwander, T.F. Stocker, S. Johnsen, A. Landals, and J. Jouzel, 2006: Isotope calibrated Greenland temperature record over Marine Isotope Stage 3 and its relation to CH_4. *Earth and Planetary Science Letters*, **243**, 504-519.

Hughen, K., J. Overpeck, R.F. Anderson, and K.M. Williams, 1996: The potential for palaeoclimate records from varved Arctic lake sediments—Baffin Island, eastern Canadian Arctic. In: *Lacustrine Environments. Palaeoclimatology and Palaeoceanography from Laminated Sediments* [Kemp, A.E.S. (ed.)]. Geological Society, London, Special Publications 116, pp. 57-71.

Hughes, M.K. and H.F. Diaz, 1994: Was there a "Medieval Warm Period" and if so, where and when? *Journal of Climatic Change*, **265**, 109-142.

Huybers, P., 2006: Early Pleistocene glacial cycles and the integrated summer insolation forcing. *Science*, **313**, 508-511.

Huybers, P. 2007: Glacial variability over the last two million years—An extended depth-derived age model, continuous obliquity pacing, and the Pleistocene progression. *Quaternary Science Reviews*, **26**, 37-55. doi:10.1016/j.quascirev.2006.07.005.

Hyvärinen, H., 1976: Flandrian pollen deposition rates and treeline history in northern Fennoscandia. *Boreas*, **5(3)**, 163-175.

Ilyashuk, E.A., B.P. Ilyashuk, D. Hammarlund, and I. Larocque, 2005: Holocene climatic and environmental changes inferred from midge records (Diptera: *Chironomidae, Chaoboridae, Ceratopogonidae*) at Lake Berkut, southern Kola Peninsula, Russia. *The Holocene*, **15**, 897-914.

Imbrie, J., A. Berger, E.A. Boyle, S.C. Clemens, A. Duffy, W.R. Howard, G. Kukla, J. Kutzbach, D.G. Martinson, A. McIntyre, A.C. Mix, B. Molfino, J.J. Morley, L.C. Peterson, N.G. Pisias, W.L. Prell, M.E. Raymo, N.J. Shackleton, and J.R. Toggweiler, 1993: On the structure and origin of major glaciation cycles. 2. The 100,000-year cycle. *Paleoceanography*, **8**, 699-735.

Imbrie, J. and N.G. Kipp, 1971: A new micropaleontological method for quantitative paleoclimatology—Application to a late Pleistocene Caribbean core. In: *The Late Cenozoic Glacial Age* [Turekian, K.K. (ed.)]. Yale Univeristy Press, New Haven, CT, pp. 71-181.

Intergovernmental Panel on Climate Change (IPCC), 1990: *Climate Change: The IPCC scientific assessment* [Houghton, J.T., G.J. Jenkins, and J.J. Ephraums (eds.)]. Cambridge University Press, Cambridge and New York, 365 pp.

Intergovernmental Panel on Climate Change (IPCC), 2007: Summary for Policymakers. In: *Climate Change 2007—The Physical Science Basis—Contribution of Working Group I to the Fourth Assessment Report of the Intergovernmental Panel on Climate Change* [Solomon, S., D. Qin, M. Manning, Z. Chen, M. Marquis, K.B. Averyt, M. Tignor, and H.L. Miller (eds.)]. Cambridge University Press, Cambridge and New York, 996 pp.

Iversen, J., 1944: *Viscum*, *Hedera* and *Ilex* as climatic indicators. A contribution to the study of past-glacial temperature climate. *Geologiska Foreningens Forhandlingar*, **66**, 463-483.

Jacoby, G.C. and R.D. D'Arrigo, 1995: Tree ring width and density evidence of climatic and potential forest change in Alaska. *Global Biogeochemical Cycles*, **9(2)**, 227-234.

Jakobsson, M. and R. Macnab, 2006: A comparison between GEBCO sheet 5.17 and the International Bathymetric Chart of the Arctic Ocean (IBCAO) version 1.0. *Marine Geophysical Researches*, **27(1)**, 35-48.

Jakobsson, M., R. Løvlie, H. Al-Hanbali, E. Arnold, J. Backman, and M. Mörth, 2000: Manganese color cycles in Arctic Ocean sediments constrain Pleistocene chronology. *Geology*, **28**, 23-26.

Jansen, E., E. Bleil, R. Henrich, L. Kringstad, and B. Slettemark, B., 1988: Paleoenvironmental changes in the Norwegian Sea and the northeast atlantic during the last 2.8 m.y.—Deep sea drilling project/ocean drilling program sites 610, 642, 643 and 644. *Paleoceanography*, **3**, 563-581.

Jansen, E., J. Overpeck, K.R. Briffa, J.-C. Duplessy, F. Joos, V. Masson-Delmotte, D. Olago, B. Otto-Bliesner, W.R. Peltier, S. Rahmstorf, R. Ramesh, D. Raynaud, D. Rind, O. Solomina, R. Villalba, and D. Zhang, 2007: Palaeoclimate. In: *Climate Change 2007—The Physical Science Basis. Contribution of Working Group I to the Fourth Assessment Report of the Intergovernmental Panel on Climate Change* [Solomon, S., D. Qin, M. Manning, Z. Chen, M. Marquis, K.B. Averyt, M. Tignor, and H.L. Miller (eds.)]. Cambridge University Press, Cambridge and New York, 996 pp.

Jenkyns, H.C., A. Forster, S. Schouten, and J.S. Sinninghe Damsté, 2004: High temperatures in the Late Cretaceous Arctic Ocean. *Nature*, **432(7019)**, 888-892.

Jennings, A., K. Knudsen, M. Hald, C. Hansen, and J. Andrews, 2002: A mid-Holocene shift in Arctic sea-ice variability on the East Greenland Shelf. *The Holocene*, **12**, 49-58.

Jiang D., H. Wang, Z. Ding, X. Lang, and H. Drange, 2005: Modeling the middle Pliocene climate with a global atmospheric general circulation model. *Journal of Geophysical Research*, **110**, D14107, doi:10.1029/2004JD005639.

Johnsen, S., H. Clausen, W. Dansgaard, K. Fuhrer, N. Gundestrup, C. Hammer, P. Iversen, J. Jouzel, B. Stauffer, and J. Steffensen, 1992: Irregular glacial interstadials recorded in a new Greenland ice core. *Nature*, **359**, 311-313.

Johnsen, S.J., D. Dahl-Jensen, W. Dansgaard, and N. Gundestrup, 1995: Greenland palaeotemperatures derived from GRIP bore hole temperature and ice core isotope profiles. *Tellus*, **47B**, 624-629.

Johnsen, S.J., W. Dansgaard, and J.W.C. White, 1989: The origin of Arctic precipitation under present and glacial conditions. *Tellus*, **41B**, 452-468. *http://www.sciencedirect.com/science/journal/02773791*.

Jones, P.D., K.R. Briffa, T.P. Barnett, and S.F.B. Tett, 1998: High-resolution palaeoclimatic records for the last millennium—Interpretation, integration and comparison with General Circulation Model control-run temperatures. *The Holocene*, **8**, 455-471.

Jones, V.J., Leng, M.J., Solovieva, N., Sloane, H.J. and Tarasov, P., 2004. Holocene climate of the Kola Peninsula—Evidence from the oxygen isotope record of diatom silica. *Quaternary Science Reviews* **23**, 833-839.

Jouzel, J., R.B. Alley, K.M. Cuffey, W. Dansgaard, P. Grootes, G. Hoffmann, S.J. Johnsen, R.D. Koster, D. Peel, C.A. Shuman, M. Stievenard, M. Stuiver, and J. White, 1997: Validity of the temperature reconstruction from water isotopes in ice cores. *Journal of Geophysical Research*, **102**, 26471-26487.

Jouzel, J., V. Masson-Delmotte, O. Cattani, G. Dreyfus, S. Falourd, G. Hoffmann, B. Minster, J. Nouet, J.M. Barnola, J. Chappellaz, H. Fischer, J.C. Gallet, S. Johnsen, M. Leuenberger, L. Loulergue, D. Luethi, H. Oerter, F. Parrenin, G. Raisbeck, D. Raynaud, A. Schilt, J. Schwander, E. Selmo, R. Souchez, R. Spahni, B. Stauffer, J.P. Steffensen, B. Stenni, T.F. Stocker, J.L. Tison, M. Werner, and E.W. Wolff, 2007: Orbital and millennial Antarctic climate variability over the past 800,000 years. *Science*, **317**, 793-796, doi:10.1126/science.1141038.

Joynt, E.H. III and A.P. Wolfe, 2001: Paleoenvironmental inference models from sediment diatom assemblages in Baffin Island lakes (Nunavut, Canada) and reconstruction of summer water temperature. *Canadian Journal of Fisheries and Aquatic Sciences*, **58**, 1222-1243.

Kaakinen, A. and M. Eronen, 2000: Holocene pollen stratigraphy indicating climatic and tree-line changes derived from a peat section at Ortino, in the Pechora lowland, northern Russia. *The Holocene*, **10**, 611-620.

Kageyama, M., O. Peyron, S. Pinot, P. Tarasov, J. Guiot, S. Joussaume, and G. Ramstein, 2001: The Last Glacial Maximum climate over Europe and western Siberia—A PMIP comparison between models and data. *Climate Dynamics*, **17**, 23-43.

Kandiano, E.S. and H.A. Bauch, 2007: Phase relationship and surface water mass change in the Northeast Atlantic during Marine Isotope Stage 11 (MIS11). *Quaternary Research*, **68(3)**, 445-455.

Kaplan, J.O., N.H. Bigelow, P.J. Bartlein, T.R. Christiansen, W. Cramer, S.P. Harrison, N.V. Matveyeva, A.D. McGuire, D.F. Murray, I.C. Prentice, V.Y. Razzhivin, B. Smith, D.A. Walker, P.M. Anderson, A.A. Andreev, L.B. Brubaker, M.E. Edwards, A.V. Lozhkin, and J.C. Ritchie, 2003: Climate change and arctic ecosystems II—Modeling paleodata-model comparisons, and future projections. *Journal of Geophysical Research*, **108(D19)**, 8171, doi:10.1029/2002JD002559.

Kaplan, M.R. and A.P. Wolfe, 2006: Spatial and temporal variability of Holocene temperature trends in the North Atlantic sector. *Quaternary Research*, **65**, 223-231.

Kapsner, W.R., R.B. Alley, C.A. Shuman, S. Anandakrishnan, and P.M. Grootes, 1995: Dominant influence of atmospheric circulation on snow accumulation in Greenland over the past 18,000 years. *Nature*, **373**, 52-54.

Karlén, W., 1988: Scandinavian glacial and climate fluctuations during the Holocene. *Quaternary Science Reviews*, **7**, 199-209.

Kaufman, D.S., T.A. Ager, N.J. Anderson, P.M. Anderson, J.T. Andrew, P.J. Bartlein, L.B. Brubakerg, L.L. Coats, L.C. Cwynar, M.L. Duvall, A.S. Dyke, M.E. Edwards, W.R. Eisner, K. Gajewski, Á. Geirsdóttir, F.S. Hu, A.E. Jennings, M.R. Kaplan, M.W. Kerwin, A.V. Lozhkin, G.M. MacDonald, G.H. Miller, C.J. Mock, W.W. Oswald, B.L. Otto-Bliesver, D.F. Porinchu, K. Ruûland, J.P. Smol, E.J. Steig, and B.B. Wolfe, 2004: Holocene thermal maximum in the western Arctic (0-180°W). *Quaternary Science Reviews*, **23**, 529-560.

Kaufman, D.S. and J. Brigham-Grette, 1993: Aminostratigraphic correlations and paleotemperature implications, Pliocene-Pleistocene high sea level deposits, northwestern Alaska. *Quaternary Science Reviews*, **12**, 21-33.

Kellogg, T.B., 1977: Paleoclimatology and paleo-oceanography of the Norwegian and Greenland seas—The last 450,000 years. *Marine Micropaleontology*, **2**, 235-249.

Kerwin, M., J.T. Overpeck, R.S. Webb, A. DeVernal, D.H. Rind, and R.J. Healy, 1999: The role of oceanic forcing in mid-Holocene northern hemisphere climatic change. *Paleoceanography*, **14**, 200-210.

Kirk-Davidov, D.B., D.P. Schrag, and J.G. Anderson, 2002: On the feedback of stratospheric clouds on polar climate. *Geophysical Research Letters*, **29(11)**, 1556 pp.

Kitoh, A. and S. Murakami, 2002: Tropical Pacific Climate at the mid-Holocene and the Last Glacial Maximum simulated by a coupled ocean-atmosphere general circulation model. *Paleoceanography*, **17**, 1-13.

Knutson, T.R., T.L. Delworth, K.W. Dixon, I.M. Held, J. Lu, V. Ramaswamy, M.D. Schwarzkopf, G. Stenchikov, and R.J. Stouffer, 2006: Assessment of twentieth-century regional surface temperature trends using the GFDL CM2 coupled models. *Journal of Climate*, **19(9)**, 1624-1651.

Koç, N. and E. Jansen, 1994: Response of the high-latitude Northern Hemisphere to climate forcing—Evidence from the Nordic Seas. *Geology*, **22**, 523-526.

Koç, N., E. Jansen, and H. Haflidason, 1993: Paleooceanographic reconstruction of surface ocean conditions in the Greenland, Iceland and Norwegian Seas through the last 14 ka based on diatoms. *Quaternary Science Reviews*, **12**, 115-140.

Koerner, R.M., 2005: Mass balance of glaciers in the Queen Elizabeth Islands, Nunavut, Canada. *Annals of Glaciology*, **42(1)**, 417-423.

Koerner, R.M. and D.A. Fisher, 1990: A record of Holocene summer climate from a Canadian high-Arctic ice core. *Nature*, **343**, 630-631.

Korhola, A., H. Olander, and T. Blom, 2000: Cladoceran and chironomid assemblages as quantitative indicators of water depth in sub-Arctic Fennoscandian lakes. *Journal of Paleolimnology*, **24**, 43-54.

Korty, R.L., K.A. Emanuel, and J.R. Scott, 2008: Tropical cyclone-induced upper-ocean mixing and climate—Application to equable climates. *Journal of Climate*, **21(4)**, 638-654.

Kremenetski, C.V., L.D. Sulerzhitsky, and R. Hantemirov, 1998: Holocene history of the northern range limits of some trees and shrubs in Russia. *Arctic and Alpine Research*, **30**, 317-333.

Kukla, G.J., 2000: The last interglacial. *Science*, **287**, 987-988.

Kump, L.R. and D. Pollard, 2008: Amplification of Cretaceous warmth by biological cloud feedbacks. *Science*, **11(5873)**, 195, doi:10.1126/science.1153883.

Kvenvolden, K.A., 1988: Methane hydrate—A major reservoir of carbon in the shallow geosphere? *Chemical Geology*, **71**, 41-51.

Kvenvolden, K.A., 1993: A primer on gas hydrates. In: *The Future of Energy Gases* [Howel, D.G. (ed.)]. U.S. Geological Survey Professional Paper 1570, pp. 279-291.

Lamb, H.H., 1977: *Climate History and the Future. Climate—Past, Present and Future*. Metheun, London, v. 2, 835 pp.

Lambeck, K., Y. Yokoyama, and T. Purcell, 2002: Into and out of the Last Glacial Maximum—Sea-level change during oxygen isotope stages 3 and 2. *Quaternary Science Reviews*, **21**, 343-360.

Larocque, I. and R.I. Hall, 2004: Holocene temperature estimates and chironomid community composition in the Abisko Valley, northern Sweden. *Quaternary Science Reviews*, **23**, 2453-2465.

Lauritzen, S.-E., 1996: Calibration of speleothem stable isotopes against historical records—A Holocene temperature curve for north Norway?. In: *Climatic Change: the Karst Record* [Lauritzen, S.-E. (ed.)]. Karst Waters Institute Special Publication, Charles Town, West Virginia, vol. 2, pp. 78-80.

Lauritzen, S.E. and J. Lundberg, 1998: Rapid temperature variations and volcanic events during the Holocene from a Norwegian speleothem record. In: *Past global changes and their significance for the future*. Volume of Abstracts, IGBP-PAGES, 88 pp.

LeGrande, A.N. and G.A. Schmidt, 2006: Global gridded data set of the oxygen isotopic composition in seawater. *Geophysical Research Letters*, **33**, L12604, doi:10.1029/2006GL026011.

Lemke, P., J. Ren, R.B. Alley, I. Allison, J. Carrasco, G. Flato, Y. Fujii, G. Kaser, P. Mote, R.H. Thomas, and T. Zhang, 2007: Observations—Changes in snow, ice and frozen ground. In: *Climate Change 2007: The Physical Science Basis—Contribution of Working Group I to the Fourth Assessment Report of the Intergovernmental Panel on Climate Change* [Solomon, S., D. Qin, M. Manning, Z. Chen, M. Marquis, K.B. Averyt, M. Tignor and H.L. Miller (eds.)]. Cambridge University Press, Cambridge and New York, 996 pp.

Leng, M.J. and J.D. Marshall, 2004: Palaeoclimate interpretation of stable isotope data from lake sediment archives. *Quaternary Science Reviews*, **23**, 811-831.

Letréguilly, A., P. Huybrechts, N. Reeh, 1991: Steady-state characteristics of the Greenland ice sheet under different climates. *Journal of Glaciology*, **37**, 149-157.

Levac, E., A. de Vernal, and W.J. Blake, 2001: Sea-surface conditions in northernmost Baffin Bay during the Holocene—Palynological evidence. *Journal of Quaternary Science*, **16**, 353-363.

Levy, L.B., D.S. Kaufman, and A. Werner, 2003: Holocene glacier fluctuations, Waskey Lake, northeastern Ahklun Mountains, southwestern Alaska. *The Holocene*, **14**, 185-193.

Ling, F. and T.J. Zhang, 2007: Modeled impacts of changes in tundra snow thickness on ground thermal regime and heat flow to the atmosphere in northernmost Alaska. *Global and Planetary Change*, **57(3-4)**, 235-246.

Lisiecki, L.E. and M.E. Raymo, 2005: A Pliocene-Pleistocene stack of 57 globally distributed benthic $\delta^{18}O$ records. *Paleoceanography*, **20**, PA1003, doi:10.1029/2004PA001071.

Loutre, M.F., 2003: Clues from MIS 11 to predict the future climate—A modeling point of view. *Earth and Planetary Science Letters*, **212(1-2)**, 213-224.

Lozhkin, A.V. and P.M. Anderson, 1995: The last interglaciation of northeast Siberia. *Quaternary Research*, **43**, 147-158.

Lozhkin, A.V. and P.M. Anderson, 1996: A late Quaternary pollen record from Elikchan 4 Lake, northeast Siberia. *Geology of the Pacific Ocean*, **12**, 609-616.

Lozhkin, A.V., P.M. Anderson, W.R. Eisner, L.G. Ravako, D.M. Hopkins, L.B. Brubaker, P.A. Colinvaux, and M.C. Miller, 1993: Late Quaternary lacustrine pollen records from southwestern Beringia. *Quaternary Research*, **9**, 314-324.

Lozhkin, A.V., P.M. Anderson, T.V. Matrosova, and P.S. Minyuk, 2007: The pollen record from El'gygytgyn Lake—Implications for vegetation and climate histories of northern Chukotka since the late middle Pleistocene. *Journal of Paleolimnology*, **37(1)**, 135-153.

Lubinski, D.J., S.L. Forman, and G.H. Miller, 1999: Holocene glacier and climate fluctuations on Franz Josef Land, Arctic Russia, 80°N. *Quaternary Science Reviews*, **18**, 85-108.

Luckman, B.H., 2007: Dendroclimatology. In: *Encyclopedia of Quaternary Science* [Elias, S. (ed.)]. **1**, 465-475.

MacDonald, G.J., 1990: Role of methane clathrates in past and future climates. *Climatic Change*, **16(3)**, 247-281.

MacDonald, G.M., T. Edwards, K. Moser, and R. Pienitz, 1993: Rapid response of treeline vegetation and lakes to past climate warming. *Nature*, **361**, 243-246.

MacDonald, G.M., B.R. Gervais, J.A. Snyder, G.A. Tarasov, and O.K. Borisova, 2000a: Radiocarbon dated *Pinus sylvestris* L. wood from beyond treeline on the Kola Peninsula, Russia. *The Holocene*, **10**, 143-147.

MacDonald, G.M., K.V. Kremenetski, and D.W. Beilman, D.W., 2007: Climate change and the northern Russian treeline zone. *Philosophical Transactions of the Royal Society B*, doi:10.1098/rstb.2007.2200.

MacDonald, G.M., A.A. Velichko, C.V. Kremenetski, O.K. Borisova, A.A. Goleva, A.A. Andreev, L.C. Cwynar, R.T. Riding, S.L. Forman, T.W.D. Edwards, R. Aravena, D. Hammarlund, J.M. Szeicz, and V.N. Gattaulin, 2000b: Holocene treeline history and climate change across northern Eurasia. *Quaternary Research*, **53**, 302-311.

Mahowald, N.M., D.R. Muhs, S. Levis, P.J Rasch, M. Yoshioka, C.S. Zender, and C. Luo, 2006: Change in atmospheric mineral aerosols in response to climate—Last glacial period, preindustrial, modern, and doubled carbon dioxide climates. *Journal of Geophysical Research*, **111**, D10202, doi:10.1029/2005JD006653.

Manabe, S. and R.J. Stouffer, 1980: Sensitivity of a global climate model to an increase of CO_2 in the atmosphere. *Journal of Geophysical Research*, **85(C10)**, 5529-5554.

Mann, M.E., R.S. Bradley, and M.K. Hughes, 1998: Global-scale temperature patterns and climate forcing over the past six centuries. *Nature*, **392**, 779-787.

Mann, D.H., D.M. Peteet, R.E. Reanier, and M.L. Kunz, 2002: Responses of an arctic landscape to Late glacial and early Holocene climatic changes—The importance of moisture. *Quaternary Science Reviews*, **21**, 997-1021, doi: 10.1016/S0277-3791(01)00116-0.

Mann, M.E. and P.D. Jones, 2003: Global surface temperatures over the past two millennia. *Geophysical Research Letters*, **30(15)**, 1820, doi:10.1029/2003GL017814.

Mann, M.E., A.Z. Zhang, M.K. Hughes, R.S. Bradley, S.K. Miller, S. Rutherford, and S. Ni, F., 2008: Proxy-based reconstructions of hemispheric and global surface temperature variations over the past two millennia. *Proceedings of the National Academy of Sciences of the United States of America*, **105(36)**, 13252-13257.

Marchant, D.R., and G.H. Denton, 1996: Miocene and Pliocene paleoclimate of the Dry Valleys region, southern Victoria Land—A geomorphological approach. *Marine Micropaleontology*, **27**, 253-271.

Marincovitch, L., Jr. and A.Y. Gladenkov, 2001: New evidence for the age of Bering Strait. *Quaternary Science Reviews*, **20(1-3)**, 329-335.

Marlowe, I.T., J.C. Green, A.C. Neal, S.C. Brassell, G. Eglinton, P.A. Course, 1984: Long-chain (N-C37-C39) alkenones in the *prymnesiophyceae*—Distribution of alkenones and other lipids and their taxonomic significance. *British Phycological Journal*, **19(3)**, 203-216.

Marotzke, J., 2000: Abrupt climate change and thermohaline circulation—Mechanisms and predictability. *Proceedings of the National Academy of Sciences of the United States of America*, **97(4)**, 1347-1350.

Marshall, S.J. and P.U. Clark, 2002: Basal temperature evolution of North American ice sheets and implications for the 100-kyr cycle. *Geophysical Research Letters*, **29(24)**, 2214 p.

Martinson, D.G. and M. Steele, 2001: Future of the Arctic sea ice cover—Implications of an Antarctic analog. *Geophysical Research Letters*, **28**, 307-310.

Masson-Delmotte, V. et al., 2006: Past and future polar amplification of climate change—Climate model intercomparisons and ice-core constraints. *Climate Dynamics*, **26**, 513-529.

Masson-Delmotte, V., J. Jouzel, A. Landais, M. Stievenard, S.J. Johnsen, J.W.C. White, M.A. Werner, A. Sveinbjornsdottir, and K. Fuhrer, 2005: GRIP deuterium excess reveals rapid and orbital-scale changes in Greenland moisture origin. *Science*, **309**, 118-121.

Mathieu, R., D. Pollard, J.E. Cole, J.W.C. White, R.S. Webb, and S.L. Thompson, 2002: Simulation of stable water isotope variations by the GENESIS GCM for modern conditions. *Journal of Geophysical Research*, **107**, doi:10.1029/2001JD900255.

Matthews, J.V., Jr., C.E. Schweger, and J. Janssens, 1990: The last (Koy-Yukon) interglaciation in the northern Yukon—Evidence from unit 4 at Chijee's Bluff, Bluefish Basin. *Géographie Physique et Quaternaire*, **44**, 341-362.

Maximova, L.N. and V.E. Romanovsky, 1988: A hypothesis of the Holocene permafrost evolution. *Proceedings of the Fifth International Conference on Permafrost*, Norwegian Institute of Technology, Trondheim, Norway, pp. 102-106.

Mayewski, P.A., L.D. Meeker, M.S. Twickler, S.I. Whitlow, Q. Yang, W.B. Lyons, and M. Prentice, 1997: Major features and forcing of high-latitude Northern Hemisphere atmospheric circulation using a 110,000-year-long glaciochemical series. *Journal of Geophysical Research*, **102**, 26345-26366.

McGhee, R., 2004: *The Last Imaginary Place; a Human History of the Arctic World*. Key Porter, Ontario, 296 pp.

McKenna, M.C., 1980. Eocene paleolatitude, climate and mammals of Ellesmere Island. *Paleogeography, Paleoclimatology and Paleoecology*, **30**, 349-362.

McLaughlin, F., E. Carmack, R. Macdonald, A.J. Weaver, and J. Smith, 2002: The Canada Basin, 1989–1995—Upstream events and far-field effects of the Barents Sea. *Journal of Geophysical Research*, **107(C7)**, 3082, doi:10.1029/2001JC000904.

McManus, J.F., 2004: A great grand-daddy of ice cores. *Nature*, **429**, 611-612.

McManus, J.F., G.C. Bond, W.S. Broecker, S. Johnsen, L. Labeyrie, and S. Higgins, 2002: High-resolution climate records from the North Atlantic during the last interglacial. *Nature*, **371**, 326-329, doi:10.1038/371326a0.

Meehl, G.A., T.F. Stocker, W.D. Collins, P. Friedlingstein, A.T. Gaye, J.M. Gregory, A. Kitoh, R. Knutti, J.M. Murphy, A. Noda, S.C.B. Raper, I.G. Watterson, A.J. Weaver and Z.-C. Zhao, 2007: Global Climate Projections. In: *Climate Change 2007: The Physical Science Basis—Contribution of Working Group I to the Fourth Assessment Report of the Intergovernmental Panel on Climate Change* [Solomon, S.,D. Qin, M. Manning, Z. Chen, M. Marquis, K.B. Averyt, M. Tignor, and H.L. Miller (eds.)]. Cambridge University Press, Cambridge and New York, pp. 747-845.

Meeker, L.D. and P.A. Mayewski, 2002: A 1400-year high-resolution record of atmospheric circulation over the North Atlantic and Asia. *The Holocene*, **12**, 257-266.

Meier, M.F., M.B. Dyurgerov, U.K. Rick, S. O'Neel, W.T. Pfeffer, R.S. Anderson, S.P. Anderson, and A.F. Glazovsky, 2007: Glaciers dominate eustatic sea-level rise in the 21st Century. *Science,* **317(5841)**, 1064-1067, doi:10.1126/science.1143906.

Miller, G.H., A.P. Wolfe, J.P. Briner, P.E. Sauer, and A. Nesje, 2005: Holocene glaciation and climate evolution of Baffin Island, Arctic Canada. *Quaternary Science Reviews*, **24**, 1703-1721.

Moberg, A., D.M. Sonechkin, K. Holmgren, N.M. Datsenko, and W. Karlen, 2005: Highly variable northern hemisphere temperatures reconstructed from low- and high-resolution proxy data. *Nature*, **433**, 613-617.

Montoya, M., H. von Storch, and T.J. Crowley, 2000: Climate simulation for 125 kyr BP with a coupled ocean-atmosphere general circulation model. *Journal of Climate,* **13**, 1057-1072.

Moran, K., J. Backman, H. Brinkhuis, S.C. Clemens, T. Cronin, G.R. Dickens, F. Eynaud, J. Gattacceca, M. Jakobsson, R.W. Jordan, M. Kaminski, J. King, N. Koç, A. Krylov, N. Martinez, J. Matthiessen, D. McInroy, T.C. Moore, J. Onodera, M. O'Regan, H. Palike, B. Rea, D. Rio, T. Sakamoto, D.C. Smith, R. Stein, K. St. John, I. Suto, N. Suzuki, K. Takahashi, M. Watanabe, M. Yamamoto, J. Farrell, M. Frank, P. Kubik, W. Jokat, and Y. Kristoffersen, 2006: The Cenozoic palaeoenvironment of the Arctic Ocean. *Nature*, **441**, 601-605.

Morison, J., K. Aagaard, and M. Steele, 2000: Recent environmental changes in the Arctic. *Arctic,* **53(4)**, 359-371.

Muhs, D.R. and J.R. Budahn, 2006: Geochemical evidence for the origin of late Quaternary loess in central Alaska. *Canadian Journal of Earth Science,* **43**, 323-337.

Muller, P.J., G. Kirst, G. Ruhland, I. von Storch, and A. Rossell-Melé, 1998: Calibration of the alkenone paleotemperature index $U_{37}^{k'}$ based on core-tops from the eastern South Atlantic and the global ocean (60°N–60°S). *Geochimica et Cosmochimica Acta,* **62**, 1757-1772.

National Research Council, 2006: *Surface temperature reconstructions for the last 2,000 years.* National Academies Press, Washington, D.C., 160 pp.

Naurzbaev, M.M., E.A. Vaganov, O.V. Sidorova, and F.H. Schweingruber, 2002: Summer temperatures in eastern Taimyr inferred from a 2427-year late-Holocene tree-ring chronology and earlier floating series. *The Holocene,* **12**, 727-736.

Nelson, R.E. and Carter, L.D. 1991. Preliminary interpretation of vegetation and paleoclimate in northern Alaska during the late Pliocene Colvillian marine transgression. In: *Geologic studies in Alaska* [Bradley, D.C. and A.B. Ford (eds.)]. U.S.Geological Survey Bulletin 1999, pp. 219-222.

Nesje, A., J. Bakke, S.O. Dahl, O. Lie, and J.A. Matthews, 2008: Norwegian mountain glaciers in the past, present and future. *Global and Planetary Change*, **60**, 10-27.

Nesje, A., J.A. Matthews, S.O. Dahl, M.S. Berrisford, and C. Andersson, 2001: Holocene glacier fluctuations of Flatebreen and winter precipitation changes in the Jostedalsbreen region, western Norway, based on glaciolacustrine records. *The Holocene,* **11**, 267-280.

Nørgaard-Pedersen, N., N. Mikkelsen, and Y. Kristoffersen, 2007a: Arctic Ocean record of last two glacial-interglacial cycles off North Greenland/Ellesmere Island—Implications for glacial history. *Marine Geology,* **244(2007)**, 93-108.

Nørgaard-Pedersen, N., N. Mikkelsen, S.J. Lassen, Y. Kristoffersen, and E. Sheldon, 2007b: Reduced sea ice concentrations in the Arctic Ocean during the last interglacial period revealed by sediment cores off northern Greenland. *Paleoceanography,* **22**, PA1218, doi:10.1029/2006PA001283.

Nørgaard-Pedersen, N., R.F. Spielhagen, H. Erlenkeuser, P.M. Grootes, J. Heinemeier, and J. Knies, 2003: The Arctic Ocean during the Last Glacial Maximum—Atlantic and polar domains of surface water mass distribution and ice cover. *Paleoceanography,* **18**, 8-1 to 8-19.

Nørgaard-Pedersen, N., R.F. Spielhagen, J. Thiede, and H. Kassens, 1998: Central Arctic surface ocean environment during the past 80,000 years. *Paleoceanography,* **13**, 193-204.

Nürnberg, D. and R. Tiedemann, 2004: Environmental changes in the Sea of Okhotsk during the last 1.1 million years. *Paleoceanography,* **19**, PA4011.

O'Brien, S.R., P.A. Mayewski, L.D. Meeker, D.A. Meese, M.S. Twickler, and S.I. Whitlow, 1995: Complexity of Holocene climate as reconstructed from a Greenland ice core. *Science,* **270**, 1962-1964.

Obata, A., 2007: Climate-carbon cycle model response to freshwater discharge into the North Atlantic. *Journal of Climate,* **20(24),** 5962-5976.

Oerlemans, J., 2001: *Glaciers and Climatic Change.* A.A. Balkema Publishers, Lisse, 148 pp.

Ogilvie, A.E.J. and T. Jónsson, 2001: "Little Ice Age" Research—A Perspective from Iceland. *Climate Change*, **48**, 9-52.

Ohkouchi, N., T.I. Eglinton, L.D. Keigwin, and J.M. Hayes, 2002: Spatial and temporal offsets between proxy records in a sediment drift. *Science*, **298**, 1224-1227.

Oswald, W.W., L.B. Brubaker, and P.M. Anderson, 1999, Late Quaternary vegetational history of the Howard Pass area, northwestern Alaska. *Canadian Journal of Botany*, **77(4)**, 570-581.

Oswald, W.W., L.B. Brubaker, F.S. Hu, and G.W. Kling, 2003: Holocene pollen records from the central Arctic Foothills, northern Alaska—Testing the role of substrate in the response of tundra to climate change. *Journal of Ecology*, **91**, 1034-1048.

Otto-Bliesner, B.L., C.D. Hewitt, T.M. Marchitto, E. Brady, A. Abe-Ouchi, M. Crucifix, S. Murakami, and S.L. Weber, 2007: Last Glacial Maximum ocean thermohaline circulation—PMIP2 model inter-comparisons and data constraints. *Geophysical Research Letters,* **34**, L12706, doi:10.1029/2007GL029475.

Otto-Bliesner, B.L., S.J. Marshall, J.T. Overpeck, G.H. Miller, A. Hu, and CAPE Last Interglacial Project members, 2006: Simulating Arctic climate warmth and icefield retreat in the Last Interglaciation. *Science,* **311**, 1751-1753, doi:10.1126/science.1120808

Overpeck, J., K. Hughen, D. Hardy, R. Bradley, R. Case, M. Douglas, B. Finney, K. Gajewski, C. Jacoby, A. Jennings, S. Lamoureux, A. Lasca, G. MacDonald, J. Moore, M. Retelle, S. Smith, A. Wolfe, and G. Zielinski, 1997: Arctic environmental change of the last four centuries. *Science,* **278**, 1251-1256.

Overpeck, Jonathan T., Bette L. Otto-Bliesner, Gifford H. Miller, Daniel R. Muhs, Richard B. Alley, Jeffrey T. Kiehl, 2006: Paleoclimatic evidence for future ice-sheet instability and rapid sea-level rise. *Science,* **311**, 1747-1750, doi:10.1126/science.1115159

Peixoto, J.P. and A.H. Oort, 1992: *Physics of Climate*. American Institute of Physics, New York, 520 pp.

Peterson, B.J., R.M. Holmes, J.W. McClelland, C.J. Vorosmarty, R.B. Lammers, A.I. Shiklomanov, I.A. Shiklomanov, and S. Rahmstorf, 2002: Increasing river discharge to the Arctic Ocean. *Science,* **298**, 2171-2173.

Pienitz, R., M.S.V. Douglas, and J.P. Smol (eds.), 2004: *Long-Term Environmental Change in Arctic and Antarctic Lakes.* Springer, Dordrecht, Germany, 579 pp.

Pienitz, R. and J.P. Smol, 1993: Diatom assemblages and their relationship to environmental variables in lakes from the boreal forest-tundra ecotone near Yellowknife, Northwest Territories, Canada. *Hydrobiologia*, **269**, 391-404.

Pienitz, R., J.P. Smol, W.M. Last, P.R. Leavitt, and B.F. Cumming, 2000: Multi-proxy Holocene palaeoclimatic record from a saline lake in the Canadian Subarctic. *The Holocene,* **10(6)**, 673-686, doi:10.1191/09596830094935.

Pierrehumbert, R.T., H. Brogniez, and R. Roca, 2007: On the relative humidity of the atmosphere. In: *The Global Circulation of the Atmosphere* [Schneider, T. and A. Sobel (eds.)]. Princeton University Press, Princeton, New Jersey, pp. 143-185.

Pinot, S., G. Ramstein, S.P. Harrison, I.C. Prentice, J. Guiot, M. Stute. and S. Joussaume, 1999: Tropical paleoclimates of the Last Glacial Maximum—Comparison of Paleoclimate Modelling Intercomparison Project (PMIP) simulations and paleodata. *Climate Dynamics*, **15**, 857-874.

Pisaric, M.F J., G.M. MacDonald, A.A Velichko, and L.C. Cwynar, 2001: The late-glacial and post-glacial vegetation history of the northwestern limits of Beringia, from pollen, stomates and tree stump evidence. *Quaternary Science Reviews*, **20**, 235-245.

Pollard, D. and S.L. Thompson, 1997: Climate and ice-sheet mass balance at the Last Glacial Maximum from the GENESIS Version 2 global climate model. *Quaternary Science Reviews*, **16**, 841-864.

Polyak, L., W.B. Curry, D.A. Darby, J. Bischof, and T.M. Cronin, 2004: Contrasting glacial/interglacial regimes in the western Arctic Ocean as exemplified by a sedimentary record from the Mendeleev Ridge. *Palaeogeography, Palaeoclimatology, Palaeoecology*, **203**, 73-93.

Porter, S.C. and G.H. Denton, 1967: Chronology of neoglaciation. American Journal of *Science,* **165**, 177-210.

Poulsen, C.J., E.J. Barron, W.H. Peterson, and P.A. Wilson, 1999: A reinterpretation of mid-Cretaceous shallow marine temperatures through model-data comparison. *Paleoceanography,* **14(6),** 679-697.

Prahl, F.G., G.J. de Lange, M. Lyle, and M.A. Sparrow, 1989: Post-depositional stability of long-chain alkenones under contrasting redox conditions. *Nature*, **341**, 434-437.

Prahl, F.G., L.A. Muelhausen, and D.L. Zahnle, 1988: Further evaluation of long-chain alkenones as indicators of paleoceanographic conditions. *Geochimica et Cosmochimica Acta,* **52**, 2303-2310.

Prentice, I.C. and T. Webb III, 1998: BIOME 6000—Reconstructing global mid-Holocene vegetation patterns from palaeoecological records. *Journal of Biogeography,* **25**, 997-1005.

Rahmstorf, S., 1996: On the freshwater forcing and transport of the Atlantic thermohaline circulation. *Climate Dynamics,* **12**, 799-811.

Rahmstorf, S., 2002: Ocean circulation and climate during the past 120,000 years. *Nature*, **419**, 207-214.

Rasmussen, S.O., K.K. Andersen, A.M. Svensson, J.P. Steffensen, B.M. Vinther, H.B. Clausen, M.L. Siggaard-Andersen, S.J. Johnsen, L.B. Larsen, D. Dahl-Jensen, M. Bigler, R. Röthlisberger, H. Fischer, K. Goto-Azuma, M.E. Hansson, and U. Ruth, 2006: A new Greenland ice core chronology for the last glacial termination. *Journal of Geophysical Research,* **111**, D061202, doi:10.1029/2005JD006079.

Raymo, M.E., 1994: The initiation of northen hemisphere glaciation. *Annual Review of Earth and Planetary Sciences*, **22**, 353-383, doi:10.1146/annurev.ea.22.050194.002033.

Raymo, M.E., 1997: The timing of major climate terminations. *Paleoceanography*, **12**, 577-585.

Raymo, M.E., B. Grant, M. Horowitz, and G.H. Rau, 1996: Mid-Pliocene warmth—Stronger greenhouse and stronger conveyer. *Marine Micropaleontology*, **27**, 313-326.

Raymo, M.E., L.E. Lisiecki, and K.H. Nisancioglu, 2006: Plio-Pleistocene ice volume, Antarctic climate, and the global $\delta^{18}O$ record. *Science*, **313**, 492-495.

Raymo, M.E., D.W. Oppo, and W. Curry, 1997: The mid-Pleistocene climate transition: A deep sea carbon isotopic perspective: *Paleoceanography*, **12**, 546-559.

Renssen, H., E. Driesschaert, M.F. Loutre, and T. Fichefet, 2006: On the importance of initial conditions for simulations of the Mid-Holocene climate. *Climate of the Past*, **2**, 91-97.

Renssen, H., H. Goosse, T. Fichefet, V. Brovkin, E. Dresschaert, and F. Wolk, 2005: Simulating the Holocene climate evolution at northern high latitudes using a coupled atmosphere-sea ice-ocean-vegetation model. *Climate Dynamics*, **24**, 23-43.

Reyes, A.V., G.C. Wiles, D.J. Smith, D.J. Barclay, S. Allen, S. Jackson, S. Larocque, S. Laxton, D. Lewis, P.E. Calkin, and J.J. Clauge, 2006: Expansion of alpine glaciers in Pacific North America in the first millennium A.D. *Geology*, **34**, 57-60.

Rignot, E. and R.H. Thomas, 2002: Mass balance of polar ice sheets. *Science*, **297**, 1502-1506.

Rind, D., 1987: Components of the Ice Age circulation. *Journal of Geophysical Research*, **92**, 4241-4281.

Rind, D., 2006: Water-vapor feedback. In: *Frontiers of Climate Modeling* [Kiehl, J.T. and V. Ramanathan (eds.)]. Cambridge University Press, Cambridge and New York, pp. 251-284, ISBN-13 978-0-521-79132-8.

Ritchie, J.C., L.C. Cwynar, and R.W. Spear, 1983: Evidence from northwest Canada for an early Holocene Milankovitch thermal maximum. *Nature*, **305**, 126-128.

Rivers, A.R. and A.H. Lynch, 2004: On the influence of land cover on early Holocene climate in northern latitudes. *Journal of Geophysical Research—Atmospheres*, **109(D21)**, D21114.

Roe, G.H. and M.R. Allen, 1999: A comparison of competing explanations for the 100,000-yr ice age cycle. *Geophysical Research Letters*, **26(15)**, 2259-2262.

Rosell-Melé, A., G. Eglinton, U. Pflaumann, and M. Sarnthein, 1995: Atlantic core top calibration of the Uk37 index as a sea-surface temperature indicator. *Geochimica et Cosmochimica Acta*, **59**, 3099-3107.

Royer, D.L., 2006: CO_2-forced climate thresholds during the Phanerozoic. *Geochimica et Cosmochimica Acta*, **70(23)**, 5665-5675.

Royer, D.L., R.A. Berner, and J. Park, 2007: Climate sensitivity constrained by CO_2 concentrations over the past 420 million years. *Nature*, **446**, 530-532.

Ruddiman, W.F., 2003: Insolation, ice sheets and greenhouse gases. *Quaternary Science Reviews*, **22**, 1597.

Ruddiman, W.F., 2006: Ice-driven CO_2 feedback on ice volume. *Climate of the Past*, **2**, 43-55.

Ruddiman, W.F., N.J. Shackleton, and A. McIntyre, 1986: North Atlantic sea-surface temperatures for the last 1.1 million years. In: *North Atlantic Paleoceanography* [Summerhayes, C.P. and N.J. Shackleton (eds.)]. Geological Society, London, Special Publication, **21**, 155-173.

Rudels, B., L.G. Anderson, E.P. Jones, and G. Kattner, 1996: Formation and evolution of the surface mixed layer and halocline of the Arctic Ocean. *Journal of Geophysical Research*, **101**, 8807-8821.

Rühland, K., A. Priesnitz, and J.P. Smol, 2003: Evidence for recent environmental changes in 50 lakes the across Canadian Arctic treeline. *Arctic, AntArctic, and Alpine Research*, **35**, 110-123.

Salvigsen, O., S.L. Forman, and G.H. Miller, 1992: Thermophilous mollusks on Svalbard during the Holocene and their paleoclimatic implications. *Polar Research*, **11**, 1-10.

Salzmann, U., A.M. Haywood, D.J. Lunt, P.J. Valdes, and D.J. Hill, 2008: A new global biome reconstruction and data-model comparison for the Middle Pliocene. *Global Ecology and Biogeography*, **17**, 432-447.

Sauer, P.E., G.H. Miller, and J.T. Overpeck, 2001: Oxygen isotope ratios of organic matter in Arctic lakes as a paleoclimate proxy—Field and laboratory investigations. *Journal of Paleolimnology*, **25**, 43-64.

Schauer, U., B. Rudels, E.P. Jones, L.G. Anderson, R.D. Muench, G. Björk, J.H. Swift, V. Ivanov, and A.-M. Larsson, 2002: Confluence and redistribution of Atlantic water in the Nansen, Amundsen and Makarov basins. *Annals of Geophysics*, **20**, 257- 273.

Schindler, D.W. and J.P. Smol, 2006: Cumulative effects of climate warming and other human activities on freshwaters of Arctic and subarctic North America. *Ambio*, **35**, 160-168.

Schlosser, P., B. Ekwurzel, S. Khatiwala, B. Newton, W. Maslowski, and S. Pfirman, 2000: Tracer studies of the Arctic freshwater budget. In: *The Freshwater Budget of the Arctic Ocean* [Lewis, E.L. (ed.)]. Kluwer Academic Publishers, Norwell, MA, pp. 453-478.

Schlosser, P., R. Newton, B. Ekwurzel, S. Khatiwala, R. Mortlock, and R. Fairbanks, 2002: Decrease of river runoff in the upper waters of the Eurasian Basin, Arctic Ocean, between 1991 and 1996—Evidence from $\delta^{18}O$ data. *Geophysical Research Letters,* **29(9)**, 1289, doi:10.1029/ 2001GL013135.

Schmidt, G.A., A.N. LeGrande, and G. Hoffman, 2007: Water isotope expressions of intrinsic and forced variability in a coupled ocean-atmosphere model. *Journal Geophyscial Research,* **112**, D10103, doi:10.1029/2006JD007781.

Schmittner, A., 2005: Decline of the marine ecosystem caused by a reduction in the Atlantic overturning circulation. *Nature,* **434**, 628-633.

Schneider, K.B. and B. Faro, 1975: Effects of sea ice on sea otters (*Enhydra lutris*). *Journal of Mammalogy,* **56**, 91-101.

Schouten, S., E.C. Hopmans, and J.S.S. Damsté, 2004: The effect of maturity and depositional redox conditions on archaeal tetraether lipid palaeothermometry. *Organic Geochemistry,* **35(5)**, 567-571.

Schrag, D.P., J.F. Adkins, K. McIntyre, J.L. Alexander, D.A. Hodell, C.D. Charles, and J.F. McManus, 2002: The oxygen isotopic composition of seawater during the Last Glacial Maximum. *Quaternary Science Reviews,* **21(1-3)**, 331-342.

Schulz, H., U. von Rad, and H. Erlenkeuser, 1998: Correlation between Arabian Sea and Greenland climate oscillations of the past 110,000 years: *Nature,* **393**, 54-57.

Scott, D.B., P.J. Mudie, V. Baki, K.D. MacKinnon, and F.E. Cole, 1989: Biostratigraphy and late Cenozoic paleoceanography of the Arctic Ocean—Foraminiferal, lithostratigraphic, and isotopic evidence. *Geological Society of America Bulletin,* **101**, 260-277.

Seager, R., D.S. Battisti, J. Yin, N. Gordon, N. Naik, A.C. Clement, and M.A. Cane, 2002: Is the Gulf Stream responsible for Europe's mild winters? *Quarterly Journal of the Royal Meteorological Society,* **128(586)**, 2563-2586.

Seppä, H., 1996: Post-glacial dynamics of vegetation and tree-lines in the far north of Fennoscandia. *Fennia,* **174**, 1-96.

Seppä, H. and H.J.B. Birks, 2001: July mean temperature and annual precipitation trends during the Holocene in the Fennoscandian tree-line area—Pollen-based climate reconstructions. *The Holocene,* **11**, 527-539.

Seppä, H. and H.J.B. Birks, 2002: Holocene climate reconstructions from the Fennoscandian tree-line area based on pollen data from Toskaljavri. *Quaternary Research,* **57**, 191-199.

Seppä, H., H.J.B. Birks, A. Odland, A. Poska, and S. Veski, 2004: A modern pollen-climate calibration set from northern Europe—Developing and testing a tool for palaeoclimatological reconstructions. *Journal of Biogeography,* **31**, 251-267.

Seppä, H., L.C. Cwynar, and G.M. MacDonald, 2003: Post-glacial vegetation reconstruction and a possible 8200 cal. Yr BP event from the low arctic of continental Nunavut, Canada. *Journal of Quaternary Science,* **18**, 621-629.

Seppä, H. and D. Hammarlund, 2000: Pollen-stratigraphical evidence of Holocene hydrological change in northern Fennoscandia supported by independent isotopic data. *Journal of Paleolimnology,* **24(1)**, 69-79.

Serreze, M.C., A.P. Barrett, A.G. Slater, M. Steele, J. Zhang, and K.E. Trenberth, 2007a: The large-scale energy budget of the Arctic. *Journal of Geophysical Research,* **112**, D11122, doi:10.1029/2006JD008230.

Serreze, M.C., A.P. Barrett, A.G. Slater, R.A. Woodgate, K. Aagaard, R.B. Lammers, M. Steele, R. Moritz, M. Meredith, and C.M. Lee, 2006: The large-scale freshwater cycle of the Arctic. *Journal of Geophysical Research—Oceans,* **111(C11)**, C11010.

Serreze, M.C. and J.A. Francis, 2006: The Arctic amplification debate. *Climatic Change,* **76**, 241-264.

Severinghaus, J.P. and E.J. Brook, 1999: Abrupt climate change at the end of the last glacial period inferred from trapped air in polar ice. *Science,* **286**, 930-934.

Severinghaus, J.P., T. Sowers, E.J. Brook, R.B. Alley, and M.L. Bender. 1998: Timing of abrupt climate change at the end of the Younger Dryas interval from thermally fractionated gases in polar ice. *Nature,* **391(6663)**, 141-146.

Sewall, J.O. and L.C. Sloan, 2001: Equable Paleogene climates—The result of a stable, positive Arctic Oscillation? *Geophysical Research Letters,* **28(19)**, 3693-3695.

Sewall, J.O. and L.C. Sloan, 2004: Disappearing Arctic sea ice reduces available water in the American west. *Geophysical Research Letters,* **31**, doi:10.1029/2003GL019133.

Shackleton, N.J., 1967: Oxygen isotope analyses and paleotemperatures reassessed. *Nature,* **215**, 15-17.

Shackleton, N.J., 1974: Attainment of isotopic equilibrium between ocean water and the benthonic foraminifera genus *Uvigerina*—Isotopic changes in the ocean during the last glacial. *Collège International du CNRS (Centre national de la Récherche scientifique),* **219**, 203-209.

Shellito, C. J., L.C. Sloan, and M. Huber, 2003: Climate model sensitivity to atmospheric CO_2 levels in the early-middle Paleogene. *Palaeogeography, Palaeoclimatology, Palaeoecology,* **193**, 113-123.

Shemesh, A., Rosqvist G., Riett-Shati, M., Rubensdotter, L., Bigler, C., Yam, R. and Karlén, W., 2001. Holocene climate change in Swedish Lapland inferred from an oxygen-isotope record of lacustrine biogenic silica. *The Holocene* **11(4)**, 447-454, doi:10.1191/095968301678302887..

Shindell, D.T., G.A. Schmidt, M.E. Mann, D. Rind, and A. Waple, 2001: Solar forcing of regional climate change during the Maunder Minimum. *Science,* **294**, 2149-2152.

Shuman, C.A., R.B. Alley, S. Anandakrishnan, J.W.C. White, P.M. Grootes and C.R. Stearns, 1995: Temperature and accumulation at the Greenland Summit—Comparison of high-resolution isotope profiles and satellite passive microwave brightness temperature trends. *Journal of Geophysical Research,* **100(D5)**, 9165-9177.

Siegenthaler, U., T.F. Stocker, E. Monnin, E. Lüthi, J. Schwander, B. Stauffer, D. Raynaud, J.M. Barnola, H. Fischer, V. Masson-Delmotte, and J. Jouzel, 2005: Stable carbon cycle–climate relationship during the late Pleistocene. *Science,* **310**, 1313-1317, doi:10.1126/science.1120130

Sloan, L.C. and E.J. Barron, 1992: A comparison of Eocene climate model results to quantified paleoclimatic interpretations. *Palaeogeography, Palaeoclimatology, Palaeoecology,* **93(3-4)**, 183-202

Sloan, L., T.J. Crowley, and D. Pollard, 1996: Modeling of middle Pliocene climate with the NCAR GENESIS general circulation model. *Marine Micropaleontology,* **27**, 51-61.

Sloan, L.C. and D. Pollard, 1998: Polar stratospheric clouds—A high latitude warming mechanism in an ancient greenhouse world. *Geophysical Research Letters,* **25(18)**, 3517-3520.

Sluijs, A., U. Rohl, S. Schouten, H.J. Brumsack, F Sangiorgi, J.S.S. Damsté, and H. Brinkhuis, 2008: Arctic late Paleocene-early Eocene paleoenvironments with special emphasis on the Paleocene-Eocene thermal maximum (Lomonosov Ridge, Integrated Ocean Drilling Program Expedition 302). *Paleoceanography,* **23(1)**, PA1S11.

Sluijs, A., S. Schouten, M. Pagani, M. Woltering, H. Brinkhuis, J.S.S. Damsté, G.R. Dickens, M. Huber, G.J. Reichart, R. Stein, J. Matthiessen, L.J. Lourens, N. Pedentchouk, J. Backman and K. Moran, 2006: Subtropical arctic ocean temperatures during the Palaeocene/Eocene thermal maximum. *Nature,* **441**, 610-613.

Smith, L.C., G.M. MacDonald, A.A. Velichko, D.W. Beilman, O.K. Borisova, K.E. Frey, K.V. Kremenetski, and Y. Sheng, 2004: Siberian peatlands a net carbon sink and global methane source since the early Holocene. *Science,* **303(5656)**, 353-356.

Smol, J.P., 1988: Paleoclimate proxy data from freshwater Arctic diatoms. *Internationale Vereinigung für Limnologie,* **23**, 837-844.

Smol, J.P., 2008: *Pollution of Lakes and Rivers—A Paleoenvironmental Perspective.* Blackwell Publishing, Oxford, U.K., 2nd ed., 280 pp.

Smol, J.P. and B.F. Cumming, 2000: Tracking long-term changes in climate using algal indicators in lake sediments. *Journal of Phycology,* **36**, 986-1011.

Smol, J.P. and M.S.V. Douglas, 2007a: From controversy to consensus—Making the case for recent climatic change in the Arctic using lake sediments. *Frontiers in Ecology and the Environment,* **5**, 466-474.

Smol, J.P. and M.S.V. Douglas, 2007b: Crossing the final ecological threshold in high Arctic ponds. *Proceedings of the National Academy of Sciences of the United States of America,* **104**, 12395-12397.

Solovieva, N., P.E. Tarasov, and G.M. MacDonald, 2005: Quantitative reconstruction of Holocene climate from the Chuna Lake pollen record, Kola Peninsula, northwest Russia. *The Holocene,* **15**, 141-148.

Sorvari, S. and A. Korhola, 1998: Recent diatom assemblage changes in subarctic Lake Saanajarvi, NW Finnish Lapland, and their paleoenvironmental implications. *Journal of Paleolimnology,* **20(3)**, 205-215.

Sorvari, S, A. Korhola, and R. Thompson, 2002: Lake diatom response to recent Arctic warming in Finnish Lapland. *Global Change Biology,* **8**, 171-181.

Sowers, T., Bender, M., Raynaud, D., 1989: Elemental and isotopic composition of occluded O_2 and N_2 in polar ice. *Journal of Geophysical Research—Atmospheres,* **94(D4)**, 5137-5150.

Spencer, M.K., R.B. Alley and J.J. Fitzpatrick, 2006: Developing a bubble number-density paleoclimatic indicator for glacier ice. *Journal of Glaciology,* **52(178)**, 358-364.

Spielhagen, R.F., K-H Baumann, H. Erlenkeuser, N.R. Nowaczyk, N. Nørgaard-Pedersen, C. Vogt, and D. Weiel, 2004: Arctic Ocean deep-sea record of northern Eurasian ice sheet history. *Quaternary Science Reviews,* **23(11-13)**, 1455-1483.

Spielhagen, R.F., G. Bonani, A. Eisenhauer, M. Frank, T. Frederichs, H. Kassens, P.W. Kubik, N. Nørgaard-Pedersen, N.R. Nowaczyk, A. Mangini, S. Schäper, R. Stein, J. Thiede, R. Tiedemann, and M. Wahsner, 1997: Arctic Ocean evidence for late Quaternary initiation of northern Eurasian ice sheets. *Geology,* **25(9)**, 783-786.

Spielhagen, R.F. and H. Erlenkeuser, 1994: Stable oxygen and carbon isotopes in planktic foraminifers from Arctic Ocean surface sediments—Reflection of the low salinity surface water layer. *Marine Geology,* **119(3/4)**, 227-250.

Spielhagen, R.F., H. Erlenkeuser, and C. Siegert, 2005: History of freshwater runoff across the Laptev Sea (Arctic) during the last deglaciation. *Global and Planetary Change,* **48(1-3)**, 187-207.

Stanton-Frazee, C., D.A. Warnke, K. Venz, D.A. Hodell, 1999: The stage 11 problem as seen from ODP site 982, In: *Marine Oxygen Isotope Stage 11 and Associated Terrestrial Records* [Poore, R.Z., L. Burckle, A. Droxler, W.E. McNulty (eds.)]. U.S. Geological Survey Open-File Report 99–312, 75 pp.

Steele, M. and T. Boyd, 1998: Retreat of the cold halocline layer in the Arctic Ocean. *Journal of Geophysical Research,* **103**, 10419-10435.

Stein, R., S.I. Nam, C. Schubert, C. Vogt, D. Fütterer, and J. Heinemeier, 1994: The last deglaciation event in the eastern central Arctic Ocean. *Science,* **264**, 692-696.

Stötter, J., M. Wastl, C. Caseldine, and T. Häberle, 1999: Holocene palaeoclimatic reconstruction in Northern Iceland—Approaches and results. *Quaternary Science Reviews*, **18**, 457-474.

Stroeve, J., M. Serreze, S. Drobot, S. Gearheard, M. Holland, J. Maslanik, W. Meier, and T. Scambos, 2008: Arctic Sea Ice Extent Plummets in 2007. *EOS, Transactions, American Geophysical Union*, **89(2)**, 13-14.

Sturm, M., T. Douglas, C. Racine, and G.E. Liston, 2005: Changing snow and shrub conditions affect albedo with global implications. *Journal of Geophysical Research,* **110**, G01004, doi:10.1029/2005JG000013.

Svendsen, J.I. and J. Mangerud, 1997: Holocene glacial and climatic variations on Spitsbergen, Svalbard. *The Holocene,* **7**, 45-57.

Teece, M.A., J.M. Getliff, J.W. Leftley, R.J. Parkes, and J.R. Maxwell, 1998: Microbial degradation of the marine *prymnesiophyte Emiliania huxleyi* under oxic and anoxic conditions as a model for early diagenesis—Long chain alkadienes, alkenones and alkyl alkenoates. *Organic Geochemistry*, **29**, 863-880.

Thomas, D.J., J.C. Zachos, T.J. Bralower, E. Thomas, and S. Bohaty, 2002: Warming the fuel for the fire—Evidence for the thermal dissociation of methane hydrate during the Paleocene-Eocene thermal maximum. *Geology,* **30(12)**, 1067-1070.

Thomsen, C., D.E. Schulz-Bull, G. Petrick, and J.C. Duinker, 1998: Seasonal variability of the long-chain alkenone flux and the effect on the Uk_{37} index in the Norwegian Sea. *Organic Geochemistry*, **28**, 311-323.

Toggweiler, J. R., 2008: Origin of the 100,000-year timescale in Antarctic temperatures and atmospheric CO_2. *Paleoceanography,* **23**, PA2211, doi:10.1029/2006PA001405.

Troitsky, S.L., 1964: Osnoviye zakonomernosti izmeneniya sostava fauny po razrezam morskikh meshmorennykh sloev usteniseyskoy vpadiny i nishne-pechorskoy depressii. *Akademia NAUK SSSR, Trudy instituta geologii i geofiziki*, **9**, 48-65 (in Russian).

Vassiljev, J., 1998: The simulated response of lakes to changes in annual and seasonal precipitation—Implication for Holocene lake-level changes in northern Europe. *Climate Dynamics*, **14**, 791-801.

Vassiljev, J., S.P. Harrison, and J. Guiot, 1998: Simulating the Holocene lake-level record of Lake Bysjon, southern Sweden. *Quaternary Research*, **49**, 62-71.

Vavrus, S., Harrison, S.P., 2003: The impact of sea-ice dynamics on the Arctic climate system. *Climate Dynamics,* **20**, 741-757.

Velichko, A.A., A.A. Andreev, and V.A. Klimanov, 1997: Climate and vegetation dynamics in the tundra and forest zone during the Late Glacial and Holocene. *Quaternary International,* **41/42**, 71-96.

Velichko, A.A. and V.P. Nechaev (eds.), 2005: Cenozoic climatic and environmental changes in Russia. [Wright, H.E. Jr., T.A. Blyakharchuk, A.A. Velichko, and Olga Borisova (eds. of English version)]. *The Geological Society of America Special Paper*, **382**, 226 pp.

Vinther, B.M., H.B. Clausen, S.J. Johnsen, S.O. Rasmussen, K.K. Andersen, S.L. Buchardt, D. Dahl-Jensen, I.K. Seierstad, M.L. Siggaard-Andersen, J.P. Steffensen, A. Svensson, J. Olsen, and J. Heinemeier, 2006: A synchronized dating of three Greenland ice cores throughout the Holocene. *Journal of Geophysical Research,* **111**, D13102, doi:13110.11029/12005JD006921.

Vörösmarty, C.J., L.D. Hinzman, B.J. Peterson, D.H. Bromwich, L.C. Hamilton, J.Morison, V.E. Romanovsky, M.Sturm, and R.S. Webb, 2001: *The Hydrologic Cycle and Its Role in Arctic and Global Environmental Change—A Rationale and Strategy for Synthesis Study*. Fairbanks, Alaska: Arctic Research Consortium of the United States, 84 pp.

Vörösmarty, C., L. Hinzman, and J. Pundsack, 2008: Introduction to special section on changes in the Arctic freshwater system—Identification, attribution, and impacts at local and global scales. *Journal of Geophysical Research—Biogeosciences,* **113(G1)**, G01S91.

Walter, K.M., M. Edwards S.A. Zimov G. Grosse F.S. Chapin, III, 2007: Thermokarst lakes as a source of atmospheric CH4 during the last deglaciation. *Science,* **318(5850)**, 633-636.

Walter, K.M., S.A. Zimov, J.P. Chanton, D. Verbyla, and F.S. Chapin, III, 2006: Methane bubbling from Siberian thaw lakes as a positive feedback to climate warming. *Nature*, **443**, 71-75.

Wang, Y.J., H. Cheng, R.L. Edwards, Z.S. An, J.Y. Wu, C.-C. Shen, and J.A. Doralé, 2001: A high-resolution absolute-dated late Pleistocene monsoon record from Hulu Cave, China. *Science,* **294**, 2345-2348.

Weckström, J., A. Korhola, P. Erästö, and L. Holmström, 2006: Temperature patterns over the past eight centuries in northern Fennoscandia inferred from sedimentary diatoms. *Quaternary Research*, **66**, 78-86.

Weijers, J.W.H., S. Schouten, O.C. Spaargaren, and J.S.S. Damsté, 2006: Occurrence and distribution of tetraether membrane lipids in solid—Implications for the use of the TEX_{86} proxy and the BIT index. *Organic Geochemistry*, **37(12)**, 1680-1693.

Weijers, J.W.H., S. Schouten, A. Sluijs, H. Brinkhuis, and J.S.S. Damsté, 2007: Warm arctic continents during the Palaeocene-Eocene thermal maximum. *Earth and Planetary Science Letters*, **261(1-2)**, 230-238.

Werner, M., U. Mikolajewicz, M. Heimann, and G. Hoffmann, 2000: Borehole versus isotope temperatures on Greenland—Seasonality does matter. *Geophysical Research Letters,* **27**, 723-726.

Whitlock, C. and M.R. Dawson, 1990: Pollen and vertebrates of the early Neogene Haughton formation, Devon Island, Arctic Canada. *Arctic,* **43(4)**, 324-330.

Wiles, G.C., D.J. Barclay, P.E. Calkin, and T.V. Lowell, 2008: Century to millennial-scale temperature variations for the last two thousand years indicated from glacial geologic records of Southern Alaska. *Global and Planetary Change*, **60**, 15-125.

Williams, C.J., A.H. Johnson, B.A. LePage, D.R. Vann and T. Sweda, 2003: Reconstruction of Tertiary Metasequoia forests II. Structure, biomass and productivity of Eocene floodplain forests in the Canadian Arctic. *Paleobiology*, **29**, 271-292.

Wohlfahrt, J., S.P. Harrison, and P. Braconnot, 2004: Synergistic feedbacks between ocean and vegetation on mid- and high-latitude climates during the Holocene. *Climate Dynamics,* **22**, 223-238.

Wohlfarth, B., G. Lemdahl, S. Olsson, T. Persson, I. Snowball, J. Ising, and V. Jones, 1995: Early Holocene environment on Bjornoya (Svalbard) inferred from multidisciplinary lake sediment studies. *Polar Research*, **14**, 253-275.

Wolfe, B.B., T.W.D. Edwards, R.Aravena, S.L. Forman, B.G. Warner, A.A. Velichko, and G. MacDonald, 2000: Holocene paleohydrology and paleoclimate at treeline, north-central Russia, inferred from oxygen isotope records in lake sediment cellulose. *Quaternary Research*, **53**, 319-329, doi:10.1006/qres.2000.2124.

Wolfe, B.B., T.W.D. Edwards, R. Aravena, and G.M. MacDonald, 1996: Rapid Holocene hydrologic change along boreal tree-line revealed by $\delta^{13}C$ and $\delta^{18}O$ in organic lake sediments, Northwest Territories, Canada. *Journal of Paleolimnology* **15(2)**, 171-181, doi:10.1007/BF00196779.

Wooller, M.J., D. Francis, M.L. Fogel, G.H. Miller, I.R. Walker, and A.P. Wolfe, 2004: Quantitative paleotemperature estimates from $\delta^{18}O$ of chironomid head capsules preserved in arctic lake sediments. *Journal of Paleolimnology,* **31(3)**, 267-274.

Wuchter, C., S. Schouten, M.J.L. Coolen, and J.S.S. Damsté, 2004: Temperature-dependent variation in the distribution of tetraether membrane lipids of marine Crenarchaeota—Implications for TEX_{86} paleothermometry. *Paleoceanography,* **19(4)**, PA4028

Zachos, J.C., Dickens, G.R., Zeebe, R.E., 2008: An early Cenozoic perspective on greenhouse warming and carbon-cycle dynamics. *Nature*, **451(7176)**, 279-283.

Zachos, Z., P. Pagani, L. Sloan, E. Thomas, and K. Billups, 2001: Trends, rhythms, and aberrations in global climate 65 Ma to present. *Science,* **292**, 686-693.

Zazula, G.D., D.G. Froese, C.E. Schweger, R.W. Mathewes, A.B. Beaudoin, A.M. Telka, C.R. Harington, and J.A. Westgate, 2003: Ice-age steppe vegetation in east Beringia—Tiny plant fossils indicate how this frozen region once sustained huge herds of mammals. *Nature*, **423**, 603.

Chapter 4. Past Rates of Climate Change in the Arctic

Abbott, M.B. and T.W.J. Stafford, 1996: Radiocarbon geochemistry of modern and ancient arctic lake systems, Baffin Island, Canada. *Quaternary Research*, **45,** 300-311.

Alley, R.B., 2000: The Younger Dryas cold interval as viewed from central *Greenland. Quaternary Science Reviews,* **19**, 213-226.

Alley, R.B., 2007: Wally was right—Predictive ability of the North Atlantic "conveyor belt" hypothesis for abrupt climate change. *Annual Review of Earth and Planetary Sciences*, **35**, 241-272.

Alley, R.B. and A.M. Ágústsdóttir, 2005: The 8k event—Cause and consequences of a major Holocene abrupt climate change. *Quaternary Science Reviews*, **24**, 1123-1149.

Alley, R.B., J. Marotzke, W.D. Nordhaus, J.T. Overpeck, D.M. Peteet, R.A. Pielke, Jr., R.T. Pierrehumbert, P.B. Rhines, T.F. Stocker, L.D. Talley, and J.M. Wallace, 2003: Abrupt climate change. *Science*, **299**, 2005-2010.

Alley, R.B., D.A. Meese, C.A. Shuman, A.J. Gow, K.C. Taylor, P.M. Grootes, J.W.C. White, M. Ram, E.D. Waddington, P.A. Mayewski, and G.A. Zielinski, 1993: Abrupt increase in snow accumulation at the end of the Younger Dryas event. *Nature*, **362**, 527-529.

Andersen, K.K., A. Svensson, S.O. Rasmussen, J.P. Steffensen, S.J. Johnsen, M. Bigler, R. Röthlisberger, U. Ruth, M.-L. Siggaard-Andersen, D. Dahl-Jensen, B.M. Vinther, and H.B. Clausen, 2006: The Greenland ice core chronology 2005, 15–42 ka. Part 1: Constructing the time scale. *Quaternary Science Reviews*, **25(23-24)**, 3246-3257.

Bard, E. and G. Delaygue, 2008: Comment on Are there connections between the Earth's magnetic field and climate? *Earth and Planetary Science Letters*, **265**, 302-307.

Bard, E., G.M. Raisbeck, F. Yiou, and J. Jouzel, 2007: Comment—Solar activity during the last 1000 yr inferred from radionuclide records. *Quaternary Science Reviews*, **26**, 2301-2304.

Bice, K.L., C.R. Scotese, D. Seidov, and E.J. Barron, 2000: Quantifying the role of geographic change in Cenozoic ocean heat transport using uncoupled atmosphere and ocean models. *Palaeogeography, Palaeoclimatology, Palaeoecology*, **161**, 295-310.

Biscaye, P.E., F.E. Grousset, M. Revel, S. VanderGaast, G.A. Zielinski, A. Vaars, and G. Kukla, 1997: Asian provenance of glacial dust (stage 2) in the Greenland Ice Sheet Project 2 Ice Core, Summit, Greenland. *Journal of Geophysical Research, Oceans*, **102(C12)**, 26765-26781.

Björck, S., O. Bennike, P. Rose, C.S. Andreson, S. Bohncke, E. Kaas, and D. Conley, 2002: Anomalously mild Younger Dryas summer conditions in southern Greenland. *Geology*, **30,** 427-430.

Björck, S., N. Koç, and G. Skot, 2003: Consistently large marine reservoir ages in the Norwegian Sea during the last deglaciation. *Quaternary Science Reviews*, **22**, 429-435.

Bond, G., B. Kromer, J. Beer, R. Muscheler, M.N. Evans, W. Showers, S. Hoffman, R. Lotti-Bond, I. Hajdas, and G. Bonani, 2001: Persistent solar influence on North Atlantic climate during the Holocene. *Science*, **294**, 2130-2136.

Bradley, R.S., 1999: *Paleoclimatology: Reconstructing Climate of the Quaternary.* Academic Press, San Diego, 613 pp.

Braun, H., L.M. Christ, S. Rahmstorf, A. Ganopolski, A. Mangini, C. Kubatzki, K. Roth, and B. Kromer, 2005: Possible solar origin of the 1,470-year glacial climate cycle demonstrated in a coupled model. *Nature*, **438**, 208-211.

Brigham-Grette, J., M. Melles, P. Minyuk, and Scientific Party, 2007: Overview and significance of a 250 ka paleoclimate record from El'gygytgyn Crater Lake, NE Russia. *Journal of Paleolimnology*, **37(1)**, 1-16.

Briner, J.P., Y. Axford, S.L. Forman, G.H. Miller, and A.P. Wolfe, 2007: Multiple generations of interglacial lake sediment preserved beneath the Laurentide Ice Sheet. *Geology*, **35,** 887-890.

Briner, J.P., N. Michelutti, D.R. Francis, G.H. Miller, Y. Axford, M.J. Wooller, and A.P. Wolfe, 2006: A multi-proxy lacustrine record of Holocene climate change on northeastern Baffin Island, Arctic Canada. *Quaternary Research*, **65**, 431-442.

Broecker, W.S., 2000: Was a change in thermohaline circulation responsible for the Little Ice Age? *Proceedings of the National Academy of Sciences of the United States of America*, **97**, 1339-1342

Broecker, W.S., D.M. Peteet, and D. Rind, 1985: Does the ocean-atmosphere system have more than one stable mode of operation? *Nature*, **315**, 21-26.

Brubaker, L.B., P.M. Anderson, M.E. Edwards, and A.V. Lozhkin, 2005: Beringia as a glacial refugium for boreal trees and shrubs—New perspectives from mapped pollen data. *Journal of Biogeography*, **32**, 833-848.

Cai, B.G., R.L. Edwards, H. Cheng, M. Tan, X. Wang, and T.S. Liu, 2008: A dry episode during the Younger Dryas and centennial-scale weak monsoon events during the early Holocene—A high-resolution stalagmite record from southeast of the Loess Plateau, China. *Geophysical Research Letters*, **35(2),** Article L02705.

Chapman, W.L. and J.E. Walsh, 2007: Simulations of Arctic temperature and pressure by global coupled models. *Journal of Climate*, **20**, 609-632.

Cronin, T.M., 1999: *Principles of Paleoclimatology.* Columbia University Press, New York, 560 pp.

Cuffey, K.M. and E.J. Brook, 2000: Ice sheets and the ice-core record of climate change. In: *Earth system science—From biogeochemical cycles to global change* [Jacobson, M.C., R.J. Charlson, H. Rodhe, and G.H. Orians (eds.)]. Academic Press, Burlington, MA, 459-497.

Cuffey, K.M. and G.D. Clow, 1997: Temperature, accumulation, and ice sheet elevation in central Greenland through the last deglacial transition. *Journal of Geophysical Research*, **102(C12)**, 26383-26396.

Cuffey, K.M., G.D. Clow, R.B. Alley, M. Stuiver, E.D. Waddington, and R.W. Saltus, 1995: Large arctic temperature change at the Wisconsin-Holocene glacial transition. *Science*, **270(5235)**, 455-458.

Cuffey, K.M. and E.J. Steig, 1998: Isotopic diffusion in polar firn—Implications for interpretation of seasonal climate parameters in ice-core records, with emphasis on central Greenland. *Journal of Glaciology*, **44(147**), 273-284.

D'Andrea, W.J. and Y. Huang, 2005: Long-chain alkenones in Greenland lake sediments—Low $\delta^{13}C$ values and exceptional abundance. *Organic Geochemistry*, **36**, 1234-1241.

Dansgaard, W., H.B. Clausen, N. Gundestrup, C.U. Hammer, S.J. Johnsen, P.M. Kristinsdottir, and N. Reeh, 1982: A new Greenland deep ice core. *Science*, **218(4579**), 1273-1277.

Dansgaard, W., S.J. Johnsen, H.B. Clausen, D. Dahl-Jensen, N.S. Gundestrup, C.U. Hammer, C.S. Hvidberg, J.P. Steffensen, A.E. Sveinbjörnsdottir, J. Jouzel, and G. Bond, 1993: Evidence for general instability of past climate from a 250-kyr ice-core record. *Nature*, **364(6434**), 218-220.

Dansgaard, W., S.J. Johnsen, H.B. Clausen, D. Dahl-Jensen, N. Gundestrup, C.U. Hammer, and H. Oeschger, 1984: North Atlantic climatic oscillations revealed by deep Greenland ice cores. In: *Climatic Processes and Climate Sensitivity* [Hansen, J.E. and T. Takahashi (eds.)]. Geophysical Monograph Series 29, American Geophysical Union, Washington, D.C., pp. 288-298.

Dansgaard, W., S.J. Johnsen, H.B. Clausen, and C.C. Langway, Jr., 1971: Climatic record revealed by the Camp Century ice core. In: *The Late Cenozoic Glacial Ages* [Turekian, K.K. (ed.)]. Yale University Press, New Bedford, CT., pp. 37-56.

Dansgaard, W., S.J. Johnsen, J. Møller, and C.C. Langway, Jr., 1969: One thousand centuries of climatic record from Camp Century on the Greenland Ice Sheet. *Science*, **166(3903)**, 377-381.

Dansgaard, W., J.W.C. White, and S.J. Johnsen, 1989: The abrupt termination of the Younger Dryas climate event. *Nature*, **339**, 532-534.

Denton, G.H., R.B. Alley, G.C. Comer, and W.S. Broecker, 2005: The role of seasonality in abrupt climate change. *Quaternary Science Reviews*, **24**, 1159-1182.

Delworth, T.L. and T.R. Knutson, 2000: Simulation of early 20th century global warming. *Science*, **287**, 2246 pp.

Donnadieu, Y., R. Pierrehumbert, R. Jacob, and F. Fluteau, 2006: Modelling the primary control of paleogeography on Cretaceous climate. *Earth and Planetary Science Letters*, **248**, 426-437.

Easterling, D.R., T.R. Karl, E.H. Mason, P.Y. Hughes, and D.P. Bowman, 1996: *United States Historical Climatology Network (U.S. HCN) Monthly Temperature and Precipitation Data.* ORNL/CDIAC-87, NDP-019/R3. Carbon Dioxide Information Analysis Center, Oak Ridge National Laboratory, U.S. Department of Energy, Oak Ridge, TN.

Eiriksson, J., G. Larsen, K.L. Knudsen, J. Heinemeier, and L.A. Simonarson, 2004: Marine reservoir age variability and water mass distribution in the Iceland Sea. *Quaternary Science Reviews*, **23**, 2247-2268.

Ellison, C.R.W., M.R. Chapman, and I.R. Hall, 2006: Surface and deep ocean interactions during the cold climate event 8200 years ago. *Science*, **312**, 1929-1932.

Francus, P., R. Bradley, M. Abbott, F. Keimig, and W. Patridge, 2002: Paleoclimate studies of minerogenic sediments using annually resolved textural parameters. *Geophysical Research Letters*, **29**, 59-1 to 59-4.

Funder, S. and L. Hansen, 1996: The Greenland Ice Sheet—A model for its culmination and decay during and after the last glacial maximum. *Bulletin of the Geological Society of Denmark*, **42**, 137-152.

Grootes, P.M. and M. Stuiver, 1997: Oxygen 18/16 variability in Greenland snow and ice with 10^{-3} to 10^{5} year time resolution. *Journal of Geophysical Research—Oceans*, **102(C12)**, 26455-26470.

Hajdas, I., G. Bonani, P. Boden, D.M. Peteet, and D.H. Mann, 1998: Cold reversal on Kodiak Island, Alaska, correlated with the European Younger Dryas by using variations of atmospheric C-14 content. *Geology*, **26(11),** 1047-1050.

Harder, S.L., D.T. Shindell, G.A. Schmidt, and E.J. Brook, 2007: A global climate model study of CH_4 emissions during the Holocene and glacial-interglacial transitions constrained by ice core data. *Global Biogeochemical Cycles*, **21**, GB1011.

Hegerl, G.C., F.W. Zwiers, P. Braconnot, N.P. Gillett, Y. Luo, J.A. Marengo Orsini, N. Nicholls, J.E. Penner, and P.A. Stott, 2007: Understanding and attributing climate change. In: *Climate Change 2007—The Physical Science Basis. Contribution of Working Group I to the Fourth Assessment Report of the Intergovernmental Panel on Climate Change* [Solomon, S., D. Qin, M. Manning, Z. Chen, M. Marquis, K.B. Averyt, M. Tignor, and H.L. Miller (eds.)]. Cambridge University Press, Cambridge and New York, pp. 663-745.

Hu, F.S., J.I. Hedges, E.S. Gorden, and L.B. Brubaker, 1999a: Lignin biomarkers and pollen in postglacial sediments of an Alaskan lake. *Geochimica et Cosmochimica Acta*, **63**, 1421-1430.

Hu, F.S., D.M. Nelson, G.H. Clarke, K.M. Ruhland, Y. Huang, D.S. Kaufman, and J.P. Smol, 2006: Abrupt climatic events during the last glacial-interglacial transition in Alaska. *Geophysical Research Letters*, **33**, L18708, doi:10.1029/2006GL027261.

Hu, F.S., D. Slawinski, H.E.J. Wright, E. Ito, R.G. Johnson, K.R. Kelts, R.F. McEwan, and A. Boedigheimer, 1999b: Abrupt changes in North American climate during early Holocene times. *Nature*, **400**, 437-440.

Huang, Y., B. Shuman, Y. Wang, and T. Webb III, 2004: Hydrogen isotope ratios of individual lipids in lake sediments as novel tracers of climatic and environmental change—A surface sediment test. *Journal Paleolimnology*, **31**, 363-375.

Huber, C., M. Leuenberger, R. Spahni, J. Flückiger, J. Schwander, T.F. Stocker, S. Johnsen, A. Landais, and J. Jouzel, 2006: Isotope calibrated Greenland temperature record over Marine Isotope Stage 3 and its relation to CH_4. *Earth and Planetary Science Letters*, **243(3-4)**, 504-519.

Hughen, H.A., J.R. Southon, S.J. Lehman, and J.T. Overpeck, 2000: Synchronous radiocarbon and climate shifts during the last deglaciation. *Science*, **290**, 1951-1954.

Hughen, K.A., M.G.L. Baillie, E. Bard, A. Bayliss, J.W. Beck, C. Bertrand, P.G. Blackwell, C.E. Buck, G. Burr, K.B. Cutler, P.E. Damon, R.L. Edwards, R.G. Fairbanks, M. Friedrich, T.P. Guilderson, B. Kromer, F.G. McCormac, S. Manning, C. Bronk Ramsey, P.J. Reimer, R.W. Reimer, S. Remmele, J.R. Southon, M. Stuiver, S. Talamo, F.W. Taylor, J. van der Plicht, and C.E. Weyhenmeyer, 2004a: Marine04 marine radiocarbon age calibration, 0-26 Cal Kyr BP. *Radiocarbon*, **46**, 1059-1086.

Hughen, K., S. Lehman, J. Southon, J. Overpeck, O. Marchal, C. Herring, and J. Turnbull, 2004b: ^{14}C Activity and global carbon cycle changes over the past 50,000 years. *Science*, **303**, 202-207.

Hughen, K., J. Overpeck, R.F. Anderson, and K.M. Williams, 1996: The potential for palaeoclimate records from varved Arctic lake sediments—Baffin Island, eastern Canadian Arctic. In: *Lacustrine Environments. Palaeoclimatology and Palaeoceanography from Laminated Sediments* [Kemp, A.E.S. (ed.)]. Geological Society, London, Special Publications 116, pp. 57-71.

Imbrie, J., A. Berger, E.A. Boyle, S.C. Clemens, A. Duffy, W.R. Howard, G. Kukla, J. Kutzbach, D.G. Martinson, A. McIntyre, A.C. Mix, B. Molfino, J.J. Morley, L.C. Peterson, N.G. Pisias, W.L. Prell, M.E. Raymo, N.J. Shackleton, and J.R. Toggweiler, 1993: On the structure and origin of major glaciation cycles. 2. The 100,000-year cycle. *Paleoceanography,* **8(6),** 699-735.

Jansen, E., J. Overpeck, K.R. Briffa, J.-C. Duplessy, F. Joos, V. Masson-Delmotte, D. Olago, B. Otto-Bliesner, W.R. Peltier, S. Rahmstorf, R. Ramesh, D. Raynaud, D. Rind, O. Solomina, R. Villalba, and D. Zhang, 2007. Palaeoclimate. In: *Climate Change 2007—The Physical Science Basis. Contribution of Working Group I to the Fourth Assessment Report of the Intergovernmental Panel on Climate Change* [Solomon, S., D. Qin, M. Manning, Z. Chen, M. Marquis, K.B. Averyt, M. Tignor and H.L. Miller (eds.)]. Cambridge University Press, Cambridge and New York, pp. 434-497.

Jennings, A.E., M. Hald, M. Smith, J.T. Andrews, 2006: Freshwater forcing from the Greenland Ice Sheet during the Younger Dryas—Evidence from southeastern Greenland shelf cores. *Quaternary Science Reviews,* **25**, 282-298.

Johnsen, S.J., 1977: Stable isotope homogenization of polar firn and ice. In: *Isotopes and Impurities in Snow and Ice.* Proceedings of International Union of Geodesy and Geophysics symposium XVI, General Assembly, Grenoble, France, August and September 1975, 210-219. IAHS-AISH Publication 118, Washington, D.C.

Johnsen, S.J., H.B. Clausen, K.M. Cuffey, G. Hoffmann, J. Schwander, and T. Creyts, 2000: Diffusion of stable isotopes in polar firn and ice—The isotope effect in firn diffusion. In: *Physics of Ice Core Records* [Hondoh, T. (ed.)]. Hokkaido University Press, Sapporo, pp. 121-140.

Johnsen, S.J., H.B. Clausen, W. Dansgaard, K. Fuhrer, N. Gundestrup, C.U. Hammer, P. Iversen, J.P. Steffensen, J. Jouzel, and B. Stauffer, 1992: Irregular glacial interstadials recorded in a new Greenland ice core. *Nature*, **359(6393)**, 311-313.

Johnsen, S., D. Dahl-Jensen, W. Dansgaard, and N. Gundestrup, 1995: Greenland palaeotemperatures derived from GRIP bore hole temperature and ice core isotope profiles. *Tellus B*, **47(5)**, 624-629.

Johnsen, S.J., W. Dansgaard, H.B. Clausen, and C.C. Langway, Jr., 1972: Oxygen isotope profiles through the Antarctic and Greenland Ice Sheets. *Nature*, **235(5339)**, 429-434.

Jouzel, J., R.B. Alley, K.M. Cuffey, W. Dansgaard, P. Grootes, G. Hoffmann, S.J. Johnsen, R.D. Koster, D. Peel, C.A. Shuman, M. Stievenard, M. Stuiver, and J. White, 1997: Validity of the temperature reconstruction from water isotopes in ice cores. *Journal of Geophysical Research*, **102(C12)**, 26471-26487.

Kaufman, D.S., T.A. Ager, N.J. Anderson, P.M. Anderson, J.T. Andrews, P.J. Bartlein, L.B. Brubaker, L.L. Coats, L.C. Cwynar, M.L. Duvall, A.S. Dyke, M.E. Edwards, W.R. Eisner, K. Gajewski, A. Geirsdóttir, F.S. Hu, A.E. Jennings, M.R. Kaplan, M.W. Kerwin, A.V. Lozhkin, G.M. MacDonald, G.H. Miller, C.J. Mock, W.W. Oswald, B.L. Otto-Bliesner, D.F. Porinchu, K. Rühland, J.P. Smol, E.J. Steig, and B.B. Wolfe, 2004: Holocene thermal maximum in the western Arctic (0–180°W). *Quaternary Science Reviews*, **23**, 529-560.

Keller, K. and D. McInerney, 2007. The dynamics of learning about a climate threshold. *Climate Dynamics*, **30**, 321-332, doi:10.1007/s00382-007-0290-5.

Kristjansdottir, G.B., 2005: *Holocene climatic and environmental changes on the Iceland shelf—$\delta^{18}O$, Mg/Ca, and tephrachronology of core MD99-2269.* PhD dissertation, Department of Geological Sciences, University of Colorado, Boulder, 423 pp.

Kristjansdottir, G.B., J.S. Stoner, A.E. Jennings, J.T. Andrews, and K. Gronvold, 2007: Geochemistry of Holocene cryptotephras from the North Iceland Shelf (MD99-2269)—Intercalibration with radiocarbon and paleomagnetic chronostratigraphies. *The Holocene*, **17(2)**, 155-176.

Landais, A., J.M. Barnola, V. Masson-Delmotte, J. Jouzel, J. Chappellaz, N. Caillon, C. Huber, M. Leuenberger, and S.J. Johnsen, 2004: A continuous record of temperature evolution over a sequence of Dansgaard-Oeschger events during Marine Isotopic Stage 4 (76 to 62 kyr BP). *Geophysical Research Letters*, **31**, L22211, doi:22210.21029/22004GL021193.

Lang, C., M. Leuenberger, J. Schwander, and S. Johnsen, 1999: 16°C rapid temperature variation in central Greenland 70,000 years ago. *Science*, **286(5441)**, 934-937.

Lauritzen, S.-E. and J. Lundberg, 2004: Isotope Stage 11, the Super-Interglacial, from a north Norwegian speleothem. In: *Studies of Cave Sediments: Physical and Chemical Records of Paleoclimate* [Sasowsky, I.D. and J. Mylroie (eds.)]. Kluwer Academic, New York, pp. 257-272.

Le Treut, H., R. Somerville, U. Cubasch, Y. Ding, C. Mauritzen, A. Mokssit, T. Peterson, and M. Prather, 2007: Historical overview of climate change. In: *Climate Change 2007—The Physical Science Basis. Contribution of Working Group I to the Fourth Assessment Report of the Intergovernmental Panel on Climate Change* [Solomon, S., D. Qin, M. Manning, Z. Chen, M. Marquis, K.B. Averyt, M. Tignor and H.L. Miller (eds.)]. Cambridge University Press, Cambridge and New York, pp. 93-127.

Leuenberger, M., C. Lang, and J. Schwander, 1999: Delta ^{15}N measurements as a calibration tool for the paleothermometer and gas-ice age differences—A case study for the 8200 B.P. event on GRIP ice. *Journal of Geophysical Research*, **104(D18)**, 22163-22170.

Lorenz, E.N., 1963: Deterministic nonperiodic flow. *Journal of the Atmospheric Sciences,* **20(2)**, 130-141.

Lozhkin, A.V. and P.M. Anderson, 1995: The last interglaciation of northeast Siberia. *Quaternary Research*, **43,** 147-158.

McConnell, J.R., R. Edwards, G.L. Kok, M.G. Flanner, C.S. Zender, E.S. Saltzman, J.R. Banta, D.R. Pasteris, M.M. Carter, and J.D.W. Kahl, 2007: 20th-century industrial black carbon emissions altered arctic climate forcing. *Science*, **317**, 1381-1384.

Meehl, G.A., T.F. Stocker, W.D. Collins, P. Friedlingstein, A.T. Gaye, J.M. Gregory, A. Kitoh, R. Knutti, J.M. Murphy, A. Noda, S.C.B. Raper, I.G. Watterson, A.J. Weaver, and Z.-C. Zhao, 2007: Global climate projections. In: *Climate Change 2007: The Physical Science Basis. Contribution of Working Group I to the Fourth Assessment Report of the Intergovernmental Panel on Climate Change* [Solomon, S., D. Qin, M. Manning, Z. Chen, M. Marquis, K.B. Averyt, M. Tignor and H.L. Miller (eds.)]. Cambridge University Press, Cambridge and New York, pp. 747-845.

Meese, D.A., A.J. Gow, R.B. Alley, G.A. Zielinski, P.M. Grootes, M. Ram, K.C. Taylor, P.A. Mayewski, and J.F. Bolzan, 1997: The Greenland Ice Sheet Project 2 depth-age scale—Methods and results. *Journal of Geophysical Research*, **102(C12)**, 26411-26423.

Miller, G.H., W.N. Mode, A.P. Wolfe, P.E. Sauer, O. Bennike, S.L. Forman, S.K. Short, and T.W.J. Stafford, 1999: Stratified interglacial lacustrine sediments from Baffin Island, Arctic Canada—Chronology and paleoenvironmental implications. *Quaternary Science Reviews*, **18,** 789-810.

Miller, G.H., A.P. Wolfe, J.P. Briner, P.E. Sauer, and A. Nesje, 2005: Holocene glaciation and climate evolution of Baffin Island, Arctic Canada. *Quaternary Science Reviews*, **24,** 1703-1721.

Monnin, E., A. Indermuhle, A. Dallenbach, J. Fluckiger, B. Stauffer, T.F. Stocker, D. Raynaud, and J.M. Barnola, 2001: Atmospheric CO_2 concentrations over the last glacial termination. *Science*, **291(5501),** 112-114.

Moran, K., J. Backman, H. Brinkhuis, S.C. Clemens, T. Cronin, G.R. Dickens, F. Eynaud, J. Gattacceca, M. Jakobsson, R.W. Jordan, M. Kaminski, J. King, N. Koç, A. Krylov, N. Martinez, J. Matthiessen, D. McInroy, T.C. Moore, J. Onodera, M. O'Regan, H. Palike, B. Rea, D. Rio, T. Sakamoto, D.C. Smith, R. Stein, K. St. John, I. Suto, N. Suzuki, K. Takahashi, M. Watanabe, M. Yamamoto, J. Farrell, M. Frank, P. Kubik, W. Jokat and Y. Kristoffersen, 2006: The Cenozoic palaeoenvironment of the Arctic Ocean. *Nature*, **441**, 601-605.

Muscheler, R., F. Joos, J. Beer, S.A. Miller, M. Vonmoos, and I. Snowball, 2007: Solar activity during the last 1000 yr inferred from radionuclide records. *Quaternary Science Reviews*, **26**, 82-97.

National Research Council, 2002: *Abrupt Climate Change, Inevitable Surprises.* National Academies Press, Washington, D.C., 230 pp.

Oeschger, H., J. Beer, U. Siegenthaler, B. Stauffer, W. Dansgaard, and C.C. Langway, Jr., 1984: Late glacial climate history from ice cores. In: *Climate Processes and Climate Sensitivity,* [Hansen, J.E. and T. Takahashi, (eds.)]. Geophysical Monograph Series 29, American Geophysical Union, Washington, D.C., pp. 299-306.

Ojala, A.E.K. and M. Tiljander, 2003: Testing the fidelity of sediment chronology—Comparison of varve and paleomagnetic results from Holocene lake sediments from central Finland. *Quaternary Science Reviews*, **22,** 1787-1803.

Oswald, W.W., P.M. Anderson, T.A. Brown, L.B. Brubaker, F.S. Hu, A.V. Lozhkin, W. Tinner, and P. Kaltenrieder, 2005: Effects of sample mass and macrofossil type on radiocarbon dating of arctic and boreal lake sediments. *The Holocene*, **15,** 758-767.

Overpeck, J., K. Hughen, D. Hardy, R. Bradley, R. Case, M. Douglas, B. Finney, K. Gajewski, G. Jacoby, A. Jennings, S. Lamoureux, A. Lasca, G. MacDonald, J. Moore, M. Retelle, S. Smith, A. Wolfe, and G. Zielinski, 1997: Arctic environmental change of the last four centuries. *Science*, **278,** 1251-1256.

Peteet, D., 1995a: Global Younger Dryas. *Quaternary International*, **28**, 3-104.

Peteet, D.M., 1995b: Global Younger Dryas. Vol. 2. Preface. *Quaternary Science Reviews*, **14**, 811 p.

Rasmussen, S.O., K.K. Andersen, A.M. Svensson, J.P. Steffensen, B.M. Vinther, H.B. Clausen, M.-L. Siggaard-Andersen, S.J. Johnsen, L.B. Larsen, D. Dahl-Jensen, M. Bigler, R. Röthlisberger, H. Fischer, K. Goto-Azuma, M.E. Hansson, and U. Ruth, 2006: A new Greenland ice core chronology for the last glacial termination. *Journal of Geophysical Research*, **111**, D06102, doi:06110.01029/02005JD006079.

Rasmussen, S.O., I.K. Seierstad, K.K. Andersen, M. Bigler, D. Dahl-Jensen, and S.J. Johnsen, 2007: Synchronization of the NGRIP, GRIP, and GISP2 ice cores across MIS 2 and palaeoclimatic implications. *Quaternary Science Review*, **27**, 18-28.

Reeh, N., 1985: Greenland ice-sheet mass balance and sea-level change. In: *Glaciers, Ice Sheets and Sea Level: Effect of a CO_2-Induced Climatic Change*. DOE/ER/60235-1, Department of Energy, Washington, D.C., pp. 155-171.

Rempel, A.W. and J.S. Wettlaufer, 2003: Segregation, transport, and interaction of climate proxies in polycrystalline ice. *Canadian Journal of Physics*, **81(1-2)**, 89-97.

Renssen, H., H. Goosse, and R. Muscheler, 2006: Coupled climate model simulation of Holocene cooling events—Oceanic feedback amplifies solar forcing. *Climate of the Past*, **2**, 79-90.

Royer, D.L., R.A. Berner, and J. Park, 2007: Climate sensitivity constrained by CO_2 concentrations over the past 420 million years. *Nature*, **446**, 530-532.

Ruddiman, W.F. and L.K. Glover, 1975: Subpolar North Atlantic circulation at 9300 yr BP— Faunal evidence. *Quaternary Research*, **5**, 361-389.

Ruddiman, W.F. and A. McIntyre, 1981: The North Atlantic Ocean during the last deglaciation. *Palaeogeography, Palaeoclimatology, Palaeoecology*, **35**, 145-214.

Saarinen, T., 1999: Paleomagnetic dating of late Holocene sediments in Fennoscandia. *Quaternary Science Reviews*, **18**, 889-897.

Sauer, P.E., T.I. Eglinton, J.M. Hayes, A. Schimmelmann, and A.L. Sessions, 2001: Compound-specific D/H ratios of lipid biomarkers from sediments as a proxy for environmental and climatic conditions. *Geochimica et Cosmochimica Acta*, **65**, 213-222.

Serreze, M.C., M.M. Holland, and J. Stroeve, 2007: Perspectives on the Arctic's shrinking sea-ice cover. *Science*, **315(5818)**, 1533, doi:10.1126/science.1139426.

Severinghaus, J.P., T. Sowers, E.J. Brook, R.B. Alley, and M.L. Bender, 1998: Timing of abrupt climate change at the end of the Younger Dryas interval from thermally fractionated gases in polar ice. *Nature*, **391**, 141-146.

Snowball, I. and P. Sandgren, 2004: Geomagnetic field intensity changes in Sweden between 9000 and 450 cal B.P.—Extending the record of archaeomagnetic jerks by means of lake sediments and the pseudo-Thellier technique. *Earth and Planetary Science Letters*, **277,** 361-376.

Snowball, I.F., L. Zillén, and M.-J. Gaillard, 2002. Rapid early Holocene environmental changes in northern Sweden based on studies of two varved lake sediment sequences. *The Holocene,* **12,** 7-16.

Snowball, I., L. Zillén, A. Ojala, T. Saarinen, and P. Sandgren, 2007: FENNOSTACK and FENNORPIS—Varve-dated Holocene palaeomagnetic secular variation and relative palaeointensity stacks for Fennoscandia. *Earth and Planetary Science Letters*, **255,** 106-115.

Steffensen, J.P., H.B. Clausen, C.U. Hammer, M. Legrand, and M. De Angelis, 1997: The chemical composition of cold events within the Eemian section of the Greenland Ice Core Project ice core from Summit, Greenland. *Journal of Geophysical Research*, **102(C12)**, 26747-26754.

Steffensen, J.P. and D. Dahl-Jensen, 1997: Modelling of alterations of the stratigraphy of ionic impurities in very old ice core strata. *Eos, Transactions,of the American Geophysical Union Fall Meeting, San Francisco,* **78**, F7 Poster U21A-22.

Steig, E.J. and R.B. Alley, 2003: Phase relationships between Antarctic and Greenland climate records. *Annals of Glaciology*, **35**, 451-456.

Stocker, T.F. and S.J. Johnsen, 2003: A minimum thermodynamic model for the bipolar seesaw. *Paleoceanography*, **18**, 1087 pp.

Stoner, J.S., A. Jennings, G.B. Kristjansdottir, G. Dunhill, J.T. Andrews, and J. Hardardottir, 2007: A paleomagnetic approach toward refining Holocene radiocarbon-based chronologies—Paleoceanographic records from the North Iceland (MD99-2269) and East Greenland (MD99-2322) margins. *Paleoceanography*, **22,** PA1209, doi:10.1029/2006PA001285.

Stuiver, M., T.F. Braziunas, P.M. Grootes, and G.A. Zielinski, 1997: Is there evidence for solar forcing of climate in the GISP2 oxygen isotope record? *Quaternary Research*, **48**, 259-266.

Stuiver, M., P.J. Reimer, E. Bard, J.W. Beck, K.A. Hughen, B. Kromer, F.G. McCormack, J. van der Plicht, and M. Spurk, 1998: INTCAL98 Radiocarbon age calibration 24,000 cal BP. *Radiocarbon*, **40**, 1041-1083.

Svendson, J.I. and J. Mangerud, 1992: Paleoclimatic inferences from glacial fluctuations on Svalbard during the last 20,000 years. *Climate Dynamics*, **6,** 213-220.

Svensson, A., K.K. Andersen, M. Bigler, H.B. Clausen, D. Dahl-Jensen, S.M. Davies, S.J. Johnsen, R. Muscheler, S.O. Rasmussen, R. Röthlisberger, J.P. Steffensen, and B.M. Vinther, 2006: The Greenland Ice Core chronology 2005, 15-42 ka. Part 2—Comparison to other records. *Quaternary Science Reviews*, **25(23-24)**, 3258-3267.

Svensson, A., S.W. Nielsen, S. Kipfstuhl, S.J. Johnsen, J.P. Steffensen, M. Bigler, U. Ruth, and R. Röthlisberger, 2005: Visual stratigraphy of the North Greenland Ice Core Project (NorthGRIP) ice core during the last glacial period. *Journal of Geophysical Research*, **110**, D02108, doi:02110.01029/02004JD005134.

Tarduno, J.A., D.B. Brinkman, P.R. Renne, R.D. Cottrell, H. Scher, and P. Castillo, 1998: Evidence for extreme climatic warmth from Late Cretaceous Arctic vertebrates. *Science*, **282**, 2241-2244.

Tilling, R.I., L. Topinka, and D.A. Swanson, 1990: *Eruptions of Mount St. Helens: Past, Present, and Future*. U.S. Geological Survey Special Interest Publication, 56 pp.

Trenberth, K.E., J.M. Caron, D.P. Stepaniak, and S. Worley, 2002: Evolution of El Niño-Southern Oscillation and global atmospheric surface temperatures. *Journal of Geophysical Research—Atmospheres*, **107(D7-8),** 4065.

Trenberth, K.E., P.D. Jones, P. Ambenje, R. Bojariu, D. Easterling, A. Klein Tank, D. Parker, F. Rahimzadeh, J.A. Renwick, M. Rusticucci, B. Soden, and P. Zhai, 2007: Observations—Surface and atmospheric climate change. In: *Climate Change 2007: The Physical Science Basis. Contribution of Working Group I to the Fourth Assessment Report of the Intergovernmental Panel on Climate Change* [Solomon, S., D. Qin, M. Manning, Z. Chen, M. Marquis, K.B. Averyt, M. Tignor, and H.L. Miller (eds.)]. Cambridge University Press, Cambridge and New York, pp. 236-336.

Vandermark, D., J.A. Tarduno, and D.B. Brinkman, 2007: A fossil champsosaur population from the High Arctic—Implications for Late Cretaceous paleotemperatures. *Palaeogeography, Palaeoclimatology, Palaeoecology*, **248**, 49-59.

Vinther, B.M., H.B. Clausen, S.J. Johnsen, S.O. Rasmussen, K.K. Andersen, S.L. Buchardt, D. Dahl-Jensen, I.K. Seierstad, M.-L. Siggaard-Andersen, J.P. Steffensen, A.M. Svensson, J. Olsen, and J. Heinemeier, 2006: A synchronized dating of three Greenland ice cores throughout the Holocene. *Journal of Geophysical Research*, **111**, D13102, doi:13110.11029/12005JD006921.

Walker, I.R., A.J. Levesque, L.C. Cwynar, and A.F. Lotter, 1997: An expanded surface-water palaeotemperature inference model for use with fossil midges from eastern Canada. *Journal of Paleolimnology*, **18,** 165-178.

Whillans, I.M. and P.M. Grootes, 1985: Isotopic diffusion in cold snow and firn. *Journal of Geophysical Research*, **90(D2)**, 3910-3918.

White, J.W.C., J.R. Lawrence, and W.S. Broecker, 1994: Modeling and interpreting D/H ratios in tree-rings—A test-case of white-pine in the northeastern United States. *Geochimica et Cosmochimica Acta*, **58(2),** 851-862.

Willemse, N.W., and T.E. Törnqvist, 1999: Holocene century-scale temperature variability from West Greenland lake records. *Geology*, **27,** 580-584.

Woillard, G.M., 1978: Grande Pile peat bog—A continuous pollen record for the last 140,000 years. *Quaternary Research*, **9**, 1-21.

Woillard, G.M., 1979: Abrupt end of the last interglacial s.s. in north-east France. *Nature*, **281(5732)**, 558-562.

Wolfe, A.P., G.H. Miller, C. Olsen, S.L. Forman, P.T. Doran, and S.U. Holmgren, 2005: Geochronology of high-latitude lake sediments. In: *Long-term Environmental Change in Arctic and Antarctic Lakes—Developments in Paleoenvironmental Research* [Pienitz, R., M.S.V. Douglas, and J.P. Smol (eds.)]. Springer, New York, pp. 19-52.

Yang, F.L., M.E. Schlesinger, 2002: On the surface and atmospheric temperature changes following the 1991 Pinatubo volcanic eruption—A GCM study. *Journal of Geophysical Research, Atmospheres*, **107(D8)**, 4073, doi:10.1029/2001JD000373.

Chapter 5. History of the Greenland Ice Sheet

Adrielsson, L. and H. Alexanderson, 2005: Interactions between the Greenland Ice Sheet and the Liverpool Land coastal ice cap during the last two glaciation cycles. *Journal of Quaternary Science*, **20**, 269-283.

Aksu, A. E., 1985: Climatic and oceanographic changes over the past 400,000 years—Evidence from deep-sea cores on Baffin Bay and David Strait. In: *Quaternary Environments—Eastern Canadian Arctic, Baffin Bay and Western Greenland* [Andrews, J.T. (ed.)]. Allen and Unwin, Boston, pp. 181-209.

Alley, R.B., 2007. Wally was right—Predictive ability of the North Atlantic "conveyor belt" hypothesis for abrupt climate change. *Annual Review of Earth and Planetary Sciences*, **35**, 241-272.

Alley, R.B. and A.M. Ágústsdóttir, 2005: The 8k event—Cause and consequences of a major Holocene abrupt climate change. *Quaternary Science Reviews*, **24**, 1123-1149, doi:10.1016/j.quascirev.2004.12.004.

Alley, R.B., A.M. Ágústsdóttir, and P.J. Fawcett, 1999: Ice-core evidence of late-Holocene reduction in north Atlantic ocean heat transport. In: *Mechanisms of Global Climate Change at Millennial Time Scales* [Clark, P.U., R.S. Webb and L.D. Keigwin (eds.)]. Geophysical Monograph 112, American Geophysical Union, Washington, D.C., pp. 301-312.

Alley, R.B., and S. Anandakrishnan, 1995: Variations in melt-layer frequency in the GISP2 ice core—Implications for Holocene summer temperatures in central Greenland. *Annals of Glaciology*, **21**, 64-70.

Alley, R.B., S. Anandakrishnan, T.K. Dupont, B.R. Parizek, and D. Pollard, 2007a: Effect of sedimentation on ice-sheet grounding-line stability. *Science*, **315(5820)**, 1838-1841.

Alley, R.B., S. Anandakrishnan, and P. Jung, 2001: Stochastic resonance in the North Atlantic. *Paleoceanography*, **16**, 190-198.

Alley, R.B., E.J. Brook and S. Anandakrishnan, 2002: A northern lead in the orbital band—North-south phasing of ice-age events. *Quaternary Science Reviews*, **21**, 431-441 (2002).

Alley, R.B., P.U. Clark, P. Huybrechts, and I. Joughin, 2005a: Ice-sheet and sea-level changes. *Science*, **310**, 456-460.

Alley, R.B. and K.M. Cuffey. 2001. Oxygen- and hydrogen-isotopic ratios of water in precipitation—Beyond paleothermometry. In: *Stable Isotope Geochemistry,* [Valle, J.W. and D. Cole (eds.)]. Reviews in Mineralogy and Geochemistry, **43**, 527-553.

Alley, R.B., T.K. Dupont, B.R. Parizek, and S. Anandakrishnan, 2005b: Access of surface meltwater to beds of sub-freezing glaciers—Preliminary insights. *Annals of Glaciology*, **40**, 8-14.

Alley, R.B., R.C. Finkel, K. Nishiizumi, S. Anandakrishnan, C.A. Shuman, G.R. Mershon, G.A. Zielinski, and P.A. Mayewski, 1995a: Changes in continental and sea-salt atmospheric loadings in central Greenland during the most recent deglaciation. *Journal of Glaciology*, **41(139)**, 503-514.

Alley, R.B., A.J. Gow, S.J. Johnsen, J. Kipfstuhl, D.A. Meese, and Th. Thorsteinsson, 1995b: Comparison of deep ice cores. *Nature*, **373**, 393-394.

Alley, R.B., A.J. Gow, D.A. Meese, J.J. Fitzpatrick, E.D. Waddington, and J.F. Bolzan, 1997: Grain-scale processes, folding, and stratigraphic disturbance in the GISP2 ice core. *Journal of Geophysical Research*, **102(C12)**, 26819-26830.

Alley, R.B. and B.R. Koci, 1990: Recent warming in central Greenland? *Annals of Glaciology*, **14**, 6-8.

Alley, R.B. and D.R. MacAyeal, 1994: Ice-rafted Debris Associated with Binge/Purge Oscillations of the Laurentide Ice Sheet. *Paleoceanography*, **9**, 503-511.

Alley, R.B., D.A. Meese, C.A. Shuman, A.J. Gow, K.C. Taylor, P.M. Grootes, J.W.C. White, M. Ram, E.D. Waddington, P.A. Mayewski, and G.A. Zielinski, 1993: Abrupt increase in snow accumulation at the end of the Younger Dryas event. *Nature*, **362**, 527-529.

Alley, R.B., M.K. Spencer, and S. Anandakrishnan, 2007b: Ice-sheet mass balance—Assessment, attribution and prognosis. *Annals of Glaciology*, **46**, 1-7.

Alley, R.B. and I.M. Whillans, 1984: Response of the East Antarctic ice sheet to sea-level rise. *Journal of Geophysical Research*, **89C**, 6487-6493.

Andersen, K.K., A. Svensson, S.J. Johnsen, S.O. Rasmussen, M. Bigler, R. Röthlisberger, U. Ruth, M.L. Siggaard-Andersen, J.P. Steffensen, D. Dahl-Jensen, B.M. Vinther, and H.B. Clausen, 2006: The Greenland ice core chronology 2005, 15-42 ka. Part 1. Constructing the time scale. *Quaternary Science Reviews*, **25**, 3246-3257.

Andresen, C.S., S. Björck, O. Bennike, and G. Bond, 2004: Holocene climate changes in southern Greenland—Evidence from lake sediments. *Journal of Quaternary Science*, **19**, 783-795.

Andrews, J.T., 2008: The role of the Iceland Ice Sheet in sediment delivery to the North Atlantic during the late Quaternary—How important was it? Evidence from the area of Denmark Strait. *Journal of Quaternary Science*, **23**, 3-20.

Andrews, J.T. and G. Dunhill, 2004: Early to mid-Holocene Atlantic water influx and deglacial meltwater events, Beaufort Sea slope, Arctic Ocean. *Quaternary Research*, **61**, 14-21.

Andrews, J.T., T.A. Cooper, A.E. Jennings, A.B. Stein, and H. Erlenkeuser, 1998a: Late Quaternary iceberg-rafted detritus events on the Denmark Strait/Southeast Greenland continental slope (about 65° N)—Related to North Atlantic Heinrich events? *Marine Geology*, **149**, 211-228.

Andrews, J.T., H. Erlenkeuser, K. Tedesco, A. Aksu, and A.J.T. Jull, 1994: Late Quaternary (Stage 2 and 3) meltwater and Heinrich events, NW Labrador Sea. *Quaternary Research*, **41**, 26-34.

Andrews, J.T., A.E. Jennings, T. Cooper, K.M. Williams, and J. Mienert, 1996: Late Quaternary sedimentation along a fjord to shelf (trough) transect, East Greenland (ca. 68°N). In: *Late Quaternary Paleoceanography of North Atlantic Margins* [Andrews, J.T., W. Austin, H. Bergsten, and A.E. Jennings (eds.)]. Geological Society, London, pp. 153-166.

Andrews, J.T., M.E. Kirby, A. Aksu, D.C. Barber, and D. Meese, 1998b: Late Quaternary detrital carbonate (DC-) events in Baffin Bay (67°–74°N)—Do they correlate with and contribute to Heinrich events in the North Atlantic? *Quaternary Science Reviews*, **17**, 1125-1137.

Andrews, J.T., L.M. Smith, R.Preston, T. Cooper, and A.E. Jennings, 1997: Spatial and temporal patterns of iceberg rafting (IRD) along the East Greenland margin, ca. 68°N, over the last 14 cal.ka. *Journal of Quaternary Science*, **12**, 1-13.

Bamber, J.L., R.B. Alley, and I. Joughin, 2007: Rapid Response of modern day ice sheets to external forcing. *Earth and Planetary Science Letters*, **257**, 1-13.

Bamber, J.L., R.L. Layberry, and S.P. Gogineni, 2001: A new ice thickness and bed data set for the Greenland Ice Sheet 1—Measurement, data reduction, and errors. *Journal of Geophysical Research*, **106**, 33773-33780.

Bard, E. and M. Frank, 2006: Climate change and solar variability—What's new under the sun? *Earth and Planetary Science Letters*, **248**, 1-14.

Bauch, H.A., H. Erlenkeuser, J.P. Helmke, and U. Struck, 2000: A paleoclimatic evaluation of marine oxygen isotope stage 11 in the high-northern Atlantic (Nordic seas). *Global and Planetary Change*, **24**, 27-39.

Bender, M.L., R.G. Fairbanks, F.W. Taylor, R.K. Matthews, J.G. Goddard, and W.S. Broecker, 1979: Uranium-series dating of the Pleistocene reef tracts of Barbados, West Indies. *Geological Society of America Bulletin, Part I*, **90**, 577-594.

Bennike, O., S. Bjorck, and K. Lambeck, 2002: Estimates of South Greenland late-glacial ice limits from a new relative sea level curve. *Earth and Planetary Science Letters*, **197**, 171-186.

Bennike, O. and J. Bocher, 1994: Land biotas of the last interglacial-glacial cycle on Jameson Land, East Greenland. *Boreas*, **23**, 479-487.

Berger, A.L., 1978: Long-term variations of caloric insolation resulting from the earth's orbital elements. *Quaternary Research*, **9**, 139-167.

Berger, A. and M.F. Loutre, 1991: Insolation values for the climate of the last 10 million years. *Quaternary Science Reviews*, **10**, 297-317.

Bigg, G.R., 1999: An estimate of the flux of iceberg calving from Greenland. *Arctic, Antarctic, and Alpine Research*, **31**, 174-178.

Björck, S., M. Rundgren, O. Ingolfsson, and S. Funder, 1997: The Preboreal oscillation around the Nordic Seas—Terrestrial and lacustrine responses. *Journal of Quaternary Science*, **12**, 455-465.

Björck, S., O. Bennike, P. Rosen, C.S. Andresen, S. Bohncke, E. Kaas, and D. Conley, 2002: Anomalously mild Younger Dryas summer conditions in southern Greenland. *Geology*, **30**, 427-430.

Blake, W. Jr., H.R. Jackson, and C.G. Currie, 1996: Seafloor evidence for glaciation, northernmost Baffin Bay. *Bulletin of the Geological Survey of Denmark*, **43**, 157-168.

Blunier, T. and E.J. Brook, 2001: Timing of millennial-scale climate change in Antarctica and Greenland during the last glacial period. *Science*, **291**, 109-112.

Bond, G., W. Broecker, S. Johnson, J. McManus, L. Labeyrie, J. Jouzel, and G. Bonani, 1993: Correlation between climate records from North Atlantic sediments and Greenland ice. *Nature*, **365**, 507-508.

Bond, G.C. and R. Lotti, 1995: Iceberg discharges into the North Atlantic on millennial time scales during the last glaciation. *Science*, **267**, 1005-1009.

Box, J.E., D.H. Bromwich, B.A. Veenhuis, L.-S. Bai, J.C. Stroeve, J.C. Rogers, K. Steffen, T. Haran, and S.-H. Wang, 2006: Greenland ice-sheet surface mass balance variability (1988–2004) from calibrated Polar MM5 output. *Journal of Climate*, **19(12)**, 2783-2800.

Bradley, R.S., 1999: *Paleoclimatology—Reconstructing Climate of the Quaternary*. Academic Press, San Diego, 613 pp.

Braun, H., M. Christl, S. Rahmstorf, A. Ganopolski, A. Mangini, C. Kubatzki, K. Roth, and B. Kromer, 2005: Possible solar origin of the 1,470-year glacial climate cycle demonstrated in a coupled model. *Nature*, **438**, 208-211.

Brodersen, K.P. and O. Bennike, 2003: Interglacial Chironomidae (Diptera) from Thule, northwest Greenland—Matching modern analogues to fossil assemblages. *Boreas*, **32**, 560-565.

Broecker, W.S., 1995: *The Glacial World According to Wally*. Eldigio Press, Palisades, NY, 89 pp.

CircumArctic PaleoEnvironments (CAPE)—Last Interglacial Project Members (P. Anderson, O. Bennike, N. Bigelow, J. Brigham-Grette, M. Duvall, M. Edwards, B. Frechette, S. Funder, S. Johnsen, J. Knies, R. Koerner, A. Lozhkin, S. Marshall, J. Matthiessen, G. Macdonald, G. Miller, M. Montoya, D. Muhs, B. Otto-Bliesner, J. Overpeck, N. Reeh, H.P. Sejrup, R. Spielhagen, C. Turner, and A. Velichko), 2006: Last Interglacial Arctic warmth confirms polar amplification of climate change. *Quaternary Science Reviews*, **25**, 1383-1400.

Cazenave, A., 2006: How fast are the ice sheets melting? *Science*, **314**, 1250-1252.

Chappellaz, J., E. Brook, T. Blunier, and B. Malaizé, 1997: CH_4 and $\delta^{18}O$ of O_2 records from Antarctic and Greenland ice—A clue for stratigraphic disturbance in the bottom part of the Greenland Ice Core Project and the Greenland Ice Sheet Project 2 ice cores. *Journal of Geophysical Research*, **102(C12)**, 26547-26557.

Chen, J.H., H.A. Curran, B. White, and G.J. Wasserburg, 1991: Precise chronology of the last interglacial period—^{234}U-^{230}Th data from fossil coral reefs in the Bahamas: *Geological Society of America Bulletin*, **103**, 82-97.

Clarke, G.K.C., N. Lhomme, and S.J. Marshall, 2005: Tracer transport in the Greenland Ice Sheet—Three-dimensional isotopic stratigraphy. *Quaternary Science Reviews*, **24**, 155-171.

Clarke, G.K.C. and S.J. Marshall, 2002: Isotopic balance of the Greenland Ice Sheet—Modeled concentrations of water isotopes from 30,000 BP to present. *Quaternary Science Reviews*, **21**, 419-430.

Cronin, T.M., 1999: *Principles of Paleoclimatology*. Columbia University Press, New York, 560 pp.

Cuffey, K.M. and E.J. Brook, 2000: Ice sheets and the ice-core record of climate change. In: *Earth System Science—From Biogeochemical Cycles to Global Change* [Jacobson, M.C., R.J. Charlson, H. Rodhe, and G.H. Orians (eds.)]. Academic Press, New York, pp. 459-497.

Cuffey, K.M. and G.D. Clow, 1997: Temperature, accumulation, and ice sheet elevation in central Greenland through the last deglacial transition. *Journal of Geophysical Research*, **102(C12)**, 26383-26396.

Cuffey, K.M. and S.J. Marshall, 2000: Substantial contribution to sea-level rise during the last interglacial from the Greenland Ice Sheet. *Nature*, **404**, 591-594.

Cuffey, K.M., R.B. Alley, P.M. Grootes, J.F. Bolzan, and S. Anandakrishnan, 1994: Calibration of the $\delta^{18}O$ isotopic paleothermometer for central Greenland, using borehole temperatures. *Journal of Glaciology*, **40(135)**, 341-349.

Cuffey, K.M., G.D. Clow, R.B. Alley, M. Stuiver, E.D. Waddington, and R.W. Saltus, 1995: Large Arctic temperature change at the glacial-Holocene transition. *Science*, **270**, 455-458.

Dahl-Jensen, D., K. Mosegaard, N. Gundestrup, G.D. Clow, S.J. Johnsen, A.W. Hansen, and N. Balling, 1998: Past temperatures directly from the Greenland Ice Sheet. *Science*, **282**, 268-271.

Dahl-Jensen, D., N. Gundestrup, S.P. Gogineni, and H. Miller, 2003: Basal melt at NorthGRIP modeled from borehole, ice-core and radio-echo sounded observations. *Annals of Glaciology*, **37**, 207-212.

Dansgaard, W., S.J. Johnsen, H.B. Clausen, D. Dahl-Jensen, N.S. Gundestrup, C.U. Hammer, C.S. Hvidberg, J.P. Steffensen, A.E. Sveinbjorndottir, J. Jouzel, and G. Bond, 1993: Evidence for general insatbility of past climate from a 250-kyr ice-core record. *Nature*, **364**, 218-220.

Darby, D.A., J. Bischof, R.F. Spielhagen, S.A. Marshall, and S.W. Herman, 2002: Arctic ice export events and their potential impact on global climate suring the late Pleistocene. *Palaeoceanography*, **17**, doi:10.1029/2001PA000639, 000615-000631 to 000615-000617.

De Abreu, C., F.F. Abrantes, N.J. Shackleton, P.C. Tzedakis, J.F. McManus, D.W. Oppo, and M.A. Hall, 2005: Ocean climate variability in the eastern North Atlantic during interglacial marine isotope stage 11—A partial analogue to the Holocene? *Paleoceanography*, **20**, Art. No. PA3009.

Denton, G.H., R.B. Alley, G.C. Comer, and W.S. Broecker, 2005: The role of seasonality in abrupt climate change. *Quaternary Science Reviews*, **24**, 1159-1182.

Dickinson, W.R., 2001: Paleoshoreline record of relative Holocene sea levels on Pacific Islands. *Earth-Science Reviews*, **55**, 191-234.

Dowdeswell, J.A., G. Uenzelmann-Neben, R.J. Whittington, and P. Marienfeld, 1994a: The Late Quaternary sedimentary record in Scoresby Sund, East Greenland. *Boreas*, **23**, 294-310.

Dowdeswell, J.A., R.J. Whittington, and P. Marienfeld, 1994b: The origin of massive diamicton facies by iceberg rafting and scouring, Scoresby Sund, East Greenland. *Sedimentology*, **41**, 21-35.

Dowdeswell, J.A., N.H. Kenyon, A. Elverhoi, J.S. Laberg, F.-J. Hollender, J. Mienert, and M.J. Siegert, 1996: Large-scale sedimentation on the glacier-influenced polar North Atlantic margins—Long-range side-scan sonar evidence. *Geophysical Research Letters*, **23**, 3535-3538.

Dowdeswell, J.A., N.H. Kenyon, and J.S. Laberg, 1997: The glacier-influenced Scoresby Sund Fan, east Greenland continental margin—Evidence from GLORIA and 3.5 kHz records. *Marine Geology*, **143**, 207-221.

Droxler, A.W., R.B. Alley, W.R. Howard, R.Z. Poore, and L.H. Burckle, 2003: Unique and exceptionally long interglacial marine isotope stage 11—Window into Earth future climate. In: *Earth's Climate and Orbital Eccentricity—The Marine Isotope Stage 11 Question* [Droxler, A.W., R.Z. Poore and L.H. Burckle (eds.)]. Geophysical Monograph 137, American Geophysical Union, pp. 1-14.

Dunhill, G., 2005: *Iceland and Greenland Margins—A Comparison of Depositional Processes Under Different Glaciological and Oceanographic Settings*. PhD dissertation, Geological Sciences, University of Colorado, Boulder.

Dupont, T.K. and R.B. Alley, 2005: Assessment of the importance of ice-shelf buttressing to ice-sheet flow. *Geophysical Research Letters*, **32**, L04503, doi:10.1029/2004GL022024.

Dupont, T.K. and R.B. Alley, 2006: Role of small ice shelves in sea-level rise. *Geophysical Research Letters*, **33**, L09503.

Dyke, A.S., J.E. Dale, and R.N. McNeely, 1996: Marine molluscs as indicators of environmental change in glaciated North America and Greenland during the last 18,000 years. *Géographie Physique et Quaternaire*, **50**, 125-184.

Dyke, A.S., J.T. Andrews, P.U. Clark, J.H. England, G.H. Miller, J. Shaw, and J.J. Veillette, 2002: The Laurentide and Innuitian ice sheets during the Last Glacial Maximum. *Quaternary Science Reviews*, **21**, 9-31.

Edwards, R.L., H. Cheng, M.T. Murrell, and S.J. Goldstein, 1997: Protactinium-231 dating of carbonates by thermal ionization mass spectrometry—Implications for Quaternary climate change: *Science*, **276**, 782-786.

Eisen, O., U. Nixdorf, F. Wilhelms, and H. Miller, 2004: Age estimates of isochronous reflection horizons by combining ice core, survey, and synthetic radar data. *Journal of Geophysical Research*, **109(B4)**, Art. No. B04106.

Eldrett, J.S., I.C. Harding, P.A. Wilson, E. Butler, and A.P. Roberts, 2007: Continental ice in Greenland during the Eocene and Oligocene. *Nature*, **446**, 176-179.

Elliot, M., L. Labeyrie, G. Bond, E. Cortijo, J.-L. Turon, N. Tiseray, and J.-C. Duplessy, 1998: Millennial-scale iceberg discharges in the Irminger Basin during the last glacial period—Relationship with the Heinrich events and environmental settings. *Paleoceanography*, **13**, 433-446.

Elverhoi, A., J.A. Dowdeswell, S. Funder, J. Mangerud, and R. Stein, 1998: Glacial and oceanic history of the polar North Atlantic margins—An overview. *Quaternary Science Reviews*, **17**, 1-10.

England, J., 1999: Coalescent Greenland and Innuitian ice during the Last Glacial Maximum—Revising the Quaternary of the Canadian High Arctic. *Quaternary Science Reviews*, **18**, 421-456.

Fairbanks, R.G., R.A. Mortlock, C. Tzu-Chien, L. Cao, A. Kaplan, T.P. Guilderson, T.W. Fairbanks, A.L. Bloom, P.M. Grootes, and M.-J. Nadeau, 2005: Radiocarbon caliubration curve spanning 0 to 50,000 years BP based on paired ^{230}Th/^{234}U/^{238}U and ^{14}C dates on pristine corals. *Quaternary Science Reviews*, **24**, 1781-1796.

Farmer, G.L., D.C., Barber, and J.T. Andrews, 2003: Provenance of late Quaternary ice-proximal sediments in the North Atlantic—Nd, Sr and Pd isotopic evidence. *Earth and Planetary Science Letters*, **209**, 227-243.

Fillon, R.H. and J.C. Duplessy, 1980: Labrador Sea bio-, tephro-, oxygen isotopic stratigraphy and late Quaternary paleoceanographic trends. *Canadian Journal of Earth Sciences*, **17**, 831-854.

Finkel, R.C. and K. Nishiizumi, 1997: Beryllium 10 concentrations in the Greenland Ice Sheet Project 2 ice core from 3–40 ka. *Journal of Geophysical Research*, **102(C12)**, 26699-26706.

Fisher, T.G., D.G. Smith, and J.T. Andrews, 2002: Preboreal oscillation caused by a glacial Lake Agassiz flood. *Quaternary Science Reviews*, **21**, 873-878.

Fleming, K. and K. Lambeck, 2004: Constraints on the Greenland Ice Sheet since the Last Glacial Maximum from sea-level observations and glacial-rebound models. *Quaternary Science Reviews*, **23**, 1053-1077.

Fruijtier, C., T. Elliot, and W. Schlager, 2000: Mass-spectrometric ^{234}U-^{230}Th ages from the Key Largo Formation, Florida Keys, United States—Constraints on diagenetic age disturbance. *Geological Society of America Bulletin*, **112**, 267-277.

Funder, S., 1989a: Quaternary geology of east Greenland. In: Chapter 13 of *Quaternary Geology of Canada and Greenland* [Fulton, R.J. (ed.)]. Geological Survey of Canada, Geology of Canada, no. 1, 839 pp.; also Geological Society of America, the Geology of North America, v. K-1, pp. 756-763.

Funder, S., 1989b: Quaternary geology of north Greenland. In: Chapter 13 of *Quaternary Geology of Canada and Greenland* [Fulton, R.J. (ed.)]. Geological Survey of Canada, Geology of Canada, no. 1, 839 pp.; also Geological Society of America, the Geology of North America, v. K-1, pp. 763-769.

Funder, S., 1989c: Quaternary geology of west Greenland. In: Chapter 13 of *Quaternary Geology of Canada and Greenland* [R.J. Fulton (ed.)]. Geological Survey of Canada, Geology of Canada, no. 1, 839 pp.; also Geological Society of America, the Geology of North America, v. K-1, pp. 749-756.

Funder, S., 1989d: Sea level history. In: Chapter 13 of *Quaternary Geology of Canada and Greenland* [R.J. Fulton (ed.)]. Geological Survey of Canada, Geology of Canada, no. 1, 839 pp.; also Geological Society of America, the Geology of North America, v. K-1, pp. 772-774.

Funder, S. and B. Fredskild, 1989: Paleofaunas and floras. In: Chapter 13 of *Quaternary Geology of Canada and Greenland* [Fulton, R.J. (ed.)]. Geological Survey of Canada, Geology of Canada, no. 1; 839 pp.; also Geological Society of America, the Geology of North America, v. **K-1**, pp. 775-783.

Funder, S. and H.C. Larsen, 1989: Quaternary geology of the shelves adjacent to Greenland. In: Chapter 13 of *Quaternary Geology of Canada and Greenland* [Fulton, R.J. (ed.)]. Geological Survey of Canada, Geology of Canada, no. 1; 839 pp.; also Geological Society of America, The Geology of North America, v. **K-1**, pp. 769-772.

Funder, S., C. Hjort, J.Y. Landvik, S.I. Nam, N. Reeh, and R. Stein, 1998: History of a stable ice margin East Greenland during the Middle and Upper Pleistocene. *Quaternary Science Reviews*, **17**, 77-123.

Funder, S., O. Bennike, J. Böcher, C. Israelson, K.S. Petersen, and L.A. Símonarson, 2001: Late Pliocene Greenland—The Kap København Formation in North Greenland. *Bulletin of the Geological Society of Denmark*, **48**, 117-134.

Funder, S., A.E. Jennings, and M.J. Kelly, 2004: Middle and late Quaternary glacial limits in Greenland. In: *Quaternary Glaciations—Extent and Chronology, Part II* [Ehlers, J. and P.L. Gibbard (eds.)]. Elsevier, Amsterdam, pp. 425-430.

Gallup, C.D., R.L. Edwards, and R.G. Johnson, 1994: The timing of high sea levels over the past 200,000 years: *Science*, **263**, 796-800.

Geirsdóttir, Á., H. Hardardottir, and J. Eirksson, 1997: The depositional history of the Younger Dryas–Preboreal Budi moraines in south-central Iceland. *Arctic and Alpine Research*, **29**, 13-23.

Geirsdóttir, Á., J. Hardardottir, and J.T. Andrews, 2000: Late Holocene terrestrial geology of Miki and I.C. Jacobsen Fjords, East Greenland. *The Holocene*, **10**, 125-134.

Gilbert, R., 1990: Rafting in glacimarine environments. In: *Glacimarine Environments—Processes and Sediments* [Dowdeswell, J.A. and J.D. Scourse (eds.)]. Geological Society, London, pp. 105-120.

Goreau, T.F., 1959: The ecology of Jamaican coral reefs I—Species composition and zonation. *Ecology*, **40(1)**, 67-90.

Gosse, J.C. and F.M. Phillips, 2001: Terrestrial in situ cosmogenic nuclides—Theory and application. *Quaternary Science Reviews*, **20**, 1475-1560.

Gregory, J.M. and P. Huybrechts, 2006: Ice-sheet contributions to future sea-level change. *Philosophical Transactions of the Royal Society A: Mathematical, Physical and Engineering Sciences*, **364(1844)**, 1709-1731.

Grootes, P.M. and M. Stuiver, 1997: Oxygen 18/16 variability in *Greenland* snow and ice with 10^{-3}- to 10^{5}-year time resolution. *Journal of Geophysical Research*, **102C**, 26455-26470.

Grousset, F.E., E. Cortijo, S. Huon, L. Herve, T. Richter, D. Burdloff, J. Duprat, and O. Weber, 2001: Zooming in on Heinrich layers. *Paleoceanography*, **16**, 240-259.

Hagen, S., 1999: North Atlantic Paleoceanography and Climate History During the Last 70 cal. ka Years. PhD dissertation, Department of Geology, University of Tromsø, Norway, 110 pp.

Hagen, S. and M. Hald, 2002: Variation in surface and deep water circulation in the Denmark Strait, North Atlantic, during marine isotope stages 3 and 2. *Paleoceaography*, **17**, 13-11 to 13-16 (10.1029/2001PA000632).

Hakansson, L., J. Briner, H. Alexanderson, A. Aldahan, and G. Possnert, 2007: ^{10}Be ages from central east Greenland constrain the extent of the Greenland Ice Sheet during the Last Glacial Maximum. *Quaternary Science Reviews*, **26(19-2),** 2316-2321**,** doi:10.1016/j.quascirev.2007.08.001.

Hald, M. and S. Hagen, 1998: Early Preboreal cooling in the Nordic seas region triggered by meltwater. *Geology*, **26**, 615-618.

Hanna, E., P. Huybrechts, I. Janssens, J. Cappelen, K. Steffens, and A. Stephens, 2005: Runoff and mass balance of the Greenland Ice Sheet—1958–2003. *Journal of Geophysical Research*, **110**, D13108, doi:10.1029/2004JD005641.

Hearty, P.J., P. Kindler, H. Cheng, and R.L. Edwards, 1999: A +20 m middle Pleistocene sea-level highstand (Bermuda and the Bahamas) due to partial collapse of Antarctic ice: *Geology*, **27**, 375-378.

Helmke, J.P., H.A. Bauch, and H. Erlenkeuser, 2003: Development of glacial and interglacial conditions in the Nordic seas between 1.5 and 0.35 Ma. *Quaternary Science Reviews*, **22**, 1717-1728.

Hemming, S.R., 2004: Heinrich events—Massive late Pleistocene detritus layers of the North Atlantic and their global climate imprint. *Reviews of Geophysics*, **42**, Art. No. RG1005.

Hemming, S.R., T.O. Vorren, and J. Kleman, 2002: Provinciality of ice rafting in the North Atlantic—Application of Ar^{40}/Ar^{39} dating of individual ice rafted hornblende grains. *Quaternary International*, **95**, 75-85.

Hooke, R. LeB., 2005: *Principles of Glacier Mechanics.* Cambridge University Press, Cambridge, 429 pp.

Hopkins, T.S., 1991: The GIN Sea—A synthesis of its physical oceanography and literature review 1972–1985. *Earth Science Reviews*, **30**, 175-318.

Huddard, A., D. Sugden, A. Dugmore, H. Norddahl, and H.G. Petersson, 2006: A modelling insight into the Icelandic Last Glacial Maximum ice sheet. *Quaternary Science Reviews*, **25**, 2283-2296.

Hughes, T.J., 1998: *Ice Sheets.* Oxford University Press, New York, 343 pp.

Hulbe, C.L., D.R. MacAyeal, G.H. Denton, J. Kleman, and T.V. Lowell, 2004: Catastrophic ice shelf breakup as the source of Heinrich event icebergs. *Paleoceanography*, **19**, Art. No. PA1004.

Huybrechts, P., 2002: Sea-level changes at the LGM from ice-dynamic reconstructions of the Greenland and Antarctic ice sheets during the glacial cycles. *Quaternary Science Reviews*, **21**, 203-231.

Huybrechts, P. and J. de Wolde, 1999: The dynamic response of the Greenland and Antarctic ice sheets to multiple-century climatic warming. *Journal of Climate*, **12**, 2169-2188.

Intergovernmental Panel on Climate Change (IPCC), 2007: Summary for policymakers. In: *Climate Change 2007—The Physical Science Basis. Contribution of Working Group I to the Fourth Assessment Report of the Intergovernmental Panel on Climate Change* [Solomon, S., D. Qin, M. Manning, Z. Chen, M. Marquis, K.B. Averyt, M.Tignor and H.L. Miller (eds.)]. Cambridge University Press, Cambridge and New York, 18 pp.

Jacobel, R.W. and B.C. Welch, 2005: A time marker at 17.5 kyr BP detected throughout West Antarctica. *Annals of Glaciology*, **41**, 47-51.

Jansen, E., J. Overpeck, K.R. Briffa, J.-C. Duplessy, F. Joos, V. Masson-Delmotte, D. Olago, B. Otto-Bliesner, W.R. Peltier, S. Rahmstorf, R. Ramesh, D. Raynaud, D. Rind, O. Solomina, R. Villalba, and D. Zhang, 2007: Palaeoclimate. In: *Climate Change 2007: The Physical Science Basis. Contribution of Working Group I to the Fourth Assessment Report of the Intergovernmental Panel on Climate Change* [Solomon, S., D. Qin, M. Manning, Z. Chen, M. Marquis, K.B. Averyt, M. Tignor and H.L. Miller (eds.)]. Cambridge University Press, Cambridge and New York, pp. 434-497.

Jennings, A.E. and N.J. Weiner, 1994: East Greenland climate change over the last 1300 years from foraminiferal evidence. *PaleoBios*, **16(Supplement to No. 2)**, 38 pp.

Jennings, A.E., K. Gronvold, R. Hilberman, M. Smith, and M. Hald, 2002a: High resolution study of Icelandic tephras in the Kangerlussuaq Trough, SE East Greenland, during the last deglaciation. *Journal of Quaternary Science*, **17**, 747-757.

Jennings, A.E., K.L. Knudsen, M. Hald, C.V. Hansen, and J.T. Andrews, 2002b: A mid-Holocene shift in Arctic sea ice variability on the East Greenland shelf. *The Holocene*, **12**, 49-58.

Jennings, A.E., M. Hald, L.M. Smith, and J.T. Andrews, 2006: Freshwater forcing from the Greenland Ice Sheet during the Younger Dryas—Evidence from southeastern Greenland shelf cores. *Quaternary Science Reviews*, **25**, 282-298.

Johannessen, O.M., K. Khvorostovsky, M.W. Miles, and L.P. Bobylev, 2005: Recent ice-sheet growth in the interior of Greenland. *Science*, **310**, 1013-1016.

Johnsen, S.J., 1977: Stable isotope homogenization of polar firn and ice. In: *Proceedings of a Symposium on Isotopes and Impurities in Snow and Ice, I.U.G.G. XVI, General Assembly, Grenoble Aug. Sept. 1975*. IAHS-AISH Publ. 118, Washington D.C., pp. 210-219.

Johnsen, S., H.B. Clausen, W. Dansgaard, N.S. Gundestrup, M. Hansson, P. Johnsson, P. Steffensen, and A.E. Sveinbjornsdottir, 1992a: A "deep" ice core from East Greenland. *Meddelelser om Gronland, Geoscience*, **29**, 3-22.

Johnsen, S.J., H.B. Clausen, W. Dansgaard, K. Fuhrer, N. Gundestrup, C.U. Hammer, P. Iversen, J. Jouzel, B. Stauffer, and J.P. Steffensen, 1992b: Irregular glacial interstadials recorded in a new Greenland ice core. *Nature*, **359**, 311-313.

Johnsen, S.J., D. Dahl-Jensen, N. Gundestrup, J.P. Steffensen, H.B. Clausen, H. Miller, V. Masson-Delmotte, A.E. Sveinbjornsdottir, and J. White, 2001: Oxygen isotope and palaeotemperature records from six Greenland ice-core stations: Camp Century, Dye-3, GRIP, GISP2, Renland and NorthGRIP. *Journal of Quaternary Science*, **16**, 299-307.

Jones, G.A. and L.D. Keigwin, 1988: Evidence from the Fram Strait (78°N) for early deglaciation. *Nature*, **336**, 56-59.

Joughin, I., S. Tulaczyk, M. Fahnestock, and R. Kwok, 1996: A mini-surge on the Ryder Glacier, Greenland, observed by satellite radar interferometry. *Science*, **274**, 228-230.

Joughin, I., S.B. Das, M.A. King, B.E. Smith, I.M. Howat, and T. Moon, 2008a: Seasonal speedup along the western flank of the Greenland Ice Sheet. *Science*, published online, 10.1126/science.1153288 (Science Express Reports).

Joughin, I., I. Howat, R.B. Alley, G. Ekstrom, M. Fahnestock, T. Moon, M. Nettles, M. Truffer, and V.C. Tsai, 2008b: Ice front variation and tidewater behavior on Helheim and Kangerdlugssuaq Glaciers, Greenland. *Journal of Geophysical Research*, **113**, F01004, doi:10.1029/2007JF000837.

Jouzel, J., R.B. Alley, K.M. Cuffey, W. Dansgaard, P. Grootes, G. Hoffmann, S.J. Johnsen, R.D. Koster, D. Peel, C.A. Shuman, M. Stievenard, M. Stuiver, and J. White, 1997: Validitiy of the temperature reconstruction from water isotopes in ice cores. *Journal of Geophysical Research*, **102(C12)**, 26471-26487.

Kahn, S.A., J. Wahr, L.A. Stearns, G.S. Hamilton, T. van Dam, K.M. Larson, and O. Francis, 2007: Elastic uplift in southeast Greenland due to rapid ice mass loss. *Geophysical Resesarch Letters*, **34**, L21701. doi:10.1029/2007GL031468.

Kamb, B., C.F. Raymond, W.D. Harrison, H. Engelhardt, K.A. Echelmeyer, N. Humphrey, M.M. Brugman, and T. Pfeffer, 1985: Glacier surge mechanism—1982–1983 surge of Variegated Glacier, Alaska. *Science*, **227**, 469-479.

Kandiano, E.S. and H.A. Bauch, 2003: Surface ocean temperatures in the north-east Atlantic during the last 500,000 years—Evidence from foraminiferal census data. *Terra Nova*, **15**, 265-271.

Kapsner, W.R., R.B. Alley, C.A. Shuman, S. Anandakrishnan, and P.M. Grootes, 1995: Dominant control of atmospheric circulation on snow accumulation in central Greenland. *Nature*, **373**, 52-54.

Kaufman, D.S. and J. Brigham-Grette, 1993: Aminostratigraphic correlations and paleotemperature implications, Pliocene-Pleistocene high sea level deposits, northwestern Alaska. *Quaternary Science Reviews*, **12**, 21-33.

Kaufman, D.S., R.C. Walter, J. Brigham-Grette, and D.M. Hopkins, 1991: Middle Pleistocene age of the Nome River glaciation, northwestern Alaska. *Quaternary Research*, **36**, 277-293.

Kelly, M., S. Funder, M. Houmark-Nielsen, K.L. Knudsen, C. Kronborg, J. Landvik, and L. Sorby, 1999: Quaternary glacial and marine environmental history of northwest Greenland—A review and reappraisal. *Quaternary Science Reviews*, **18**, 373-392.

Kiehl, J.T. and P.R. Gent, 2004: The Community Climate System model, version two. *Journal of Climate*, **17**, 3666-3682.

Kindler, P. and P.J. Hearty, 2000: Elevated marine terraces from Eleuthera (Bahamas) and Bermuda—Sedimentological, petrographic and geochronological evidence for important deglaciation events during the middle Pleistocene. *Global and Planetary Change*, **24**, 41-58.

Kleiven, H.F., E. Jansen, T. Fronval, and T.M. Smith, 2002: Intensification of Northern Hemisphere glaciations in the circum Atlantic region (3.5–2.4 Ma)—Ice-rafted detritus evidence. *Palaeogeography, Palaeoclimatology, Palaeoecology*, **184**, 213-223.

Kobashi, T., J.P. Severinghaus, and J.-M. Barnola. 2008: 4±1.5°C abrupt warming 11,270 years ago identified from trapped air in Greenland ice. *Earth and Planetary Science Letters*, **268**, 397-407.

Koerner, R.M., 1989: Ice core evidence for extensive melting of the Greenland Ice Sheet in the last interglacial. *Science*, **244**, 964-968.

Koerner, R.M. and D.A. Fisher, 2002: Ice-core evidence for widespread Arctic glacier retreat in the last interglacial and the early Holocene. *Annals of Glaciology*, **35**, 19-24.

Kuijpers, A., N. Abrahamsen, G. Hoffmann, V. Huhnerbach, P. Konradi, H. Kunzendorf, N. Mikkelsen, J. Thiede, and W. Weinrebe, scientific party of RV Poseidon, surveyors of the Royal Danish Administration for Navigation, and Hydrology, 1999: Climate change and the Viking-age fjord environment of the Eastern Settlement, South Greenland. *Geology of Greenland Survey Bulletin*, **183**, 61-67.

Lambeck, K., C. Smither, and P. Johnston, 1998: Sea-level change, glacial rebound and mantle viscosity for northern Europe. *Geophysical Journal International*, **134**, 102-144.

Lemke, P., J. Ren, R.B. Alley, I. Allison, J. Carrasco, G. Flato, Y. Fujii, G. Kaser, P. Mote, R.H. Thomas, and T. Zhang, 2007: Observations—Changes in snow, ice and frozen ground. In: *Climate Change 2007: The Physical Science Basis. Contribution of Working Group I to the Fourth Assessment Report of the Intergovernmental Panel on Climate Change* [Solomon, S., D. Qin, M. Manning, Z. Chen, M. Marquis, K.B. Averyt, M. Tignor and H.L. Miller (eds.)]. Cambridge University Press, Cambridge and New York, 996 pp.

Lhomme, N., G.K.C. Clarke, and S.J. Marshall, 2005: Tracer transport in the Greenland Ice Sheet—Constraints on ice cores and glacial history. *Quaternary Science Reviews*, **24**, 173-194.

Lie, O. and O. Paasche, 2006: How extreme was northern hemisphere seasonality during the Younger Dryas? *Quaternary Science Reviews*, **25**, 404-407, doi:10.1016/j.quascirev.2005.11.003.

Lisitzin, A.P., 2002: *Sea-Ice and Iceberg Sedimentation in the Ocean—Recent and Past.* Springer-Verlag, Berlin, 563 pp.

Lloyd, J.M., 2006: Late Holocene environmental change in Disko Bugt, West Greenland—Interaction between climate, ocean circulation and Jakobshavn Isbrae. *Boreas*, **35**, 35-49.

Lloyd, J.M., L.A. Park, A. Kuijpers, and M. Moros, 2005: Early Holocene palaeoceanography and deglacial chronology of Disko Bugt, West Greenland. *Quaternary Science Reviews*, **24**, 1741-1755.

Ljung, K. and S. Björck, 2004: A lacustrine record of the Pleistocene–Holocene boundary in southernmost Greenland. *GFF (Geological Society of Sweden)*, **126[part 3]**, 273-278.

Locke, W.W.I., J.T. Andrews, and P.J. Webber, 1979: *A Manual for Lichenometry.* British Geomorphological Research Group, Technical Bulletin 26, 47 pp.

Long, A.J., D.H. Roberts, and S. Dawson, 2006: Early Holocene history of the west Greenland Ice Sheet and the GH-8.2 event. *Quaternary Science Reviews*, **25**, 904-922.

Mangerud, J. and Funder, S., 1994: The interglacial-glacial record at the mouth of Scoresby Sund, East Greenland. *Boreas*, **23**, 349-358.

Marienfeld, P., 1992a: Recent sedimentary processes in Scoresby Sund, East Greenland. *Boreas*, **21**, 169-186.

Marienfeld, P., 1992b: Postglacial sedimentary history of Scoresby Sund, East Greenland. *Polarforschung*, **60**, 181-195.

Marshall, S.J. and K.M. Cuffey, 2000: Peregrinations of the Greenland Ice Sheet divide in the last glacial cycle—Implications for central Greenland ice cores. *Earth and Planetary Science Letters*, **179**, 73-90.

Marwick, P.J., 1998: Fossil crocodilians as indicators of Late Cretaceous and Cenozoic climates—Implications for using palaeontological data in reconstructing palaeoclimate. *Palaeogeography, Palaeoclimatology, Palaeoecology*, **137**, 205-271.

McCave, I.N. and B.E. Tucholke, 1986: Deep current-controlled sedimentation in the western North Atlantic. In: *The Geology of North America: The Western North Atlantic Region* [Vogt, P.R. and B.E. Tucholke (eds.)]. Geological Society of America, pp. 451-468.

McManus, J.F., D.W. Oppo, and J.L. Cullen, 1999: A 0.5-million-year record of millennial-scale climate variability in the North Atlantic. *Science*, **283**, 971-975.

Meier, M.F. and A. Post, 1987: Fast tidewater glaciers. *Journal of Geophysical Research*, **92(B9)**, 9051-9058.

Mienert, J., J.T. Andrews, and J.D. Milliman, 1992: The East Greenland continental margin (65°N) since the last deglaciation—Changes in sea floor properties and ocean circulation. *Marine Geology*, **106**: 217-238.

Milne, G.A., J.X. Mitrovica, H.G. Scherneck, J.L. Davis, J.M. Johansson, H. Koivula, and M. Vermeer, 2004: Continuous GPS measurements of postglacial adjustment in Fennoscandia—2. Modeling results. *Journal of Geophysical Research*, **109(B2)**, Art. No. B02412.

Mitrovica, J.X., 1996: Haskell [1935] Revisited. *Journal Geophysical Research*, **101**, 555-569.

Mitrovica, J.X. and G.A. Milne, 2002: On the origin of Late Holocene highstands within equatorial ocean basins. *Quaternary Science Reviews*, **21**, 2179-2190.

Mitrovica, J.X. and W.R. Peltier, 1991: On post-glacial geoid relaxation over the equatorial oceans. *Journal Geophysical Research*, **96**, 20053-20071.

Mitrovica, J.X., M.E. Tamisiea, J.L. Davis, and G.A. Milne, 2001: Recent mass balance of polar ice sheets inferred from patterns of global sea-level change. *Nature*, **409**, 1026-1029.

Mitrovica, J.X., J. Wahr, I. Matsuyama, A. Paulson, and M.E. Tamisiea, 2006: Reanalysis of ancient eclipse, astronomic and geodetic data—A possible route to resolving the enigma of global sea-level rise. *Earth and Planetary Science Letters*, **243**, 390-399.

Moran, K., J. Backman, H. Brinkhuis, S.C. Clemens, T. Cronin, G.R. Dickens, F. Eynaud, J. Gattacceca, M. Jakobsson, R.W. Jordan, M. Kaminski, J. King, N. Koç, A. Krylov, N. Martinez, J. Matthiessen, D. McInroy, T.C. Moore, J. Onodera, M. O'Regan, H. Palike, B. Rea, D. Rio, T. Sakamoto, D.C. Smith, R. Stein, K. St. John, I. Suto, N. Suzuki, K. Takahashi, M. Watanabe, M. Yamamoto, J. Farrell, M. Frank, P. Kubik, W. Jokat and Y. Kristoffersen, 2006: The Cenozoic palaeoenvironment of the Arctic Ocean. *Nature*, **441**, 601-605.

Moros, M., K.G. Jensen, and A. Kuijpers, 2006: Mid- to late-Holocene hydrological variability and climatic variability in Disko Bugt, central West Greenland. *The Holocene*, **16**, 357-367.

Mudie, P.J., A. Rochon, M.A. Prins, D. Soenarjo, S.R. Troelstra, E. Levac, D.B. Scott, L. Roncaglia, and A. Kuijpers, 2006: Late Pleistocene-Holocene marine geology of Nares Strait region—Palaeoceanography from foraminifera and dinoflagellate cysts, sedimentology and stable istopes. *Polarforshung*, **74(1/3)**, 169-183. hdl:10013/epic.29931.d001

Muhs, D.R., 2002: Evidence for the timing and duration of the last interglacial period from high-precision uranium-series ages of corals on tectonically stable coastlines. *Quaternary Research*, **58**, 36-40.

Muhs, D.R. and B.J. Szabo, 1994: New uranium-series ages of the Waimanalo Limestone, Oahu, Hawaii—Implications for sea level during the last interglacial period: *Marine Geology*, **118**, 315-326.

Muhs, D.R., K.R. Simmons, and B. Steinke, 2002: Timing and warmth of the last interglacial period—New U-series evidence from Hawaii and Bermuda and a new fossil compilation for North America. *Quaternary Science Reviews*, **21**, 1355-1383.

Muhs, D.R., J.F. Wehmiller, K.R. Simmons, and L.L. York, 2004: Quaternary sea level history of the United States. In: *The Quaternary Period in the United States* [Gillespie, A.R., S.C. Porter, and B.F. Atwater, (eds.)]. Elsevier, Amsterdam, pp. 147-183.

Multer, H.G., E. Gischler, J. Lundberg, K.R. Simmons, and E.A. Shinn, 2002: Key Largo Limestone revisited—Pleistocene shelf-edge facies, Florida Keys, USA: *Facies*, **46**, 229-272.

Munk, W., 2002: Twentieth century sea level—An enigma. *Proceedings of the National Academy of Sciences of the United States of America*, **99**, 6550-6555.

Muscheler, R., R. Beer, P.W. Kubik, and H.A. Synal, 2005: Geomagnetic field intensity during the last 60,000 years based on Be-10 and Cl-36 from the Summit ice cores and ^{14}C. *Quaternary Science Reviews*, **24**, 1849-1860.

Nam, S. I. and R. Stein, 1999: Late Quaternary variations in sediment accumulation rates and their paleoenvironment implications—A case study from the East Greenland continental margin. *GeoResearch Forum*, **5**, 223-240.

Nishiizumi, K., R.C. Finkel, K.V. Ponganis, T. Graf, C.P. Kohl, and K. Marti, 1996: In situ produced cosmogenic nuclides in GISP2 rock core from Greenland Summit (Abstract, Fall Meeting 1996). *Eos, Transactions of the American Geophysical Union*, **77(46) Supplement**, F428, Abstract OS41B-10.

North Greenland Ice Core Project Members. 2004. High-resolution record of Northern Hemisphere climate extending into the last interglacial period. *Nature*, **431**, 147-151.

O'Cofaigh, C., J. Taylor, J.A. Dowdeswell, and C.J. Pudsey, 2003: Palaeo-ice streams, trough mouth fans and high-latitude continental slope sedimentation. *Boreas*, **32**, 37-55.

Oerlemans, J., 1994: Quantifying global warming from the retreat of glaciers. *Science*, **264**, 243-245.

Oerlemans, J., 2001: *Glaciers and Climatic Change*. A.A. Balkema Publishers, Lisse, 148 pp.

Otto-Bliesner, B.L., J.T. Overpeck, S.J. Marshall, and G. Miller, 2006: Simulating Arctic climate warmth and icefield retreat in the Last Interglaciation. *Science*, **311**, 1751-1753.

Overpeck, J.T., B.L. Otto-Bliesner, G.H., Miller, D.R. Muhs, R.B. Alley, and J.T. Kiehl, 2006: Paleoclimatic evidence for future ice-sheet instability and rapid sea-level rise, *Science*, **311**, 1747-1750.

Parizek, B.R. and R.B. Alley, 2004: Implications of increased Greenland surface melt under global-warming scenarios—Ice-sheet simulations. *Quaternary Science Reviews*, **23**, 1013-1027.

Parnell, J., S. Bowden, C. Taylor, and J.T. Andrews, 2007: Biomarker determination as a provenance tool for detrital carbonate events (Heinrich events?)—Fingerprinting Quaternary glacial sources in Baffin Bay. *Earth and Planetary Science Letters*, **257**, 71-82.

Paterson, W.S.B., 1994: *The Physics of Glaciers*. Pergamon, Oxford, 3rd ed., 480 pp.

Payne, A.J., A. Vieli, A.P. Shepherd, D.J. Wingham, and E. Rignot, 2004: Recent dramatic thinning of largest West Antarctic ice stream triggered by oceans. *Geophysical Research Letters*, **31**, L23401., doi:10.1029/2004GL021284.

Peltier, W.R., 2004: Glocal glacial isostasy and the surface of the ice-age Earth—The ICE-5G (VM2) Model and GRACE. *Annual Review of Earth and Planetary Sciences*, **32**, 111-149, doi:10.1146/annurev.earth.32.082503.144359.

Peltier, W.R. and R.G. Fairbanks, 2006: Global glacial ice volume and Last Glacial Maximum duration from an extended Barbados sea level record. *Quaternary Science Reviews*, **25**, 3322-3337.

Petrenko, V.V., J.P. Severinghaus, E.J. Brook, N. Reeh, and H. Schaefer, 2006: Gas records from the West Greenland ice margin covering the Last Glacial Termination—A horizontal ice core. *Quaternary Science Reviews*, **25**, 865-875.

Plag, H.-P. and H.-U. Juttner, 2001: Inversion of global tide gauge data for present-day ice load changes. In: *Proceedings of the International Symposium on Environmental Research in the Arctic and Fifth Ny-Alesund Science Seminar* [Yamanouchi, T. (ed.)]. *Memoirs of the National Institute of Polar Research (Japan)*, **54**, 301-317.

Poore, R.Z. and H.J. Dowsett, 2001: Pleistocene reduction of polar ice caps—Evidence from Cariaco Basin marine sediments. *Geology*, **29**, 1-74.

Raynaud, D., J. Chappellaz, C. Ritz, and P. Martinerie, 1997: Air content along the Greenland Ice Core Project core—A record of surface climatic parameters and elevation in central Greenland. *Journal of Geophysical Research*, **102(C12)**, 26607-26613.

Reeh, N., 1984: Reconstruction of the glacial ice covers of Greenland and the Canadian Arctic islands by 3-dimensional, perfectly plastic ice-sheet modeling. *Annals of Glaciology*, **5**, 115-121.

Reeh, N., 1985: Greenland Ice-Sheet Mass Balance and Sea-Level change. In: *Glaciers, Ice Sheets, and Sea Level—Effect of a CO_2 induced climatic change*. National Academies Press, Washington D.C., 155-171.

Reeh, N., 2004: Holocene climate and fjord glaciations in Northeast Greenland—Implications for IRD deposition in the North Atlantic. *Sedimentary Geology*, **165**, 333-342.

Reeh, N., C. Mayer, H. Miller, H.H. Thomsen, and A. Weidick, 1999: Present and past climate control on fjord glaciations in Greenland—Implications for IRD-deposition in the sea. *Geophysical Research Letters*, **26**, 1039-1042.

Ridley, J.K., P. Huybrechts, J.M. Gregory, and J.A. Lowe, 2005: Elimination of the Greenland Ice Sheet in a high CO_2 climate. *Journal of Climate*, **18**, 3409-3427.

Rignot, E. and P. Kanagaratnam, 2006: Changes in the velocity structure of the Greenland Ice Sheet. *Science*, **311**, 986-990.

Rohling, E.J., M. Fenton, F.J. Jorissen, P. Bertrand, G. Ganssen, and J.P. Caulet. 1998. Magnitudes of sea-level lowstands of the past 500,000 years. *Nature*, **394**, 162-165.

Schofield, J.E., K.J. Edwards, and J.A. McMullen, 2007: Modern pollen-vegetation relationships in subarctic southern Greenland and the interpretation of fossil pollen data from the Norse landnam. *Journal of Biogeography*, **34**, 473-488.

Severinghaus, J.P., T. Sowers, E.J. Brook, R.B. Alley, and M.L. Bender, 1998: Timing of abrupt climate change at the end of the Younger Dryas interval from thermally fractionated gases in polar ice. *Nature*, **391**, 141-146.

Shackleton, N.J., J. Backman, H. Zimmerman, D.V. Kent, M.A. Hall, D.G. Roberts, D. Schnitker, J.G. Baldauf, A. Desprairies, R. Homrighausen, P. Huddlestun, J.B. Keene, A.J. Kaltenback, K.A.O. Krumsiek, A.C. Morton, J.W. Murray, and J. Westberg-smith, 1984: Oxygen isotope calibration of the onset of ice-rafting and history of glaciation in the North-Atlantic region. *Nature*, **307**, 620-623.

Sluijs, A., S. Schouten, M. Pagani, M. Woltering, H. Brinkhuis, J.S.S. Damsté, G.R. Dickens, M. Huber, G.J. Reichart, R. Stein, J. Matthiessen, L.J. Lourens, N. Pedentchouk, J. Backman, and K. Moran, 2006: Subtropical arctic ocean temperatures during the Palaeocene/Eocene thermal maximum. *Nature*, **441**, 610-613.

Smith, J.M.B. and T.P. Bayliss-Smith, 1998: Kelp-plucking—Coastal erosion facilitated by bull-kelp *Durvillaea antarctica* at subantarctic Macquarie Island. *Antarctic Science*, **10**, 431-438.

Souchez, R., A. Bouzette, H.B. Clausen, S.J. Johnsen, and J. Jouzel, 1998: A stacked mixing sequence at the base of the Dye 3 core, Greenland. Geophysical *Research Letters*, **25**, 1943-1946.

Sowers, T., M. Bender, D. Raynaud, and Y.S. Korotkevich, 1992: Delta-N-15 of N2 in air trapped in polar ice—A tracer of gas-transport in the firn and a possible constraint on ice age/gas age-differences. *Journal of Geophysical Research*, **97(D14)**, 15683-15697.

Sparrenbom, C.J., O. Bennike, S. Bjõrck, and K. Lambeck, 2006a: Holocene relative sea-level changes in the Qaqortoq area, southern Greenland. *Boreas*, **35**, 171-187.

Sparrenbom, C.J., O. Bennike, S. Bjorck, and K. Lambeck, 2006b: Relative sea-level changes since 15,000 cal. yr BP in the Nanortalik area, southern Greenland. *Journal of Quaternary Science*, **21**, 29-48.

Spencer, M.K., R.B. Alley, and J.J. Fitzpatrick, 2006: Developing a bubble number-density paleoclimatic indicator for glacier ice. *Journal of Glaciology*, **52(178)**, 358-364.

Stastna, M. and Peltier, W.R., 2007: On box models of the North Atlantic thermohaline circulation—Intrinsic and extrinsic millennial timescale variability in response to deterministic and stochastic forcing. *Journal of Geophysical Research*, **112,** C10023.

St. John, K., 2008: Cenozoic ice-rafting history of the central Arctic Ocean—Terrigenous sands on the Lomonosov Ridge. *Paleoceanography*, **23**, PA1SO5, doi:10.1029/2007PA001483.

St. John, K.E. and Krissek, L.A., 2002: The late Miocene to Pleistocene ice-rafting history of southeast Greenland. *Boreas*, **31**, 28-35.

Stein, A.B., 1996: Seismic stratigraphy and seafloor morphology of the Kangerlugssuaq region, East Greenland—Evidence for glaciations to the Continental shelf break during the late Weischelian and earlier. MSc thesis, University of Colorado, Boulder, 293 pp.

Stein, R., Nam, S.-I., Grobe, H., and Hubberten, H., 1996: Late Quaternary glacial history and short-term ice-rafted debris fluctuations along the East Greenland continental margin. In: *Late Quaternary Paleoceanography of North Atlantic Margins* [Andrews, J.T., W.A. Austen, H. Bergsetn, and A.E. Jennings (eds.)]. Geological Society, London, pp. 135-151.

Stirling, C.H., T.M. Esat, M.T. McCulloch, and K. Lambeck, 1995: High-precision U-series dating of corals from Western Australia and implications for the timing and duration of the last interglacial. *Earth and Planetary Science Letters*, **135**, 115-130.

Stirling, C.H., T.M. Esat, K. Lambeck, and M.T. McCulloch, 1998: Timing and duration of the last interglacial—Evidence for a restricted interval of widespread coral reef growth. *Earth and Planetary Science Letters,* **160**, 745-762.

Stirling, C.H. T.M., Esat, K. Lambeck, M.T. McCulloch, S.G. Blake, D.-C. Lee, and A.N. Halliday, 2001: Orbital forcing of the marine isotope stage 9 interglacial. *Science*, **291**, 290-293.

Stocker, T.F. and S.J. Johnsen, 2003: A minimum thermodynamic model for the bipolar seesaw. *Paleoceanography*, **18**, Art. No. 1087.

Stoner, J.S., J.E.T. Channell, and C. Hillaire-Marcel, 1995: Magnetic properties of deep-sea sediments off southwest Greenland—Evidence for major differences between the last two deglaciations. *Geology*, **23**: 241-244.

Stuiver, M., T.F. Braziunas, P.M. Grootes, and G.A. Zielinski, 1997: Is there evidence for solar forcing of climate in the GISP2 oxygen isotope record? *Quaternary Research*, **48**, 259-266.

Stuiver, M., P.J. Reimer, E. Bard, J.W. Beck, K.A. Hughen, B. Kromer, F.G. McCormack, J. van der Plicht, and M. Spurk, 1998: INTCAL98 Radiocarbon age calibration 24,000–0 cal BP. *Radiocarbon*, **40**, 1041-1083.

Sugden, D. and B. John, 1976: *Glaciers and Landscape*. Arnold, London, and Halsted, New York, 376 pp.

Suwa, M., J.C. von Fischer, M.L. Bender, A. Landais, and E.J. Brook, 2006: Chronology reconstruction for the disturbed bottom section of the GISP2 and the GRIP ice cores—Implications for Termination II in Greenland. *Journal of Geophysical Research*, **111(D2)**, Art. No. D02101.

Syvitski, J.P.M., J.T. Andrews, and J.A. Dowdeswell, 1996: Sediment deposition in an iceberg-dominated glacimarine environment, East Greenland—Basin fill implications. *Global and Planetary Change*, **12**, 251-270.

Syvitski, J.P.M., A. Stein, J.T. Andrews, and J.D. Milliman, 2001: Icebergs and seafloor of the East Greenland (Kangerdlussuaq) continental margin. *Arctic, Antarctic and Alpine Research*, **33**, 52-61.

Tang, C.C.L., C.K. Ross, and T. Yao, 2004: The circulation, water masses and sea-ice of Baffin Bay. *Progress in Oceanography*, **63**, 183-228.

Tarasov, L. and W. Richard Peltier, 2002: Greenland glacial history and local geodynamic consequences. *Geophysical Journal International*, **150(1)**, 198-229, doi:10.1046/j.1365-246X.2002.01702.x

Tarasov, L. and W.R. Peltier, 2003: Greenland glacial history, borehole constraints, and Eemian extent. *Journal of Geophysical Research*, **108(B3)**, Art. No. 2143.

Tarduno, J.A., D.B. Brinkman, P.R. Renne, R.D. Cottrell, H. Scher and P. Castillo, 1998: Evidence for extreme climatic warmth from Late Cretaceous Arctic vertebrates. *Science,* **282**, 2241-2244.

Thiede, J., A. Winkler, T. Wolf-Welling, O. Eldholm, A.M. Myhre, K.H. Baumann, R. Henrick, and R. Stein, 1998: Late Cenozoic history of the Polar North Atlantic—Results from ocean drilling. *Quaternary Science Reviews*, **17**, 185-208.

Thomas, R.H. and PARCA Investigators, 2001: Program for Arctic Regional Climate Assessment (PARCA)—Goals, key findings, and future directions. *Journal of Geophysical Research*, **106(D24),** 33691-33705.

Thomas, R., W. Abdalati, E. Frederick, W. Krabill, S. Manizade, and K. Steffen, 2003: Investigation of surface melting and dynamic thinning on Jakobshavn Isbrae, Greenland. *Journal of Glaciology*, **49**, 231-239.

Thomas, R., E. Frederick, W. Krabill, S. Manizade, and C. Martin, 2006: Progressive increase in ice loss from Greenland. *Geophysical Research Letters*, **33**, L10503, doi:10.1029/2006GL026075.

Thompson, W.G. and S.L. Goldstein, 2005: Open-system coral ages reveal persistent suborbital sea-level cycles. *Science*, **308**, 401-404.

Toniazzo, T., J.M. Gregory, and P. Huybrechts, 2004: Climatic impact of a Greenland deglaciation and its possible irreversibility. *Journal of Climate*, **17**, 21-33.

Vandermark, D., J.A. Tarduno, and D.B. Brinkman, 2007: A fossil champsosaur population from the High Arctic—Implications for Late Cretaceous paleotemperatures. *Palaeogeography, Palaeoclimatology, Palaeoecology,* **248**, 49-59.

van der Plicht, J., B. Van Geel, S.J.P. Bohncke, J.A.A. Bos, M. Blaauw, A.O.M. Speranza, R. Muscheler, and S. Björck, 2004: The Preboreal climate reversal and a subsequent solar-forced climate shift. *Journal of Quaternary Science*, **19**, 263-269.

van der Veen, C.J.,1999: *Fundamentals of Glacier Dynamics.* Balkema, Rotterdam, 462 pp.

van Kreveld, S., M. Sarthein, H. Erlenkeuser, P. Grootes, S. Jung, M.J. Nadeau, U. Pflaumann, and A. Voelker, 2000: Potential links between surging ice sheets, circulation changes, and the Dansgaard-Oeschger cycles in the Irminger Sea, 60–18 ka. *Paleoceanography,* **15**, 425-442.

Velicogna, I. and J. Wahr, 2006: Acceleration of Greenland ice mass loss in spring 2004. *Nature*, **443**, 329-331.

Vorren, T.O. and J.S. Laberg, 1997: Trough mouth fans—Palaeoclimate and ice-sheet monitors. *Quaternary Science Reviews*, **16**, 865-886.

Weidick, A. 1993: Neoglacial change of ice cover and the related response of the Earth's crust in West Greenland. *Grønlands Geologiske Undersøgelse Rapport*, **159**, 121-126.

Weidick, A., 1996: Neoglacial changes of ice cover and sea level in Greenland—A classical enigma. In: *The Paleo-Eskimo Cultures of Greenland* [Gronnow, B. and J. Pind (eds.)]. Danish Polar Center, Copenhagen, pp. 257-270.

Weidick, A., H. Oerter, N. Reeh, H.H. Thomsen, and L. Thorning, 1990: The recession of the Inland Ice margin during the Holocene Climatic Optimum in the Jakobshavn-Isfjord area of West Greenland. *Global and Planetary Change*, **82**, 389-399.

Weidick, A., M. Kelly, and O. Bennike, 2004: Later Quaternary development of the southern sector of the Greenland Ice Sheet, with particular reference to the Qassimiut lobe. *Boreas*, **33**, 284-299.

Whillans, I.M., 1976: Radio-echo layers and recent stability of West Antarctic ice sheet. *Nature*, **264**, 152-155.

Wilken, M. and J. Mienert, 2006: Submarine glacigenic debris flows, deep-sea channels and past ice-stream behaviour of the East Greenland continental margin. *Quaternary Science Reviews*, **25**, 784-810.

Willerslev, E., E. Cappellini, W. Boomsma, R. Nielsen, M.B. Hebsgaard, T.B. Brand, M. Hofreiter, M. Bunce, H.N. Poinar, D. Dahl-Jensen, S. Johnsen, J.P. Steffensen, O. Bennike, J.L. Schwenninger, R. Nathan, S. Armitage, C.J. de Hoog, V. Alfimov, M. Christi, J. Beer, R. Muscheler, J. Barker, M. Sharkp, K.E.H. Penkman, J. Haile, P. Taberlet, M.T.P. Gilbert, A. Casoli, E. Campani, and M.J. Collins, 2007: Ancient biomolecules from deep ice cores reveal a forested southern Greenland. *Science*, **317**, 111-114.

Zielinski, G.A., P.A. Mayewski, L.D. Meeker, S. Whitlow, M.S. Twickler, M. Morrison, D.A. Meese, A.J. Gow, and R.B. Alley, 1994: Record of volcanism since 7000 B.C. from the GISP2 *Greenland* ice core and implications for the volcano-climate system. *Science*, **264**, 948-950.

Zwally, H.J., W. Abdalati, T. Herring, K. Larson, J. Saba, and K. Steffen, 2002: Surface melt-induced acceleration of Greenland ice-sheet flow. *Science*, **297**, 218-222.

Chapter 6. History of Arctic Sea Ice

Andersen, C., N. Koç, and M. Moros, 2004: A highly unstable Holocene climate in the subpolar North Atlantic: evidence from diatoms. *Quaternary Science Reviews*, **23**, 2155-2166.

Anderson, R.K., G.H. Miller, J.P. Briner, N.A. Lifton, and S.B. DeVogel, 2008: A millennial perspective on Arctic warming from ^{14}C in quartz and plants emerging from beneath ice caps. *Geophysical Research Letters*, **35**, L01502, doi:10.1029/2007GL032057.

Andrews, J.T., 2000: Icebergs and iceberg rafted detritus (IRD) in the North Atlantic—Facts and assumptions. *Oceanography*, **13**, 100-108.

Andrews, J.T., 2007: A moderate resolution, definitive(?) record for Holocene variations in ice rafting around Iceland. *37th Arctic Workshop Program and Abstracts*, pp. 34-35.

Andrews, J.T. and D.D. Eberl, 2007: Quantitative mineralogy of surface sediments on the Iceland shelf, and application to down-core studies of Holocene ice-rafted sediments. *Journal Sedimentary Research*, **77**, 469-479.

Andrews, J.T., A.E. Jennings, M. Moros, C. Hillaire-Marcel, and D.D. Eberl, 2006: Is there a pervasive Holocene ice-rafted debris (IRD) signal in the northern North Atlantic? The answer appears to be either no, or it depends on the proxy! *PAGES Newsletter*, **14**, 7-9.

Backman, J., M. Jakobsson, M. Frank, F. Sangiorgi, H. Brinkhuis, C. Stickley, M. O'Regan, R. Løvlie, H. Pälike, D. Spofforth, J. Gattacecca, K. Moran, J. King, and C. Heil, 2008: Age model and core-seismic integration for the Cenozoic Arctic Coring Expedition sediments from the Lomonosov Ridge. *Paleoceanography*, **23**, PA1S03, doi:10.1029/2007PA001476.

Backman, J., M. Jakobsson, R. Løvlie, L. Polyak, and L.A. Febo, 2004: Is the central Arctic Ocean a sediment starved basin? *Quaternary Science Reviews*, **23**, 1435-1454.

Backman, J., K. Moran, D.B. McInroy, L.A. Mayer, and the Expedition 302 scientists, 2006: *Proceedings of IODP, 302*. Edinburgh (Integrated Ocean Drilling Program Management International, Inc.), doi:10.2204/iodp.proc.302.2006.

Belt, S.T., G. Masse, S.J. Rowland, M. Poulin, C. Michel, and B. LeBlanc, 2007: A novel chemical fossil of palaeo sea ice: IP25. *Organic Geochemistry*, **38**, 16-27.

Bennike, O., 2004: Holocene sea-ice variations in Greenland—Onshore evidence. *The Holocene*, **14**, 607-613.

Bennike, O. and J. Böcher, 1990: Forest-tundra neighboring the North Pole—Plant and insect remains from Plio-Pleistocene Kap København Formation, North Greenland. *Arctic*, **43(4)**, 331-338.

Berger, A. and M.F. Loutre, 2004: An exceptionally long interglacial ahead? *Science*, **297(5585)**, 1287-1288.

Bergthorsson, P., 1969: An estimate of drift ice and temperature in Iceland in 1000 years. *Jokull*, **19**, 94-101.

Blake, W., Jr., 1975: Radiocarbon age determination and postglacial emergence at Cape Storm, southern Ellesmere Island, Arctic Canada. *Geografiska Annaler*, **57**, 1-71.

Blake, W., Jr., 2006: Occurrence of the Mytilus edulis complex on Nordaustlandet, Svalbard— Radiocarbon ages and climatic implications. *Polar Research*, **25(2)**, 123-137.

Bond, G., B. Kromer, J. Beer, R. Muscheler, M.N. Evans, W. Showers, S. Hoffman, R. Lotti-Bond, I. Hajdas, G. Bonani, 2001: Persistent solar influence on North Atlantic climate during the Holocene. *Science*, **294**, 2130-2136.

Bond, G., W. Showers, M. Cheseby, R. Lotti, P. Almasi, P. deMenocal, P. Priore, H. Cullen, I. Hajdas, and G. Bonani, 1997: A pervasive millennial-scale cycle in North Atlantic Holocene and glacial climates. *Science*, **278**, 1257-1265.

Brigham-Grette, J. and L.D. Carter, 1992: Pliocene marine transgressions of northern Alaska—Circumarctic correlations and paleoclimate. *Arctic*, **43(4)**, 74-89.

Brigham-Grette, J. and D.M. Hopkins, 1995: Emergent-marine record and paleoclimate of the last interglaciation along the northwest Alaskan coast. *Quaternary Research*, **43**, 154-173.

Brigham-Grette, J., D.M. Hopkins, V.F. Ivanov, A. Basilyan, S.L. Benson, P. Heiser, and V. Pushkar, 2001: Last interglacial (Isotope Stage 5) glacial and sea level history of coastal Chukotka Peninsula and St. Lawrence Island, western Beringia. *Quaternary Science Reviews*, **20(1-3)**, 419-436.

CircumArctic PaleoEnvironments (CAPE)—Last Interglacial Project Members, 2006: Last Interglacial Arctic warmth confirms polar amplification of climate change. *Quaternary Science Reviews*, **25**, 1383-1400.

Carter, L.D., J. Brigham-Grette, L. Marincovich, Jr., V.L. Pease, and U.S. Hillhouse, 1986: Late Cenozoic Arctic Ocean sea ice and terrestrial paleoclimate. *Geology*, **14**, 675-678.

Cavalieri, D.J., C.L. Parkinson, and K.Y. Vinnikov, 2003: 30-Year satellite record reveals contrasting Arctic and Antarctic sea ice variability, *Geophysical Research Letters*, **30(18)**, 1970, doi:10.1029/2003GL018031.

Clark, D.L., L.A. Chern, J.A. Hogler, C.M. Mennicke, and E.D. Atkins, 1990: Late Neogene climatic evolution of the central Arctic Ocean. Marine *Geology*, **93**, 69-94.

Comiso, J.C., C.L. Parkinson, R. Gersten, and L. Stock, 2008: Accelerated decline in the Arctic sea ice cover. *Geophysical Research Letters*, **35**, L01703, doi:10.1029/2007GL031972.

Cronin, T.M., T.R. Holtz, Jr., R. Stein, R. Spielhagen, D. Fütterer, and J. Wollenburg, 1995: Late Quaternary paleoceanography of the Eurasian Basin, Arctic Ocean. *Paleoceanography*, **10**, 259-281.

Cronin, T.M., S.A. Smith, F. Eynaud, M. O'Regan, and J. King, 2008: Quaternary paleoceanography of the central Arctic based on IODP ACEX 302 foraminiferal assemblages. *Paleoceanography*, **23**, PA1S18, doi:10.1029/2007PA001484.

Curran, M.A.J., T. van Ommen, V. I. Morgan, K.L. Phillips, and A.S. Palmer, 2003: Ice core evidence for Antarctic sea ice decline since the 1950s. *Science*, **302**, 1203-1206.

Darby, D.A., 2003: Sources of sediment found in the sea ice from the western Arctic Ocean—New insights into processes of entrainment and drift patterns. *Journal of Geophysical Research*, **108**, 13-1 to 13-10, doi: 10, 1111029/1112002JC1001350, 1112003.

Darby, D.A., 2008: Arctic perennial ice cover over the last 14 million years. *Paleoceanography*, **23**, PA1S07, doi:10.1029/2007PA001479.

Darby, D.A. and J. Bischof, 2004: A Holocene record of changing Arctic Ocean ice drift analogous to the effects of the Arctic Oscilation. *Paleoceanography*, **19(1 of 9)**, 002004, doi:10.1029/2003PA000961.

Darby, D.A., M. Jakobsson, and L. Polyak, 2005: Icebreaker expedition collects key Arctic seafloor and ice data. *Eos, Transactions of the American Geophysical Union*, **86(52)**, 549-556.

Darby, D.A., L. Polyak, and H. Bauch, 2006: Past glacial and interglacial conditions in the Arctic Ocean and marginal seas—A review. In: Structure and function of contemporary food webs on Arctic shelves—A Pan-Arctic comparison, [Wassman, P. (ed.)]. *Progress in Oceanography*, **71**, 129-144.

de Vernal, A. and C. Hillaire-Marcel, 2000: Sea-ice cover, sea-surface salinity and halo/thermocline structure of the northwest North Atlantic—Modern versus full glacial conditions. *Quaternary Science Reviews*, **19**, 65-85.

Delworth, T.L., S. Manabe, and R.J. Stouffer, 1997: Multidecadal climate variability in the Greenland Sea and surrounding regions: A coupled model simulation. *Geophysical Research Letters*, **24**, 257-260.

Devaney, J.R., 1991: Sedimentological highlights of the Lower Triassic Bjorne Formation, Ellesmere Island, Arctic Archipelago. In: *Current Research Part B, Geological Survey of Canada Paper 91-1B*, pp. 33-40.

Dowdeswell, J.A., R.J. Whittington, and P. Marienfeld, 1994: The origin of massive diamicton facies by iceberg rafting and scouring, Scorsby Sund, East Greenland. *Sedimentology*, **41**, 21-35.

Duk-Rodkin, A., R.W. Barendregt, D.G. Froese, F. Weber, R.J. Enkin, I.R. Smith, G.D. Zazula, P. Waters, and R. Klassen, 2004: Timing and extent of Plio-Pleistocene glaciations in north-western Canada and east-central Alaska. In: *Quaternary Glaciations—Extent and Chronology, Part II, North America*, [Ehlers, J. and P.L. Gibbard (eds.)]. Elsevier, Amsterdam, pp. 313-345.

Dunhill, G., 1998: *Comparison of Sea-Ice and Glacial-Ice Rafted Debris—Grain Size, Surface Features, and Grain Shape*. U.S. Geological Survey Open-File Report 98–0367, 74 pp.

Dyke, A.S., J. England, E. Reimnitz, and H. Jetté, 1997: Changes in driftwood delivery to the Canadian Arctic Archipelago—The hypothesis of postglacial oscillations of the Transpolar Drift. *Arctic,* **50**, 1-16.

Dyke, A.S., J. Hooper, C.R. Harington, and J.M. Savelle, 1999: The Late Wisconsinan and Holocene record of walrus (*Odobenus rosmarus*) from North America—A review with new data from Arctic and Atlantic Canada. *Arctic,* **52**, 160-181.

Dyke, A.S., J. Hooper, and J.M. Savelle, 1996: A history of sea ice in the Canadian Arctic Archipelago based on the postglacial remains of the bowhead whale (*Balaena mysticetus*). *Arctic,* **49**, 235-255.

Eggertsson, O., 1993: Origin of the driftwood on the coasts of Iceland—A dendrochronological study. *Jokull*, **43**, 15-32.

Eiriksson, J., K.L. Knudsen, H. Haflidason, and P. Henriksen, 2000: Late-glacial and Holocene paleoceanography of the North Iceland Shelf. Journal of Quaternary *Science,* **15**, 23-42.

Eldrett, J.S., I.C. Harding, P.A. Wilson, E. Butler, and A.P. Roberts, 2007: Continental ice in Greenland during the Eocene and Oligocene. *Nature*, **446**, 176-179.

Feyling-Hanssen, R.W., S. Funder, and K.S. Petersen, 1983: The Lodin Elv Formation—A Plio/Pleistocene occurrence in Greenland. *Bulletin of the Geological Society of Denmark*, **31**, 81-106.

Fischer, H., 2001: Imprint of large-scale atmospheric transport patterns on sea-salt records in northern Greenland ice cores. *Journal of Geophysical Research,* **106**, 23977-23984.

Fischer, H. and B. Mieding, 2005: A 1,000-year ice core record of interannual to multidecadal variations in atmospheric circulation over the North Atlantic. *Climate Dynamics,* **25**, 65-74.

Fischer, H., F. Fundel, U. Ruth, B. Twarloh, A. Wegner, R. Udisti, S. Becagli, E. Castellano, et al., 2007a: Reconstruction of millennial changes in transport, dust emission and regional differences in sea ice coverage using the deep EPICA ice cores from the Atlantic and Indian Ocean sector of Antarctica, *Earth and Planetary Science Letters*, **260**, 340-354.

Fischer, H., M.L. Siggaard-Andersen, U. Ruth, R. Röthlisberger, and E.W. Wolff, 2007b: Glacial-interglacial changes in mineral dust and sea salt records in polar ice cores—Sources, transport, deposition. Reviews of Geophysics, 45, RG1002, Eldrett, J.S., I.C. Harding, P.A. Wilson, E. Butler, and A.P. Roberts, 2007: Continental ice in Greenland during the Eocene and Oligocene. *Nature*, **446**, 176-179.

Fisher, D.A., R.M. Koerner, J.C. Bourgeois, G. Zielinski, C. Wake, C.U. Hammer, H.B. Clausen, N. Gundestrup, S. Johnsen, K. Goto-Azuma, T. Hondoh, E. Blake, and M. Gerasimoff, 1998: Penny Ice Cap cores, Baffin Island, Canada, and the Wisconsinan Foxe Dome connection—Two states of Hudson Bay ice cover. *Science,* **279**, 692-695.

Francis, J.A. and E. Hunter, 2006: New insight into the disappearing Arctic sea ice. *Eos, Transactions of the American Geophysical Union*, **87**, 509-511.

Francis, J.E., 1988: A 50-million-year-old fossil forest from Strathcona Fiord, Ellesmere Island, Arctic Canada—Evidence for a warm polar climate. *Arctic,* **41(4)**, 314-318.

Funder, S., N. Abrahamsen, O. Bennike, and R.W. Feyling-Hansen, 1985: Forested Arctic—Evidence from North Greenland, *Geology* **13**, 542-546.

Funder, S., O. Bennike, J. Böcher, C. Israelson, K.S. Petersen, and L.A. Simonarson, 2001: Late Pliocene Greenland—The Kap København formation in North Greenland. *Bulletin of the Geological Society of Denmark*, **48**, 77-134.

Funder, S., I. Demidov, and Y. Yelovicheva, 2002: Hydrography and mollusc faunas of the Baltic and the White Sea-North Sea seaway in the Eemian. *Palaeogeography, Palaeoclimatology, Palaeoecology*, **184**, 275-304.

Funder, S. and K. Kjær, 2007: Ice free Arctic Ocean, an early Holocene analogue. *Eos, Transactions of the American Geophysical Union*, **88(52)**, Fall Meeting Supplement, Abstract PP11A-0203.

Fyles, J.G., 1990: Beaufort Formation (late Tertiary) as seen from Prince Patrick Island, arctic Canada. *Arctic,* **43**, 393-403.

Fyles, J.G., L.V. Hills, J.V. Mathews, Jr., R.W. Barendregt, J. Baker, E. Irving, and H. Jetté, 1994: Ballast Brook and Beaufort Formations (Late Tertiary) on northern Banks Island, Arctic Canada. *Quaternary International*, **22/23**, 41-171.

Fyles, J.G., L. Marincovich Jr., J.V. Mathews Jr., and R. Barendregt, 1991: Unique mollusc find in the Beaufort Formation (Pliocene) Meighen Island, Arctic Canada. In: *Current Research, Part B, Geological Survey of Canada, Paper 91-1B*, pp. 461-468.

Fyles, J.G., D.H. McNeil, , J.V. Matthews, R.W. Barendregt, L. Marincovich, Jr., E. Brouwers, J. Bednarski, J. Brigham-Grette, L.O. Ovenden, J. Baker, and E. Irving, 1998: Geology of the Hvitland Beds (Late Pliocene), White Point Lowland, Ellesmere Island, Arctic Canada. *Geological Survey of Canada Bulletin*, **512**, 1-35.

Giraudeau, J., A.E. Jennings, and J.T. Andrews, 2004: Timing and mechanisms of surface and intermediate water circulation changes in the Nordic Seas over the last 10,000 cal years—A view from the North Iceland shelf. *Quaternary Science Reviews*, **23**, 2127-2139.

Goosse, H., E. Driesschaert, T. Fichefet, and M.-F. Loutre, 2007: Information on the early Holocene climate constrains the summer sea ice projections for the 21st century. *Climate of the Past Discussions*, **2**, 999-1020.

Gow, A.J. and W.B. Tucker III, 1987: Physical properties of sea ice discharge from Fram Strait. *Science*, **236**, 436-439.

Grumet, N.S., C.P. Wake, P.A. Mayewski, G.A. Zielinski, S.I. Whitlow, R.M. Koerner, D.A. Fisher, and J.M. Woollett, 2001: Variability of sea-ice extent in Baffin Bay over the last millennium. *Climatic Change*, **49**, 129-145.

Guelle, W., M. Schulz, Y. Balkanski, and F. Dentener, 2001: Influence of the source formulation on modeling the atmospheric global distribution of sea salt aerosol. *Journal of Geophysical Research*, **106**, 27509-27524.

Haggblom, A., 1982: Driftwood in Svalbard as an indicator of sea ice conditions. *Geografiska AnnalerSeries A, Physical Geography*, **64A**, 81-94.

Harington, C.R., 2003: *Annotated Bibliography of Quaternary Vertebrates of Northern North America With Radiocarbon Dates*. University of Toronto Press, Toronto, 539 pp.

Hastings, A.D., 1960: Environment of Southeast Greenland. Quartermaster Research and Engineering Command U.S. Army Technical Report EP-140, 48 pp.

Hebbeln, D., 2000: Flux of ice-rafted detritus from sea ice in the Fram Strait. *Deep-Sea Research II*, **47**, 1773-1790.

Helland, P.E. and M.A. Holmes, 1997: Surface textural analysis of quartz sand grains from ODP Site 918 off the southeast coast of Greenland suggests glaciation of 24 southern Greenland at 11 Ma. *Palaeogeography, Palaeoclimatology, Palaeoecology*, **135**, 109-121.

Herman, Y., 1974: Arctic Ocean sediments, microfauna, and the climatic record in late Cenozoic time. In: *Marine Geology and Oceanography of the Arctic Seas* [Herman, Y. (ed.)]. Springer-Verlag, Berlin, pp. 283-348.

Holland, M.M., C.M. Bitz, M. Eby, and A.J. Weaver, 2001: The role of ice-ocean interactions in the variability of the north Atlantic thermohaline circulation. *Journal of Climate*, **14**, 656-675.

Holland, M.M., C.M. Bitz, and B. Tremblay, 2006a: Future abrupt reductions in the summer Arctic sea ice. *Geophysical Research Letters*, **33**, L23503, doi: 10.1029/2006GL028024.

Holland, M.M., J. Finnis, and M.C. Serreze, 2006b: Simulated Arctic Ocean freshwater budgets in the 20th and 21st centuries. *Journal of Climate*, **19**, 6221-6242.

Hutterli, M.A., T. Crueger, H. Fischer, K.K. Andersen, C.C. Raible, T.F. Stocker, M.L. Siggaard-Andersen, J.R. McConnell, R.C. Bales, and J. Burkhardt, 2007: The influence of regional circulation patterns on wet and dry mineral dust and sea salt deposition over Greenland. *Climate Dynamics*, **28**, 635-647.

Isaksson, E., T. Kekonen, J.C. Moore, and R. Mulvaney, 2005: The methanesulfonic acid (MSA) record in a Svalbard ice core. *Annals of Glaciology*, **42**, 345-351.

Jakobsson, M., J. Backman, B. Rudels, J. Nycander, M. Frank, L. Mayer, W. Jokat, F. Sangiorgi, M. O'Regan, H. Brinkhuis, J. King, and K. Moran, 2007: The Early Miocene onset of a ventilated circulation regime in the Arctic Ocean. *Nature*, **447(21)**, 986-990. doi:10.1038/Nature05924.

Jennings, A.E., K. Grönvold, R. Hilberman, M. Smith, and M. Hald, 2002: High-resolution study of Icelandic tephras in the Kangerlussuq Trough, Southeast Greenland, during the last deglaciation. *Journal of Quaternary Science*, **7**, 747-757.

Jennings, A.E. and N.J. Weiner, 1996: Environmental change in eastern Greenland during the last 1300 years—Evidence from foraminifera and lithofacies in Nansen Fjord, 68°N. *The Holocene*, **6**, 179-191.

Jennings, A.E., N.J. Weiner, G. Helgadottir, and J.T. Andrews, 2004: Modern foraminiferal faunas of the Southwest to Northern Iceland shelf—Oceanographic and environmental controls. *Journal of Foraminiferal Research*, **34**, 180-207.

Jones, P.D., T.J. Osborn, and K.R. Briffa, 2001: The evolution of climate over the last millennium. *Science*, **292**, 662-667.

Kaufman, D.S., 1991: *Pliocene-Pleistocene Chronostratigraphy, Nome, Alaska*. Ph.D. dissertation, University of Colorado, Boulder, 297 pp.

Kaufman, D.S., T.A. Ager, N.J. Anderson, P.M. Anderson, J.T. Andrews, P.J. Bartlein, L.B. Brubaker, L.L. Coats, L.C. Cwynar, M.L. Duvall, A.S. Dyke, M.E. Edwards, W.R. Eisner, K. Gajewski, A. Geirsdóttir, F.S. Hu, A.E. Jennings, M.R. Kaplan, M.W. Kerwin, A.V. Lozhkin, G.M. MacDonald, G.H. Miller, C.J. Mock, W.W. Oswald, B.L. Otto-Bliesner, D.F. Porinchu, K. Rühland, J.P. Smol, E.J. Steig, and B.B. Wolfe, 2004. Holocene thermal maximum in the western Arctic (0-180°W). *Quaternary Science Reviews*, **23**, 529-560.

Kaufman, D.S. and J. Brigham-Grette, 1993: Aminostratigraphic correlations and paleotemperature implications, Pliocene-Pleistocene high sea level deposits, northwestern Alaska. *Quaternary Science Reviews*, **12**, 21-33.

Kinnard, C., C.M. Zdanowicz, D.A. Fisher, and C.P. Wake, 2006: Calibration of an ice-core glaciochemical (sea-salt) record with sea-ice variability in the Canadian Arctic. *Annals of Glaciology*, **44**, 383-390.

Kinnard, C., C.M. Zdanowicz, R. Koerner, and D.A. Fisher, 2008: A changing Arctic seasonal ice zone—Observations from 1870–2003 and possible oceanographic consequences. *Geophysical Research Letters*, **35**, L02507, doi:10.1029/2007GL032507.

Knies, J. and C. Gaina, 2008: Middle Miocene ice sheet expansion in the Arctic—Views from the Barents Sea. *Geochemistry, Geophysics, Geosystems*, **9**, Q02015, doi:10.1029/2007GC001824.

Knies, J., J. Matthiessen, C. Vogt, and R. Stein, 2002: Evidence of "mid-Pliocene (similar to 3 Ma) global warmth" in the eastern Arctic Ocean and implications for the Svalbard/Barents Sea ice sheet during the late Pliocene and early Pleistocene (similar to 3–1.7 Ma). *Boreas*, **31(1)**, 82-93.

Koç, N. and E. Jansen, 1994: Response of the high-latitude Northern Hemisphere to orbital climate forcing—Evidence from the Nordic Seas. *Geology*, **22**, 523-526.

Koch, L., 1945: The East Greenland Ice. *Meddelelser om Grønland*, **130(3)**, 1-375.

Krylov, A.A., I.A. Andreeva, C. Vogt, J. Backman, V.V. Krupskaya, G.E. Grikurov, K. Moran, and H. Shoji, 2008: A shift in heavy and clay mineral provenance indicates a middle Miocene onset of a perennial sea-ice cover in the Arctic Ocean. *Paleoceanography*, **23**, PA1S06, doi:10.1029/2007PA001497.

LePage, B.A., H. Yang, and M. Matsumoto, 2005: The evolution and biogeographic history of Metasequoia. In: *The Geobiology and Ecology of Metasequoia* [LePage, B.A., C.J. Williams, and H. Yang (eds.)]. Springer, New York, Chapter 1, pp. 4-81.

Levac, E., A. de Verna, and W.J. Blake, 2001: Sea-surface conditions in northernmost Baffin Bay during the Holocene—Palynological evidence. Journal of Quaternary *Science*, **16**, 353-363.

Levermann, A., J. Mignot, S. Nawrath, and S. Rahmstorf, 2007: The role of Northern sea ice cover for the weakening of the thermohaline circulation under global warming. *Journal of Climate*, doi:10.1175/JCLI4232.1.

Lisiecki, L.E. and M.E. Raymo, 2005: A Pliocene-Pleistocene stack of 57 globally distributed benthic $\delta^{18}O$ records. *Paleoceanography*, **20**, doi:10.1029 2004PA001071.

Lisitzin, A.P., 2002: *Sea-Ice and Iceberg Sedimentation in the Ocean, Recent and Past.* Springer-Verlag, Berlin, 563 pp.

Lowenstein, T.K. and Demicco, R.V., 2006: Elevated Eocene atmospheric CO_2 and its subsequent decline. *Science*, **313**, 1928.

Magnusdottir, G., C. Deser, and R. Saravanan, 2004: The effects of North Atlantic SST and sea ice anomalies on the winter circulation in CCSM3, Part I—Main features and storm track characteristics of the response, *Journal of Climate*, **17**, 857-876.

Mann, M.E., R.S. Bradley, and M.K. Hughes, 1999: Northern Hemisphere temperatures during the millennium—Inferences, uncertainties, and limitations. *Geophysical Research Letters*, **26**, 759-764.

Maslanik, J., S. Drobot, C. Fowler, W. Emery, and R. Barry, 2007a: On the Arctic climate paradox and the continuing role of atmospheric circulation in affecting sea ice conditions. *Geophysical Research Letters*, 10.1029/2006GL028269.

Maslanik, J.A, C. Fowler, J. Stroeve, S. Drobot, and J. Zwally, 2007b: A younger, thinner Arctic ice cover—Increased potential for rapid, extensive ice loss, *Geophysical Research Letters*, **34**, L24501, doi:10.1029/2007GL032043.

Matthews, J.V., Jr., 1987: Plant macrofossils from the Neogene Beaufort Formation on Banks and Meighen islands, District of Franklin. In: *Current Research, Part A*, Geological Survey of Canada Paper 87-1A, pp. 73-87.

Matthews, J.V., Jr. and L.E. Ovenden, 1990: Late Tertiary plant macrofossils from localities in northern North America (Alaska, Yukon, and Northwest Territories). *Arctic*, **43(2)**, 364-392.

Matthews, J.V., Jr. and A. Telka, 1997. Insect fossils from the Yukon. In: *Insects of the Yukon* [Danks, H.V. and J.A. Downes (eds.)]. Biological Survey of Canada (Terrestrial Arthopods), Ottawa, pp. 911-962.

Mauritzen, C. and S. Hakkinen, 1997: Influence of sea ice on the thermohaline circulation in the Arctic-North Atlantic Ocean. *Geophysical Research Letters*, **24**, 3257-3260.

Mayewski, P.A., L.D. Meeker, S. Whitlow, M.S. Twickler, M.C. Morrison, P. Bloomfield, G.C. Bond, R.B. Alley, A.J. Gow, P.M. Grootes, D.A. Meese, M. Ram, K.C. Taylor, and W. Wumkes, 1994: Changes in atmospheric circulation and ocean ice cover over the North Atlantic during the last 41,000 years. *Science*, **263**, 1747-1751.

McKenna, M.C., 1980. Eocene paleolatitude, climate and mammals of Ellesmere Island. *Paleogeography, Paleoclimatology and Paleoecology*, **30**, 349-362.

McNeil, D.H., 1990: Tertiary marine events of the Beaufort-Mackenzie Basin and correlation of Oligocene to Pliocene marine outcrops in Arctic North America. *Arctic*, 1990, **43(4)**, 301-313.

Miller, G.H., 1985: Aminostratigraphy of Baffin Island shell-bearing deposits. In: *Late Quaternary Environments—Eastern Canadian Arctic, Baffin Bay and West Greenland*, [Andrews, J.T. (ed.)]. Allen and Unwin Publishers, Boston, pp. 394-427.

Moran, K., J. Backman, H. Brinkhuis, S.C. Clemens, T. Cronin, G.R. Dickens, F. Eynaud, J. Gattacceca, M. Jakobsson, R.W. Jordan, M. Kaminski, J. King, N. Koç, A. Krylov, N. Martinez, J. Matthiessen, D. McInroy, T.C. Moore, J. Onodera, A.M. O'Regan, H. Pälike, B. Rea, D. Rio, T. Sakamoto, D.C. Smith, R. Stein, K. St. John, I. Suto, N. Suzuki, K. Takahashi, M. Watanabe, M. Yamamoto, J. Farrell, M. Frank, P. Kubik, W. Jokat, and Y. Kristoffersen, 2006: The Cenozoic palaeoenvironment of the Arctic Ocean. *Nature*, **441**, 601-605.

Moros, M., J.T. Andrews, D.D. Eberl, and E. Jansen, 2006: The Holocene history of drift ice in the northern North Atlantic—Evidence for different spatial and temporal modes. *Palaeoceanography*, **21(1 of 10)**, doi:10.1029/2005PA001214.

Moros, M., K. Emeis, B. Risebrobakken, I. Snowball, A. Kuijpers, J. McManus, E. Jansen, 2004: Sea surface temperatures and ice rafting in the Holocene North Atlantic—Climate influences on northern Europe and Greenland. *Quaternary Science Reviews*, **23**, 2113-2126.

Mosher, B.W., P. Winkler, and J.-L. Jaffrezo, 1993: Seasonal aerosol chemistry at Dye 3, Greenland. *Atmospheric Environment*, **27A**, 2761-2772.

Mudie, P.J., A. Rochon, M.A. Prins, D. Soenarjo, S.R. Troelstra, E. Levac, D.B. Scott, L. Roncaglia, and A. Kuijpers, 2006: Late Pleistocene-Holocene marine geology of Nares Strait region—Palaeoceanography from foraminifera and dinoflagellate cysts, sedimentology and stable istopes. *Polarforshung*, **74**, 169-183.

Mullen, M.W. and D.H. McNeil, 1995: Biostratigraphic and paleoclimatic significance of a new Pliocene foraminiferal fauna from the central Arctic Ocean. *Marine Micropaleontology*, **26(1)**, 273-280.

Nørgaard-Pedersen, N., N. Mikkelsen, and Y. Kristoffersen, 2007a: Arctic Ocean record of last two glacial-interglacial cycles off North Greenland/Ellesmere Island—Implications for glacial history. Marine *Geology*, **244**, 93-108.

Nørgaard-Pedersen, N., N. Mikkelsen, S.J. Lassen, Y. Kristoffersen, and E. Sheldon, 2007b: Arctic Ocean sediment cores off northern Greenland reveal reduced sea ice concentrations during the last interglacial period. *Paleoceanography*, **22**, PA1218, doi:10.1029/2006PA001283.

O'Brien, S.R., P.A. Mayewski, L.D. Meeker, D.A. Meese, M.S. Twickler, and S.I. Whitlow, 1995: Complexity of Holocene climate as reconstructed from a Greenland ice core. *Science*, **270**, 1962-1964.

O'Regan, M., K. Moran, J. Backman, M. Jakobsson, F. Sangiorgi, H., Brinkhuis, R. A. Pockalny, A. Skelton, C. Stickley, N. Koç, and H. Brumsack, 2008: Mid-Cenozoic tectonic and paleoenvironmental setting of the central Arctic Ocean, *Paleoceanography*, **23**, PA1S20, doi:10.1029/2007PA001559.

Ogilvie, A., 1996: Sea-ice conditions off the coasts of Iceland A.D. 1601–1850 with special reference to part of the Maunder Minimum period (1675–1715). In: *North European climate data in the latter part of the Maunder Minimum period A.D. 1675–1715*. Stavanger Museum of Archaeology, Norway AmS-Varia 25, pp. 9-12.

Ogilvie, A.E., L.K. Barlow, and A.E. Jennings, 2000: North Atlantic Climate c. A.D. 1000—Millennial reflections on the Viking discoveries of Iceland, Greenland and North America. *Weather*, **55**, 34-45.

Ogilvie, A.E.J., 1984: The past climate and sea-ice record from Iceland, Part I—Data to A.D. 1780. *Climatic Change*, **6**, 131-152.

Oleinik, A., L. Marincovich, P. Swart, and R. Port, 2007: Cold Late Oligocene Arctic Ocean—Faunal and stable isotopic evidence. *EOS, Transactions of the American Geophysical Union*, **88(52)**, Abstract PP43D-05.

Otto-Bliesner, B.L., S.J. Marshall, J.T. Overpeck, G.H. Miller, A. Hu, and CAPE Last Interglacial Project members, 2006, Simulating Arctic climate warmth and icefield retreat in the Last Interglaciation. *Science*, **311**, 1751-1753. doi:10.1126/science.1120808

Overpeck, J., K. Hughen, D. Hardy, R. Bradley, R. Case, M. Douglas, B. Finney, K. Gajewski, G. Jacoby, A. Jennings, S. Lamoureux, A. Lasca, G. MacDonald, J. Moore, M. Retelle, S. Smith, A. Wolfe, and G. Zielinski, 1997: Arctic environmental changes of the last four centuries. *Science*, **278**, 1251-1256.

Pearson, P.N. and M.R. Palmer, 2000: Atmospheric carbon dioxide concentrations over the past 60 million years. *Nature*, **406**, 695-699.

Peterson, B.J., J. McClelland, R. Curry, R.M. Holmes, J.E. Walsh, and K. Aagaard, 2006: Trajectory shifts in the Arctic and Subarctic freshwater cycle. *Science*, **313**, 1061-1066.

Petit, J.R., J. Jouzel, D. Raynaud, N.I. Barkov, J.-M. Barnola, I. Basile, M. Bender, J. Chappellaz, M. Davis, G. Delaygue, M. Delmotte, V.M. Kotlyakov, M. Legrand, V.Y. Lipenkov, C. Lorius, L. Pepin, C. Ritz, E. Saltzman, and M. Stievenard, 1999: Climate and atmospheric history of the past 420,000 years from the Vostok ice core, Antarctica. *Nature*, **399**, 429-436.

Polyak, L., W.B. Curry, D.A. Darby, J. Bischof, and T.M. Cronin, 2004: Contrasting glacial/interglacial regimes in the western Arctic Ocean as exemplified by a sedimentary record from the Mendeleev Ridge. *Paleogeography, Paleoclimatology and Paleoecology*, **203**, 73-93.

Polyak, L., S. Korsun, L.A. Febo, V. Stanovoy, T. Khusid, M. Hald, B.E. Paulsen, and D.J. Lubinski, 2002: Benthic foraminiferal assemblages from the southern Kara Sea, a river influenced Arctic marine environment. *Journal of Foraminiferal Research*, **32**, 252-273.

Polyakov, I.V., A. Beszczynska, E.C. Carmack, I.A. Dmitrenko, E. Fahrbach, I.E. Frolov, R. Gerdes, E. Hansen, J. Holfort, V.V. Ivanov, M.A. Johnson, M. Karcher, F. Kauker, J. Morison, K.A. Orvik, U. Schauer, H.L. Simmons, Ø. Skagseth, V.T. Sokolov, M. Steele, L.A. Timokhov, D. Walsh, and J.E. Walsh, 2005: One more step toward a warmer Arctic. *Geophysical Research Letters,* **32**, L17605, doi:10.1029/2005GL023740.

Rankin, A.M., E.W. Wolff, and S. Martin, 2002: Frost flowers—Implications for tropospheric chemistry and ice core interpretation. *Journal of Geophysical Research,* **107**, 4683, doi:10.1029/2002JD002492.

Rankin, A.M., E.W. Wolff, and R. Mulvaney, 2005: A reinterpretation of sea salt records in Greenland and Antarctic ice cores. *Annals of Glaciology,* **39**, 276-282.

Rayner, N.A., D.E. Parker, E.B. Horton, C.K. Folland, L.V. Alexander, D.P. Rowell, E.C. Kent, and A. Kaplan, 2003: Global analysis of sea surface temperature, sea ice, and night marine air temperature since the late nineteenth century, *Journal of Geophysical Research,* **108(D14)**, 4407. doi 4410.1029/2002JD002670.

Repenning, C.A., E.M. Brouwers, L.C. Carter, L. Marincovich, Jr., and T.A. Ager, 1987: *The Beringian Ancestry of Phenacomys (Rodentia: Critetidae) and the Beginning of the Modern Arctic Ocean Borderland Biota.* U.S. Geological Survey Bulletin 1687, 31 pp.

Rigor, I.G. and J.M. Wallace, 2004: Variations in the age of Arctic sea-ice and summer sea-ice extent. *Geophysical Research Letters,* **31**, L09401. doi:10.1029/2004GL019492

Risebrobakken, B., E. Jansen, C. Andersson, E. Mjelde, and K. Hevroy, 2003: A high-resolution study of Holocene paleoclimatic and paleoceanographic changes in the Nordic Seas. *Paleoceanography,* **18**, 1017-1031.

Rothrock, D.A. and J. Zhang, 2005: Arctic Ocean sea ice volume: What explains its recent depletion? *Journal of Geophysical Research,* **110**, C01002, doi:10.1029/2004JC002282.

Savelle, J.M., A.A. Dyke, and A.P. McCartney, 2000: Holocene bowhead whale (*Balaena mysticetus*) mortality patterns in the Canadian Arctic Archipelago. *Arctic,* **53(4)**, 414-421.

Seager, R., D.S. Battisti, J. Yin, N. Gordon, N. Naik, A.C. Clement, M.A. Cane, 2002: Is the Gulf Stream responsible for Europe's mild winters? *Quarterly Journal of the Royal Meteorological Society,* **128B**, 2563-2586.

Seidenkrantz, M.-S., S. Aagaard-Sørensen, H. Sulsbrück, A. Kuijpers, K.G. Jensen, and H. Kunzendorf, 2007: Hydrography and climate of the last 4400 years in a SW Greenland fjord—Implications for Labrador Sea palaeoceanography. *The Holocene,* **17**, 387-401.

Serreze, M.C., A.P. Barrett, A.J. Slater, M. Steele, J. Zhang, and K.E. Trenberth, 2007a: The large-scale energy budget of the Arctic. *Journal of Geophysical Research,* **112**, D11122, doi:10.1029/2006JD008230.

Serreze, M.C., M.M. Holland, and J. Stroeve, 2007b: Perspectives on the Arctic's shrinking sea ice cover. *Science,* **315**, 1533-1536.

Sewall, J.O. and L.C. Sloan, 2004: Disappearing Arctic sea ice reduces available water in the American west, *Geophysical Research Letters,* **31**, doi:10.1029/2003GL019133.

Sher, A.V., T.N. Kaplina, Y.V. Kouznetsov, E.I. Virina, and V.S. Zazhigin, 1979: Late Cenozoic of the Kolyma Lowland. In: *XIV Pacific Science Congress. Tour Guide XI.* Academy of Sciences of USSR, Moscow, 115 pp.

Shimada, K., T. Kamoshida, M. Itoh, S. Nishino, E. Carmack, F. McLaughlin, S. Zimmermann, A. Proshutinsky, 2006: Pacific Ocean inflow—Influence on catastrophic reduction of sea ice cover in the Arctic Ocean. *Geophysical Research Letters,* **33**, L08605, doi:10.1029/2005GL025624.

Singarayer, J.S., J. Bamber, and P.J. Valdes, 2006: Twenty-first century climate impacts from a declining Arctic sea ice cover. *Journal of Climate,* **19**, 1109-1125.

Smith, L.M., G.H. Miller, B. Otto-Bliesner, and S.-I. Shin, 2003: Sensitivity of the Northern Hemisphere climate system to extreme changes in Arctic sea ice. *Quaternary Science Reviews,* **22**, 645-658.

Solignac, S., J. Giraudeau, and A. de Vernal, 2006: Holocene sea surface conditions in the western North Atlantic: Spatial and temporal heterogeneities. *Palaeoceanography,* **21**, 1-16.

Steele, M., W. Ermold, J. Zhang, 2008: Arctic Ocean surface warming trends over the past 100 years. *Geophysical Research Letters,* **35**, L02614, doi:10.1029/2007GL031651.

St. John, K.E., 2008: Cenozoic ice-rafting history of the central Arctic Ocean—Terrigenous sands on the Lomonosov Ridge. *Paleoceanography,* **23**, PA1S05, doi:10.1029/2007PA001483.

Stroeve, J., M.M. Holland, W. Meier, T. Scambos, and M. Serreze, 2007: Arctic sea ice decline: Faster than forecast. *Geophysical Research Letters,* **34**, doi:10.1029/2007GL029703.

Stroeve, J.C., T. Markus, and W.N. Meier, 2006: Recent changes in the Arctic melt season, *Annals of Glaciology,* **44**, 367-374.

Stroeve, J., M. Serreze, S. Drobot, S. Gearheard, M. Holland, J. Maslanik, W. Meier, T. Scambos, 2008: Arctic sea ice extent plummets in 2007. *Eos, Transactions of the American Geophysical Union,* **89**, 13-14.

Thompson, D.W.J. and J.M. Wallace, 1998: The Arctic Oscillation signature in the wintertime geopotential height and temperature fields. *Geophysical Research Letters,* **25(9)**, 1297-1300.

Thorndike, A.S., 1986: Kinematics of sea ice. In: *The Geophysics of Sea Ice* [Untersteiner, N. (ed.)]. NATO ASI Series, Series B, Physics, Vol. 146, Plenum Press, New York, pp. 489-549.

Tremblay, L.B., L.A. Mysak, and A.S. Dyke, 1997: Evidence from driftwood records for century-to-millennial scale variations of the high latitude atmospheric circulation during the Holocene. *Geophysical Research Letters,* **24**,: 2027-2030.

Tsukernik, M., D.N. Kindig, and M.C. Serreze, 2007: Characteristics of winter cyclone activity in the northern North Atlantic—Insights from observations and regional modeling, *Journal of Geophysical Research,* **112**, D03101, doi:10.1029/2006JD007184.

Turney, C., M. Baillie, S. Clemens, D. Brown, J. Palmer, J. Pilcher, P. Reimer, and H.H. Leuschner, 2005: Testing solar forcing of pervasive Holocene climate cycles. *Journal of Quaternary Science,* **20**, 511-518.

Van Loon, H. and J.C. Rogers, 1978: The seesaw in winter temperature between Greenland and northern Europe. Part I—General Description. *Monthly Weather Review,* **106**, 296-310.

Vincent, J.-S., 1990: Late Tertiary and early Pleistocene deposits and history of Banks Island, southwestern Canadian Arctic Archipelago. *Arctic,* **43**, 339-363.

Vinje, T., 1999: Barents Sea ice edge variation over the past 400 years. *Extended Abtracts, Workshop on Sea-Ice Charts of the Arctic, Seattle, WA.* World Meteorological Organization, WMO/TD, **949**, 4-6.

Vinje, T., 2001: Anomalies and Trends of Sea-Ice Extent and Atmospheric Circulation in the Nordic Seas during the Period 1864-1998. *Journal of Climate,* **14**, 255-267.

Walsh, J.E., 1978: A data set on Northen Hemisphere sea ice extent. World Data Center-A for Glaciology, Glaciological Data, Report GD-2, part 1, pp. 49-51.

Walsh, J.E. and W.L. Chapman, 2001: 20th-century sea-ice variations from observational data. *Annals of Glaciology,* **33**, 444-448.

White, J.M. and T.A. Ager, 1994: Palynology, paleoclimatology and correlation of middle Miocene beds from Porcupine River (Locality 90-1), Alaska. *Quaternary International,* **22/23**, 43-78.

White, J.M., T.A. Ager, D.P. Adam, E.B. Leopold, G. Liu, H. Jetté, and C.E. Schweger, 1997: An 18-million-year record of vegetation and climate change in northwestern Canada and Alaska—Tectonic and global climatic correlates. *Palaeogeography, Paleoclimatology, Palaeoecology,* **130**, 293-306.

Whitlock, C. and M.R. Dawson, 1990: Pollen and vertebrates of the early Neogene Haughton Formation, Devon Island, Arctic Canada. *Arctic,* **43(4)**, 324-330.

Whitlow, S., P.A. Mayewski, and J.E. Dibb, 1992: A comparison of major chemical species seasonal concentration and accumulation at the South Pole and Summit, Greenland. *Atmospheric Environment,* **26A**, 2045-2054.

Williams, C.J. 2006. Paleoenvironmental reconstruction of Polar Miocene and Pliocene Forests from the Western Canadian Arctic. 19th Annual Keck Symposium; *http://keckgeology.org/publications.*

Williams, C.J., A.H. Johnson, B.A. LePage, D.R. Vann and T. Sweda, 2003: Reconstruction of Tertiary metasequoia forests II. Structure, biomass and productivity of Eocene floodplain forests in the Canadian Arctic. *Paleobiology,* **29(2)**, 238-274.

Wolfe, J.A., 1980: Tertiary climates and floristic relationships at high latitudes in the Northern Hemisphere. *Palaeogeography, Palaeoclimatology, and Palaeoecology,* **30**, 313-323.

Wolfe, J.A., 1997: Relations of environmental change to angiosperm evolution during the Late Cretaceous and Tertiary. In: *Evolution and Diversification of Land Plants,* [Iwatsuki, K. and P.H. Raven (eds.)]. Springer-Verlag, Tokyo, pp. 269-290.

Wolff, E.W., H. Fischer, F. Fundel, U. Ruth, B. Twarloh, G.C. Littot, R. Mulvaney, R. Röthlisberger, M. de Angelis, C.F. Boutron, M. Hansson, U. Jonsell, M. A. Hutterli, F. Lambert, P. Kaufmann, B. Stauffer, T.F. Stocker, J.P. Steffensen, M. Bigler, M.L. Siggaard-Andersen, R. Udisti, S. Becagli, E. Castellano, M. Severi, D. Wagenbach, C. Barbante, P. Gabrielli, and V. Gaspari, 2006: Southern Ocean sea-ice extent, productivity and iron flux over the past eight glacial cycles. *Nature,* **440**, 491-496.

Wolff, E.W., A.M. Rankin, and R. Röthlisberger, 2003: An ice core indicator of Antarctic sea ice production? *Geophysical Research Letters,* **30**, 2158, doi:10.1029/2003GL018454.

Wollenburg, J.E. and W. Kuhnt, 2000: The response of benthic foraminifers to carbon flux and primary production in the Arctic Ocean. *Marine Micropaleontology,* **40**, 189-231.

Zachos, J., M. Pagani, L. Sloan, E. Thomas, and K. Billups, 2001: Trends, rhythms, and aberrations in global climate 65 Ma to present. *Science,* **292(5517)**, 686-693.

Zhang, X. and J.E. Walsh, 2006: Toward a seasonally ice-covered Arctic Ocean: Scenarios from the IPCC AR4 model simulations. *Journal of Climate,* **19**, 1730-1747.

Chapter 7. Key Findings and Recommendations

Jansen, E., J. Overpeck, K.R. Briffa, J.-C. Duplessy, F. Joos, V. Masson-Delmotte, D. Olago, B. Otto-Bliesner, W.R. Peltier, S. Rahmstorf, R. Ramesh, D. Raynaud, D. Rind, O. Solomina, R. Villalba, and D. Zhang, 2007: Palaeoclimate. In: *Climate Change 2007—The Physical Science Basis. Contribution of Working Group I to the Fourth Assessment Report of the Intergovernmental Panel on Climate Change* [Solomon, S., D. Qin, M. Manning, Z. Chen, M. Marquis, K.B. Averyt, M. Tignor and H.L. Miller (eds.)]. Cambridge University Press, Cambridge and New York, pp. 434-497.

PHOTOGRAPHY CREDITS

Cover/Title Page/Table of Contents:

EXECUTIVE SUMMARY Rushing water; *Richard Alley, Pennsylvania State University*

CHAPTER 1 White Dryas in flower; *Richard Alley, Pennsylvania State University*

CHAPTER 2 Ice thin section; *Joan Fitzpatrick, U.S. Geological Survey*

CHAPTER 3 Lake coring; *Gifford Miller, INSTAAR, University of Colorado*

CHAPTER 4 Greenland camp with flags; *James White, INSTAAR, University of Colorado*

CHAPTER 5 Glacier confluence; *Richard Alley, Pennsylvania State University*

CHAPTER 6 Shipboard operations in sea ice; *Leonid Polyak, Byrd Polar Research Center, Ohio State University*

CHAPTER 7 Perched erratic boulder; *Gifford Miller, INSTAAR, University of Colorado*

Page 6:

Greenland oblique; *image courtesy of National Aeronautics and Space Administration*

Page 7:

Northwest Passage, open; *image courtesy of National Aeronautics and Space Administration*

Page 22:

Benthic foraminifers; *Harry Dowsett, U.S. Geological Survey*

Page 29:

Igaliku, southwest Greenland; *Hans Oerter, Alfred Wegener Institute, Germany*

Page 32:

Elephant Foot Glacier; *Hans Oerter, Alfred Wegener Institute, Germany*

Page 42:

Shipboard operations; *Leonid Polyak, Byrd Polar Research Center, Ohio State University*

Page 58:

Cottongrass, Jakobshaven, Greenland; *Richard Alley, Pennsylvania State University*

Page 68:

Melting ice; *Hans Oerter, Alfred Wegener Institute, Germany*

Page 85:

Polar bear; *Hans Oerter, Alfred Wegener Institute, Germany*

Page 86:

Bearberry, Milne Land, Greenland; *Richard Alley, Pennsylvania State University*

Page 92:

Melt pond, Greenland; *James White, INSTAAR, University of Colorado*

Page 98:

Alpe Fjord, Greenland; *Richard Alley, Pennsylvania State University*

Page 105:

Meltwater; *Hans Oerter, Alfred Wegener Institute, Germany*

Page 107:

Walrus and iceberg; *Hans Oerter, Alfred Wegener Institute, Germany*

Page 108:

Broad-leaved willowherb; *Hans Oerter, Alfred Wegener Institute, Germany*

Page 115:

Harebell, Jakobshaven, Greenland; *Richard Alley, Pennsylvania State University*

Page 116:

Icebergs, Bjorn Islands, Scoresby Sund, East Greenland; *Richard Alley, Pennsylvania State University*

Page 120:

Arctic poppy; *Hans Oerter, Alfred Wegener Institute, Germany*

Page 122:

Rock ptarmigan; *Richard Alley, Pennsylvania State University*

Page 126:

Macro-photograph of air bubbles trapped in an Antarctic ice core—the scale at the top is a millimeter scale; *Joan Fitzpatrick, U.S. Geological Survey*

Page 127:

Iceberg, Scoresby Sund, East Greenland; *Richard Alley, Pennsylvania State University*

Page 128:

Cottongrass, Jakobshaven, Greenland; *Richard Alley, Pennsylvania State University*

Page 131:

Musk oxen, Kong Oscar Fjord area, northeast Greenland; *Hannes Grobe, Alfred Wegener Institute, Germany*

Page 137:

Snowshoe hare; *Hans Oerter, Alfred Wegener Institute, Germany*

Page 138:

Iceberg; *Hans Oerter, Alfred Wegener Institute, Germany*

Page 146:

Ship in sea ice; *Leonid Polyak, Byrd Polar Research Center, Ohio State University*

Page 150:

Meltwater lake; *Hans Oerter, Alfred Wegener Institute, Germany*

Page 160:

Ship in sea ice; *Leonid Polyak, Byrd Polar Research Center, Ohio State University*

Page 165:

Mosquito sunset; *Julie Brigham-Grette, University of Massachusetts*

Page 170:

Snow bunting, Jakobshaven, Greenland; *Richard Alley, Pennsylvania State University*

Page 173:

Clyde Inlet, Baffin Island; *Gifford Miller, INSTAAR, University of Colorado*

Page 174:

Humpback whale; *Hans Oerter, Alfred Wegener Institute, Germany*

Page 178:

Jakobshaven Isfjord, Greenland; *Hans Oerter, Alfred Wegener Institute, Germany*

Contact Information

Global Change Research Information Office
c/o Climate Change Science Program Office
1717 Pennsylvania Avenue, NW
Suite 250
Washington, DC 20006
202-223-6262 (voice)
202-223-3065 (fax)

The Climate Change Science Program incorporates the U.S. Global Change Research Program and the Climate Change Research Initiative.

To obtain a copy of this document, place an order at the Global Change Research Information Office (GCRIO) web site: *http://www.gcrio.org/orders.*

Climate Change Science Program and the Subcommittee on Global Change Research

www.ingramcontent.com/pod-product-compliance
Lightning Source LLC
Chambersburg PA
CBHW080635180526
45168CB00008B/3181
* 9 7 8 1 5 0 7 8 0 9 5 3 2 *